Fungi in Bioremediation

Bioremediation is an expanding area of environmental biotechnology and may be defined as the application of biological processes to the treatment of pollution. Much bioremediation work has concentrated on organic pollutants, although the range of substances that can be transformed or detoxified by microorganisms includes both natural materials and inorganic pollutants, such as toxic metals. The majority of applications developed to date involve bacteria, and there is a distinct lack of appreciation of the potential roles and involvement of fungi in bioremediation, despite clear evidence of their metabolic and morphological versatility. This volume highlights the potential of filamentous fungi, including mycorrhizas, in bioremediation and discusses the physiology and biochemistry of pollutant transformations.

Membership of the British Mycological Society is open to all with an interest in fungi, whether professional or amateur. It is an international society with members throughout the world. Further details regarding membership activities and publications can be obtained from the British Mycological Society, Joseph Banks Building, Royal Botanic Gardens, Kew, Richmond, Surrey TW9 3AB, UK, http://www.britmycolsoc.org.uk.

GEOFF GADD, Professor of Microbiology, is Head of the Division of Environmental and Applied Biology in the School of Life Sciences at the University of Dundee.

To my family, Julie, Katy and Richard Gadd, who have been a constant source of irritation throughout this project

Fungi in Bioremediation

EDITED BY

G. M. GADD

Published for the British Mycological Society

CAMBRIDGE
UNIVERSITY PRESS

PUBLISHED BY THE PRESS SYNDICATE OF THE UNIVERSITY OF CAMBRIDGE
The Pitt Building, Trumpington Street, Cambridge, United Kingdom

CAMBRIDGE UNIVERSITY PRESS
The Edinburgh Building, Cambridge CB2 2RU, UK
40 West 20th Street, New York NY 10011–4211, USA
10 Stamford Road, Oakleigh, VIC 3166, Australia
Ruiz de Alarcón 13, 28014 Madrid, Spain
Dock House, The Waterfront, Cape Town 8001, South Africa

http://www.cambridge.org

First published 2001

Printed in the United Kingdom at the University Press, Cambridge

Typeface Times 10/13pt   *System* Poltype® [v n]

A catalogue record for this book is available from the British Library

Library of Congress Cataloguing in Publication data

Fungi in Bioremediation/[edited by] G.M. Gadd.
   p.   cm. – (British Mycological Society symposium series; 23)
Includes bibliographical references (p.   ).
ISBN 0 521 78119 1 (hb)
1. Fungal remediation.   I. Gadd, Geoffrey M.   II. Series.
TD192.72.F86 2001
628.5–dc21   2001025609

ISBN 0 521 78119 1 hardback

Contents

List of contributors

Michelle Barclay
Oxford Centre for Environmental Biotechnology at the NERC Centre for Ecology and Hydrology, Mansfield Rd, Oxford OX1 3SR, UK

Joan W. Bennett
Department of Cell and Molecular Biology, ARS, USDA, Tulane University, New Orleans, LA 70118, USA

Helmut Brandl
University of Zürich, Institute of Environmental Sciences, Winterthurerstrasse 190, CH-8057 Zürich, Switzerland

John A. Buswell
Department of Biology, The Chinese University of Hong Kong, Shatin, New Territories, Hong Kong SAR, China

Marta Noemí Cabello
Universidad Nacional de La Plata, Instituto de Botanica 'Spegazzini', 53 No 477, 1900 La Plata, Argentina

Carl E. Cerniglia
National Center for Toxicological Research, Food and Drug Administration, Jefferson AR 72079, USA

Roni Cohen
Department of Plant Pathology and Microbiology, and the Otto Warburg Center for Biotechnology in Agriculture, Faculty of Agricultural, Food and Environmental Quality Sciences, the Hebrew University of Jerusalem, PO Box 12, Rehovot 76100, Israel

William J. Connick, Jr
Southern Regional Research Center, 1100 Robert E. Lee Boulevard, New Orleans, LA 70179, USA

Ronald L. Crawford
Department of Microbiology, Molecular Biology and Biochemistry and the Environmental Biotechnology Institute, University of Idaho, Moscow, ID 83844-1052, USA

Donald Daigle
Southern Regional Research Center, 1100 Robert E. Lee Boulevard, New Orleans, LA 70179, USA

Christine S. Evans
Fungal Biotechnology Group, University of Westminster, London W1M 8JS, UK

Geoffrey M. Gadd
Division of Environmental and Applied Biology, Biological Sciences Institute, School of Life Sciences, University of Dundee, Dundee DD1 4HN, UK

Yitzhak Hadar
Department of Plant Pathology and Microbiology, and the Otto Warburg Center for Biotechnology in Agriculture, Faculty of Agriculural, Food and Environmental Quality Sciences, the Hebrew University of Jerusalem, PO Box 12, Rehovot 76100, Israel

Patricia J. Harvey
School of Chemical and Life Sciences, University of Greenwich, Wellington St, London SE18 6PF, UK

John N. Hedger
Fungal Biotechnology Group, University of Westminster, London W1M 8JS, UK

Jeremy S. Knapp
Division of Microbiology, School of Biochemistry and Molecular Biology, University of Leeds, Leeds LS2 9JT, UK

Christopher J. Knowles
Oxford Centre for Environmental Biotechnology at the Department of Engineering Sciences, University of Oxford,

Parks Rd, Oxford OX1 3PJ, UK

Sarah E. Maloney
Centre for Applied Microbiology and Research, Porton Down, Salisbury SP4 0JG, UK

Zacharia Mathew
Department of Microbiology and Molecular Genetics and the Center for Microbial Ecology, Michigan State University, East Lansing, MI 48824-1101, USA

Andrew A. Meharg
Department of Plant and Soil Science, University of Aberdeen, Cruickshank Building, St Machar Drive, Aberdeen AB24 3UU, UK

David A. Newcombe
Department of Microbiology, Molecular Biology and Biochemistry and the Environmental Biotechnology Institute, University of Idaho, Moscow, ID 83844-1052, USA

C. Adinarayana Reddy
Department of Microbiology and Molecular Genetics and the Center for Microbial Ecology, Michigan State University, East Lansing, MI 48824-1101, USA

Ian Singleton
Department of Agricultural and Environmental Science, King George VI Building, University of Newcastle upon Tyne, Newcastle upon Tyne, NE1 7RU, UK

John B. Sutherland
National Center for Toxicological Research, Food and Drug Administration, Jefferson, AR 72079, USA

Christopher F. Thurston
Division of Life Sciences, King's College London, 150 Stamford St, London SE1 8WA, UK

John M. Tobin
School of Biotechnology, Dublin City University, Dublin 9, Ireland

Eli J. Vantoch-Wood
Division of Microbiology, School of Biochemistry and Molecular Biology, University of Leeds, Leeds LS2 9JT, UK

Kenneth Wunch
Department of Cell and Molecular Biology, ARS, USDA, Tulane University, New Orleans, LA 70118, USA

Fuming Zhang
Division of Microbiology, School of Biochemistry and Molecular Biology, University of Leeds, Leeds LS2 9JT, UK

Preface

Bioremediation is an expanding area of environmental biotechnology and may simply be considered to be the application of biological processes to the treatment of pollution. The metabolic versatility of microorganisms underpins practically all bioremediation applications and most work to date has concentrated on organic pollutants, although the range of substances which can be transformed or detoxified by microorganisms includes solid and liquid wastes, natural materials and inorganic pollutants such as toxic metals and metalloids. However, the majority of applications developed to date involve bacteria and there is a distinct lack of appreciation of the potential roles, involvement and possibilities of fungi in environmental bioremediation despite clear and growing evidence of their metabolic and morphological versatility. The fundamental importance of fungi in the environment with regard to decomposition and transformation of both organic and inorganic substrates and resultant cycling of elements is of obvious relevance to the treatment of wastes, while the branching, filamentous mode of growth can allow efficient colonization and exploration of, for example, contaminated soil and other solid substrates. This, together with the growing importance of fungi as model systems in eukaryotic cell and molecular biology, physiology and biochemistry, provides the rationale for this work.

The prime objective of this book is to highlight the potential of filamentous fungi in bioremediation, and to discuss the physiology, chemistry and biochemistry of organic and inorganic pollutant transformations. The chapters are written by leading international authorities in their fields and represent the latest and most complete synthesis of this subject area. Organic and inorganic pollutants are covered, although it is intriguing that, as in bacterial research, these two areas are largely segregated, unlike the real nature of environmental pollution in many cases. Perhaps combined research on both organic and inorganic pollutants and organism response is a worthy topic for future research. Another point worth emphasizing is that virtually all of the transformation processes described revolve around intrinsic properties of fungi that underpin fungal growth and survival, and are integral to environmental function. Thus, mechanisms for breakdown of recalcitrant plant residues can also act on synthetic pollutants and this has led to much interest in ligninolytic fungi, especially the white rots exemplified by *Phanerochaete chrysosporium*. The metabolic versatility of this organism provides a theme and foundation for several chapters, and much organic bioremediation knowledge arises from work with this organism. However, as is also evident, there are many other fungi with interesting properties and these may be applicable in other specific contexts: interesting isolates may perhaps be unearthed from the rain forest or, alternatively, extremely polluted locations! Additionally, appreciation of the important environmental roles of mycorrhizal fungi is increasing and these fungi, so intimately associated with the flow of carbon and other essential elements in the biosphere, may have wider significance in conjunction with revegetation and phytoremediation initiatives. The range of organic molecules degraded, decomposed or transformed by fungi includes recalcitrant plant biomolecules, polycyclic aromatic hydrocarbons, nitroaromatics, chlorinated aromatics, BTEX compounds, as well as miscellaneous dyes, pesticides, effluent components, and even cyanide. Unlike organic molecules, metals cannot be destroyed, but fungi, like other microorganisms, can effect transformations between mobile and immobile forms. This is not only of bioremedial significance but also underpins important environmental roles including the solubilization of essential metals and associated anionic components such as phosphate, so important for plant (and microbial) productivity. I hope this book succeeds in providing a fascinating insight into an important area of fungal biology, but I would stress that the field has plenty of room for expansion. Many areas remain poorly understood with some yet to receive detailed application of modern techniques in cell and molecular biology, while the interface between chemistry and biology is particularly important and with considerable reciprocity. The reactions and processes catalysed by fungi, of dynamic cellular and environmental significance, pose a continual analytical challenge. Who knows what future treasures lie hidden within the vastness of fungal biodiversity?

I would like to thank all the authors who have contributed to this work in an enthusiastic and professional manner, and all at Cambridge University Press who have facilitated progress. In Dundee, special thanks go to

Angela Nicoll who greatly assisted collation, editing and formatting of chapters and Karen Kinnear for editing queries and index preparation. Their help was indispensable and greatly appreciated. Finally, I would like to thank the British Mycological Society for wholeheartedly supporting this project, and my family, Julie, Katy and Richard Gadd.

Geoffrey Michael Gadd
2001

1

Degradation of plant cell wall polymers

CHRISTINE S. EVANS AND JOHN N. HEDGER

Introduction

Processes of natural bioremediation of lignocellulose involve a range of organisms, but predominantly fungi (Hammel, 1997). Laboratory studies on the degradation of lignocellulose, including wood, straw, and cereal grains, have focused mainly on a few fungal species that grow well in the laboratory and can be readily manipulated in liquid culture to express enzymes of academic interest. Our current understanding of the mechanism of lignocellulose degradation stems from such studies. Although some of these enzymes have economic potential in a range of industries, for example pulp and paper manufacture and the detergent industry, it is frequently expensive and uneconomic to use them for bioremediation of pollutants in soils and water columns. In the successful commercial bioremediation processes developed, whole organisms have been used in preference to their isolated enzymes (Lamar & Dietrich, 1992; Bogan & Lamar, 1999; Jerger & Woodhull, 1999).

Most fungi are robust organisms and are generally more tolerant to high concentrations of polluting chemicals than are bacteria, which explains why fungi have been investigated extensively since the mid-1980s for their bioremediation capacities. However, the species investigated have been primarily those studied extensively under laboratory conditions, which may not necessarily represent the ideal organisms for bioremediation. Fungi in little-explored forests of the world, for example tropical forests, may yet prove to have even better bioremediation capabilities than the temperate organisms currently studied, exhibiting more tolerance to temperature and specialist environments. This chapter discusses the current state of knowledge on the degradation of lignocellulose and how this relates with the ecology of lignocellulolytic fungi. This knowledge is important to modify and enhance the mechanisms of degradation of industrial pollutants such as chlorophenols, nitrophenols and polyaromatic hydrocarbons by these fungi.

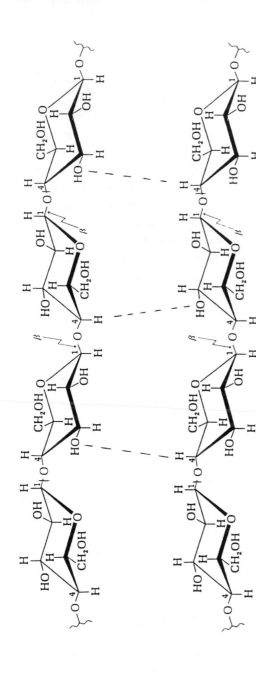

Fig. 1.1. Structure of cellulose formed from β-1,4-linked cellobiose units, with hydrogen bonding between parallel chains.

Structure and function of plant cell walls

Plant cell walls offer several benefits to the growing and mature plant. They provide an exo-skeleton giving rigidity and support, enabling the water column to reach the plant apex, and they serve as a protective barrier against predators and pathogens. As the wall is an extracytoplasmic product, it was not considered to be a living part of the cell (Newcomb, 1980), although this view is now challenged and many metabolic processes are now known to occur within the cell wall structure for maintenance and in response to attack by pathogens (Dey, Brownleader & Harborne, 1997a). Every cell of a plant is surrounded by a primary wall that undergoes plastic extension as the cell grows. Some cells such as those of the parenchyma keep a primary wall throughout their lifespan. The primary wall is composed of cellulose fibrils, hemicellulose and protein with large amounts of pectin forming a viscous matrix that cements the wall together. The precise molecular composition varies between cell types, tissues and plant species, although an approximate dry weight ratio would be 30% cellulose, 25% hemicellulose, 35% pectin, and 10% protein (Taiz & Zeiger, 1991).

Cellulose is formed by polymerization of D-glucose molecules linked in the β-1,4 position, resulting in flat, linear chains (Fig. 1.1). Hydrogen bonding between chains leads to the formation of a microfibril up to 3 nm in diameter. These crystalline microfibrils are laid down in different orientations within the primary wall, providing the structural support for the wall. The surrounding wall matrix is composed of hemicellulose, protein and pectin in which the microfibrils are embedded. Hemicelluloses are mixed polymers of different neutral and acidic polysaccharides. They adhere to the surface of the cellulose microfibrils by hydrogen bonding, through OH groups on the sugars, and enhance the strength of the cell wall. Pectic polysaccharides are covalently bound to the hemicelluloses. The protein components of the primary cell wall are hydroxyproline-rich glycoproteins, named extensins, that are involved in cell wall architecture and plant disease resistance (Brownleader *et al.*, 1996; Dey *et al.*, 1997b). As some cell types mature, a secondary cell wall is deposited between the primary wall and the plasmalemma that is more dense than primary walls and binds less water. Wood contains cell walls of the greatest maturity of all plant cell types, with several layers (S1, S2, S3) in the secondary wall (Fig. 1.2). These are composed of cellulose, hemicellulose and lignin but with less pectin than primary cell walls. An approximate ratio of these components would be 35% cellulose, 25% hemicellulose including pectin,

Fig. 1.2. Transmission electron micrograph of an ultra-thin section of beech wood showing the wood cell wall structure. M, middle lamella; P, primary wall layer; S1, S2 and S3, secondary wall layers. Magnification × 28 000.

and up to 35% lignin, depending on the plant species. The cellulose microfibrils are orientated at different angles in each layer of the three secondary wall layers to provide increased strength. Lignin gives rigidity to the wall (Cowling & Kirk, 1976; Montgomery, 1982). Strength is related to these structural components, particularly the orientation and crystallinity of the cellulose microfibrils (Preston, 1974), whereas toughness is a reflection of the elastic component of the cell wall, giving an indication of the potential for plasticity (Lucas *et al.*, 1995).

Lignin is a three-dimensional aromatic polymer that surrounds the microfibrils, with some covalent attachment to the hemicellulose. It is composed of up to three monomeric units of cinnamyl alcohols: coumaryl alcohol, mainly confined to grasses; coniferyl alcohol, the major monomeric unit in gymnosperm wood; and sinapyl alcohol, predominant in angiosperm wood (Fig. 1.3). Polymerization of these monomers is by a free radical reaction catalysed by peroxidase. This results in a variety of bonds

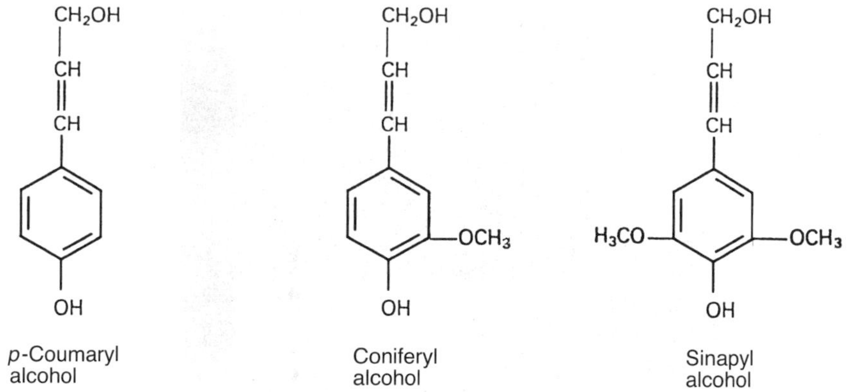

Fig. 1.3. Monomeric phenylpropanoid units that polymerize to form lignin.

in the polymer, with β-O-4 bonds between the C_4 carbon of the aromatic ring and the β-carbon of the side chain being the predominant bond linkage. Other common bonds are α-O-4, C_3 or C_5 aryl-O-4 linkages with some biphenyl linkages. Substitution on the aromatic rings of the monomers determines the type of lignin as syringyl (made from coniferyl and sinapyl alcohols) or guaiacyl lignin (made from only coniferyl alcohol) (Adler, 1977). Side chains of lignin are composed of cinnamyl alcohols, aldehydes and hydroxylated substitutions on the α- and β-carbons. It is the high proportion of ether bonds in the polymer, from the methoxy groups and polymerizing bonds, that gives lignin its unique structure and properties as a strong resistant polymer. It has been found that the polysaccharides in the cell wall can influence the structure of lignin during synthesis. The motion of coniferyl alcohol (one of the lignin monomers) and its oligomers near a cellulose surface can change the course of dehydrogenation polymerization into lignin, as electrostatic forces restrict the motion of the monomer and oligomers (Houtman & Atalla, 1995). This is consistent with experimental observations of the lignin–polysaccharide alignment in cell walls as observed by Raman microprobe studies (Atalla & Agarwal, 1985).

Hemicelluloses in the secondary wood wall vary between plant species, with xylan the predominant polymer. The simple β-1,4-linked-D-xylopyranosyl main chain carries a variable number of neutral or uronic acid monosaccharide substituents, or short oligosaccharide side chains (Fig. 1.4). Hardwood (angiosperms) xylans are primarily of the glucuroxylan type, while softwood (gymnosperms) xylans are glucuronoarabinoxylans (Joseleau, Comtat & Ruel, 1992). The structure

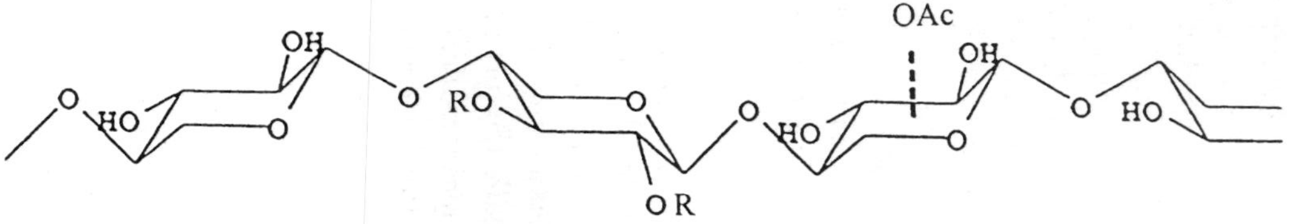

R can be any of the following substituents:

4-OMe-α-D-GlcA (1,2)Xyl   Feruloyl
α-D-GlcA (1,2)Xyl   p-Coumaryl
α-L-Ara (1,3)Xyl   Lignin
β-D-Xyl (1,2) α-L-Ara (1,3)Xyl
β-D-Gal (1,5) α-L-Ara (1,3)Xyl

Fig. 1.4. Basic structure of xylan. Me, methyl; GlcA, gluconic acids; Xyl, xylose; Ara, arabinose; Gal, galactose.

of xylans in the wood cell wall is difficult to characterize but it is clear that many covalent linkages occur with other wall polymers, as complete extraction of xylan from wood requires drastic alkaline conditions. Stable lignin–xylan complexes remain in wood pulp after Kraft pulping, and may involve carbon–carbon bonding while other bonds forming acetals or glycosides include oxygen. Other cross-linking polymers in the cell wall are ferulic and *p*-coumaric acids, providing important structural components. Dehydrodiferulate oligosaccharide esters have been extracted from wheat bran but not the free dehydrodiferulate acids, indicating that cross-linkages are formed with hemicellulose components (Kroon *et al.*, 1999).

The secondary wood cell wall structure has been visualized using transmission electron microscopy, revealing distinct layers in the secondary wall (Fig. 1.2). The dense wall structure makes the cells impenetrable to microorganisms without prior degradation of the wall polymers. The size of the pores within the wood cell wall, and hence water-holding capacity of the wall, is low; permeability studies with indicator molecules and dyes indicate that molecules above 2000 Da are unable to penetrate (Cowling, 1975; Srebotnik, Messner & Foisner, 1988; Flournoy, Kirk & Highley, 1991). In wheat cell walls, pores with radii of 1.5–3 nm (measured by gas adsorption) predominate, which are below the size that would allow free penetration by degrading enzymes (Chesson, Gardner & Wood, 1997). Casual predators and pathogens are deterred from establishing an ecological niche in such substrates.

The ecophysiology of lignin degradation

Most reviews of lignocellulose degradation have focused on the mechanisms of the process rather than the ecophysiology of the organisms involved. The Basidiomycota and Ascomycota, mostly in the orders 'Aphyllophorales', Agaricales and Sphaeriales, are considered by Cooke & Rayner (1984) to be responsible for decomposition of a high proportion of the annual terrestrial production of 100 gigatonnes of lignocellulose-rich plant cell wall material, of which lignin alone accounts for 20 gigatonnes (Kirk & Fenn, 1982). The basis of most studies on lignocellulose-degrading fungi has been economic rather than ecological, with focus on the applied aspects of lignin decomposition, including biodeterioration, bioremediation and bioconversion. This has led to an overemphasis on a few fungi as model organisms, particularly in the study of lignin decomposition, without any attempt to decide if they represent the spectrum of lignin-degrading systems in the fungi as a whole. Awareness and understanding of a wider number of species with good potential for economic use will lead to improvements in bioremediation technology.

Another gap in our knowledge of ligninolytic fungi is that not only have comparatively few taxa been studied but nearly all of them originate from the northern temperate forest and taiga biomes. This is in spite of the fact that the biodiversity of decomposer fungi is much higher in tropical ecosystems, especially tropical forest. In tropical forest, 74% of the primary production is deposited as woody litter and 8% as small litter, 10–35 tonnes litter $ha^{-1} yr^{-1}$, illustrating the enormous quantities of lignocellulose processed by decomposer fungi and termites in tropical forest (Swift, Heal & Anderson, 1976). The tropical forest biome contains 400–450 gigatonnes of plant biomass compared with 120–150 gigatonnes for temperate and boreal forest biomes (Dixon *et al.*, 1994). It is estimated that there are three times more taxa of higher fungi in tropical ecosystems than in other forest ecosystems, of which a much higher proportion are decomposers (Hedger, 1985). In spite of this, the isolation and screening of wood- and litter-decomposing fungi from tropical forests has yet to be systematically commenced (Lodge & Cantrell, 1995).

Lignocellulose degradation by fungi used in bioconversion of lignocellulose wastes

Most world mushroom production is from *Agaricus bisporus*, *Pleurotus ostreatus*, and *Lentinula edodes* and related species (Stamets & Chilton, 1983; Stamets, 1993), all grown on a range of substrates prepared from lignocellulose wastes such as straw and sawdust. These taxa have been widely used in physiological studies of cellulose and lignin decomposition in order to determine the role of lignocellulolytic systems in bioconversion of lignocellulose wastes to fruit bodies. Detailed studies on the ligninases of these taxa provided the early understanding of the ligninase systems (Wood, 1980; Kirk & Farrell, 1987; Hatakka, 1994). However, surprisingly little is known of the ecophysiology of the mycelia of these fungi in their natural environments: soil and litter in the case of *Agaricus* spp. and wood in the case of *Pleurotus* spp. and *L. edodes*. These are very different resources, and the published contrasts in the enzyme systems of these two groups of cultivated fungi may be related to the autecology of their mycelia (Wood, Matcham & Fermor, 1988). However there are no *in vivo* studies of lignocellulose degradation by these fungi. When used for bioremediation, it is the fungal mycelia and not their fruit bodies that transform lignocellulosic wastes and transform aromatic pollutants.

The ecology of wood-rotting fungi

Another source of isolates for the study of lignin decomposition has been the higher fungi, which cause significant economic losses to the timber industry. Pathogens of trees are an obvious example and include fungi such as *Armillaria* spp. and *Heterobasidion annosum*, where pathogenicity includes white rot exploitation of the lignocellulose resource by mycelia of these fungi (Stenlid & Redfern, 1998). Their ligninolytic systems have been studied (Asiegbu *et al.*, 1998; Rehman & Thurston 1992), findings showing that their role in pathogenicity is much less important than in the subsequent phases of saprotrophic exploitation and inoculum production (Rishbeth, 1979).

Decay of timber in-service has also yielded information on the lignocellulose-degrading enzymes produced by fungi like *Serpula lacrymans, Lenzites trabea*, and *Fibroporia vaillentii*. These basidiomycetes are all brown rot fungi, a physiological group that probably coevolved with the Coniferales in the northern taiga and temperate forests (Watling, 1982) and which are important because most in-service timber in Europe and the USA is softwood. Few studies have been made of these economically important fungi in the natural environment. It is salutary to realise that *S. lacrymans*, the dry rot fungus, although the subject of many papers on the nature of its mode of decomposition of cellulose (Montgomery, 1982; Kleman-Leyer *et al.*, 1992), has never been found outside the built environment, although recent studies indicate that it may have its origins in North Indian forests (White *et al.*, 1997). Another well-studied wood-degrading taxon little known outside the laboratory is the white rot thermophilic basidiomycete *Phanerochaete chrysosporium*, which was first considered as a problem in the 1970s in self-heating wood chip piles in its anamorphic state, *Sporotrichum pulverulentum* (Burdsall, 1981). Although this fungus has been the subject of many investigations of cellulases and ligninases because of their potential in bioremediation (Johnsrud, 1988), its natural 'niche' remains unknown.

The ecology of Trametes versicolor *and the dynamics of wood decay*

Up to the early 1980s, most detailed studies on lignin-degrading enzymes were on the 'economically important' fungi discussed above. However, since then the search for ligninolytic systems has been extended to include species of little economic importance but of applied potential because of their rate of growth and high enzymic activity. The most obvious example is *Trametes (Coriolus) versicolor*, the ligninases of which were first studied by Dodson *et al.* (1987), which causes white rot decay of broad-leaved tree species in temperate forest ecosystems. This fungus has been widely used in bioremediation programmes and characterization of its ligninases is now well understood, providing information that can be related to its ecophysiology in the natural environment. A good example is the regulatory effect of nitrogen on ligninolytic enzyme expression, a reflection of the inductive effect of low nitrogen levels (C:N 200:1 to 1000:1) found in wood (Swift 1982; Leatham & Kirk, 1983).

Unfortunately, laboratory results of this type have led to the simplistic view that 'success' of fungi in the natural environment can be simply related to the physiology of their mycelia in culture. It might be assumed that active ligninases and cellulases and fast mycelial growth in culture can explain the ubiquity of *T. versicolor* in broad-leaved forest. Fortunately, studies on the ligninases of this fungus coincided with studies on the population dynamics of communities of wood decay fungi, including *T. versicolor* (Rayner & Todd, 1979). The fungus causes a rapid white rot invasion of moribund or fallen trees of species such as birch, beech and oak. Rayner & Webber (1985) have shown that the outcome of primary resource capture by fungi like *T. versicolor* is a result of mechanisms that operate in the early stages of colonization. Early phases of expansion are by a rapidly extending mycelium, which utilizes free sugars in the wood of the tree. Entry into broken or cut ends of the wood from the spore rain means that an individual mycelium is usually restricted to an elongated form because of the faster rates of expansion of mycelia along vessels and tracheids. Following this resource capture, contact between mycelia of genetically distinct individuals of *T. versicolor*, and with mycelia of other species of wood-rotting fungi, results in combative behaviour. *T. versicolor* is typical of early colonizers of wood, an assemblage of fungi characterized by Cooke & Rayner (1984) as disturbance tolerant, with a combative mycelial strategy and active lignocellulose exploitation. The wood volume retained by the mycelium is covered by a melanized pseudosclerotial plate, resulting in a mosaic of individuals – the 'spalted' wood of the turners. White rot exploitation of the wood within these volumes by lignocellulolytic enzymes produced by the mycelia may then take place for a number of years. However, the initial phase of occupation and retention of the resource has little to do with the lignocellulolytic potential of the fungus. Comparison of lignocellulose decomposition by common white rot competitors of *T. versicolor*, for example *Stereum hirsutum* and *Hy-*

poxylon multiforme, show them to produce less ligninases and cellulases; however, they are equally successful colonizers since the initial outcome of competition is solely determined by occupation and retention of substrate (Rayner & Todd, 1979). *T. versicolor* and other 'primary resource capturers' may, in fact, be eventually replaced by mycelia of other more combative wood decomposers: 'secondary resource capturers', for example *Lenzites betulina* (Rayner, Boddy & Dowson, 1987; Holmer & Stenlid, 1997).

The strategy of many wood-rotting fungi is to exploit the retained wood relatively slowly, their mycelia being characterized as slow growing, stress tolerant, combative and defensive (Rayner & Boddy, 1988; Holmer & Stenlid, 1997). These fungi may persist in wood much longer than *T. versicolor*, although they appear to be less active degraders of lignocellulose under laboratory conditions. Their success is related to slow growth combined with retention of the wood, and tolerance to the developing nutrient stress in the wood as it decays and to extractives in heartwood. Many of these fungi are members of the 'Aphyllophorales', good examples being the genera *Ganoderma*, *Fomes* and *Inonotus*, which may persist for decades on fallen trees. Lignocellulose degradation by such fungi has been little studied, mostly because of their slow growth, difficulties in culturing and little apparent biotechnological potential. However, the later stages in decomposition of wood offer different physiological challenges to mycelia, for example the presence of complex recalcitrant aromatic compounds; consequently their degradative systems may well be of interest.

Another life strategy group of wood decomposer fungi is contained within the ascomycete order Sphaeriales. Genera in this order (e.g. *Xylaria, Daldinea* and *Hypoxylon*) have been studied by Rayner & Boddy (1988), who showed that they occupy and retain volumes of wood in the way described above for combative white rotters like *T. versicolor* but are characterized by a relatively slow white rot and a reduction of the water content of the wood. Physiological studies showed that the mycelia of species in these genera were tolerant of water stress and able to grow at potentials as low as 10–12 Mpa, explaining their successful retention of dry lignocellulose resources (Boddy, Gibbon & Grundy, 1985; Bravo-Velasquez & Hedger, 1988). The operation of ligninases under such low water potentials is of applied interest and the few studies carried out on these fungi have revealed unexpectedly low laccase and manganese peroxidase (MnP) activities, in spite of their *in vivo* abilities to cause extensive white rot (Ullah, 2000).

Litter-decomposing fungi

Studies on lignocellulolytic systems have mostly been limited to wood-rotting fungi, while litter-decomposing fungi that colonize small debris such as leaves and twigs have received little attention. Except for the mushrooms *A. bisporus* and *Volvariella volvacea*, which are both grown commercially on composted lignocellulose, the ligninolytic abilities of other litter-decomposing higher fungi are poorly understood, yet studies on their ecology have shown that they are major processors of lignocellulose in forest ecosystems (Hedger & Basuki, 1982). Frankland (1984), in a study of litter decomposition in Meathop Wood, UK, showed that the agaric *Mycena galopus* was responsible for a large proportion of the breakdown of the leaf litter of oak and other trees.

It is to be expected that the ecophysiologies of other litter-decomposing fungi may be different from those of wood decomposers, given the much lower lignin content of small litter, which consists of leaves, small twigs, seeds and fruits (Swift *et al.*, 1976). What effect this might have on the ligninolytic systems discussed in this chapter is difficult to predict, but needs study. An exception is a study of isolates of litter- and wood-decomposing fungi from a forest in Ecuador in which laccase and MnP activities of 27 different taxa were compared (Ullah, 2000). The 19 wood-rotting fungi had significantly greater laccase activity than 11 isolates from leaf litter, two being close to those of *T. versicolor* control cultures. However, the litter decomposer fungi had significantly higher titres of MnP activity than did the wood decomposers. Such results underline the need for an ecological perspective in the selection of fungal isolates for studies of cellulases and ligninolytic systems, in order to interpret the value of the different components of the system to the ecology of the organisms and perhaps to find novel ligninolytic systems.

Mechanisms of degradation

White rot basidiomycetes that degrade all cell wall polymers are generally considered to be the most effective lignocellulose degraders (Crawford, 1981; Hammel, 1997). However, from the perspective of an individual fungus, this may not be the case. The only reason for fungi to attack lignocellulose is to obtain sufficient carbon and nitrogen for survival. Unless energy used to obtain glucose is less than that resulting from its uptake and metabolism, there is no advantage to the organism in colonizing lignocellulose substrates. In fact, brown rot rather than white rot basidiomycetes may have the most efficient mechanisms for obtaining glucose from lignocellulose as they are able to extract glucose from the cellulose without expending energy on lignin degradation (Highley, 1977; Micales, 1995). They modify the lignin by methylation but no depolymerization occurs.

The soft rot fungi are a specialized group of organisms that grow in a localized niche within the secondary wood cell wall and degrade the cell wall polymers slowly (Hale & Eaton, 1985; Daniel & Nilsson, 1989). Characteristic patterns of decay in wood are channels within the secondary wall wherein the hyphae lie, degrading polymers immediately around the hyphal surface.

So what are the mechanisms employed by microorganisms to break down lignocellulose? Our understanding of lignocellulose degradation is based on laboratory studies with white rot basidiomycetes, using a selected number of organisms such as *P. chrysosporium* and *T. versicolor*, and, as mentioned above, it is probable that more effective organisms operate in the ecosystem that have not been characterized in the laboratory. All of these organisms have potential for use in bioremediation processes, although to date *P. chrysosporium, T. versicolor* and *P. ostreatus* have been the primary species used in pilot and field bioremediation trials.

Lignocellulolytic enzymes

The usual biological answer to breaking down a polymer is to use highly specific enzymes. This is normally extremely effective as a minimum amount of protein (enzyme) can be synthesized by the organism to cleave a regular repeating bond between units of the polymer. Examples of some of these enzymes are those that catalyse hydrolytic reactions to attack carbohydrates such as cellulose and hemicellulose (Walker & Wilson, 1991; Goyal, Ghosh & Eveleigh, 1991). They tend to be specific to a particular bond chemistry, for instance, β-1,4 specificity is required to hydrolyse cellulose, and α-1,4 specificity for starch hydrolysis.

Cellulases

Our knowledge of the biochemistry of enzymes that depolymerize cellulose is based mainly on studies of *Trichoderma* spp.: the most prolific sources of cellulases known (Mandels & Steinberg, 1976). When enzymes from wood-rotting basidiomycetes have been screened for cellulases, their composition has closely resembled those of cellulases from *Trichoderma reesei*, with five endoglucanases, one exoglucanase and two β-1,4-glucosidases identified (Eriksson & Pettersson, 1975). Cellulase is a complex of enzyme activities that includes exo- (cellobiohydrolases) and endocellulases, with β-glucosidases; these act in synergy to depolymerize cellulose fibrils, releasing glucose and cellobiose (a glucose dimer). Glucose is readily taken up by the fungus, providing carbon for energy and growth. Cellobiose is converted to glucose by the action of β-1,4-glucosidase (Evans, 1985; Gallagher & Evans, 1990). There are several controlling feedback mechanisms on production of the specific components of the cellulase complex by fungi, for instance glucose represses and cellobiose or cellulose stimulates the production of exo- and endocellulases. Cellobiohydrolases are composed of a cellulose-binding domain linked to a catalytic domain through a proline and a OH-amino acid linker region (Ong et al., 1989). Genes for exo- and endocellulases have been isolated and the gene products characterized, providing an improved understanding of the biochemical mechanisms involved (Beguin, 1990; Covert, Wymelenberg & Cullen, 1992). Feedback control of exo- and endocellulase production is exerted by glucose and sucrose with catabolite repression at concentrations of $1\,\mathrm{g}\,\mathrm{l}^{-1}$, but induction of cellulase occurs with $1\,\mathrm{mg}\,\mathrm{l}^{-1}$ cellobiose or cellulose (Eveleigh, 1987). Figure 1.5 shows a scheme for cellulolysis that represents the biological activities of the cellulase complex.

Although cellulases have been isolated from cultures of brown rot and white rot fungi, there is a fundamental difference in the mechanism of cellulose hydrolysis in the two fungal rots. Brown rots produce complete breakage of amorphous cellulose fibrils while white rots cause a progressive decay from the fibril surfaces (Klemen-Layer et al., 1992; Gilardi, Abis & Cass, 1995). The mechanism that brown rots use to access the cellulose in the wood cell wall is thought to be by generation of hydroxyl radicals from the reaction of H_2O_2 with Fe^{2+} in the Fenton reaction (Koenigs, 1974; Hyde & Wood, 1997). There has been generally less interest in the mechanisms of cellulose degradation by brown rots compared with white rots, as industrial usage of residual lignin is limited. In contrast, removal of lignin but leaving the cellulose intact has great potential in industries such as pulp and paper production.

Enzymes for hydrolysis of hemicellulose have also been identified in many wood-rotting fungi. Hard woods contain xylan, and fungal species colonizing them produce xylanases. Other hemicelluloses are hydrolysed by mannanases, galactosidases and glucosidases. These enzymes have very similar characteristics to the cellulase complex in that different enzymes attack exo- and endohemicellulose (Visser et al., 1992).

ENZYMES   CELLULOSE

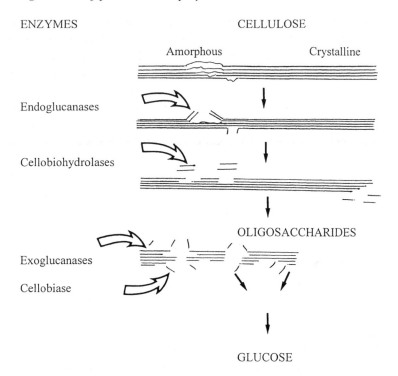

Fig. 1.5. Diagrammatic representation of the enzymic digestion of cellulose.

Ligninases

As previously described, lignin is not a symmetrical well-ordered polymer
with a single repeating bond that links the monomeric units. It is difficult,
therefore, to envisage a single enzyme that has the capability to depolymerize such an irregular structure. Although many different enzymes could be
produced by a degrading organism, this would prove energetically waste-
ful, as more energy would be expended in synthesis and secretion of
proteins than would be gained from metabolism of the end products. This
is particularly so for lignin as there is very little calorific value in the
polymer for the fungus, which is unable to survive on lignin as sole carbon
source (Kirk, Connors & Zeikus, 1976). It is assumed that white rot species
only degrade lignin as a means to access the cellulose in the wood cell wall.
The fungal enzymes that are used to degrade lignin are therefore non-
specific with respect to substrate. They function mainly by the production
of free radicals that are able to attack a wide range of organic molecules.

Peroxidases using H_2O_2, and laccases (polyphenol oxidases) using molecular oxygen are the enzymes responsible for attack on lignin (Field *et al.*, 1993; Evans *et al.*, 1994).

Peroxidases

The first enzyme shown to attack lignin-type compounds (model dimers, trimers and later polymers) was lignin peroxidase or LiP, isolated from *P. chrysosporium* (Tien & Kirk, 1984). LiP is a haem-containing peroxidase ($\sim 42\,000\ M_r$) with an unusually high redox potential. It is highly glycosylated, as are most enzymes secreted for extracellular action. Most but not all white rot species produce it (Hatakka, 1994). It is distinctive in its ability to oxidize methoxyl substituents on non-phenolic aromatic rings by the generation of cation radicals that undergo further reactions. The pH optimum of LiP is below 3.0 but the enzyme shows signs of instability if kept in such acidic environments (Tien & Kirk, 1988). Natural environments in wood cells are acidic, approximately pH 4.0, but localized pockets may occur with lower pH because of secretion of oxalic acid by the fungi (Dutton *et al.*, 1993; Dutton & Evans, 1996). *In vitro* peroxidases can be stimulated by low nitrogen stress on the fungi. Veratryl alcohol is a substrate for LiP and is used in a spectrophotometric assay to monitor its activity by measuring veratraldehyde production in the presence of H_2O_2 (Tien & Kirk, 1984). Veratryl alcohol is produced extracellularly as a secondary metabolite by many white rot fungi and enhances LiP activity (Collins & Dobson, 1995) probably by protecting LiP from inactivation by excess H_2O_2 (Chung & Aust, 1995) (Fig. 1.6).

MnP is also a haem-containing enzyme and is generally considered to be essential for lignin degradation *in vivo*. The catalytic cycle of MnP is similar to that of LiP, in addition to Mn^{2+} being converted to Mn^{3+} during the reaction. In reaction with some substrates such as dimethoxyphenol, MnP activity is independent of manganese (Archibald, 1992). Increased bio-bleaching of Kraft pulps has correlated with purified MnP activity from *T. versicolor* (Paice *et al.*, 1993; Reid & Paice, 1998). The white rot species that have been examined have all shown MnP activity, whereas not all have LiP activity. LiP and MnP both require H_2O_2, which must be generated by the fungus. Fungal enzymes producing H_2O_2 include glucose oxidase (Eriksson *et al.*, 1986), glyoxal oxidase (Kersten & Kirk, 1987) and aryl alcohol oxidase (Guillen, Martinez & Martinez, 1990; Guillen & Evans, 1994). In addition, MnP can generate H_2O_2 through oxidation of organic acids (Urzua *et al.*, 1995), while MnP-chelates can oxidize a range of phenolic compounds, including a variety of synthetic lignins (Masaphy,

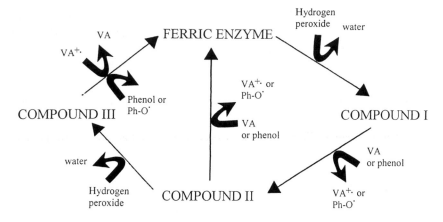

Fig. 1.6. Catalytic cycle of lignin peroxidase (LiP). VA, veratryl alcohol; Ph, phenol.

Henis & Levanon, 1996; Hofrichter *et al*, 1998). Manganese(III)-chelates, such as with oxalate, are small molecules that are able to diffuse into the pores of the wood cell walls that are inaccessible to enzymes (Archibald & Roy, 1992; Evans *et al.*, 1994). The rate of lignin degradation *in vivo* is controlled by the slowest reaction step, which in the case of MnP activity may be availability of H_2O_2 or Mn^{2+} rather than the catalytic rate of MnP. As excess H_2O_2 can destroy the catalytic site of MnP, the rate of production of H_2O_2 may have important effects on the rate of lignin degradation (Palma, Moreira & Feijoo Lema, 1997).

Laccases

The ability to oxidize phenolic compounds extracellularly is used to differentiate white rot fungi from brown rot species. A rapid screening test for white rots based on polyphenol oxidase activity – the Bavendamm test – monitors the development of a brown colouration on agar plates containing guaiacol or gallic acid. Polyphenol oxidase activity is shown by several enzymes including tyrosinase (oxidation of monophenols) and laccase, which oxidizes mono- and diphenols. The majority of white rot fungi produce laccase, frequently as the dominant extracellular enzyme in liquid cultures in the laboratory (e.g. for *T. versicolor* and *P. ostreatus*). Specific isomers of laccase can be induced *in vitro* by addition of compounds such as 2,5-dimethylaniline – a lignin mimic compound. *P. chrysosporium* generally does not produce laccase under artificial growth conditions though it has been reported during growth on a defined medium containing cellulose (Srinivasan *et al.*, 1995).

In the presence of oxygen, laccase converts mono- and diphenolic groups to quinone radicals then quinones in a multistep oxidation process (Thurston, 1994). Fungal laccase is a copper-containing enzyme found in several isoforms. Two are blue isomers with different isoelectric points and different abilities in binding to ion-exchange gels. Yellow laccases, also containing copper centres, have been isolated from cultures of *T. versicolor* and *Panus tigrinus* grown on solid substrates (Leontievsky *et al.*, 1997). Their colour is thought to result from binding of soil phenolics by the enzyme. Fungal laccases were first thought to be enzymes involved in the oxidation of phenolics to effect their removal from the fungal environment, though it was observed that mutant strains of *S. pulverulentum* (*P. chrysosporium*) without laccase were unable to degrade lignin (Ander & Eriksson, 1976). There is now general acceptance that laccases as well as LiP and MnP are involved in lignin degradation through attack on free phenolic groups in lignin and the generation of free radicals. Laccase and MnP both react with free phenolic groups, although in the presence of some low-molecular-mass mediators, laccase can also react with non-phenolic substituents on the aromatic rings. This permits laccase to operate at a higher redox level than normally (Leontievsky *et al.*, 1997).

Different white rot species produce various combinations of LiP, MnP and laccase depending on growth substrates. For example, *P. chrysosporium* secretes mainly LiP and MnP (Glenn & Gold, 1985); *Phlebia radiata* secretes laccase and MnP (Hatakka, 1994); *T. versicolor* synthesizes all three ligninolytic enzymes (Kadhim *et al.*, 1999). Enzymes produced in liquid cultures in laboratories are considered atypical of enzymes produced *in vivo*.

How these enzymes operate *in vivo* has been partially demonstrated using electron microscopy of sections of substrates colonized with fungal hyphae (reviewed by Evans *et al.*, 1991; Daniel, 1994). The technique used was immunogold-cytochemical labelling, which visualized specific enzyme molecules under the electron beam of the transmission electron microscope (Fig. 1.7). These studies have shown that enzymes such as LiP, MnP, laccases and cellulases are too large to penetrate into intact, secondary wood cell walls and remain close to the surface of the fungal hyphae or adhere to the inner surface of the wood cell wall (Evans *et al.*, 1991). Degradation of lignocellulose occurs by surface interaction between cell wall and enzymes, but initiation of decay can occur at a distance from the fungal hyphae, probably involving diffusible small molecular mass molecules such as oxalate, H_2O_2, $Fe^{2+/3+}$ and $Mn^{2+/3+}$ (Evans *et al.*, 1994). A predominant theory on the mechanism of cellulose degradation by brown

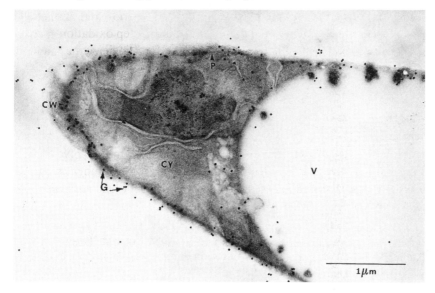

Fig. 1.7. Transmission electron micrograph showing localization of laccase on a hypha of *Trametes versicolor* by immunogold-cytochemical labelling. CW, cell wall; V, vacuole; CY, cytoplasm; G, gold-labelled laccase; P, plasmalemma.

rot fungi (which do not degrade lignin) is that Fe^{2+} reacts with H_2O_2 producing destructive hydroxyl radicals (Koenigs, 1974). Similar radicals can also be released from reactions between oxalate and H_2O_2 (Wood, 1994).

Conclusions

The ligninolytic capacity of white rot fungi makes them the most interesting taxa of fungi for use in bioremediation. Without an understanding of the mechanisms of lignin degradation, it would not be possible to address degradation of pollutants such as chlorophenols, nitrophenols and polyaromatic hydrocarbons in a practical way. These compounds can all be transformed using ligninolytic enzymes because of the free radical reactions, and development of industrial treatments using these fungi are proving successful (Bogan & Lamar, 1999; Jerger & Woodhull, 1999). Much lignocellulose material remains as waste products from industries such as forestry, agriculture and paper manufacture and use, which can be disposed of by accelerated natural biodegradation. The only products of value to accrue from such waste treatment are mushroom production and composts (Wood, 1989). New opportunities for commercial products and processes in fungal treatment of lignocellulosic materials are likely to arise in the future, and screening for novel organisms should contribute to the development of new technologies.

References

Adler, E. (1977). Lignin chemistry – past, present and future. *Wood Science Technology*, **11**, 169–218.

Ander, P. & Eriksson, K. E. (1976). The importance of phenol oxidase activity in lignin degradation by the white rot fungus *Sporotrichum pulverulentum*. *Archives in Microbiology*, **109**, 1–8.

Archibald, F. S. (1992). Lignin peroxidase activity is not important in biological bleaching and delignification of unbleached Kraft pulp by *Trametes versicolor*. *Applied and Environmental Microbiology*, **58**, 3101–3109.

Archibald, F. S. & Roy, B. (1992). Production of manganic chelates by laccase from the lignin degrading fungus *Trametes (Coriolus) versicolor*. *Applied and Environmental Microbiology*, **58**, 1496–1499.

Asiegbu, F. O., Johansson, M., Woodward, S. & Huttermann, A. (1998). Biochemistry of the host-parasite interaction. In Heterobasidion annosum: *Biology, Ecology, Impact and Control*, eds. S. Woodward, J. Stenlid, R. Karjalainen & A. Huttermann, pp. 167–194. Wallingford, UK: CAB International.

Atalla, R. H. & Agarwal, U. P. (1985). Raman microprobe evidence for lignin orientation in the native state of woody tissue. *Science*, **227**, 636–638.

Beguin, P. (1990). Molecular biology of cellulose degradation. *Annual Reviews of Microbiology*, **44**, 219–248.

Boddy, L., Gibbon D. M. & Grundy, M. A. (1985). Ecology of *Daldinia concentrica*: effect of abiotic variables on mycelial extension and interspecific interaction. *Transactions of the British Mycological Society*, **85**, 201–211.

Bogan, B. W. & Lamar, R. T. (1999). Surfactant enhancement of white rot fungal PAH soil remediation. In *Bioremediation Technologies for Polycyclic Aromatic Hydrocarbon Compounds*, eds. A. Leeson & B. C. Alleman, pp. 81–86. Columbus, OH: Battelle Press.

Bravo-Velasquez E. & Hedger, J. N. (1988). The effect of ecological disturbance on competition between *Crinipellis perniciosa* and other tropical fungi. *Proceedings of the Royal Society of Edinburgh Series B*, **94**, 159–166.

Brownleader, M. D., Byron, O., Rowe, A., Trevan, M., Welham, K. & Dey, P. M. (1996). Investigations into the molecular size and shape of tomato extensin. *Biochemical Journal*, **320**, 577–583.

Burdsall, H. H. (1981). The taxonomy of *Sporotrichum pruinosum* and *Sporotrichum pulverulentum/Phanerochaete chrysosporium*. *Mycologia*, **73**, 675–680.

Chesson, A., Gardner, P. T. & Wood, T. J. (1997). Cell wall porosity and available surface area of wheat straw and wheat grain fractions. *Journal of Science, Food and Agriculture*, **75**, 289–295.

Chung, N. & Aust, S. D. (1995). Inactivation of lignin peroxidase by hydrogen peroxide during the oxidation of phenols. *Archives of Biochemistry and Biophysics*, **316**, 851–855.

Collins, P. J. & Dobson, A. D. W. (1995). Extracellular lignin and manganese peroxidase production by the white rot fungus *Coriolus versicolor*. *Biotechnology Letters*, **17**, 989–992.

Cooke, R. G. & Rayner, A. D. M. (1984). *Ecology of saprotrophic fungi*. New York: Longman.

Covert, S. F., Wymelenberg, A. V. & Cullen, D. (1992). Structure, organisation and transcription of a cellobiohydrolase gene cluster from *Phanerochaete chrysosporium*. *Applied and Environmental Microbiology*, **58**, 2168–2175.

Cowling, E. B. (1975). Physical and chemical constraints in the hydrolysis of cellulose and lignocellulosic materials. *Biotechnology and Bioengineering Symposium*, **5**, 163–181.

Cowling, E. B. & Kirk, T. K. (1976). Properties of lignocellulosic materials as substrates for enzymatic conversion processes. *Biotechnology and Bioengineering Symposium*, **6**, 95–123.

Crawford, R. L. (1981). *Lignin Biodegradation and Transformation*. New York: Wiley & Sons.

Daniel, G. (1994). Use of electron microscopy for aiding our understanding of wood biodegradation. *FEMS Microbiology Reviews*, **13**, 199–233.

Daniel, G. F. & Nilsson, T. (1989). Interactions between soft rot fungi and CCA preservatives in *Betula verrucosa*. *Journal of the Institute of Wood Science*, **11**, 162–171.

Dey, P. M., Brownleader, M. D. & Harborne, J. B. (1997a). The plant, the cell and its molecular components. In *Plant Biochemistry*, eds. P. M. Dey & J. B. Harborne, pp. 1–47. New York: Academic Press.

Dey, P. M., Brownleader, M. D., Pantelides, A. T., Trevan, M., Smith, J. J. & Saddler, G. (1997b). Extensin from suspension-cultured potato cells: a hydroxyproline-rich glycoprotein devoid of agglutinin activity. *Planta*, **202**, 179–187.

Dixon, R. K., Brown, S., Houghton, R. A., Solomon, A. M., Trexler, M. C. & Wisniewski, J. (1994). Carbon pools and flux of global forest ecosystems. *Science*, **26**, 185–190.

Dodson, P. J., Evans, C. S., Harvey, P. J. & Palmer, J. M. (1987). Production and properties of an extracellular peroxidase from *Coriolus versicolor* which catalyses C_α–C_β cleavage in a lignin model compound. *FEMS Microbiology Letters*, **42**, 17–22.

Dutton, M. V. & Evans, C. S. (1996). Oxalate production by fungi: its role in pathogenicity and ecology in the soil environment. *Canadian Journal of Microbiology*, **42**, 881–895.

Dutton, M. V., Evans, C. S., Atkey, P. T. & Wood, D. A. (1993). Oxalate production by basidiomycetes including the white-rot species *Coriolus versicolor* and *Phanerochaete chrysosporium*. *Applied Microbiology and Biotechnology*, **39**, 5–10.

Eriksson, K. E. & Pettersson, B. (1975). Extracellular enzyme system utilized by the fungus *Sporotrichum pulverulentum* for the breakdown of cellulose. *European Journal of Biochemistry*, **51**, 213–218.

Eriksson, K. E., Pettersson, B., Volc, J. & Musilek, V. (1986). Formation and partial characterisation of glucose-2-oxidase, a hydrogen peroxide producing enzyme in *Phanerochaete chrysosporium*. *Applied Microbiology and Biotechnology*, **23**, 253–257.

Evans, C. S. (1985). Properties of the β-D-glucosidase (cellobiase) from the wood-rotting fungus *Coriolus versicolor*. *Applied Microbiology and Biotechnology*, **22**, 128–131.

Evans, C. S., Gallagher, I. M., Atkey, P. T. & Wood, D. A. (1991). Localisation of degradative enzymes in white-rot decay of lignocellulose. *Biodegradation*, **2**, 25–31.

Evans, C. S., Dutton, M. V., Guillen, F. & Veness, R. G. (1994). Enzymes and small molecular mass agents involved with lignocellulose degradation. *FEMS Microbiology Reviews*, **13**, 235–240.

Eveleigh, D. (1987). Cellulase: a perspective. *Philosophical Transactions of the Royal Society, London*, **321**, 435–437.

Field, J. A., de Jong, E., Feijoo-Costa, G. & de Bont, J. A. M. (1993). Screening for ligninolytic fungi applicable to the biodegradation of xenobiotics. *Trends in Biotechnology*, **11**, 44–49.

Flournoy, D. S., Kirk, T. K. & Highley, T. L. (1991). Wood decay by brown-rot fungi: changes in pore structure and cell wall volume. *Holzforschung*, **45**, 383–388.

Frankland, J. C. (1984). Autoecology and the mycelium of a woodland litter decomposer. In *Ecology and Physiology of the Fungal Mycelium*, eds. D. H. Jennings & A. D. M. Rayner, pp. 383–417. Cambridge, UK: Cambridge University Press.

Gallagher, I. M. & Evans, C. S. (1990). Immunogold-cytochemical labelling of β-glucosidase in the white-rot fungus *Coriolus versicolor*. *Applied Microbiology and Biotechnology*, **32**, 588–593.

Gilardi, G., Abis, L. & Cass, A. E. G. (1995). Carbon-13 CP/MAS solid state NMR and FT-IR spectroscopy of wood cell wall biodegradation. *Enzyme and Microbial Technology*, **17**, 268–275.

Glenn, J. K. & Gold M. H. (1985). Purification and characterisation of an extracellular Mn(II)-dependent peroxidase from the lignin degrading basidiomycete, *Phanerochaete chrysosporium*. *Archives of Biochemistry and Biophysics*, **242**, 329–341.

Goyal, A., Ghosh, B. & Eveleigh, D. (1991). Characteristics of fungal cellulases. *Bioresource Technology*, **36**, 37–50.

Guillen, F. & Evans, C. S. (1994). Anisaldehyde and veratraldehyde acting as redox cycling agents for hydrogen-peroxide-production by *Pleurotus eryngii*. *Applied and Environmental Microbiology*, **60**, 2811–2817.

Guillen, F., Martinez, A. T. & Martinez, M. J. (1990). Production of hydrogen peroxide by aryl-alcohol oxidase from the ligninolytic fungus *Pleurotus eryngii*. *Applied Microbiology and Biotechnology*, **32**, 465–469.

Hale, M. D. & Eaton, R. A. (1985). The ultrastructure of soft-rot fungi. Cavity-forming hyphae in wood cell walls. *Mycologia*, **77**, 594–605.

Hammel, K. E. (1997). Fungal degradation of lignin. In *Driven by Nature: Plant Litter Quality and Decomposition*, eds. G. Cadisch and K. E. Giller, pp. 33–45. Wallingford, UK: CAB International.

Hatakka, A. (1994). Lignin-modifying enzymes from selected white-rot fungi: production and role in lignin degradation. *FEMS Microbiology Reviews*, **13**, 125–135.

Hedger, J. N. (1985). Tropical agarics: resource relations and fruiting periodicity. In *Developmental Biology of Higher Fungi*, eds. D. Moore, L. A. Casselton, D. A. Wood & J. C. Frankland, pp. 41–86. Cambridge, UK: Cambridge University Press.

Hedger, J. N. & Basuki, T. (1982). The role of basidiomycetes in composts, a model system for decomposition studies. In *Decomposer Basidiomycetes: their Biology and Ecology*, eds. J. C. Frankland, J. N. Hedger & M. J. Swift, pp. 307–337. Cambridge, UK: Cambridge University Press.

Highley, T. L. (1977). Requirements for cellulose degradation by a brown rot fungus. *Material und Organismen*, **12**, 25–36.

Hofrichter, M., Scheibner, K., Schneegab, I. & Fritsche, W. (1998). Enzymatic combustion of aromatic and aliphatic compounds by manganese peroxidase from *Nematoloma frowardii*. *Applied and Environmental Microbiology*, **64**, 399–404.

Holmer, L. & Stenlid, J. (1997). Competitive hierarchies of wood decomposing basidiomycetes in artificial systems based on variable inoculum sizes. *Oikos*, **79**, 77–84.

Houtman, C. J. & Atalla, R. H. (1995). Cellulose–lignin interactions. *Plant Physiology*, **107**, 977–984.

Hyde, S. M. & Wood, P. M. (1997). A mechanism for production of hydroxyl radicals by the brown rot fungus *Coniophora puteana*: Fe(III) reduction by cellobiose dehydrogenase and Fe(II) oxidation at a distance from the hyphae. *Microbiology*, **143**, 259–266.

Jerger, D. E. & Woodhull, P. (1999). Technology development for biological treatment of explosives-contaminated soils. In *Bioremediation of Nitroaromatic and Haloaromatic Compounds*, eds. B. C. Alleman & A. Leeson, pp. 33–44. Columbus, OH: Battelle Press.

Johnsrud, S. C. (1988). Selection and screening of white rot fungi for de-lignification and upgrading of lignocellulosic materials. In *Treatment of Lignocellulosics with White Rot Fungi*, eds. F. Zadrazil & P. Reiniger, pp. 50–55. London: Elsevier.

Joseleau, J. P., Comtat, J. & Ruel, K. (1992). Chemical structure of xylans and their interaction in the plant cell walls. In *Xylans and Xylanases*, Vol. 7, *Progress in Biotechnology*, eds. J. Visser, G. Beldman, M. A. Kusters-van Someren and A. G. J. Voragen. pp. 1–16. London: Elsevier.

Kadhim, H., Graham, C., Barratt, P., Evans, C. S. & Rastall, R. A. (1999). Removal of phenolic compounds in water using *Coriolus versicolor* grown on wheat bran. *Enzyme and Microbial Technology*, **24**, 303–307.

Kersten, P. J. & Kirk, K. T. (1987). Involvement of a new enzyme, glyoxal oxidase in extracellular hydrogen peroxide production by *Phanerochaete chrysosporium*. *Journal of Bacteriology*, **169**, 2195–2201.

Kirk, T. K. & Farrell, R. L. (1987). Enzymatic 'combustion': the microbial degradation of lignin. *Annual Reviews in Microbiology*, **41**, 465–505.

Kirk, T. K. & Fenn, P. (1982). Formation and action of the ligninolytic system in basidiomycetes. In *Decomposer Basidiomycetes: their Biology and Ecology*, eds. J. C. Frankland, J. N. Hedger & M. J. Swift, pp. 67–90. Cambridge, UK: Cambridge University Press.

Kirk, T. K., Connors, W. J. & Zeikus, J. G. (1976). Requirements for a growth substrate during lignin decomposition by two wood rotting fungi. *Applied and Environmental Microbiology*, **32**, 192–194.

Klemen-Leyer, K., Agosin, E., Conner, A. H. & Kirk, T. K. (1992). Changes in molecular size distribution of cellulose during attack by white and brown rot fungi. *Applied and Environmental Microbiology*, **58**, 1266–1270.

Koenigs, J. W. (1974). Hydrogen peroxide and iron: a proposed system for decomposition of wood by brown-rot basidiomycetes. *Wood Fiber*, **6**, 66–80.

Kroon, P. A. Garcia-Conesa, M. T., Fillingham, I. J., Hazlewood, G. P. & Williamson G. (1999). Release of ferulic acid dehydrodimers from plant cell walls by feruloyl esterases. *Journal of Science, Food and Agriculture*, **79**,

428–434.

Lamar, R. T. & Dietrich, D. M. (1992). Use of lignin-degrading fungi in the disposal of pentachlorophenol-treated wood. *Journal of Industrial Microbiology*, **9**, 181–191.

Leatham, G. F. & Kirk, T. K. (1983). Regulation of ligninolytic activity by nutrient nitrogen in white rot basidiomycetes. *FEMS Microbiology Letters*, **16**, 65–67.

Leontievsky, A. A., Vares, T., Lankinen, P., Shergill, J. K., Pozdnyyakova, N. N., Myasoedova, N. M., Kalkkinen, N., Golovleva, L. A., Cammack R., Thurston C. F. & Hatakka, A. (1997). Blue and yellow laccases of ligninolytic fungi. *FEMS Microbiology Letters*, **156**, 9–14.

Lodge, J. & Cantrell, S. (1995). Fungal communities in wet tropical forests, variation in time and space. *Canadian Journal of Botany*, **73**, 1391–1398.

Lucas, P. W., Darvell, B. W., Lee, P. K. D., Yuen, T. D. B. & Choong, M. F. (1995). The toughness of plant cell walls. *Philosophical Transactions of the Royal Society of London Series B*, **348**, 363–372.

Mandels, M. & Steinberg, D. (1976). Recent advances in cellulase technology. *Fermentation Technology*, **54**, 267–286.

Masaphy, S., Henis, Y. & Levanon, D. (1996). Manganese-enhanced biotransformation of atrazine by the white rot fungus *Pleurotus pulmonarius* and its correlation with oxidation activity. *Applied and Environmental Microbiology*, **62**, 3587–3593.

Micales, J. A. (1995). In vitro oxalic acid production by the brown rot fungus *Postia placenta*. *Material und Organismen*, **28**, 197–207.

Montgomery, R. A. P. (1982). The role of polysaccharase enzymes in the decay of wood by basidiomycetes. In *Decomposer Basidiomycetes: their Biology and Ecology*, eds. J. C. Frankland, J. N. Hedger & Swift, pp. 51–65. Cambridge, UK: Cambridge University Press.

Newcomb, E. H. (1980). *The Biochemistry of Plants*, ed. N. E. Tolbert, pp. 2–50. New York: Academic Press.

Ong, E., Greenwood, J. M., Gilkes, N. R., Kilburn, D. G., Miller, R. C. Jr. & Warren, R. A. J. (1989). The cellulose-binding domain of cellulases: tools for biotechnology. *Trends in Biotechnology*, **7**, 239–243.

Paice, M. G., Reid, I. D., Bourbonnais, R., Archibald, F. S. & Jurasek, L. (1993). Manganese peroxidase produced by *Trametes versicolor* during pulp bleaching demethylates and delignifies Kraft pulp. *Applied and Environmental Microbiology*, **59**, 260–265.

Palma, C., Moreira, M. T. & Feijoo Lema, J. M. (1997). Enhanced catalytic properties of MnP by exogenous addition of manganese and hydrogen peroxide. *Biotechnology Letters*, **19**, 263–267.

Preston, R. M. (1974). *The Physical Biology of Plant Cell Walls*. London: Chapman & Hall.

Rayner, A. D. M. & Boddy, L. A. (1988). *Fungal Decomposition of Wood*, Chichester, UK: Wiley & Sons.

Rayner, A. D. M. & Todd, N. K. (1979). Population and community structure, and dynamics of fungi in decaying wood. *Advances in Botanical Research*, **7**, 333–420.

Rayner, A. D. M. & Webber, J. (1985). Interspecific mycelial interactions – an overview. In *Ecology and Physiology of the Fungal Mycelium*, eds. D. H. Jennings & A. D. M. Rayner, pp. 383–417. Cambridge, UK: Cambridge University Press.

Rayner, A. D. M., Boddy, L. & Dowson, C. G. (1987). Temporary parasitism of *Coriolus* spp. by *Lenzites betulina*, a strategy for domain capture in wood decay fungi. *FEMS Microbiology Ecology*, **45**, 53–58.

Rehman, A. U. & Thurston, C. F. (1992). Purification of laccase I from *Armillaria mellea. Journal of General Microbiology*, **138**. 1251–1257.

Reid, I. D. & Paice, M. G. (1998). Effects of manganese peroxidase on residual lignin of softwood Kraft pulp. *Applied and Environmental Microbiology*, **64**, 2273–2274.

Rishbeth, J. (1979). Modern aspects of the biological control of *Fomes* and *Armillaria. European Journal of Forest Pathology*, **9**, 331–340.

Srebotnik, E., Messner, K. & Foisner, R. (1988). Penetrability of white-rot degraded pine wood by the lignin peroxide of *Phanerochaete chrysosporium. Applied and Environmental Microbiology*, **54**, 2608–2614.

Srinivasin, C., D'Souza, T. M., Boominnathan, K. & Reddy, C. A. (1995). Demonstration of laccase in the white rot basidiomycete *Phanerochaete chrysosporium* BKM-F1767. *Applied and Environmental Microbiology*, **61**, 4274–4277.

Stamets, P. (1993). *Growing gourmet and medicinal mushrooms*, Berkeley, CA: Ten Speed Press.

Stamets, P. & Chilton, J. S. (1983). *The mushroom cultivator*, Olympia, WA: Agarikon Press.

Stenlid, J. & Redfern, D. (1998). Spread within the tree and stand. In Heterobasidion annosum: *Biology, Ecology, Impact and Control*, eds. S. Woodward, J. Stenlid, R. Karjalainen & A. Huttermann, pp. 125–142. Wallingford, UK: CAB International.

Swift, M. J. (1982). Basidiomycetes as components of forest ecosystems. In *Decomposer Basidiomycetes: their Biology and Ecology*, eds. J. C. Frankland, J. N. Hedger & M. J. Swift, pp. 307–337. Cambridge, UK: Cambridge University Press.

Swift, M. J., Heal, O. W. & Anderson, J. M. (1976). *Decomposition in Terrestrial Ecosystems*. Oxford, UK: Blackwell.

Taiz, L. & Zeiger, E. (1991). *Plant Physiology*. Redwood City, CA: Benjamin Cummings.

Thurston, C. F. (1994). The structure and function of fungal laccases. *Microbiology*, **140**, 19–26.

Tien, M. & Kirk, T. K. (1984). Lignin-degrading enzyme from *Phanerochaete chrysosporium*: purification, characterization and catalytic properties of a unique hydrogen peroxide-requiring oxygenase. *Proceedings of the National Academy of Sciences USA*, **81**, 2280–2284.

Tien, M. & Kirk, T. K. (1988). Lignin peroxidase of *Phanerochaete chrysosporium. Methods in Enzymology*, **161**, 238–249.

Ullah, M. A. (2000). Biotreatment of pentachlorophenol using wood rotting fungi. PhD thesis. University of Westminster, London.

Urzua, U., Lorrondo, S., Lobos, S., Larrain, J. & Vicupa, R. (1995). Oxidation reactions catalysed by manganese peroxidase isoenzymes from *Ceriporiopsis subvermispora. FEBS Letters*, **371**, 132–136.

Visser, J., Beldman, G., Kusters-van Someren, M. A. & Vorragen, A. G. J. (1992). *Xylans and Xylanases*, Vol. 7, *Progress in Biotechnology*. London: Elsevier.

Walker, L. P. & Wilson, D. B. (1991). *Enzymatic Hydrolysis of Cellulose*, pp. 3–14. London: Elsevier.

Watling, R. (1982). Taxonomic status and ecological identity in the basidiomycetes. In *Decomposer Basidiomycetes: their Biology and Ecology*, eds. J. C. Frankland, J. N. Hedger & M. J. Swift, pp. 1–32. Cambridge, UK: Cambridge University Press.

White, N. A., Low, G. A., Singh, J., Staines, H. & Palfreyman, J. W. (1997). Isolation and environmental study of wild *Serpula lachrymans* and *S. himantioides* from the Himalayan forests. *Mycological Research*, **101**, 580–584.

Wood, D. A. (1980). Production, purification and properties of laccase of *Agaricus bisporus. Journal of General Microbiology*, **117**, 327–338.

Wood, D. A. (1989). Mushroom biotechnology. *International Industrial Biotechnology*, **9**, 5–8.

Wood, D. A., Matcham S. E. & Fermor T. R. (1988). Production and function of enzymes during lignocellulose degradation. In *Treatment of Lignocellulosics with White Rot Fungi*, eds. F. Zadrazil & P. Reiniger, pp. 43–49. London: Elsevier.

Wood, P. M. (1994). Pathways for production of Fenton's reagent by wood rotting fungi. *FEMS Microbiology Reviews*, **13**, 313–320.

2

The biochemistry of ligninolytic fungi

PATRICIA J. HARVEY
AND CHRISTOPHER F. THURSTON

Introduction

The principal relevance of ligninolytic fungi to the field of bioremediation lies in their ability to degrade aromatic compounds. There are three groups of aromatics that constitute substantial pollutants: polyaromatic hydro-carbons (PAHs), benzene/toluene/ethyl benzene/xylene (BTEX) and the synthetic substituted aromatics typified by the chlorophenols. It may well be that ligninolytic fungi can play a useful role in bioremediation of all three types of pollutant, but the most interest is in degradation of the first and last groups, as BTEX remediation can exploit bacterial populations that promise to be efficient contributors to the process. We will largely be concerned with systems that are of possible direct application to PAH degradation as the halogenated hydrocarbons are degraded by increasingly well-understood biochemical pathways (see Reddy, Gelpke & Gold, 1998; Reddy & Gold, 1999). One of the main difficulties in the development of practical bioremediation processes rests in bringing metabolically active organisms into contact with the pollutant (see Field et al., 1995; Boyle, Wiesner & Richardson, 1998; Head, 1998; Novotny et al., 1999). The secreted enzyme systems of ligninolytic fungi may prove to be a powerful tool for PAH removal, and it is this aspect of their biochemistry to which this chapter is directed.

PAHs are a class of carcinogenic chemical that are formed whenever organic materials are burned; the amount of PAHs in soils coming from atmospheric fall-out have been rising steadily over the twentieth century. They appear in particularly high concentrations in many industrial sites, particularly those associated with petroleum and gas production. They are also found in high concentrations in the wood-preserving industry, which has relied heavily on creosote and anthracene oil as wood protectants (creosote contains 85% by weight of PAHs). PAHs have two or more fused benzene rings and are insoluble and stable, with angularly arranged ring structures (phenanthrene, benzo[a]pyrene, pyrene) being more stable than linear arrangements (anthracene, benz[*a*]anthracene) and with the two- and three-ringed structures proving to be much more biodegradable than four-, five- and six-membered rings. PAH degradation depends on the ability of microbes to introduce oxygen into the rings, which has the effect of increasing both PAH solubility and chemical reactivity (Sutherland, 1992; Wilson & Jones, 1993; Meulenberg *et al.*, 1997).

In the case of bacteria, dioxygenases typically catalyse the introduction of two oxygen atoms into the substrate to form dioxethanes, which are then further oxidized to dihydroxy products (Butler & Mason, 1997). Catechol, protocatechuic acid and gentisic acid are the usual dihydroxy products and are, in turn, ring-opened to succinic, fumaric, pyruvic and acetic acids, all of which are used for energy and cell protein synthesis. Alternatively, cytochrome P450-type monooxygenases may be employed to catalyse ring epoxidation as the first step in a pathway leading to PAH detoxification via the formation of various conjugates (Ferris *et al.*, 1976). These are not generally degraded (see Cerniglia, 1997). Both dioxygenases and cytochromes P450, however, are intracellular enzymes. PAH structures larger than two or three rings are extremely insoluble in water and cannot be taken up through the microbial cell wall. They, therefore, cannot be degraded by microorganisms that use dioxygenases or cytochromes P450 for PAH activation (see Wilson & Jones, 1993). The 'white rot' basidiomycetous fungi, however, possess an extracellular oxidative enzyme system that is used for the initial stages of attack on polyaromatic lignin. The extracellular reactions that they catalyse include lignin depolymerization as well as demethoxylation, decarboxylation, hydroxylation and aromatic ring opening. Many of the reactions result in oxygen activation, creating oxygen radicals that perpetuate the oxidative attack (Schoemaker *et al.*, 1985; Kirk & Farrell, 1987; Schoemaker, 1990). These features have sparked considerable research interest in the potential of white rot fungi to degrade the higher-molecular-weight, polymeric xenobiotics that are not degraded by bacteria.

The parallels between lignin and PAHs as targets for degradation

In the degradation of both lignin and higher-molecular-weight PAHs, initial extracellular depolymerization or extracellular aromatic ring-opening reactions must take place before the constituent rings can be metabolized. However, whereas PAHs consist of fused benzene rings, lignin is a much larger, more heterogeneous and amorphous polymer made up of phenylpropane subunits with a high content of β-O ether linkages and abundant carbon–carbon side chains (Adler, 1977). The problems that beset a microorganism in catalysing degradation of PAHs compared with lignin are therefore different. Both lignin and PAHs however, are hydrophobic and highly insoluble and pose similar problems for catalysis by enzymes, which tend to be water soluble and usually highly stereospecific. Available evidence suggests that many white rot fungi tackle the problem of lignin degradation with small diffusible oxidizing agents generated by their extracellular enzymes (see below). The same may be true in their degradation of PAHs (Bumpus *et al.*, 1985; Haemmerli *et al.*, 1986; Hammel, Kalyanaraman & Kirk, 1986a; Bogan & Lamar, 1995; Collins *et al.*, 1996).

The wood-rotting fungi

The process of degradation of woody plant material remains incompletely understood, notwithstanding its central place in terrestrial carbon cycling (the overwhelming majority of carbon fixed by land plants is found in lignocellulose). No easily isolated bacteria can completely degrade lignin, although some actinomycetes are able to achieve extensive modification of this material (McCarthy, 1987) and there are several lines of evidence suggesting that some bacteria can at least solubilize lignin. In contrast, a relatively large number of basidiomycete fungi are found naturally as wood or leaf-litter degraders and the majority of these – the white rot fungi – are capable of breaking down lignin in those examples that have been tested. It should be noted however, that the bulk conversion of radiolabelled lignin to $^{14}CO_2$ has only been demonstrated with a handful of species, while less direct evidence exists for a very much larger selection. The initial steps in this process involve oxidase and peroxidase enzymes secreted by these fungi. Different fungi appear to be able to achieve essentially the same effect with different combinations of enzymes. The most common components are the multicopper oxidase (laccase), and manganese peroxidase (MnP) (de Jong *et al.*, 1992). Lignin peroxidase (LiP) is a crucial component in some fungi and there are some less well-studied peroxidases that have atypical substrate specificities. The properties of these enzymes, the problem of H_2O_2 generation for activity of the peroxidases and the likely limitations on practical systems will be discussed.

Laccase

It is a paradox that there is still so much that is not understood about such an intensively studied enzyme. Laccase (benzenediol : oxygen oxidoreductase, EC 1.10.3.2) is a member of the small group of proteins known as the blue multicopper oxidases (Messerschmidt, 1997). These proteins (laccase, ascorbate oxidase and ceruloplasmin) all contain four or more copper atoms and have the property of reducing dioxygen completely to water. Laccase has been the focus of much spectroscopic and kinetic analysis because it was thought to be smaller and less complex than ascorbate oxidase and ceruloplasmin (Reinhammar & Malmstrom, 1981). Elucidation of the crystal structures of ascorbate oxidase (Messerschmidt *et al.*, 1989), ceruloplasmin (Lindley *et al.*, 1997) and type 2 copper-depleted laccase (Ducros *et al.*, 1998) have shown that laccase and ascorbate oxidase are remarkably similar, except that most (but not all) of the laccases so far analysed are monomeric, whereas the well-characterized ascorbate oxidases of the Cucurbitaceae are dimers of identical subunits. In general, laccase possesses a highly specific binding pocket for oxygen, but the binding pocket for reducing substrates appears to be shallow and relatively non-stereospecific. In fact, the governing feature of whether a compound will or will not be oxidized seems to be dictated to a very large extent by the redox potential differences between the reducing substrate and type I copper in the active site of the protein. This property endows laccase with the ability to oxidize a broad range of substrates provided their redox potentials are not too high (<1 V/NHE (normal hydrogen electrode)) (Reinhammer, 1972; Xu, 1996).

The laccases of ligninolytic fungi are secreted glycoproteins with the ability to catalyse the one-electron oxidation of a wide range of dihydroxy and diamino aromatic compounds (Thurston, 1994; Smith, Thurston & Wood, 1997). The role of laccase in lignin degradation has been a puzzle for many years. Briefly, the historical position is as follows. Until the 1980s, laccase was the only enzyme known to be secreted by fungi that had the ability to oxidize (poly)phenolic materials, but (i) not all ligninolytic fungi produced this enzyme, (ii) not all laccase-minus mutants showed reduced ligninolytic activity, (iii) purified laccases generally polymerized lignin-like materials rather than depolymerizing them and (iv) when tested with model compounds representing typical bond structures found in lignin, laccases were not able to oxidize the non-phenolic compounds, which generally had higher redox potentials than phenolic compounds (Meyer, 1987; Thurston, 1994; Smith *et al.*, 1997). The discovery of the ligninolytic peroxidases secreted by the white rot fungus *Phanerochaete chrysosporium* in 1983 (Tien & Kirk, 1983; see below) further diminished the likelihood that laccase was involved in lignin breakdown. This conclusion may need to be modified. Study of several fungal laccases showed that they vary quite markedly in their ability to oxidize substrates of different redox potentials (Xu, 1996), with some laccases being able to oxidize compounds with a standard redox potential up to 0.8 V/SCE (saturated calomel electrode) (1.042 V/NHE) (Kersten *et al.*, 1990; Xu *et al.*, 1996). Second, study of the distribution of the enzymes amongst the ligninolytic fungi suggests that secretion of laccase with MnP may be a common pattern (de Jong *et al.*, 1992).

Reactions catalysed by laccase

In common with MnP and LiP, laccase catalyses the single-electron oxidation of phenolic compounds to (cation) radicals. The radicals then react further in a manner that depends on the nature of the substituent groups and reaction conditions. With aromatic, lignin-like phenolics, their oxidation frequently results in carbon-to-carbon and carbon-to-oxygen coupling reactions between radicals, yielding products with a higher range of molecular size than the original (Gierer & Opara, 1973). However, as pointed out by Gianfreda, Xu & Bollag (1999), the tendency of laccase to catalyse polymerization is not necessarily a disadvantage for bioremediation purposes as sequestration of a pollutant by oxidative polymerization is an acceptable method of pollutant removal. Using lignin itself, which has a relatively low number of phenolic side chains (15 phenolic OH per 100 C_9 units (Adler, 1977)), laccase catalyses both polymerization and depolymerization reactions. Depolymerization comes about because the single-electron oxidation of some phenolic compounds can result in (depolymerizing) alkyl–arene cleavage (Kirk, Harkin & Cowling, 1968). Again, dependent on the nature of the phenol selected and reaction conditions, C_α-oxidation, demethoxylation and decarboxylation reactions are possible (Krisnangkura & Gold, 1979; Schoemaker, 1990). Reactions of this nature would be able to increase the redox status of PAHs, leading to their incorporation into cellular material. Clearly, the ability of laccase to catalyse non-specific one-electron oxidations is likely to be of relevance in PAH remediation. However, the influence of co-substrates in laccase-catalysed reactions is of even greater interest in the context of improving PAH accessibility for remediation.

Influence of co-substrates

Three independent lines of enquiry lead to the conclusion that laccase is capable of an extended range of reactions if a co-substrate is provided. First, the laccases of *Trametes versicolor* can oxidize certain non-phenolic substrates and depolymerize Kraft lignin *in vitro*, if given 2,2′-azinobis(3-ethylbenzthiazoline-6-sulfonate) (ABTS) as a co-substrate (Bourbonnais & Paice, 1990; Bourbonnais *et al.*, 1995; Collins *et al.*, 1996). Note, however, that in these reactions, the mechanism of degradation by the laccase–ABTS couple involves hydrogen atom abstraction rather than single-electron oxidation (Muheim *et al*, 1992). 3,4-Dimethoxybenzyl alcohol (veratryl alcohol, VA), for example, has a standard redox potential of 1.36 ± 0.01 V/NHE (Bietti, Baciocchi & Steenken 1998), placing it out of the substrate range for oxidation by laccase alone (Kersten *et al.*, 1990). However, when both laccase and ABTS are present in the reaction mixture, VA is oxidized to veratraldehyde, but no other products are detectable. This contrasts with the situation with LiP, for which VA serves as an assay substrate (Tien & Kirk, 1988). With LiP, which catalyses the single-electron oxidation of VA, quinones and ring-opened products are also produced (Leisola *et al.*, 1985a; Haemmerli *et al.*, 1987; Bietti *et al.*, 1998). Second, laccase will delignify Kraft pulps with 1-hydroxybenzotriazole as co-substrate (Call, 1994). Third, the white rot fungus *Pycnoporus cinnabarinus*, which is strongly ligninolytic, secretes 3,4-hydroxyanthranilic acid along with laccase. 3,4-Hydroxyanthranilic acid, acting as a co-substrate for laccase, enables the cleavage of a range of non-phenolic model compounds and the depolymerization of 'soluble lignin' (Eggert *et al.*, 1995, 1996; Eggert, Temp & Eriksson, 1997).

Enzyme reactions that produce small diffusible oxidizing agents able to penetrate into substrate matrices not otherwise permeable to the enzymes themselves provide a route for increasing the availability of PAHs for microbial metabolism and are implicated in many of the reactions catalysed by both LiP and MnP (see below). Co-substrate-assisted reactions of laccase may extend its potential in PAH degradation, as has now been demonstrated by the laccase–ABTS-coupled oxidation of anthracene to anthraquinone and oxidation of benzo[*a*]pyrene (Collins *et al.*, 1996).

Manganese peroxidase

MnP is an extracellular glycosylated heme enzyme secreted by a variety of white rot fungi that uses H_2O_2 to oxidize Mn^{II} to a Mn^{III}-chelate; this in turn, oxidizes phenolic substrates as a freely diffusible, non-specific oxidant (for a review, see Gold & Alic, 1993). The ability to generate oxidizing Mn-chelates that might ultimately increase the availability of PAHs for their degradation underpins the expectation of the usefulness of MnP in bioremediation (see for example, Bogan, Lamar & Hammel, 1996a).

MnP belongs to class II of the peroxidase family, which is designated for extracellular fungal peroxidases. These have only limited homology with sequences from other peroxidases but share a striking level of structural similarity within the class, which is related to the envelopment of the protein around a heme moiety. The heme is the site of oxidation of the protein by H_2O_2, which is essential in creating the catalytic intermediates, termed Compound I and Compound II, that are required for catalysis (Dunford, 1999; see Equations 2.1 and 2.2). MnP has been crystallized (Sundaramoorthy *et al.*, 1997) and from X-ray crystallographic studies as well as DNA sequence comparisons, its heme environment is very similar to that of other plant and fungal peroxidases (see Banci, 1997; Smith & Veicht, 1998; Dunford, 1999). However, MnP seems unable to exploit the oxidizing power of H_2O_2 to oxidize organic substrates with redox potentials beyond 1.12 V/SCE (1.362 V/NHE) (Popp & Kirk, 1991). It is unique though, in being able to oxidize Mn^{II} to Mn^{III}. This specificity relates not only to the redox potential that can be sustained by the oxidized protein and the influence of chelators but also to the structural properties of the protein. In MnP, for example, there is a unique binding site for Mn^{II} that involves the carboxylate side chains of three acidic amino acid residues. In LiP, with which MnP shares a high degree of sequence homology, neutral or positive residues replace these and preclude Mn^{II} from binding.

Catalytic cycle

Reaction of the native enzyme (E, see Equation 2.1 below) with H_2O_2 yields Compound I, which contains an oxyferryl heme with a porphyrin cation radical. Two steps of single electron reduction by Mn^{II} restore the native enzyme via the intermediate Compound II.

$$E + H_2O_2 \rightarrow \text{Compound I} + H_2O \tag{2.1}$$
$$\text{Compound I} + Mn^{II} \rightarrow \text{Compound II} + Mn^{III} \tag{2.2}$$
$$\text{Compound II} + Mn^{II} \rightarrow E + Mn^{III} + H_2O \tag{2.3}$$
$$\text{Compound II} + H_2O_2 \rightarrow \text{Compound III (inactive)} \tag{2.4}$$

Importantly for catalysis, the supply of H_2O_2 relative to Mn^{II} needs to be poised to ensure that the competing reaction of Compound II with H_2O_2 does not take place (Equation 2.4), since this has the effect of driving the enzyme into a catalytically inactive mode (Wariishi, Akileswaran & Gold, 1988).

Requirement for chelators

Organic acids such as malonate, citrate, glyoxylate and oxalate are essential in chelating and stabilizing Mn^{III} (Glenn & Gold, 1985) and are common secondary metabolites of wood-rotting basidiomycetes, secreted at the same time as MnP. Among these, oxalate shows unique effects in chelating and stabilizing Mn^{III} (Kuan & Tien, 1993; Kishi *et al.*, 1994) and may bind quite closely to the heme during catalysis. Manganese(II) reacts with the oxidized forms of MnP as a monochelated complex (Kuan & Tien, 1993) but is released from the enzyme in its dichelated form (Kishi *et al.*, 1994).

Oxidation of phenolic compounds

Unchelated Mn^{III} has a high standard redox potential of 1.5 V/NHE but its potential is reduced by chelation with organic acids to the order of 1.12 V/SCE (1.362 V/NHE) (Popp & Kirk, 1991). According to the Nernst equation, this means that a very broad range of organic substrates might be oxidized by the MnP oxidation system if the concentration of chelated Mn^{III} could be maintained sufficiently high. In practice, the rate of generation of Mn^{III} by MnP seems to be a limiting factor in restricting the substrate range of MnP to the more easily oxidizable phenolic-containing lignin substructures. Manganese(III)-chelates that are generated by MnP oxidize monomeric phenols, phenolic lignin dimers (Wariishi, Valli & Gold, 1989a; Tuor *et al.*, 1992) and synthetic lignin (Wariishi, Valli & Gold, 1991) to phenoxy radicals, with the effect that similar reactions as described for laccase ensue. Consequently, while MnP differs from laccase in being highly specific for the nature of its reducing substrate (Mn^{II}), the fact that its reactions are mediated by a small, diffusible and non-specific redox agent (Mn^{III}-chelate) ensures it has a role in pathways leading to the metabolism of the more inaccessible PAHs.

Influence of co-substrates

In a similar manner to laccase, the range of substrates that can be attacked by MnP is increased in the presence of co-substrates. For example, in the presence of glutathione, VA, which is not normally oxidized by the MnP-Mn-chelate system, can be oxidized to veratraldehyde via thiol radicals generated from the oxidation of glutathione by Mn-chelates (Wariishi *et al.*, 1989b). Similarly there is evidence for the degradation of non-phenolic lignin (Bao *et al.*, 1994; Jensen *et al.*, 1996) and phenanthrene (Moen & Hammel, 1994) in the presence of unsaturated lipids. Most probably, lipid peroxidation by Mn^{III}-chelates creates lipid peroxyl and alkoxyl radicals, which initiate degradation either via radical cations or *via* oxy radicals that add to the aromatic rings (Jensen *et al.*, 1996).

Lignin peroxidase

LiP is a water-soluble, glycosylated enzyme secreted by white rot fungi and, like MnP, is also dependent on H_2O_2 for catalysis (see Kirk & Farrell, 1987; Harvey, Schoemaker & Palmer, 1987a; Gold & Alic, 1993; Reddy & D'Souza, 1994). LiP, however, is unique in being able to produce radical cations from the one-electron oxidation of non-phenolic aromatic compounds such as VA or 1,4-dimethoxybenzene (DMB), which have redox potentials beyond the reach of either MnP or laccase (Kersten *et al.*, 1990; Popp & Kirk, 1991). Radical cations of VA or DMB are able to act as non-specific redox mediators, with the effect that both the substrate range and redox capacity of LiP can be extended (see below). LiP comprises a family of isozymes, and its major isozymes have now been crystallized (Edwards *et al.*, 1993; Piontek, Glumoff & Winterhalter, 1993; Choinowski, Blodig & Winterhalter, 1999), the genes identified (Cullen, 1997) and site-directed mutagenesis programmes are under way (Doyle *et al.*, 1998; Ambert-Balay, Fuchs & Tien, 1998).

Catalytic mechanism

LiP has the same heme and similar active site residues as MnP and the same catalytic cycle: initial oxidation with H_2O_2 yields the two-electron oxidized enzyme (Compound I) with Fe^{IV} and a porphyrin cation radical. Two steps of single-electron reduction restore the enzyme to the native state (Marquez *et al.*, 1988). Compound II can also react with H_2O_2 to form the catalytic dead-end intermediate Compound III, depending on the relative concentrations of H_2O_2 to reducing substrate; see Equation 2.4 above (Wariishi *et al.*, 1990; Cai & Tien, 1992). Compound III will either react with further H_2O_2 to be destroyed completely, or it will break down slowly to the native state by as yet unclear mechanisms.

Redox potential, in part, determines whether an aromatic nucleus is a substrate for LiP. Strong electron-withdrawing groups such as an α-carbonyl group, or nitro groups on the benzene ring tend to deactivate

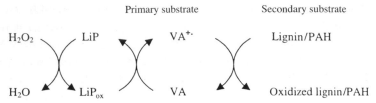

Fig. 2.1. The oxidation of lignin or polyaromatic hydrocarbons (PAH) mediated by veratryl alcohol (VA). The radical $VA^{+\cdot}$ is generated as the primary product of oxidation of H_2O_2 catalysed by lignin peroxidase (LiP).

aromatic nuclei, whereas alkoxy groups tend to activate them. On the surface of it, LiP would appear to be relatively non-specific towards the nature of the reductant and will accept electrons, directly, from a broad range of substrates including sulfides (Baciocchi *et al.*, 2000). However, in much the same way that Mn^{II} is a crucial substrate for MnP, small dimethoxylated non-phenolic aromatics such as VA or DMB serve as crucial substrates for LiP (Harvey & Candeias, 1995). The products of the oxidation of dimethoxylated aromatics are radical cations (Harvey *et al.*, 1985; Kersten *et al.*, 1985; Schoemaker *et al.*, 1985; Hammel, Kalyanaraman & Kirk, 1986b). As with phenolic (cation) radicals, they will react without further enzyme involvement in a variety of different ways depending on the nature of the reaction conditions and the pattern of substitution around the aromatic ring. The radical $VA^{+\cdot}$, for example, breaks down by carbon–hydrogen cleavage to give a carbon-centred radical and, ultimately, veratraldehyde (Candeias & Harvey, 1995; Bietti *et al.*, 1998). Radical cations, however, are also able to act as one-electron oxidants themselves, creating reactive radical cations in secondary substrates by charge transfer (Harvey, Schoemaker & Palmer, 1986, Harvey *et al.*,1992) (Fig. 2.1). So, for example, when the secondary substrates are phenylpropanoid structures that typify lignin, carbon–carbon and C_β–O-ether side chain cleavage reactions of the respective radical cations lead to lignin depolymerization in the secondary substrates. Note that under these circumstances, the primary oxidant is restored following oxidation of the secondary substrate.

Redox mediation with $VA^{+\cdot}$

VA is a fungal metabolite produced at the same time as LiP (Lundquist & Kirk, 1978) and numerous different reactions of the LiP–redox mediator oxidizing system with VA as the mediator have now been documented for a broad range of compounds. They include benzo[*a*]pyrene oxidation

(Haemmerli *et al.*, 1986); lignin depolymerization (Hammel & Moen, 1991; Hammel *et al.*, 1993); aromatic ring-opening and ring hydroxylation reactions (Schoemaker, 1990). However, the mechanism by which VA enhances the catalysis of LiP has been highly contentious (see Valli, Wariishi & Gold, 1990). In 1986, Harvey *et al.* proposed that $VA^{+\cdot}$ was generated by LiP specifically to act as a small diffusible mediator in the process of lignin oxidation – akin to the Mn^{III}-chelates described above for MnP. On oxidizing lignin aromatic moieties to radical cations, $VA^{+\cdot}$ was reduced back to VA. Recognizing the tendency of LiP to form Compound III from Compound II with substrates that did not form stable radical cations, this model was later extended. VA was proposed to reduce Compound I and form a bound Compound II–$VA^{+\cdot}$ intermediate. By forming the Compound II–$VA^{+\cdot}$ intermediate, reduction by a further VA molecule was facilitated over the alternative reaction of Compound II with H_2O_2. On completion of the catalytic cycle, two VA radical cations were released (Harvey, Schoemaker & Palmer, 1987b; Harvey *et al.*, 1989; Harvey & Candeias, 1995). A wealth of data support this model (Gilardi *et al.*, 1990; Edwards *et al.*, 1993; Candeias & Harvey, 1995; Ambert-Balay *et al.*, 1998; Bietti *et al.*, 1998; Doyle *et al.*, 1998). The most significant of these, which has also been the subject of dispute, relates to the rate of decay of $VA^{+\cdot}$, since this determines whether or not $VA^{+\cdot}$ is capable of acting as a diffusible redox mediator. Aust, for example, reported a value of 1.2×10^3 s^{-1} (Khindaria, Yamazaki & Aust, 1996), implying that $VA^{+\cdot}$ could not act as a diffusible mediator. However, Candeias & Harvey (1995) reported that the rate of decay of $VA^{+\cdot}$ is 17 ± 1 s^{-1} at pH ≤ 5 and this value has now been confirmed by Bietti *et al.* (1998). Therefore, depending on reaction conditions, $VA^{+\cdot}$ would indeed be able to act as a diffusible redox mediator. The model of Sheng & Gold (1999) may need to be revised in the light of these data. The ability of $VA^{+\cdot}$ to act as a diffusible mediator means that, according to the Nernst equation, compounds with a higher standard redox potential than the $VA^{+\cdot}/VA$ couple could also be oxidized by the LiP–VA system, provided $VA^{+\cdot}$ can be generated in a sufficiently high (local) concentration. Such seems to be the case for the oxidation of the dye poly R (Candeias & Harvey, 1995) and isoeugenol (ten Have *et al.*, 1999). Isoeugenol, for example, has an ionization potential (IP), of 9.0 eV (ten Have *et al.*, 1999). The IP value gives a measure of the ease with which one electron is abstracted from the highest occupied molecular orbital. The IP for VA, by comparison is 8.67 ± 0.06 eV (ten Have *et al.*, 1998, 1999). In mixed substrate oxidations with VA functioning as a redox mediator, isoeugenol can now be oxidized (ten Have *et al.*, 1998, 1999).

Requirement for a pH gradient

The deprotonation reaction of $VA^{+\cdot}$ is induced by OH^- ($k = 1.3 \times 10^9 \, mol \, l^{-1} \, s^{-1}$; Bietti *et al.*, 1998). As pointed out by these authors, this means that for an optimal rate of polymer oxidation catalysed by the LiP-generated $VA^{+\cdot}$ system, it is important to keep the pH of the enzyme environment as low as possible, yet maintain a high local pH in the vicinity of the lignin macromolecule to favour base-induced carbon–carbon fragmentation. LiP catalysis prevails over an acid range of pH values (up to pH 4.5) with an optimum at around pH 2.75, the lower limit being dictated by the opposing tendency for protein precipitation. Many of these wood-degrading fungi secrete oxalate during secondary metabolism (Dutton *et al.*, 1993), which may have the desired effect of creating a pH gradient extending from the fungus (acid), which represents the source of enzyme, to the polymeric substrate (alkaline). The implications of this analysis are that for biodegradation purposes, both VA and LiP will be required, as well as a carefully regulated pH gradient.

Oxygen activation in LiP-catalysed reactions

Activated oxygen species (superoxide anion ($O_2^{-\cdot}$); perhydroxyl radical ($HOO\cdot$); H_2O_2 and hydroxyl radical ($OH\cdot$)) may play a role in the oxidation processes initiated by LiP (see, for example, Forney *et al.*, 1982; Bes, Ranjeva & Boudet, 1983; Renganathan, Miki & Gold, 1986; Palmer, Harvey & Schoemaker, 1987). Carbon-centred radicals formed from the fragmentation of radical cations react with oxygen at a diffusion-limited rate. Since, during growth of LiP-secreting fungi, an acid environment is created (Dutton *et al.*, 1993), the appearance of $HOO\cdot$ ($pK_a = 4.8$) may be favoured, depending on the pH value attained. The perhydroxyl radical has a high standard redox potential (1.5 V/NHE) and could well perpetuate the process of oxidative attack on aromatic structures. It is also able to dismutate with superoxide ($k = 1.5 \times 10^7 \, M^{-1} \, s^{-1}$) to provide H_2O_2 to drive further peroxidative reactions. Indeed, many examples exist indicating superstoichiometric oxidation by LiP with respect to the initial amount of H_2O_2 supplied to start the reaction (Hammel *et al.*, 1985), either by supplying H_2O_2 or by forming $HOO\cdot$. Activated oxygen will also react with radical cations to bring about aromatic ring-opening (Haemmerli *et al.*, 1987; Schoemaker, 1990). The ability to catalyse extracellular aromatic ring-opening reactions without requiring NAD(P)H, which is needed for the cytochromes P450 or bacterial dioxygenases (Butler &

Mason, 1997), is likely to be important in the degradation of high-molecular-weight PAHs.

Oxidation of phenolic compounds

Since the reactivity of LiP towards substrates depends largely on redox potential, it is to be expected that phenolic compounds will also be oxidized by LiP. Phenolic compounds can be oxidized by the enzyme directly, or by radical cations such as $VA^{+\cdot}$ generated as products of LiP catalysis. In mixed substrate experiments with VA and phenolics, the reaction products lie very strongly in the direction of oxidized phenolics (Harvey & Palmer, 1990; Harvey *et al.*, 1993; Chung & Aust, 1995a,b). This outcome mimics the reactivity of MnP or laccase towards phenols and the dissipation of the oxidizing power of LiP with H_2O_2. In the absence of VA, however, reduction of Compound I by phenolics yields an enzyme-associated phenoxy radical with Compound II, and, unlike the situation with the $VA^{+\cdot}$-associated enzyme intermediate, there seems now to be no kinetic pathway to enable Compound II to be reduced as effectively to the native state. Consequently, a reaction with H_2O_2 ensues forming the catalytically inactive Compound III intermediate of LiP. When LiP was used to synthesize lignin from phenolic precursors (Sarkanen *et al.*, 1991), it was supplied in stoichiometric, not catalytic amount with respect to the phenolic substrate, because of its inactivation when oxidizing phenolic compounds. A further corollary is that experiments aimed at recovering active LiP activity from phenolic-containing materials are not generally successful (see Khazaal *et al.*, 1993; Schutzendubel *et al.*, 1999).

In vivo 'feedback control' of the LiP enzyme system by phenolic compounds in conjunction with the phenol-oxidizing activities of other enzymes may be biologically relevant in lignin breakdown by the fungus (Harvey *et al.*, 1993). However, for an enzyme-based technology dealing with phenolic-containing material and requiring an active (LiP) enzyme system, it must be ensured that the redox mediator is oxidized instead of the phenolic compounds. One (technologically complex) solution lies in compartmentation of the enzyme–redox mediator system away from the phenolic contaminants. This would have the added advantage of minimizing the risk of dissipating the unique oxidizing capability of the LiP–redox mediator system with H_2O_2 towards non-phenolic compounds.

Enzyme systems

The oxidizing enzyme systems of white rot fungi offer unique opportunities for bioremediation of the higher-molecular-weight PAHs by increasing their bioavailability, water solubility and redox status for subsequent metabolism. However, from the foregoing analysis, it is clear that successfully emulating the action of the extracellular enzymes produced by ligninolytic fungi in an enzyme-based technology will be both expensive and technically complex, for a number of reasons.

1. LiP and MnP have obligatory requirements for cofactors: Mn^{II} and an organic acid chelator in the case of MnP, and an organic redox mediator in the case of LiP. Manganese(II) is relatively abundant, but the organic cofactors will represent a significant cost to the system.

2. Redox mediators like VA produce radical cations with only a limited lifetime ($17 \pm 1 \, s^{-1}$ at pH < 5) and, in the absence of redox mediation, they fragment and are lost to the system. They may need to be continuously replenished.

3. Cosubstrates (e.g. lipids) that activate oxygen and increase the substrate range of all enzymes might be required, adding a further cost to the system.

4. A pH gradient may be necessary, extending from the enzyme (acid) to the PAHs (alkaline).

5. Oxidation of phenolics by LiP would have to be avoided to prevent enzyme inactivation. This might be achieved by compartmentation or with combinations of laccase and MnP, although we do not presently know what amounts of the different enzymes would make up an optimal mixture. Even in *P. chrysosporium*, there are profound differences of expression of individual *lip* and *mnp* genes under different environmental conditions that currently have no mechanistic explanation (Bogan *et al.*, 1996b; Janse *et al.*, 1998).

6. Hydrogen peroxide is required for LiP and MnP catalysis. This compound is highly reactive and toxic and would need to be supplied at a regulated rate to prevent enzyme inactivation, either recoverable (Compound III formation) or permanent (heme destruction). It is impossible to conceive an economically viable procedure for direct application of this chemical throughout bulk contaminated soil such that its concentration is maintained within the range required for peroxidase activity over periods of months to years. Addition of FAD-containing oxidases (cellobiose dehydrogenase (Wilson, Hogg & Jones,

1990), glucose oxidase (Green, 1977)) that reduce O_2 to H_2O_2 may represent a way forward but would add considerable complexity to the system.

7. Oxidation of phenolics tends toward their polymerization to higher-molecular-weight compounds. To prevent recondensations between phenoxy radicals, further enzyme mixtures might need to be introduced, for example glucose oxidase (Green, 1977) cellobiose dehydrogenase (Ander *et al.*, 1990; Wilson *et al.*, 1990; Samejima & Eriksson, 1991) or VA oxidase (Marzullo *et al.*, 1995). Many of these reduce quinoids or radical compounds and may shift the depolymerization–repolymerization equilibrium towards degradation, but their inclusion again adds complexity.

8. Addition of oxygen-consuming enzymes, such as laccase, glucose, cellobiose or VA oxidases, may reduce the supply of oxygen necessary for aromatic ring opening and ring-activation reactions.

9. Extracellular turnover is not well characterized for any fungus (see Wood, 1980), but polluted soils are certain to be a less-than-optimal environment for maintenance of any enzyme activity.

Implications for bioremediation strategies: biofarming

There is much work devoted to the development of enzyme-based bio-remediation processes, particularly for the more water-soluble pollutants (discussed in detail for laccase in Gianfreda *et al.*, 1999). If, however, bioremediation of large-scale pollution such as former sites of coal gas production is to be practicable, it is unlikely that enzyme technology will be competitive with whole organism-based methods. Specifically, we predict that the use of fungi expressing a complete ligninolytic system will be the most effective enhancer of PAH removal from contaminated soil. Already there are good indications for this approach, with sequential breakdown by white rot fungi followed by indigenous bacteria being advocated as a method for effective PAH bioremediation (Meulenberg *et al.*, 1997). The process will be a 'biofarming' procedure and involve digging up the contaminated soil so that it can be mixed with organism(s) and substrate(s). White rot fungi normally express ligninolysis in wood, so these organisms will need to be introduced as inocula into a contaminated site. The mixture will have to support ligninolytic growth of the fungus and will probably require the presence of a solid substrate such as straw, wood chips/dust or a compost of some sort, although the use of a soluble humic fraction, (as can be isolated from compost; Smith, 1994), deserves further study. Solid substrates such as straw or wood chips may also be advantageous as a means to distribute fungal inoculum evenly in large volumes of soil. Consideration will need to be given to extant PAH-degrading soil microorganisms that antagonize fungal growth by antibiotic production or by competition for nutrients (Tucker *et al.*, 1995), and vice versa, if a synergistic outcome between white rot fungi and soil microorganisms is to be attained (Gramss, Voigt & Kirsche, 1999). It should be noted that some seminal experiments have already been conducted in this area by Lamar and his colleagues (Lamar & Dietrich, 1992; Lamar, Evans & Glaser, 1993) and there are also studies on bioremediation of PAH-treated wood (Majcherczyk & Hutterman, 1998) and related soil contamination (Borazjani & Diehl, 1998)

Control of expression of the laccase and peroxidase genes: how to keep the necessary genes turned on

The control of expression of the genes for ligninolytic enzymes is yet another aspect of the problem that must influence conditions for efficient PAH removal. In *P. chrysosporium*, the induction of LiP activity and MnP activity in laboratory cultures is typically achieved by stressing cultures that have grown up with glucose as the carbon source (see Leisola, Thanei-Wyss & Fiechter, 1985b). The stress usually involves exposure to a pure oxygen atmosphere under conditions of nutrient (either nitrogen or carbon) starvation. Studies using cellulose as the carbon source, however, have indicated that freely available glucose at the outset of spore germination is sufficient to switch oxidative metabolism completely away from that participating in ligninolysis (Zacchi *et al.*, 2000a; Zacchi, Morris & Harvey, 2000b; Zacchi, Palmer & Harvey, 2000c). Cultures maintained on free glucose, even under so-called 'carbon-limited conditions', contained severely impaired mitochondria and used anaerobic metabolism for ATP synthesis. In the absence of freely available glucose, cultures did not need to be exposed to a pure oxygen atmosphere to express the ligninolytic system and retained functional mitochondria (Zacchi *et al.*, 2000c). The practice of cultivating cultures on glucose-containing medium has led to the commonly held misconception that peroxidase induction and lignin degradation are 'idiophasic' or part of secondary metabolism. These processes are subject to catabolite repression, being switched off by high concentrations of readily metabolizable carbon source, but this has no relevance to the normal habitat of these fungi (dead wood is effectively glucose free; Cote, 1977). It will, however, have relevance in considering the nature of any amendments that are made to the contaminated site, and microbial communities that effectively remove free glucose may, therefore, play a key role in permitting expression of the ligninolytic system and consequent xenobiotic degradation. By the same token, in many white rot fungi, laccases are only secreted at high levels in laboratory cultures if induced with compounds such as 2,5-xylidine, although there is a wide variation in their behaviour in respect to laccase control (Gianfreda *et al.*, 1999). When white rot fungi degrade wood or leaf litter, however, they secrete the necessary enzymes along with the necessary co-substrates and other factors, and this would suggest that fungi growing on a lignocellulose substrate would be suitable for PAH removal, but quite possibly not optimal. If the regulation of gene expression for ligninolytic enzymes is to be changed, it is likely that this will be done by gene engineering as none of the white rot fungi have facile conventional genetic systems. This, in turn, will require careful handling if the necessary regulatory hurdles are to be crossed. Obtaining permissions for release of such manipulated organisms should not be an insuperable problem as (i) the intention of environmental clean-up is generally popular, (ii) the manipulated organisms would be most unlikely to degrade wood faster than existing organisms (which have the benefit of millions of years of evolution), and (iii) changing the levels of secretion of extracellular enzymes does not allow some of the wilder scenarios claimed for recombinants containing genes specifying products from other organisms. It would, nevertheless, be highly desirable to set up procedures for gene engineering that were not dependent on antibiotic selection systems.

Conclusions

Wood-degrading white rot fungi show remarkable versatility in being able to degrade a wide spectrum of recalcitrant organopollutants. They are in possession of extracellular oxidizing enzyme systems that will increase the bioavailability, water solubility, and redox status of PAHs in preparation for their subsequent metabolism. Enzyme technology, however, is unlikely to be competitive with whole organism-based methods. Current limitations to the widespread use of white rot fungi in bioremediation include a full understanding of factors that regulate expression of the ligninolytic system and much remains to be achieved. However, the development of a technology that employs lignin-degrading fungi to remediate soils is promising.

References

Adler, E. (1977). Lignin chemistry – past, present and future. *Wood Science Technology*, **11**, 169–218.

Ambert-Balay, K., Fuchs, S. & Tien, M. (1998). Identification of the veratryl alcohol binding site in lignin peroxidase by site-directed mutagenesis. *Biochemical and Biophysical Research Communications*, **251**, 283–286.

Ander, P., Mishra, C., Farrell, R. & Eriksson, K.-E. (1990). Redox reactions in lignin degradation: interaction between laccase, different peroxidases and cellobiose: quinone oxidoreductase. *Journal of Biotechnology*, **13**, 189–198.

Baciocchi, E., Gerini, M. F., Harvey, P. J., Lanzalunga, O. & Manccinelli, S. (2000). Oxidation of aromatic sulfides by lignin peroxidase from *Phanerochaete chrysosporium. European Journal of Biochemistry*, **267**, 2705–2710.

Banci, L. (1997). Structural properties of peroxidases. *Journal of Biotechnology*, **53**, 253–263.

Bao, W., Fukushima, Y., Jensen, K. A., Moen, M. A. & Hammel, K. E. (1994). Oxidative degradation of non-phenolic lignin during lipid peroxidation by fungal manganese peroxidase. *FEBS Letters*, **354**, 297–300.

Bes, B., Ranjeva, R. & Boudet, A. M. (1983). Evidence for the involvement of activated oxygen in fungal degradation of lignocellulose. *Biochimie*, **65**, 283–289.

Bietti, M., Baciocchi, E. & Steenken, S. (1998). Lifetime, reduction potential and base induced fragmentation of the veratryl alcohol radical cation in aqueous solution. Pulse radiolysis studies on a ligninase 'mediator'. *Journal of Physical Chemistry A*, **102**, 7337–7342.

Bogan, B. W. & Lamar, R. T. (1995). One-electron oxidation in the degradation of creosote polycyclic aromatic hydrocarbons by *Phanerochaete chrysosporium. Applied and Environmental Microbiology*, **61**, 2631–2635.

Bogan, B. W., Lamar, R. T. & Hammel, K. (1996a). Fluorene oxidation *in vivo* by *Phanerochaete chrysosporium* and *in vitro* during manganese peroxidase-dependent lipid peroxidation. *Applied and Environmental Microbiology*, **62**, 1788–1792.

Bogan, B. W., Schoenike, B., Lamar, R. T. & Cullen, D. (1996b). Expression of LiP genes during growth in soil and oxidation of anthracene by *Phanerochaete chrysosporium. Applied and Environmental Microbiology*, **62**, 3697–3703.

Borazjani, A. & Diehl, S. V. (1998). Bioremediation of soils contaminated with organic wood preservatives. In *Forest Product Biotechnology*, eds. A. Bruce & J. W. Palfreyman, pp. 117–127. London: Taylor and Francis.

Bourbonnais, R. & Paice, M. G. (1990). Oxidation of non-phenolic substrates. An expanded role for laccase in lignin biodegradation. *FEBS Letters*, **267**, 99–102.

Bourbonnais, R. , Paice, M. G., Reid, I. D., Lanthier, P. & Yaguchi, M. (1995). Lignin oxidation by laccase isozymes from *Trametes versicolor* and role of the mediator 2,2′-azinobis(3-ethylbenzthiazoline-6-sulphonate) in kraft lignin depolymerisation. *Applied and Environmental Biotechnology*, **61**, 1876–1880.

Boyle, D., Wiesner, C. & Richardson, A. (1998). Factors affecting the degradation of polyaromatic hydrocarbons in soil by white-rot fungi. *Soil Biology and Biochemistry*, **30**, 873–882.

Bumpus, J. A., Tien, M., Wright, D. & Aust, A. D. (1985). Oxidation of persistent environmental pollutants by a white-rot fungus. *Science*, **228**, 1434–1436.

Butler, C. S. & Mason, J. R. (1997). Structure-function analysis of the bacterial aromatic ring-hydroxylating dioxygenases. *Advances in Microbial Physiology*, **38**, 47–84.

Cai, D, & Tien, M. (1992). Kinetic studies on the formation and decomposition of compounds II and III. Reactions of lignin peroxidase with H_2O_2. *Journal of Biological Chemistry*, **267**, 11149–11155.

Call, H. P. (1994). Process for modifying, breaking down or bleaching lignin, materials containing lignin or like substances. World Patent Application WO 94/29510.

Candeias, L. P. & Harvey, P. J. (1995). Lifetime and reactivity of the veratryl alcohol radical cation. *Journal of Biological Chemistry*, **270**, 16745–16748.

Cerniglia, C. E. (1997). Fungal metabolism of polycyclic aromatic hydrocarbons: past, present and future applications in bioremediation. *Journal of Industrial Microbiology and Biotechnology*, **19**, 324–333.

Choinowski, T., Blodig, W. & Winterhalter, K. H. (1999). The crystal structure of lignin peroxidase at 1.70 Å resolution reveals a hydroxy group on the C_α of tryptophan 171: a novel radical site formed during the redox cycle. *Journal of Molecular Biology*, **286**, 809–827.

Chung, N. & Aust, S. D. (1995a). Veratryl alcohol-mediated indirect oxidation of phenol by lignin peroxidase. *Archives of Biochemistry and Biophysics*, **316**, 733–737.

Chung, N. & Aust S. D. (1995b). Inactivation of lignin peroxidase by hydrogen peroxide during the oxidation of phenols. *Archives of Biochemistry and Biophysics*, **316**, 851–855.

Collins, P. J., Kotterman, M. J. J., Field, J. A. & Dobson, A. D. W. (1996). Oxidation of anthracene and benzo[*a*]pyrene by laccases from *Trametes versicolor*. *Applied and Environmental Microbiology*, **62**, 4563–4567.

Cote, W. A. (1977). Wood ultrastructure in relation to chemical composition. In *Recent Advances in Phytochemistry*, Vol. II, *The Structure, Biosynthesis and Degradation of Wood*, eds. F. A. Loewus & V. C. Runeckles, pp. 1–44. New York: Plenum Press.

Cullen, D. (1997). Recent advances on the molecular genetics of ligninolytic fungi. *Journal of Biotechnology*, **53**, 273–289.

de Jong, E., de Vries, F. P., Field, J. A., van de Zwan, R. P. & de Bont, J. A. M. (1992). Isolation of basidiomycetes with high peroxidative activity. *Mycological Research*, **96**, 1098–1104.

Doyle, W. A., Blodig, W., Veicht, N. C., Piontek, K. & Smith, A. T. (1998). Two substrate interaction sites in lignin peroxidase revealed by site-directed mutagenesis. *Biochemistry*, **37**, 15097–15105.

Ducros, V., Brzozowski, Wilson, K. S., Brown, S. H., Østergaard, P., Schneider, P., Yaver, D. S., Pedersen, A. H. & Davies, G. J. (1998). Crystal structure of the type-2 Cu depleted laccase from *Coprinus cinereus* at 2.2 Å resolution. *Nature Structural Biology*, **5**, 310–316.

Dunford, H. B. (1999). *Heme Peroxidase*. New York: Wiley & Sons.

Dutton, M. V., Evans, C. S., Atkey, P. T. & Wood, D. A. (1993). Oxalate production by Basidiomycetes, including the white-rot species *Coriolus versicolor* and *Phanerochaete chrysosporium*. *Archives of Microbiology and Biotechnology*, **39**, 5–10.

Edwards, S. L., Raag, R., Wariishi, H., Gold, M. H. & Poulos, T. L. (1993). Crystal structure of lignin peroxidase. *Proceedings of the National Academy of Sciences USA*, **90**, 750–754.

Eggert, C., Temp, U., Dean, J. F. D. & Eriksson, K.-E. L. (1995). Laccase-mediated formation of the phenoxazinone derivative, 3-hydroxyanthranilic acid. *FEBS Letters*, **376**, 202–206.

Eggert, C., Temp, U., Dean, J. F. D. & Eriksson, K.-E. L. (1996). A fungal metabolite mediates oxidation of non-phenolic structures and synthetic lignin by laccase. *FEBS Letters*, **391**, 144–148.

Eggert, C., Temp, U. & Eriksson, K.-E. L. (1997). Laccase is essential for lignin degradation by the white-rot fungus *Pycnoporus cinnabarinus*. *FEBS Letters*, **407**, 89–92.

Ferris, J. P., MacDonald, L. H., Patrie, M. A. & Martin, M. A. (1976). Aryl hydrocarbon hydroxylase activity in the fungus *Cunninghamella bainieri*: evidence for the presence of cytochrome P-450. *Archives of Biochemistry and Biophysics*, **175**, 443–452.

Field, J. A., Boelsma, F., Baten, H. & Rulkens, W. H. (1995). Oxidation of anthracene in water/solvent mixtures by the white-rot fungus, *Bjerkandera* sp. strain BOS55. *Applied Microbiology and Biotechnology*, **44**, 234–240.

Forney, L. J., Reddy, C. A., Tien, M. & Aust, S. D. (1982). The involvement of hydroxyl radical derived from H_2O_2 in lignin degradation by the white-rot fungus *Phanerochaete chrysosporium*. *Journal of Biological Chemistry*, **257**, 11455–11462.

Gianfreda, L., Xu, F. & Bollag, J.-M. (1999). Laccases: a useful group of oxidoreductive enzymes. *Bioremediation Journal*, **3**, 1–25.

Gierer, J. & Opara, A. E. (1973). Studies of the enzymatic degradation of lignin. *Acta Chemika Scandanavica*, **27**, 2909–2922.

Gilardi, G., Harvey, P. J., Cass, A. E. G. & Palmer, J. M. (1990). Radical intermediates in veratryl alcohol oxidation by ligninase. NMR evidence. *Biochimica et Biophysica Acta*, **1041**, 129–132.

Glenn J. K. & Gold, M. H. (1985). Purification and characterisation of an extracellular Mn(II)-dependent peroxidase from the lignin-degrading basidiomycete, *Phanerochaete chrysosporium*. *Archives of Biochemistry and Biophysics*, **242**, 329–341.

Gold, M. H. & Alic, M. (1993). Molecular biology of the lignin-degrading basidiomycete *Phanerochaete chrysosporium*. *Microbiological Reviews*, **57**, 605–622.

Gramss, G., Voigt, K.-D. & Kirsche, B. (1999). Degradation of polycyclic aromatic hydrocarbons with three to seven aromatic rings by higher fungi in sterile and unsterile soils. *Biodegradation*, **10**, 51–62.

Green, T. R. (1977). Significance of glucose oxidase in lignin degradation. *Nature*, **268**, 78–80

Haemmerli, S. D., Leisola, M. S. A., Sanglard, D. & Fiechter, A. (1986). Oxidation of benzo[*a*]pyrene by extracellular ligninases of *Phanerochaete chrysosporium*. *Journal of Biological Chemistry*, **261**, 6900–6903.

Haemmerli, S. D., Schoemaker, H. E., Schmidt, H. W. H. & Leisola, M. S. A.

(1987). Oxidation of veratryl alcohol by the lignin peroxidase of *Phanerochaete chrysosporium*. *FEBS Letters*, **220**, 149–154.

Hammel, K. E. & Moen, M. A. (1991). Depolymerisation of a synthetic lignin *in vitro* by lignin peroxidase. *Enzyme Microbial Technology*, **13**, 15–18.

Hammel, K. E., Tien, M., Kalyanaraman, B. & Kirk, T. K. (1985). Mechanism of $C_{\alpha-\beta}$ cleavage of a lignin model dimer by *Phanerochaete chrysosporium* ligninase. *Journal of Biological Chemistry*, **260**, 8348–8353.

Hammel, K. E., Kalyanaraman, B. & Kirk, T. K. (1986a). Oxidation of polycyclic aromatic hydrocarbons and dibenzo(p)-dioxins by *Phanerochaete chrysosporium* ligninase. *Journal of Biological Chemistry*, **261**, 6948–6952.

Hammel, K. E., Kalyanaraman, B. & Kirk, T. K. (1986b). Substrate free radicals are intermediates in ligninase catalysis. *Proceedings of the National Academy of Sciences USA*, **83**, 3708–3712.

Hammel, K. E., Jensen, K. A., Mozuch, M. D., Landucci, L. L., Tien, M. & Pease, E. A. (1993). Ligninolysis by a purified lignin peroxidase. *Journal of Biological Chemistry*, **268**, 12274–12281.

Harvey, P. J. & Candeias, L. (1995). Radical cation cofactors in lignin peroxidase catalysis. *Biochemical Society Transactions*, **23**, 261–267.

Harvey, P. J. & Palmer, J. (1990). Oxidation of phenolic compounds by ligninase. *Journal of Biotechnology*, **13**, 169–179.

Harvey, P. J., Schoemaker, H. E., Bowen, R. M. & Palmer, J. M. (1985). Single-electron transfer processes and the reaction mechanism of enzymic degradation of lignin. *FEBS Letters*, **183**, 13–16.

Harvey, P. J., Schoemaker, H. E. & Palmer, J. M. (1986). Veratryl alcohol as a mediator and the role of radical cations in lignin biodegradation by *Phanerochaete chrysosporium*. *FEBS Letters*, **195**, 242–246.

Harvey, P. J., Schoemaker, H. E. & Palmer, J. M. (1987a). Lignin degradation by white-rot fungi. *Plant, Cell and Environment*, **10**, 709–714.

Harvey, P. J., Schoemaker, H. E. & Palmer, J. M. (1987b). Mechanisms of ligninase catalysis. In *Les Colloques d' INRA*, Vol. 40, *Lignin Enzymic and Microbial Degradation*, ed. E. Odier, pp. 145–150. Paris: INRA Publications.

Harvey, P. J., Palmer, J. M., Schoemaker, H. E., Dekker, H. L. & Wever, R. (1989). Pre-steady-state kinetic study on the formation of Compound I and II of ligninase. *Biochimica et Biophysica Acta*, **994**, 59–63.

Harvey, P. J., Floris, R., Lundell, T., Palmer, J., Schoemaker, H. E. & Wever, R. (1992). Catalytic mechanisms and regulation of lignin peroxidase. *Biochemical Society Transactions*, **20**, 345–349.

Harvey, P. J., Gilardi, G.-F., Goble, M. & Palmer, J. M. (1993). Charge transfer reactions and feedback control of lignin peroxidase by phenolic compounds: significance in lignin degradation. *Journal of Biotechnology*, **30**, 57–69.

Head, I. M. (1998). Bioremediation: towards a credible technology. *Microbiology*, **144**, 599–608.

Janse, B. J. H., Gaskell, J., Akhtar, M. & Cullen, D. (1998). Expression of *Phanerochaete chrysosporium* genes coding lignin peroxidases, manganese peroxidases and glyoxal oxidase in wood. *Applied and Environmental Microbiology*, **64**, 3536–3538.

Jensen, K. A., Bao, W., Kawai, S., Srebotnik, E. & Hammel, K. E. (1996). Manganese-dependent cleavage of nonphenolic lignin structures by *Ceriporiopsis subvermispora* in the absence of lignin peroxidase. *Applied and Environmental Microbiology*, **62**, 3679–3686.

Kersten, P. J., Tien, M., Kalyanaraman, B. & Kirk, T. K. (1985). The ligninase of

Phanerochaete chrysosporium generates cation radicals from methoxybenzenes. *Journal of Biological Chemistry*, **260**, 2609–2612.

Kersten, P. J., Kalyanaraman, B., Hammel, K. E., Reinhammer, B. & Kirk, T. K. (1990). Comparison of lignin peroxidase, horseradish peroxidase and laccase in the oxidation of methoxybenzenes. *Biochemical Journal*, **268**, 475–480.

Khazaal, K. A., Owen, E., Dodson, A. P., Palmer, J. M. & Harvey, P. J. (1993). Treatment of barley straw with ligninase: effect on activity and fate of the enzyme shortly after being added to straw. *Animal Feed Science and Technology*, **41**, 15–21.

Khindaria, A., Yamazaki, I. & Aust, S. D. (1996). Stabilisation of the veratryl alcohol radical cation by lignin peroxidase. *Biochemistry*, **35**, 6418–6424.

Kirk, T. K. & Farrell, R. L. (1987). Enzymatic 'combustion': the microbial degradation of lignin. *Annual Review of Microbiology*, **41**, 465–505.

Kirk, T. K., Harkin, J. M. & Cowling, E. B. (1968). Oxidation of guaiacyl- and veratryl-glycerol *β*-guaiacyl ether by *Polyporus versicolor* and *Stereum frustulatum. Biochimica et Biophysica Acta*, **168**, 145–163.

Kishi, K., Wariishi, H., Marquez, L., Dunford, H. B. & Gold, M. H. (1994). Mechanism of manganese peroxidase compound II reduction. Effect of organic acid chelators and pH. *Biochemistry*, **33**, 8694–8701.

Krisnangkura, K. & Gold, M. H. (1979). Peroxidase catalysed oxidative decarboxylation of vanillic acid to methoxy-*p*-hyroquinone. *Phytochemistry*, **18**, 2019–2021.

Kuan, I.-C. & Tien, M. (1993). Stimulation of Mn peroxidase activity: a possible role for oxalate in lignin biodegradation. *Proceedings of the National Academy of Sciences USA*, **90**, 1242–1246.

Lamar, R. T. & Dietrich, D. M. (1992). Use of lignin-degrading fungi in the disposal of pentachlorophenol-treated wood. *Journal of Industrial Microbiology*, **9**, 181–191.

Lamar, R. T., Evans, J. W. & Glaser, J. A. (1993). Solid-phase treatment of a pentachlorophenol-contaminated soil using lignin-degrading fungi. *Environmental Science Technology*, **27**, 2572–2576.

Leisola, M. S. A., Schmidt, B., Thanei-Wyss, U. & Fiechter, A. (1985a). Aromatic ring cleavage of veratryl alcohol by *Phanerochaete chrysosporium. FEBS Letters*, **189**, 267–270.

Leisola, M. S. A., Thanei-Wyss, U. & Fiechter, A. (1985b). Strategies for production of high ligninase activities by *Phanerochaete chrysosporium. Journal of Biotechnology*, **3**, 97–107.

Lindley, P. F., Zaitseva, I., Zaitsev, V., Card, G., Moshkov, K. & Bax, B. (1997). The structure of human ceruloplasmin at 3.1 Å resolution. In *Multi-copper Oxidases*, ed. A. Messerschmidt, pp. 81–102. Singapore: World Scientific Publishing.

Lundquist, K. & Kirk, T. K. (1978). *De novo* synthesis and decomposition of veratryl alcohol by a lignin-degrading basidiomycete. *Phytochemistry*, **17**, 1676.

Majcherczyk, A. & Hutterman, A. (1998). Bioremediation of wood treated with preservatives using white-rot fungi. In *Forest Product Biotechnology*, eds. A. Bruce & J. W. Palfreyman, pp. 129–140. London: Taylor and Francis.

Marquez, L., Wariishi., H., Dunford, H. B. & Gold, M. H. (1988). Spectroscopic and kinetic properties of the oxidized intermediates of lignin peroxidase from *Phanerochaete chrysosporium. Journal of Biological Chemistry*, **263**, 10549–10552.

Marzullo L., Cannio, R., Giardina, P. & Santini, M. T. (1995). Veratryl alcohol oxidase from *Pleurotus ostreatus* participates in lignin biodegradation and prevents polymerisation of laccase-oxidized substrates. *Journal of Biological Chemistry*, **270**, 3827–3828.

McCarthy, A. J. (1987). Lignocellulose-degrading actinomycetes. *FEMS Microbiology Reviews*, **46**,145–163.

Messerschmidt, A. (1997). *Multi-copper Oxidases*, ed. A. Messerschmidt, 465 pp. Singapore: World Scientific Publishing.

Messerschmidt, A., Rossi, A., Ladenstein, R., Huber, R., Bolognesi, M., Guiseppina, G., Marchesini, A., Petruzzelli, R. & Finazzi-agro, A. (1989). X-ray crystal structure of the blue oxidase ascorbate oxidase from zucchini. *Journal of Molecular Biology*, **206**, 513–529.

Meulenberg, R., Rijnaarts, H. H. H. M., Doddema, H. J. & Field, J. A. (1997). Partially oxidised polycyclic aromatic hydrocarbons show an increased bioavailability and biodegradability. *FEMS Microbiology Letters*, **152**, 45–49.

Meyer, A. M. (1987). Polyphenol oxidases in plants – recent progress. *Phytochemistry*, **26**, 11–20.

Moen, M. A. & Hammel, K. E. (1994). Lipid-peroxidation by the manganese peroxidase of *Phanerochaete chrysosporium* is the basis for phenanthrene oxidation by the intact fungus. *Applied and Environmental Microbiology*, **60**, 1956–1960.

Muheim, A., Fiechter, A., Harvey, P. J. & Schoemaker, H. E. (1992). On the mechanism of oxidation of non-phenolic lignin model compounds by the laccase-ABTS couple. *Holzforschung*, **46**, 121–126.

Novotny, C., Erbanova, P., Dsasek, V., Kubatova, A., Cajthaml, T., Lang, E., Krahl, J. & Zadrazil, F. (1999). Extracellular oxidative enzyme production and PAH removal in soil by exploratory mycelium of white-rot fungi. *Biodegradation*, **10**, 159–168.

Palmer, J. M., Harvey, P. J. & Schoemaker, H. E. (1987). The role of peroxidases, radical cations and oxygen in the degradation of lignin. *Philosophical Transactions of the Royal Society of London Series A*, **321**, 495–505.

Piontek, K., Glumoff, T. & Winterhalter, K. (1993). Low pH crystal structure of glycosylated lignin peroxidase from *Phanerochaete chrysosporium* at 2.5 Å resolution. *FEBS Letters*, **315**, 119–124.

Popp, J. L. & Kirk, T. K. (1991). Oxidation of methoxybenzenes by manganese peroxidase and by Mn^{3+}. *Archives of Biochemistry and Biophysics*, **288**, 145–148.

Reddy C. A. & D'Souza, T. M. (1994). Physiology and molecular biology of the lignin peroxidases of *Phanerochaete chrysosporium*. *FEMS Microbiology Reviews*, **13**, 137–152.

Reddy, G. V. B. & Gold, M. H. (1999). A two-component tetrachlorohydroquinone reductive dehalogenase system from the lignin-degrading basidiomycete *Phanerochaete chrysosporium*. *Biochemical and Biophysical Research Communications*, **257**, 901–905.

Reddy, G. V. B., Gelpke, M. D. S. & Gold, M. H. (1998). Degradation of 2,4,6-trichlorophenol by *Phanerochaete chrysosporium*: involvement of reductive chlorination. *Journal of Bacteriology*, **180**, 5159–5164.

Reinhammer, B. (1972). Oxidation-reduction potentials of the electron acceptors I laccases and stellacyanin. *Biochimica et Biophysica Acta*, **275**, 245–259.

Reinhammar, B. & Malmstrom, B. G. (1981). Blue copper-containing oxidases.

In *Copper Proteins*, 2nd edn, ed. T. G. Spiro, pp. 109–149. New York: Wiley & Sons.

Renganathan, V., Miki, K. & Gold, M. H. (1986). Role of molecular oxygen in lignin peroxidase reactions. *Archives of Biochemistry and Biophysics*, **246**, 155–161.

Samejima, M. & Eriksson, K.-E. (1991). Mechanisms of redox interactions between lignin peroxidase and cellobiose:quinone oxidoreductase. *FEBS Letters*, **292**, 151–153.

Sarkanen, S., Razal, R. A., Piccariello, T., Yamamoto, E. & Lewis, N. G. (1991). Lignin peroxidase: toward a clarification of its role *in vivo. Journal of Biological Chemistry*, **268**, 3636–3643.

Schoemaker, H. E. (1990). On the chemistry of lignin biodegradation. *Recueil des Travaux Chimiques des Pays-Bas*, **109**, 255–272.

Schoemaker, H. E., Harvey, P. J., Bowen, R. M. & Palmer, J. M. (1985). On the mechanisms of enzymatic lignin breakdown. *FEBS Letters*, **183**, 7–12.

Schutzendubel, A., Majcherczyk, A., Johannes, C. & Hutterman, A. (1999). Degradation of fluorene, anthracene, phenanthrene, fluoranthene and pyrene lacks connection to the production of extracellular enzymes by *Pleurotus ostreatus* and *Bjerkandera adusta. International Biodeterioration and Biodegradation*, **43**, 93–100.

Sheng, D. & Gold, M. H. (1999). Oxidative polymerization of ribonuclease A by lignin peroxidase from *Phanerochaete chrysosporium.* Role of veratryl alcohol in polymer oxidation. *European Journal of Biochemistry*, **259**, 626–634.

Smith, A. T. & Veicht, N. (1998). Substrate binding and catalysis in heme peroxidases. *Current Opinions in Chemistry and Biology*, **2**, 269–278.

Smith, J. F. (1994). Factors affecting the selectivity of composts suitable for the cultivation of *Agaricus* species. PhD Thesis. University of London, UK.

Smith, M., Thurston, C. F. & Wood, D. A. (1997). Fungal laccases: role in delignification and possible industrial applications. In *Multi-copper Oxidases*, ed. A. Messerschmidt, pp. 201–224. Singapore: World Scientific Publishing.

Sundaramoorthy, M., Kishi, K., Gold, M. H. & Poulos, T. (1997). Crystal structures of substrate binding site mutants of manganese peroxidase. *Journal of Biological Chemistry*, **272**, 17574–17580.

Sutherland, J. B. (1992). Detoxification of polycyclic aromatic hydrocarbons by fungi. *Journal of Industrial Microbiology*, **9**, 53–62.

ten Have, R., Rietjens, I. M. C. M., Hartmans, S., Swarts, H. J. & Field, J. (1998). Calculated ionisation potentials determine the oxidation of vanillin precursors by lignin peroxidase. *FEBS Letters*, **430**, 390–392.

ten Have, R., de Thouars, R. G., Swarts, H. J. & Field, J. (1999). Veratryl alcohol-mediated oxidation of isoeugenyl acetate by lignin peroxidase. *European Journal of Biochemistry*, **265**, 1–8.

Thurston, C. F. (1994). The structure and function of fungal laccases. *Microbiology*, **140**, 19–26.

Tien, M. & Kirk, T. K. (1983). Lignin-degrading enzyme from the Hymenomycete *Phanerochaete chrysosporium* Burds. *Science*, **221**, 661–663.

Tien, M. & Kirk, T. K. (1988). Lignin peroxidase of *Phanerochaete chrysosporium. Methods in Enzymology*, **161**, 238–249.

Tucker, B., Radtke, C., Kwon, S.-I. & Anderson, A. J. (1995). Suppression of bioremediation of *Phanerochaete chrysosporium* by soil factors. *Journal of*

Hazardous Materials, **41**, 251–265.

Tuor, U., Wariishi, H., Schoemaker, H. & Gold, M. H. (1992). Oxidation of phenolic arylglycerol β-aryl ether lignin model compounds by manganese peroxidase from *Phanerochaete chrysosporium*: oxidative cleavage of and α-carbonyl model compound. *Biochemistry*, **31**, 4986–4995.

Valli, K., Wariishi, H. & Gold, M. H. (1990). Oxidation of monomethoxylated aromatic compounds by lignin peroxidase: role of veratryl alcohol in lignin biodegradation. *Biochemistry*, **29**, 8535–8539.

Wariishi, H., Akileswaran, L. & Gold, M. H. (1988). Manganese peroxidase from the basidiomycete *Phanerochaete chrysosporium*: spectral characterisation of the oxidised states and the catalytic cycle. *Biochemistry*, **27**, 5365–5370.

Wariishi, H., Valli, K. & Gold, M. H. (1989a). Oxidative cleavage of a phenolic diarylpropane lignin model dimer by manganese peroxidase from *Phanerochaete chrysosporium*. *Biochemistry*, **28**, 6017–6023.

Wariishi, H., Valli, K., Renganathan, V. & Gold, M. H. (1989b). Thiol-mediated oxidation of nonphenolic lignin model compounds by manganese peroxidase of *Phanerochaete chrysosporium*. *Journal of Biological Chemistry*, **264**, 14185–14191.

Wariishi, H., Marquez, L., Dunford, H. B. & Gold, M. H. (1990). Lignin peroxidase Compounds II and III. Spectral and kinetic characterisation of reactions with peroxides. *Journal of Biological Chemistry*, **19**, 11137–11142.

Wariishi, H., Valli, K. and Gold, M. H. (1991). *In vitro* depolymerisation of lignin by manganese peroxidase of *Phanerochaete chrysosporium*. *Biochemical and Biophysical Research Communications*, **176**, 269–275.

Wilson, M. T., Hogg, N. & Jones, G. D. (1990). Reactions of reduced cellobiose oxidase with oxygen. Is cellobiose oxidase primarily an oxidase? *Biochemical Journal*, **270**, 265–267.

Wilson, S. & Jones, K. C. (1993). Bioremediation of soil contaminated with polynuclear aromatic hydrocarbons (PAHs): a review. *Environmental Pollution*, **81**, 229–249.

Wood, D. A. (1980). Inactivation of extracellular laccase during fruiting of *Agaricus bisporus*. *Journal of General Microbiology*, **117**, 339–345.

Xu, F. (1996). Oxidation of phenols, anilines, and benzenethiols by fungal laccases: correlation between activity and redox potentials as well as halide inhibition. *Biochemistry*, **35**, 7608–7614.

Xu, F., Shin, W., Brown, S. H., Wahleithner, J., Sundaram, U. M. & Solomon, E. (1996). A study of a series of recombinant fungal laccases and bilirubin oxidase that exhibit significant differences in redox potential, substrate specificity and stability. *Biochimica et Biophysica Acta*, **1292**, 303–311.

Zacchi, L., Burla, G., Zuolong, D. & Harvey, P. J. (2000a). Metabolism of cellulose by *Phanerochaete chrysosporium* in continuously agitated cultures is associated with enhanced production of lignin peroxidase. *Journal of Biotechnology*, **78**, 185–192.

Zacchi, L., Morris, I. & Harvey, P. J. (2000b). Disordered ultrastructure in lignin-peroxidase-secreting hyphae of the white-rot fungus *Phanerochaete chrysosporium*. *Microbiology*, **146**, 759–765.

Zacchi, L., Palmer, J. M. & Harvey, P. J. (2000c). Respiratory pathways and oxygen toxicity in *Phanerochaete chrysosporium*. *FEMS Microbiology Letters*, **183**, 153–157.

3

Bioremediation potential of white rot fungi

C. ADINARAYANA REDDY
AND ZACHARIA MATHEW

Introduction

Environmental pollutants are a serious concern worldwide because of the hazards they pose to the health of humans and animals. An estimated 80 billion pounds of hazardous organopollutants are produced annually by the chemical, agricultural, oil, paper, textile, aerospace, and other industries in the USA alone (Aust, 1990). Only about 10% of these wastes are believed to be disposed of in an environmentally safe manner (EPA, 1988; Fernando & Aust, 1994). Traditional methods of disposing of hazardous wastes (physical, chemical, and thermal treatments and land filling) have not always been efficacious. It has been estimated that it costs about one trillion dollars to decontaminate toxic waste sites in the USA alone using traditional waste disposal methods (Barr & Aust, 1994). Considering these staggering costs for cleaning up the environment, an alternative, rapid, efficacious and cost-effective method is needed. One method that has become increasingly popular for decontamination of the environment has been bioremediation. The use of indigenous or suitable introduced micro-organisms at contamination sites often provides an efficient and economically attractive solution to the pollution problem. One of the early reports indicated that lignin-degrading white rot fungi, as exemplified by *Phanerochaete chrysosporium*, can degrade an extremely diverse group of environmental pollutants (Bumpus *et al.*, 1985). Since then, there has been intense worldwide research to unravel the potential of white rot fungi in bioremediation. This ability of white rot fungi to degrade a wide spectrum of environmental pollutants sets them apart from many other microbes used in bioremediation.

In order to understand the non-specific ability of white rot fungi to degrade a wide variety of pollutants, one should consider their ecological niche. White rot fungi are wood-degrading basidiomycetes and are among the most active degraders of lignin, the key structural polymer of woody plants. Lignin is a highly complex, three-dimensional, amorphous, hetero-polymer and consists of phenylpropanoid monomer units that are ran-domly linked to each other in a variety of C–C and C–O linkages. Further-more, the chiral carbons in lignin occur in both L and D configurations. Therefore, lignin is one of the most difficult biopolymers to be degraded by microbial enzymes. The complexity of the lignin polymer and its stereochemical irregularity results, at least in part, from the free radical mechanism of lignin synthesis seen in woody plants (Kirk & Farrell, 1987; Boominathan & Reddy, 1992; Barr & Aust, 1994). It has been hy-pothesized, therefore, that lignin degradation must also involve a non-specific and non-stereoselective mechanism. Extensive research since the early 1980s has shown that the white rot fungi have developed unique non-specific enzyme systems with the ability to attack not only lignin but also a broad spectrum of halogenated and non-halogenated aromatic compounds as well as some non-aromatic organopollutants (Table 3.1). Even complex mixtures of pollutants such as Aroclors are degraded effi-ciently by white rot fungi (Reddy, 1995).

White rot fungi offer a number of advantages for use in bioremediation. The key enzymes of the lignin degradation system (LDS) are extracellular, obviating the need to internalize the substrates and allowing substrates of low solubility to be oxidized. Furthermore, the extracellular enzyme sys-tem of the white rot fungi enables these organisms to tolerate a relatively higher concentration of toxic pollutants than would otherwise be possible. White rot fungi catalyse degradation of lignin as well as pollutants using a non-specific free radical mechanism and are, therefore, capable of degrad-ing a wide variety of pollutants. The constitutive nature of the key enzymes involved in the LDS obviates the need (in most cases) for these organisms to be adapted to the chemical being degraded. White rot fungi are also ubiquitous in nature. Although they degrade lignin, they cannot utilize it as a source of energy for growth and instead require cosubstrates such as cellulose or other carbon sources. The preferred substrates for growth of white rot fungi in nature are lignocellulosic substrates. Therefore, inexpen-sive lignocellulosics such as corn cobs, straw, peanut shells and sawdust can be added as nutrients to the contaminated sites to obtain enhanced degradation of pollutants by these organisms. Finally, white rot fungi grow by hyphal extension and thus can reach pollutants in the soil in ways that other organisms cannot. This chapter deals mostly with the bi-oremediation potential of *P. chrysosporium*, which has been extensively studied as the model organism for bioremediation, and only briefly with

Table 3.1. *Environmental pollutants degraded by white rot fungi*

Type	Examples
Polycyclic aromatic hydrocarbons	Anthracene, 2-methyl anthracene, 9-methyl anthracene, benzo[*a*]pyrene, fluorene, naphthalene, acenaphthene, acenaphthylene, phenanthrene, pyrene, biphenylene
Chlorinated aromatic compounds	Chlorophenols (e.g, pentachlorophenols (PCP), trichlorophenols (TCP), and dichlorophenols (DCP)); chlorolignols, 2,4-dichlorophenoxyacetic acid (2,4-D), 2,4,5-trichlorophenoxyacetic acid (2,4,5-T), polychlorinated biphenyls (PCBs), dioxins, Chlorobenzenes
Dyes	Azure B, Congo Red , Disperse Yellow 3 (DY3), Orange II, Poly R, Reactive Black 5, Reactive Orange 96, Reactive Violet 5, Remazol Brilliant Blue R (RBBR), Solvent yellow 14, Tropaeolin
Nitroaromatics	TNT (2,4,6-trinitrotoluene), 2,4-dinitrotoluene, 2-amino-4,6-dinitrotoluene, 1-chloro-2,4-dinitrobenzene, 2,4-dichloro-1-nitrobenzene, 1,3-dinitrobenzene
Pesticides	Alachlor, Aldrin, Chlordane, 1,1,1-trichloro-2,2-bis(4-chlorophenyl)ethane (DDT), Heptachlor, Lindane, Mirex, Atrazine
Other environmental pollutants	Benzene, toluene, ethylbenzene, *o*-, *m*-, *p*-xylenes (BTEX compounds), linear alkylbenzene sulfonate (LAS), trichloroethylene

the other white rot fungi. There have been several recent reviews on bioremediation by white rot fungi (Hammel, 1992; Lamar, 1992; Bumpus, 1993; Barr & Aust, 1994; Fernando & Aust, 1994; Crawford, 1995; Reddy, 1995).

Lignin-degrading enzymes

The (LDS) of white rot fungi consists of a battery of enzymes that catalyse oxidation of xenobiotics in addition to their ability to degrade lignin. The

LDS cleaves the carbon–carbon and carbon–oxygen bonds of the lignin molecule regardless of the chiral conformations of the lignin molecule (Fernando & Aust, 1994). This manner of bond fission may result partially from the free radical mechanism of lignin degradation employed by white rot fungi (Kirk & Farrell, 1987; Aust, 1990; Boominathan & Reddy, 1992; Fernando & Aust, 1994). In addition, free radical species generated during the degradation process (of either lignin or organopollutants) may serve as secondary oxidants, which may, in turn, mediate the oxidation of other compounds away from the active sites of the enzymes (Barr & Aust, 1994). Nitrogen deficiency was observed to initiate the degradation of lignin, while nitrogen-rich cultures suppressed the degradation of pollutants by *P. chrysosporium* (Bumpus *et al.*, 1985; Barr & Aust, 1994; Reddy, 1995).

LDS in *P. chrysosporium* is expressed during secondary metabolism in response to starvation for nutrients such as nitrogen and carbon. Ligninolytic peroxidases, which are believed to be involved in lignin degradation by this organism, are completely suppressed in media containing high levels of nitrogen or carbon. The three major families of lignin-modifying enzymes (LMEs) that are believed to be involved in lignin degradation are laccases, lignin peroxidases (LiPs) and manganese-dependent peroxidases (MnPs). Some white rot fungi produce all three classes of LME while the others produce different combinations of the three. Important physiological and biochemical features of LMEs have been reviewed (Cullen & Kersten, 1992; Reddy, 1993; Reddy & D'Souza, 1994; Thurston, 1994).

Laccase (benzenediol:oxygen oxidoreductase, EC 1.10.3.2) of white rot fungi is a glycosylated copper-containing enzyme that catalyses the four-electron reduction of dioxygen to water by substrate molecules of phenolic origin without the generation of H_2O_2. Besides phenolic compounds they also can attack non-phenolic aromatics with high redox potentials in the presence of small aromatic compounds such as 2,2′-azinobis(-3-ethyl-benz-thiazoline-6-sulfonic acid) (Hatakka, 1994; Thurston, 1994).

LiPs (EC 1.11.1.7) are extracellular, glycosylated heme proteins that catalyse H_2O_2-dependent one-electron oxidation of lignin-related aromatic compounds to aryl cation radicals, leading to a variety of end products through non-enzymic reactions. LiPs have a higher redox potential than do most peroxidases and appear to oxidize a greater range of chemicals than many other peroxidases. MnPs (EC 1.11.1.7) are extracellular glycosylated heme proteins that catalyse H_2O_2-dependent oxidation of Mn^{2+} to Mn^{3+}. It is the Mn^{3+} state of the enzyme that actually mediates the oxidation of phenolic substrates, while non-phenolic compounds are oxidized via cation radicals (Kirk & Farrell, 1987; Hammel, 1992; Barr & Aust, 1994; Reddy and D'Souza, 1994).

It is important to realize that the key step in lignin degradation by laccase or the ligninolytic peroxidases (LiP and MnP) involves the formation of free radical intermediates, which are formed when one electron is removed or added to the ground state of a chemical. Such free radicals are highly reactive and rapidly give up or abstract an electron from another chemical. This free radical mechanism provides the basis for the non-specific nature of degradation of a variety of structurally diverse pollutants (Barr & Aust, 1994).

X-ray crystallographic structures of LiP and MnP from *P. chrysosporium* have been determined (reviewed in Reddy, 1995). LiP, similar to cytochrome *c* peroxidase, was shown to have histidine as the proximal ligand that accepts a proton from H_2O_2 while the distal arginine facilitates oxygen–oxygen bond cleavage. These studies further revealed close structural similarities between the LiPs and MnPs except that MnP had five disulfide bonds rather than the four disulfide bonds seen in LiP. A new cation binding site has also been located in MnP.

The *lip*, *mnp* and laccase gene families from a variety of white rot fungi have been cloned and sequenced (Cullen & Kersten, 1992; Gold & Alic, 1993; Reddy & D'Souza, 1994; Mansur *et al.*, 1997). Both LiP and MnP are regulated at the mRNA level by nitrogen. In *P. chrysosporium*, the gene transcription of MnPs is also regulated by Mn^{2+} and by heat shock (Gold & Alic, 1993). Both homologous and heterologous expression of *lip* and *mnp* have been reported (Gold & Alic, 1993; Mayfield *et al.*, 1994; Reddy & D'Souza, 1994).

Polycyclic aromatic hydrocarbons

Polycyclic aromatic hydrocarbons (PAHs) are widespread, hazardous environmental pollutants that are released into the air, soil, water and marine environments by the burning of fossil fuels and wood, coal mining and oil drilling (Fernando & Aust, 1994). Several of these PAHs are mutagenic and carcinogenic (Zhang & Jenssen, 1994; Clonfero *et al.*, 1996). At least 22 of the PAHs undergo 70–100% breakdown in 27 days in nitrogen-limited cultures of *P. chrysosporium*. Earlier studies have also demonstrated that some PAHs, such as benzo[*a*]pyrene, benz[*a*]anthracene, anthracene, pyrene and perylene, are directly oxidized by the LiPs of *P. chrysosporium* to quinone-type products (Reddy, 1995). Purified LiP of *P. chrysosporium* oxidizes PAHs to corresponding quinones (Haemmerli *et al.*, 1986; Hammel, Kalyanaraman & Kirk, 1986). Both ligninolytic and non-ligninolytic cultures of *P. chrysosporium* degraded radiolabelled phenathrene to $^{14}CO_2$, suggesting that the ligninolytic enzymes as well as other non-ligninolytic enzymes may also be involved in the degradation pathway (Sutherland *et al.*, 1991; Dhawale, Dhawale & Dean-Ross, 1992; Hammel *et al.*, 1992; Sutherland, 1992). In nitrogen-limited cultures of *Phanerochaete laevis*, both MnPs and laccases were synthesized but no LiP was detected (Bogan & Lamar, 1996).

Several white rot fungi have been reported to degrade anthracene (Vyas *et al.*, 1994). The predominant ligninolytic enzyme produced during the degradation of anthracene and benzo[*a*]pyrene by *P. laevis* was reported to be MnP (Bogan & Lamar, 1996). The MnP levels in *P. laevis* were stimulated by Mn^{2+} in the culture medium. *In vitro*, the MnPs were shown to produce small amounts of quinones as intermediates in the degradation of anthracene and benzo[*a*]pyrene (Bogan & Lamar, 1996). The laccase from *Coriolopsis gallica* has been shown to be involved in the oxidation of benzo[*a*]pyrene, 9-methylanthracene, 2-methylanthracene, anthracene, biphenylene, acenaphthene and phenanthrene (Pickard *et al.*, 1999). Laccases of *Trametes versicolor* have also been shown to carry out the oxidation of the PAHs, acenaphthene, acenaphthylene, anthracene and fluorene, mediated by small-molecular-weight aromatic compounds such as phenol, aniline and 4-hydroxybenzyl alcohol (Johannes & Majcherczyk, 2000). Mediators such as 2,2'-azinobis(3-ethylbenzthiazoline-6-sulfonic acid) (ABTS) and 1-hydroxybenzotriazole mediate the transformation of anthracene by laccase of *T. versicolor* (Johannes, Majcherczyk & Hutterman, 1996). Laccases are also capable of degrading phenathrene to give phenathrene-9,10-quinone and 2,2'-diphenic acid as the major products (Bohmer, Messner & Srebotnik, 1998). Two- to fivefold augmentation of degradation of anthracene, pyrene, and benzo[*a*]pyrene in the presence of non-ionic surfactants such as Tween 80 by *Bjerkandera* sp. suggested the possibilities for further optimization to obtain enhanced degradation of pollutants by white rot fungi (Kotterman, Rietberg & Field, 1998).

Bjerkandera sp. strain BOS55 removed 38.5% of benzo[*a*]pyrene from soil after 56 days of incubation (Field *et al.*, 1994). However, the anthracene biodegradation rate was not repressed by nitrogen levels when *Bjerkandera* sp. BOS55 was grown on a glucose-BII medium (Field *et al.*, 1994). The ability of *P. chrysosporium* to degrade PAHs has led to its application in the treatment of coal- and creosote-contaminated soils supplemented with wood chips, corn cobs or sawdust (Reddy, 1995). The PAH constituents were reduced after treatment to 10–20% of the original levels (Reddy, 1995). However, field-scale experiments in which PAH-contaminated soils were mixed with corn cobs, sawdust or bark chips and heavily inoculated with *P. chrysosporium* (10–30% w/w), did not show significant changes in the concentration of PAHs. Various studies have shown practical difficulties in bioremediation by white rot fungi because of the competition offered by indigenous organisms in the soil being treated and the difficulty in growing the fungi to sufficient biomass (Reddy, 1995). Indigenous soil bacteria have been known to antagonize the growth of *P. chrysosporium* depending on the pH and the nitrogen/carbon sources available (Radtke, Cook & Anderson, 1994). Detection and monitoring the growth of fungi in soils by a polymerase chain reaction (PCR) procedure may well prove to be useful in assessing the survival of the fungus applied to the contaminated soil and in deducing its relative contribution to bioremediation in soil (Johnston & Aust, 1994).

Dioxins

The halogenated dioxins and dioxin-like compounds, for example poly-chlorinated dibenzo-*p*-dioxins (PCDD), polychlorinated dibenzofurans (PCDF) and polychlorinated diphenyl ethers (PCDE), are released in the environment in the form of paper mill effluents, ash formed from combustion processes and as contaminants of chemicals such as chlorophenols. They are relatively chemically stable, lipophilic in nature and highly toxic when released into the environment (Valli *et al.*, 1992a; Witiich, 1998). A multistep pathway for the degradation of 2,7-dichlorodibenzo-dioxin involving LiP and MnP in *P. chrysosporium* has been proposed (Valli, Wariishi & Gold, 1992b). In the first step, 2,7-dichlorobenzo-*p*-dioxin is oxidatively cleaved, catalysed by LiP, to yield 4-chloro-1,2-benzo-quinone, 2-hydroxy-1,4-benzoquinone and chloride. Then 4-chloro-1,2-benzoquinone is reduced to 1-chloro-3,4-dihydroxybenzene, followed by the methylation of the latter intermediate to yield 1-chloro-3,4-dimethoxy-benzene. This product, in turn, is oxidized to produce 2-methoxy-1,4-benzoquinone and chloride. The intermediate product 2-methoxy-1,4-benzoquinone is reduced to 2-methoxy-1,4-dihydroxybenzene by LiP. In the succeeding step, 2-methoxy-1,4-dihydroxybenzene is further oxidized to yield 4-hydroxy-1,2-benzoquinone, to be reduced later to yield 1,2,4-trihydroxybenzene by LiP or MnP. The product of reduction of 2-hydroxy-1,4-benzoquinone, one of the early intermediates, is also 1,2,4-trihydroxybenzene. This key intermediate is converted to *β*-ketoadipic acid after reduction and ring cleavage (Valli *et al.*, 1992b). Up to

60% degradation of 2,3,7,8-tetrachlorodibenzo-*p*-dioxin (TCDD) by *Phanerochaete sordida* has also been reported (Takada *et al.*, 1996).

Polychlorinated biphenyls

Because of their thermal and electrical properties, polychlorinated biphenyls (PCBs) were used at one time in dielectric fluids, heat-transfer fluids, hydraulic fluids, flame retardants, adhesives, solvent extenders, textiles and printing (Robinson & Lenn, 1994). PCBs are marketed as complex mixtures under the trade names of Aroclor, Clophen and Delor. Three of the commonly used Aroclors are 1242, 1254, and 1260, which contain 42, 54 and 60% chlorine by weight with an average of 3, 5, and 6 chlorine atoms per biphenyl molecule. The inherent chemical inertness, owing to a stable molecular structure and hydrophobicity, and the presence of a mixture of a large number of congeners in the commercially available PCBs is at the heart of the problem of their low biodegradation in ecosystems and persistence in the environment (Robinson & Lenn, 1994). Mutagenic effects of PCBs in rodents are well known (Robinson & Lenn, 1994). There is a great deal more known about bacterial degradation than about fungal degradation of PCBs. The extent of degradation of PCBs by basidiomycetes seems to be dependent on the level of chlorination and the fungal strains employed. For instance, *Aspergillus niger* has been shown to dechlorinate Aroclor 1242 efficiently but not 1254 from contaminated soils (Murado, Tejedor & Baluja, 1976). The degradation of PCBs by *P. chrysosporium* decreases in the following order: biphenyl (23%), 2-chlorobiphenyl (16%), 2,2',4,4'-tetrachlorobiphenyl (TeCB) (10%) (Thomas, Carlswell & Georgiou, 1992). *P. chrysosporium* was shown to degrade 11% of 4,4'-dichlorobiphenyl (DCB) and 10% of 2,2',4,4'-tetrachlorobiphenyl but only negligible amounts of 3,3',4,4'-tetrachlorobiphenyl; this further supported the idea that degradation of PCB congeners is dependent on the chlorine substitution pattern on the biphenyl ring (Dietrich, Hickey & Lamar, 1995). *P. chrysosporium, Coriolopsis polyzona* and *T. versicolor*, respectively, caused 25, 41 and 50% degradation of the PCBs present in a commercial Delor 106 mixture (Novotny *et al.*, 1997). Yadav *et al.* (1995a) reported 82, 31, and 18% degradation, respectively, of Aroclor 1242, 1254 and 1260 by *P chrysosporium*. Degradation of Aroclor reported by Yadav *et al.* (1995a) is particularly significant because this was the first conclusive demonstration of substantial degradation of Aroclor 1260 by a fungus in pure culture. Congeners with varying numbers of *o*-, *m*- and *p*-chlorines were extensively degraded, indicating relative non-specificity for the position of chlorine substitutions on the biphenyl ring. In addition, degradation does not require induction by biphenyl and occurs in high nitrogen or malt-extract media, in which LiPs and MnPs are not known to be produced. Further studies showed that 4-chlorobenzoic acid and 4-chlorobenzoyl alcohol were metabolic intermediates in the PCB degradation pathway of *P. chrysosporium* (Dietrich *et al.*, 1995).

Other white rot fungi such as *T. versicolor* and *Pleurotus ostreatus* were also shown to degrade more than 95% of the mono- and dichlorobiphenyls added to cultures (Zeddel, Majcherczyk & Hutterman, 1993) but few other details are available. *Bjerkandera adusta*, *P. ostreatus* and *T. versicolor* were shown to be more efficient than *P. chrysosporium* in degrading six PCB congeners: 2,3-DCB, 4,4'-DCB, 2,4',5'-TCB, 2,2'4,4'-TeCB, 2,2',5,5'-TeCB, and 2,2',4,4',5,5'-hexachlorobiphenyl (Beaudette *et al.*, 1998). Clearly, more work needs to be done regarding the biochemistry of the PCB degradation pathway, identification of the enzymes involved, and optimization of the culture conditions to obtain enhanced degradation of PCBs.

Chlorophenols

Chlorophenols over the years have been generated for applications in agriculture and are important constituents of paper-mill effluents (Huynh et al., 1985; Aust, 1990). Large-scale use of pentachlorophenol (PCP) as a wood preservative and as a fungicide/herbicide has led to the contamination of terrestrial and aquatic ecosystems and it is one of the priority pollutants listed by the US Environmental Protection Agency (EPA, 1988). A number of studies have shown that PCP is rapidly degraded by *P. chrysosporium* under nitrogen-limiting secondary metabolic conditions (i.e. ligninolytic conditions) while degradation was inhibited in high-nitrogen media (i.e. non-ligninolytic conditions), suggesting the involvement of LDS in PCP degradation by *P. chrysosporium* (Mileski *et al.*, 1988). PCP degradation of 20–50% was reported in nitrogen-limited static cultures (Reddy, 1995). Subsequent studies by Reddy & Gold (2000) showed that PCP degradation is initiated by a LiP- or MnP-catalysed oxidative dechlorination reaction to produce tetrachloro-1,4-benzoquinone (TCBQ). The quinone was further reduced to tetrachlorodihydrobenzene (TCDB), which undergoes successive dechlorinations to produce 1,4-hydroquinone. This was hydroxylated to form 1,2,4-trihydroxybenzene (THB). In an alternative pathway, TCBQ can undergo enzymic or non-enzymic conversion to produce 2,3,5-trichlorotrihydroxybenzene (TCTB), which undergoes successive reductive dechlorinations to produce THB. Presumably

THB undergoes ring cleavage with subsequent degradation to produce carbon dioxide. Reddy & Gold (1999) also showed that tetrachloro-1,4-hydroquinone (TCHQ) is dechlorinated to trichlorohydroxyquinone by cell extracts of *P. chrysosporium*. *T. versicolor* cultures grown under conditions conducive for laccase production, but with no detectable LiP, catalysed degradation of PCP (Ricotta, Unz & Bollag, 1996). Addition of purified extracellular laccase to such cultures enhanced PCP breakdown in the first few days of incubation. These studies clearly established a role for laccase in the degradation of PCP and possibly other chlorophenols.

The pathway for degradation of 2,4,6-trichlorophenol (TCP) by *P. chrysosporium* has also been elucidated (Joshi & Gold, 1993; Armenante, Pal & Lewandowski, 1994). Degradation of TCP was shown to involve cycles of peroxidase-catalysed oxidative dechlorination reactions followed by quinone reduction reactions to yield the key intermediate 1,2,4,5-tetrahydroxybenzene, which undergoes further degradation to carbon dioxide. It is noteworthy that in the proposed pathway all the three chlorines of TCP are removed prior to ring cleavage. 2,4-Dichlorophenol (DCP), 2,4,5-trichlorophenol and TCP were all oxidized by *P. chrysosporium* to give the corresponding 1,4-benzoquinones (Hammel, 1992). Valli and Gold (1991) showed that DCP degradation by *P. chrysosporium* involved LiP and MnP not only in the initial oxidation of DCP but also at multiple stages in the pathway.

Several species of *Phanerochaete* are moderately sensitive to PCP (Lamar, Larsen & Kirk, 1990). *P. sordida* and *P. chrysosporium* were able to grow at 25 ppm PCP with lower growth rates compared with those in media containing 5 ppm PCP (Lamar *et al.*, 1990). Comparison of the toxicity of PCP to several *Phanerochaete* species and selected other white rot fungi showed that *T. versicolor* was the fastest growing species that remained viable at high levels of PCP (40 mg l^{-1}). It was reported that there was enhanced production of laccases by *T. versicolor* in the presence of 2-chlorophenol (Grey, Hofer & Schlosser, 1998). When *T. versicolor* was grown in wheat straw cultures, over 40% of the added [^{14}C]-DCP and [^{14}C]-PCP was broken down to $^{14}CO_2$ (Fahr *et al.*, 1999). Immobilizing *P. chrysosporium* on porous polystyrene–divinylbenzene carriers resulted in increased production of LiP by both batch and repeated batch shake cultures of this organism (Ruckenstein & Wang, 1994). Immobilized spores exhibited a higher activity than immobilized 1-day-old mycelial pellets in the degradation of 2-chlorophenol (Ruckenstein & Wang, 1994).

In the early field studies where soil was augmented with peat as a source of carbon, PCP was 88–91% depleted (reviewed in Reddy, 1995). While most of the PCP was converted to non-extractable soil-bound products, only a small amount of the PCP was degraded. In PCP-contaminated soils inoculated with *Lentinula edodes*, there was 99% biotransformation of PCP within 10 weeks, with the depletion occurring rapidly in the first 4 weeks and declining thereafter (Okeke *et al.*, 1997). The biotransformation of PCP by *L. edodes* when competing with indigenous soil microorganisms was markedly lower (< 50%). During the rapid degradation of PCP, production of laccases and MnPs was maximal (Okeke *et al.*, 1997). Ten weeks after inoculation, both PCP and pentachloroanisole were almost completely degraded in monocultures of *L. edodes*. Pentachloroanisole, an intermediate in the degradation of PCP, as well as other chloroanisoles have also been shown to be toxic pollutants (Kennedy, Aust & Bumpus, 1990; Okeke *et al.*, 1993, 1994).

The above studies form the basis for further studies on bioremediation of chlorophenols by white rot fungi. These include selection of the appropriate strains, selection of suitable inocula, optimization of the growth parameters in accordance with peak enzyme production, estimation of the tolerance of the fungus and the toxicity of the chlorophenol, and the levels of degradation of the chlorophenol.

Nitroaromatics

Nitroaromatics are used in the manufacture of explosives, pesticides, pharmaceuticals, dyes and plastics and are often found in ground water and soil near production sites. Nitrotoluenes and residues of related explosives are common pollutants at a number of military facilities. Some of the nitroaromatics are highly toxic while a few others are carcinogens or mutagens (Crawford, 1995). Regular exposure to TNT (2,4,6-trinitrotoluene) was shown to cause liver damage and anaemia in workers. *P. chrysosporium* has been shown to degrade 2,4-dinitrotoluene by a proposed pathway involving oxidative, reductive and methyl transfer reactions (Valli *et al.*, 1992a). *P. chrysosporium* was also shown to degrade 30 to 50% of added TNT, a highly oxidized compound compared with many other environmental pollutants, when the TNT concentration was less than $20\,\mathrm{mg\,l^{-1}}$ (Fernando, Bumpus & Aust, 1990). Higher concentrations of TNT affected its degradation and intermediates such as 2-hydroxylamino-4,6-dinitrotoluene, 4-hydroxylamino-2,6-dinitrotoluene and 4-hydroxylamino-4,6-dinitrotoluene accumulated; these intermediates, not TNT itself, inhibited TNT degradation (Fernando & Aust, 1994; Barr & Aust, 1994). Addition of TNT to ligninolytic cultures resulted in the disappearance of LiP activity (Valli *et al.*, 1992a), which was shown also to be an effect of metabolites not TNT itself: such as 2-hydroxylamino-4,6-dinitrotoluene and 4-hydroxyl amino-2,6-dinitrotoluene 2-aminodinitrotoluene or 4-aminodinitrotoluene were also not inhibitory to LiP (Bumpus & Tatarko, 1994; Michels & Gottschalk, 1994). The involvement of an aromatic nitroreductase in the catalytic reduction of nitro groups of 1,3-dinitrobenzene, 2,4-dinitrotoluene, 2,4,6-trinitrotoluene, 1-chloro-2,4-dinitrobenzene and 2,4-dichloro-1-nitrobenzene, converting them to their corresponding hydroxylamino or amino congeners, has also been reported (Rieble, Joshi & Gold, 1994). Studies with *P. chrysosporium* have further demonstrated the occurrence of degradation of TNT in agitated cultures as opposed to stationary cultures if TNT was added at the beginning instead of after 6 days (Hawari *et al.*, 1999). In this study, MnP from *P. chrysosporium* was shown to appear almost immediately after the disappearance of TNT. These results further suggest that TNT inhibits MnP production and that MnP is not necessarily important for TNT degradation in this organism.

Among other white rot fungi, *Phlebia radiata* was shown to degrade TNT and 2-amino-4,6-dinitrotoluenes (22% and 76%, respectively), suggesting the potential of this fungus for use in bioremediation of munitions-contaminated sites (Van *et al.*, 1999). *P. ostreatus* and *T. versicolor* have also been reported to degrade TNT in solid-state systems (Majcherczyk, Zeddel & Huttermann, 1994). A wider search for other TNT-degrading white rot fungi may provide more efficient strains that, unlike *P. chrysosporium*, are not inhibited by intermediary metabolites in the nitroaromatic degradation pathway. An MnP from *Nematoloma frowardii* has been implicated in the degradation of uniformly ring-labelled [^{14}C]-2-amino-4,6-dinitrotoluene to $^{14}CO_2$, in the presence of reduced glutathione (Scheibner, Hofrichter & Fritsche, 1997). Glutathione and L-cysteine enhanced the degradation of TNT by *Nematoloma frowardii* (Scheibner & Hofrichter, 1998).

Dyes

Synthetic dyes (azo dyes, anthraquinone dyes, triarylmethane dyes and phthalocyanine dyes) are widely used in textile dyeing, paper printing, colour photography and in petroleum products. These industrial dyes are released into the environment primarily from dye-manufacturing and dye-using industries. It has been estimated that approximately 10–15% of the dyes produced end up in the industrial effluents (Spadaro, Gold &

Renganathan, 1992). Azo dyes are the predominant group and account for 50% of all the industrial dyes produced worldwide. Azo dyes and several other groups of dyes are recalcitrant to conventional wastewater treatments and persist in the environment. Azo dyes are reduced in mammals to carcinogenic aromatics, which are oxidized to N-hydroxy derivatives and finally give rise to electrophiles capable of forming covalent linkages with DNA amines.

P. chrysosporium has been reported to decolorize azo dyes Congo Red, Orange II and Tropaeolin (Cripps, Bumpus & Aust, 1990). Crude LiP decolorized all the dyes tested except Congo Red, suggesting the involvement of enzymes other than LiP in the degradation of that dye (Cripps *et al.*, 1990). The azo dyes 4-phenylazophenol, 4-phenylazo-2-methoxyphenol, Disperse Yellow 3 (DY3; 2-(4′-acetamidophenylazo)-4-methylphenol), 4-phenylazoaniline, N,N-dimethyl-4-phenylazoaniline, Disperse Orange 3 (4-(4′-nitrophenylazo)-aniline), and Solvent Yellow 14 (1-phenylazo-2-naphthol) were extensively degraded by *P. chrysosporium* under nitrogen-limiting conditions, as shown by radiolabelling of ring carbons (Spadaro *et al.*, 1992). Some dyes were, however, degraded under nitrogen-sufficient non-ligninolytic conditions as well, suggesting the involvement of non-LDS enzymes in the degradation of at least some of the azo dyes (Spadaro *et al.*, 1992).

DY3, a carcinogenic azo dye, was oxidized by LiPs and MnPs to yield 4-methyl-1-2-benzoquinone, acetanilide and a dimer of DY3 (Spadaro & Renganathan, 1994). Utilization of wheat straw during solid-state fermentation in *P. ostreatus* produced an enzyme that could decolorize Remazol Brilliant Blue R (RBBR) (Vyas & Molitoris, 1995). Furthermore, it was found that MnP and LiP were not responsible for the decolorization of RBBR in this fungus (Vyas & Molitoris, 1995). The RBBR-degrading activity was independent of Mn^{2+} and was not influenced by veratryl alcohol, but it was inhibited by $Na_2S_2O_5$, NaCN, NaN_3 and depletion of oxygen. These results suggested that this enzyme may be an oxygenase with a metal centre and is distinct from LiP and MnP (Vyas & Molitoris, 1995). In contrast to this, LiP from *B. adusta* and the MnPs from *B. adusta* and *Pleurotus eryngii* were reported to be involved in the decolorization of the industrial dyes Reactive Violet 5, Reactive Black 5, Reactive Orange 96, Reactive Red 198 and Reactive Blue 38 and 15 (Heinfling *et al.*, 1998). Efficient decolorization of azo and phthalocyanine dyes by LiP from *B. adusta* occurred in the presence of veratryl alcohol and not in its absence (Heinfling *et al.*, 1998). It was of interest that MnP from *B. adusta* oxidized dyes in an Mn^{2+}-independent manner, whereas Mn^{2+} was shown to be critical for the activity of MnPs from a number of other organisms. MnP from *P. chrysosporium* showed low activity towards industrial dyes in the presence or absence of Mn^{2+} (Heinfling *et al.*, 1998). Rodriguez, Pickard & Vázquez-Duhalt (1999) studied the decolorization of a large number of industrial dyes by *P. ostreatus* and *Trametes hispida* and observed that only laccase activity was correlated with the decolorization activity of the crude extracts of these two organisms.

Recently, Raghukumar, D'Souza & Reddy (1999) reported that *Flavodon flavus*, a basidiomycete isolated from the coastal marine environment, produces laccases, MnPs, and LiPs, and that it efficiently degrades the dyes poly R, poly B, azure B and RBBR. However, decolorization of Brilliant Green was relatively less efficient. Better degradation of the dyes was seen in the presence of salts (simulating marine environment), suggesting the potential of this organism for bioremediation of pollutants in the marine environment.

Much remains to be done in elucidating the biochemistry of dye decolourization and in identifying optimal organisms and culture conditions for dye decolourization by white rot fungi. Selected dye decolorizing enzymes from white rot fungi should be good candidates for immobilization and use in bioremediation applications.

Decolorization of industrial effluents

The pulp and paper industry releases large volumes of intensely coloured bleach plant effluents (BPEs), which contain chlorophenols, chlorolignols and other pollutants. The BPE-decolourizing activity of the ligninolytic white rot fungi *P. chrysosporium* and *T. versicolor* has been known for some time (Boominathan & Reddy, 1992), but the enzyme systems used by these organisms to degrade BPEs have only recently been elucidated. MnPs were shown to play the primary role in BPE decolorization by *P. chrysosporium* (Michel *et al.*, 1991). Lackner, Srebotnik & Messner, (1991) independently confirmed the importance of MnP in BPE decolorization and showed that the oxidation of BPE was mediated by Mn^{2+}. Purified *P. chrysosporium* MnPs also catalysed BPE decolorization in the presence of lactate, Mn^{2+} and H_2O_2. These results indicate that Mn^{2+} chelated to lactate or other organic acids is primarily responsible for BPE decolorization *in vivo*. A further report by Jaspers, Jiminez & Penninckx (1994) independently confirmed the findings of Michel *et al.* (1991) and Lackner *et al.* (1991), showing that MnP of *P. chrysosporium*, but not purified LiP, is able to decolorize BPEs.

Laccases appear to play the primary role in BPE decolorization by *T. versicolor* (Archibald, Paice & Jurasek, 1990). *T. versicolor* laccases, in the presence of phenolic substrates, were able to generate Mn^{III} chelates similar to those produced by MnP. Furthermore, several laccases of *T. versicolor* were shown to dechlorinate a number of toxic polychlorinated phenols which are major constituents of BPEs (Roy-Arcand & Archibald, 1991). The results obtained so far on BPE decolorization by *P. chrysosporium* and other white rot fungi may lead to the design of effective biomimetic systems that are able to generate chelated Mn^{III} for the degradation of BPEs.

One common group of compounds found in pulp mill effluents that are non-chlorinated are resins, a group of diterpenoid carboxylic acids that are constituents of wood from pines, spruce and firs. Resin acids may account for up to 0.2–0.8% of the total weight of the wood and may be released into the water during pulping by chemical and mechanical treatments. Resins have been implicated in the toxicity of the effluents (Kovacs & Voss, 1992). In order to degrade resin acids, fungi seem to require an additional carbon source (metabolic susbstrate). Many of the detoxification reactions by fungi occur by hydroxylation reactions (Liss, Bicho & Saddler, 1997). For example, *Mortierella isabellina* can hydroxylate dehydroabietic acid, abietic acid and isopimaric acids while *Chaetomium cochliodes* can transform dehydroabietic acid (Yano *et al.*, 1995). However, transformations by several of these fungi may be incomplete. A pretreatment of wood chips with fungi may prove to be a useful method for removing toxic resins before pulping (Wang *et al.*, 1995).

Guaiacols, the by-products of the bleaching process employed in the paper industry, are also one of the persistent pollutants of terrestrial and aquatic ecosystems. It has been proposed that the first step in the dechlorination pathway of tetrachloroguaiacol by laccase of *Coriolus versicolor* is the demethylation step that results in tetrachlorocatechol, which subsequently is dechlorinated to give 2,3,5-trichloro-6-hydroxy-*p*-benzoquinone, 2,5-dichloro-3,6-dihydroxy-*p*-benzoquinone and dichloro-6-hydroxy-*p*-benzoquinone (Imura, Hartikainen & Tatsumi, 1996).

Pesticides

Despite the ban or restrictions placed on alkyl halide insecticides such as aldrin, heptachlor, chlordane, lindane and mirex, bioaccumulation and toxicity arising from their persistence in water, soils and sediments pose serious environmental hazards. *P. chrysosporium* was shown to degrade extensively a variety of pesticides. It showed up to 23% degradation of [^{14}C]-lindane and [^{14}C]-chlordane to $^{14}CO_2$ in 30 days in liquid cultures and in 60 days in soil-corn cob cultures (Kennedy, Aust & Bumpus, 1990). However, aldrin, heptachlor and mirex did not undergo appreciable degradation but underwent substantial biotransformation, as indicated by the disappearance of the starting substrate and the appearance of intermediary metabolites. Degradation of lindane and chlordane by *P. chrysosporium* was attributed to P450-monooxygenase rather than the LMEs (Kennedy *et al.*, 1990; Mougin *et al.*, 1996). Arisoy (1998) reported extensive degradation of heptachlor and moderate degradation of lindane by *P. chrysosporium, P. eryngi, Pleurotus florida* and *Pleurotus sajor-caju*.

DDT (1,1,1-trichloro-2,2-bis(4-chlorophenyl)ethane) is one of the most persistent environmental pollutants in the environment. The recalcitrance of DDT to microbial degradation is generally attributed to trichloromethyl group. DDT was one of the earliest chlorinated aromatic compounds shown to be degraded by *P. chrysosporium* (Bumpus *et al.*, 1985). Substantial degradation of [^{14}C]-DDT to $^{14}CO_2$ was observed (Bumpus & Aust, 1987). The first metabolite produced was DDD (1,1-dichloro-2,2-bis(4-chlorophenyl)ethane) which disappeared from the cultures on continued incubation (Boominathan & Reddy, 1992; Barr & Aust, 1994). An amended soil system that contained ground corn cobs was shown to support growth and [^{14}C]-DDT degradation to $^{14}CO_2$. DDT degradation by *P. chrysosporium*, similar to lignin degradation, required the presence of another carbon source such as cellulose to serve as a growth substrate.

Alachlor (2-chloro-*N*-(2,6-diethylphenyl)-*N*-(methoxymethyl)-acetamide) and the related acetanilide herbicides Metalochlor and Propachlor, which are considered to be potential carcinogens, are transformed by white rot fungi; *Ceriporiopsis subvermispora, Phlebia tremellosa,* and *P. chrysosporium* degraded Alachlor, after 122 days of incubation, by 14, 12, and 6.3%, respectively (Ferrey *et al.*, 1994). *Fomitopsis pinicola*, a brown rot fungus, did not break down Alachlor under these conditions.

Atrazine (2-chloro-4-ethylamine-6-isopropylamino-1,3,4-triazine) is a chorinated triazine and is one of the most extensively used herbicides worldwide. It is known to undergo relatively slow biotransformation in soils and persists in the environment. *P. chrysosporium* degraded 48% of the atrazine after 14 days of incubation in a nitrogen-limited medium (Mougin *et al.*, 1994). The [^{14}C]-ethyl carbons of atrazine were degraded to $^{14}CO_2$ while very little of [^{14}C]-ring-labelled atrazine was affected (Mougin *et al.*, 1994). Hydroxylated and/or N-dealkylated metabolites of atrazine were the main products observed in the spent medium. Similar observations were made with atrazine degradation by *Pleurotus pulmonarius* except that 2-chloro-4-ethylamino-6-(1-hydroxyisopropyl) amino-1,3,5-triazine, a novel metabolite, was also produced.

Chlorophenoxyacetic acids are one of the most common herbicides, used for selective weed control, defoliation and as plant growth regulators (Loos, 1975; EPA, 1988). Chlorophenoxyacetic acids are known to be teratogenic and mutagenic, cause damage to the nervous system and suppress the immune system (Hileman, 1996). LiPs and MnPs were not required for the degradation of either 2,4-D (2,4-dichlorophenoxyacetic acid), or 2,4,5-T (2,4,5-trichlorophenoxyacetic acid) by *P. chrysosporium* (Yadav & Reddy, 1993a). A mixture of 2,4-D and 2,4,5-T was degraded at a higher rate by *P. chrysosporium* than when these compounds were present individually (Yadav & Reddy, 1993a). Yadav & Reddy (1993a) had reported a concomitant increase in degradation of 2,4-D by *P. chrysosporium* with an increase in the level of nitrogen or carbon. These results are at variance with those of Ryan & Bumpus (1989), who reported suppression of the degradation of 2,4,5-T under nitrogen-sufficient conditions.

Effects of Mn^{2+} and nitrogen limitation on the degradation of ring and side chain carbons labelled with ^{14}C-2,4,5-T by *Dichomitus squalens* and *P. chrysosporium* suggested that, in both fungi, side chain cleavage was catalysed by a mechanism independent of the LDS (Reddy, Joshi & Gold, 1997) but degradation of the aromatic ring was dependent on the LDS. These investigators further elucidated the pathway for 2,4,5-T degradation by *D. squalens* and showed that it involves chlorophenol intermediates that were further metabolized in a manner similar to that previously reported for *P. chrysosporium* (Valli & Gold, 1991; Joshi & Gold, 1993). Nerve agents VX and Russian VX (RVX), and the insecticide analog diisopropyl-amiton, which contain phosphothiolate bonds (P–S), were rapidly and completely oxidized by *P. ostreatus* in the presence of ABTS (Amitai *et al.*, 1998).

BTEX compounds

BTEX compounds (<u>b</u>enzene, <u>t</u>oluene, <u>e</u>thylbenzene, and *o*-, *m*-, and *p*-<u>x</u>ylenes) are a family of priority environmental pollutants listed by the US EPA (EPA, 1988). BTEX compounds are components of gasoline and aviation fuels and enter soil, sediments and ground water from <u>l</u>eaking <u>u</u>nderground <u>s</u>torage <u>t</u>anks (LUST) and pipelines, accidental spills, and inadequate waste disposal practices. BTEX components were shown to be efficiently degraded by *P. chrysosporium* when these components were added individually or as mixtures (Yadav & Reddy, 1993b). There was much greater degradation of BTEX compounds in malt extract medium or in defined high-nitrogen medium (in which LiP and MnP expression was blocked) than in defined low-nitrogen medium. It was remarkable that the fungus was shown to carry out substantial degradation of [^{14}C]-ring-labelled benzene and toluene to $^{14}CO_2$. *P. chrysosporium* was also shown to degrade high concentrations of *p*-cresol ($150\,mg\,l^{-1}$) and phenol ($50\,mg\,l^{-1}$), which are often found in effluents from petroleum-related industries, individually or in combination (Kennes & Lema, 1994).

Other environmental pollutants

Chlorobenzenes

Chlorobenzenes are significant environmental pollutants. They are used in the synthesis of various halogenated pesticides and dyes and are also used as degreasers and solvents. *P. chrysosporium* was shown to degrade both chlorobenzenes and *o*-, *m*, and *p*-dichlorobenzenes extensively (Yadav, Wallace & Reddy, 1995b). Furthermore, simultaneous degradation of chloro- and methyl-substituted benzenes was observed.

Trichloroethylene

Trichloroethylene (TCE) is a volatile aliphatic halocarbon compound that is commonly used as an industrial degreasing solvent and as a precursor in the synthesis of various industrial chemicals. TCE, known to be one of the more important pollutants of soils, air and aquifers in the USA, is a suspected carcinogen and exposure to it is also known to cause cardiac and neurological problems. Yadav, Bethea & Reddy (2000) showed that *P. chrysosporium* degrades TCE under nutrient-rich conditions and that TCE degradation is not linked to LiP or MnP production. Instead, TCE degradation appears to involve an alternative enzyme system that is probably upregulated under nutrient-rich conditions. Comparison of the values for total TCE removal (46.2%) and degradation (38.5%) suggests that most of the TCE is converted to carbon dioxide (Yadav *et al.*, 2000).

Linear alkylbenzene sulfonate

Linear alkylbenzene sulfonate (LAS) is an anionic surfactant that is used in laundry detergents worldwide and accounts for 28% of the total annual production of synthetic surfactants in the USA, Western Europe and Japan. Commercial LAS is a mixture of homologues with alkyl side chains ranging from 10 to 15 carbons in length. Yadav showed recently in our laboratory that *P. chrysosporium*, in contrast to its known ability to cleave and degrade aromatic rings with or without substitution, shows negligible degradation of LAS (J. S. Yadav, unpublished data). Instead, this organism extensively transforms LAS into polar metabolites, primarily sulfophenyl carboxylates of varying chain length. Our results further showed that transformation of LAS may involve processes other than or in addition to β-oxidative shortening of the side chain, which has been observed in bacteria.

Conclusions and future perspectives

White rot fungi appear to have a great potential for bioremediation applications because of their ability to degrade a wide range of structurally diverse chloroaromatic, nitroaromatic and polyaromatic compounds. They owe this at least in part to their ability to produce extracellular peroxidases and laccases that catalyse the breakdown of organic pollutants through free radical-mediated reactions. Most laboratory scale studies to date have been done using *P. chrysosporium*. However, there has been growing interest in screening a number of other genera of white rot fungi hoping to identify organisms that produce higher levels of LMEs and/or LMEs with a higher degree of specific activity and greater ability to degrade various xenobiotic compounds. Though considerable success has been achieved in the laboratory in demonstrating extensive degradation by *P. chrysosporium* of important pollutants such as chlorophenols, PAHs and PCBs, field-scale studies have not achieved a similar degree of success because of a variety of factors. These include the inability of the fungus to compete with native microbes in soils and other ecosystems; the concentration and nature of the organopollutant present at a given site; inadequate understanding and/or inability to meet the nutrient requirements of the fungus to enable it to thrive at the contaminated site; unfavourable local conditions, such as the pH, temperature and moisture; and local unavailability of inexpensive nutrients such as corn cobs, sawdust etc., which need to be added to the site to enhance the growth of the fungus. Recently, there have been several promising field-scale bioremediation studies using white rot fungi, and startup companies such as Intec One-eighty (Logan, Utah) and Mycotech (Butte, Montana) are hoping to prosper by selling the technology to interested industries.

Progress is being made in obtaining a better understanding of the comparative biology of the LDS in a broad group of white rot fungi in order to identify organisms that may be superior to the commonly used model fungus *P. chrysosporium*. Much information is becoming available on the enzymology and molecular biology of LDS. The crystal structure of both LiP and MnP have been published. The major genes encoding LiPs, MnPs and laccases have also been cloned and sequenced, and studies on the regulation of expression of key enzymes of the LDS are in progress. Continuing progress in this area should lead to the successful genetic engineering of white rot fungi to enable improved design and application for optimal bioremediation strategies for treating contaminated sites.

Acknowledgements

Research from the author's laboratory reported here was supported in part by grant DE-FG02-85ER 13369 from the US Department of Energy and BIR 912-006 from the MSU Center for Microbial Ecology.

References

Amitai, G., Adani, R., Sod-Moriah, G., Rabinovitz, I., Vincze, A., Leader, A., Chefetz, B., Leibovitz-Persky, L., Friesem, D. & Hadar, Y. (1998). Oxidative biodegradation of phosphorothiolates by fungal laccase. *FEBS Letters*, **438**, 195–200.

Archibald, F. S., Paice, M. G. & Jurasek, L. (1990). Decolorization of kraft bleachery effluent chromophores by *Coriolus* (*Trametes*) *versicolor. Enzyme and Microbial Technology*, **12**, 846–853.

Arisoy, M. (1998). Biodegradation of chlorinated organic compounds by white-rot fungi. *Bulletin of Environmental Contamination and Toxicology*, **60**, 872–876.

Armenante, P. M., Pal, N. & Lewandowski, G. (1994). Role of mycelium and extracellular protein in the biodegradation of 2,4,6-trichlorophenol by *Phanerochaete chrysosporium. Applied and Environmental Microbiology*, **60**, 1711–1718.

Aust, S. D. (1990). Degradation of environmental pollutants by *Phanerochaete chrysosporium. Microbial Ecology*, **20**, 197–209.

Barr, D. P. & Aust, S. D. (1994). Pollutant degradation by white rot fungi. *Reviews in Environmental Contamination and Toxicology*, **138**, 49–72.

Beaudette, L. A., Davies, S., Fedorak, P. M., Ward, O. P. & Pickard, M. A. (1998). Comparison of gas chromatography and mineralization for measuring loss of selected polychlorinated biphenyl congeners in cultures of white rot fungi. *Applied and Environmental Microbiology*, **64**, 2020–2025.

Bogan, B. W. & Lamar, R. T. (1996). Polycyclic aromatic hydrocarbon-degrading capabilities of *Phanerochaete laevis* HHB-1625 and its extracellular ligninolytic enzymes. *Applied and Environmental Microbiology*, **62**, 1597–1603.

Bohmer, S., Messner, K. & Srebotnik, E. (1998). Oxidation of phenanthrene by a fungal laccase in the presence of 1-hydroxybenzotriazole and unsaturated lipids. *Biochemical and Biophysical Research Communications*, **244**, 233–238.

Boominathan, K. & Reddy, C. A. (1992). Fungal degradation of lignin: biotechnological applications. In *Handbook of Applied Mycology*, eds. D. K. Arora, R. P. Elander & K. G. Mukerji, Vol. 4, pp. 763–822. New York: Marcel Dekker.

Bumpus, J. A. (1993). White-rot fungi and their potential use in soil bioremediation processes. In *Soil Biochemistry*, eds. J. M. Bollag & G. Stotzky, pp. 65–100. New York: Marcel Dekker.

Bumpus, J. A. & Aust, S. D. (1987). Biodegradation of DDT [1,1,1-trichloro-2,2-bis (4-chlorophenyl) ethane] by the white-rot fungus *Phanerochaete chrysosporium*. *Applied and Environmental Microbiology*, **53**, 2001–2003.

Bumpus, J. A. & Tatarko, M. (1994). Biodegradation of 2,4,6-trinitrotoluene by *Phanerochaete chrysosporium*: identification of initial degradation products and the discovery of a TNT metabolite that inhibits lignin peroxidases. *Current Microbiology*, **28**, 185–190.

Bumpus, J. A., Tien, M., Wright, D. & Aust, S. D. (1985). Oxidation of persistent environmental pollutants by a white-rot fungus. *Science*, **228**,1434–1436.

Clonfero, E., Nardini, B., Marchioro, M., Bordin, A. & Gabbani, G. (1996). Mutagenicity and contents of polycyclic aromatic hydrocarbons in used and recycled motor oils. *Mutation Research*, **368**, 283–291.

Crawford, R. L. (1995). The microbiology and treatment of nitroaromatic compounds. *Current Biology*, **6**, 329–336.

Cripps, C., Bumpus, J. A. & Aust, S. D. (1990). Biodegradation of azo and heterocyclic dyes by *Phanerochaete chrysosporium*. *Applied and Environmental Microbiology*, **56**, 1114–1118.

Cullen, D. & Kersten, P. (1992). Fungal enzymes for lignocellulose degradation. In *Applied Molecular Genetics of Filamentous Fungi*, eds. J. R. Kinghorn & G. Turner. Glasgow: Blackie Academic and Professional (Chapman & Hall).

Dhawale, S. W., Dhawale, S. S. & Dean-Ross, D. (1992). Degradation of phenanthrene by *Phanerochaete chrysosporium* occurs under ligninolytic as well as non-ligninolytic conditions. *Applied and Environmental Microbiology*, **58**, 3000–3006.

Dietrich, D., Hickey, W. J. & Lamar, R. (1995). Degradation of 4,4′-dichlorobiphenyl, 3,3′,4,4′-tetrachlorobiphenyl, and 2,2′,4,4′,5,5′-hexachlorobiphenyl by the white rot fungus *Phanerochaete chrysosporium*. *Applied and Environmental Microbiology*, **61**, 3904–3909.

EPA (US Environmental Protection Agency) (1988). 2,4-D. *Reviews in Environmental Contamination and Toxicology*, **104**, 63–72.

Fahr, K., Wetzstein, H. G., Grey, R. & Schlosser, D. (1999). Degradation of 2,4-dichlorophenol and pentachlorophenol by two brown rot fungi. *FEMS Microbiology Letters* **175**, 127–132.

Fernando, T. & Aust, S. D. (1994). Biodegradation of toxic chemicals by white rot fungi. In *Biological Degradation and Bioremediation of Toxic Chemicals*, ed. G. R. Chaudhry, pp. 386–402. London: Chapman & Hall.

Fernando, T., Bumpus, J. A. & Aust, S. D. (1990). Biodegradation of TNT (2,4,6-trinitrotoluene) by *Phanerochaete chrysosporium*. *Applied and Environmental Microbiology*, **56**, 1666–1671.

Ferrey, M. L., Koskinen, W. C., Blanchette, R. A. & Burnes, T. A. (1994). Mineralization of alachlor by lignin-degrading fungi. *Canadian Journal of Microbiology*, **40**, 795–798.

Field, J. A., Heessels, E., Wijngaarde, R., Kotterman, M., de Jong, E. & de Bont, J. A. M. (1994). The physiology of polycyclic aromatic hydrocarbon biodegradation by the white rot fungus, *Bjerkandera* sp. strain BOS55. In *Applied Biotechnology for Site Remediation*, eds. R. E. Hinchee, D. B. Andersson, F. B. Metting, G. D. Sayler, pp. 143–151. Columbus, OH: CRC Press.

Gold, M. H. & Alic, M. (1993). Molecular biology of the lignin-degrading basidiomycete *Phanerochaete chrysosporium*. *Microbiological Reviews*, **57**, 605–622.

Grey, R., Hofer, C. & Schlosser, D. (1998). Degradation of 2-chlorophenol and formation of 2-chloro-1,4-benzoquinone by mycelia and cell-free crude culture liquids of *Trametes versicolor* in relation to extracellular laccase activity. *Journal of Basic Microbiology*, **38**, 371–382.

Haemmerli, S. D., Leisola, M. S. A., Sanglard, D. & Fiechter, A. (1986). Oxidation of benzo[*a*]pyrene by extracellular ligninase of *Phanerochaete chrysosporium*. *Journal of Biological Chemistry*, **261**, 6900–6903.

Hammel, K. E. (1992). Oxidation of aromatic pollutants by lignin-degrading fungi and their extracellular peroxidases. In *Metal Ions in Biological Systems*, eds. H. Siegel & A. Siegel, pp. 41–60. New York: Marcel Dekker.

Hammel, K. E., Kalyanaraman, B. & Kirk, T. K. (1986). Oxidation of polycyclic aromatic hydrocarbons and dibenzo[*p*]dioxins by *Phanerochaete chrysosporium* ligninase. *Journal of Biological Chemistry*, **261**, 16948–16952.

Hammel, K. E., Gai, W. Z., Green, B. & Moen, M. A. (1992). Oxidative degradation of phenanthrene by the ligninolytic fungus *Phanerochaete chrysosporium*. *Applied and Environmental Microbiology*, **58**, 1832–1838.

Hatakka, A. (1994). Lignin-modifying enzymes from selected white-rot fungi: production and role in lignin degradation. *FEMS Microbiology Reviews*, **13**, 125–135.

Hawari, J., Halasz, A., Beaudet, S., Paquet, L., Ampleman, G. & Thiboutot, S. (1999). Biotransformation of 2,4,6-trinitrotoluene with *Phanerochaete chrysosporium*. *Applied and Environmental Microbiology*, **65**, 2977–2986.

Heinfling, A., Martinez, M. J., Martinez, A. T., Bergbauer, M. & Szewzyk, U. (1998). Transformation of industrial dyes by manganese peroxidases from *Bjerkandera adusta* and *Pleurotus eryngii* in a manganese-independent reaction. *Applied and Environmental Microbiology*, **64**, 2788–2793.

Hileman, B. (1996). Immune system suppression linked to widely used pesticides. *Chemical and Engineering News*, **74**, 23.

Huynh, V. B., Chang, H. M., Joyce, T. W. & Kirk, K. (1985). Dechlorination of chloroorganics by a white rot fungus. *Technical Association of the Pulp Paper Industry*, **68**, 98–102.

Imura, Y., Hartikainen, P. & Tatsumi, K. (1996). Dechlorination of tetraguaiacol by laccase of white-rot basidiomycete *Coriolus versicolor*. *Applied Microbiology and Biotechnology*, **45**, 434–439.

Jaspers, C. J., Jimenez, G. & Penninckx, M. J. (1994). Evidence for a role of manganese peroxidase in the decolorization of kraft pulp bleach effluent by *Phanerochaete chrysosporium*: effects of initial culture conditions on enzyme production. *Journal of Biotechnology*, **37**, 229–234.

Johannes, C. & Majcherczyk, A. (2000). Natural mediators in the oxidation of polycyclic aromatic hydrocarbons by laccase mediator systems. *Applied and Environmental Microbiology*, **66**, 524–528.

Johannes, C., Majcherczyk, A. & Hutterman, A. (1996). Degradation of anthracene by laccase of *Trametes versicolor* in the presence of different mediator compounds. *Applied Microbiology and Biotechnology*, **46**, 313–317.

Johnston, C. G. & Aust, S. D. (1994). Detection of *Phanerochaete chrysosporium* in soil by PCR and restriction enzyme analysis. *Applied and Environmental Microbiology*, **60**, 2350–2354.

Joshi, D. K. & Gold, M. H. (1993). Degradation of 2,4,5-trichlorophenol by lignin-degrading basidiomycete *Phanerochaete chrysosporium. Applied and Environmental Microbiology*, **59**, 1779–1785.

Kennedy, D. W., Aust, S. D. & Bumpus, J. A. (1990). Comparative biodegradation of alkyl halide insecticides by the white rot fungus, *Phanerochaete chrysosporium* (BKM-F-1767). *Applied and Environmental Microbiology*, **56**, 2347–2353.

Kennes, C. & Lema, J. M (1994). Simultaneous biodegradation of *p*-cresol and phenol by the basidiomycete *Phanerochaete crysosporium. Journal of Industrial Microbiology*, **13**, 311–314.

Kirk T. K. and Farrell, R. L. (1987). Enzymatic 'combustion': the microbial degradation of lignin. *Annual Review of Microbiology*, **41**, 465–505.

Kotterman, M. J., Rietberg, H. J. & Field, J. A. (1998). Polycyclic aromatic hydrocarbon oxidation by the white-rot fungus *Bjerkandera* sp. strain BOS55 in the presence of nonionic surfactants. *Biotechnology and Bioengineering*, **57**, 220–227.

Kovacs, T. G. & Voss, R. H. (1992). Biological and chemical characterization of newsprint/speciality mill effluents. *Water Research*, **26**, 771–780.

Lackner, R., Srebotnik, E. & Messner, K. (1991). Oxidative degradation of high molecular weight chlorolignin by manganese peroxidase of *Phanerochaete chrysosporium. Biochemistry and Biophysics Research Communications*, **178**, 1092–1098.

Lamar, R. T. (1992). The role of fungal lignin-degrading enzymes in xenobiotic degradation. *Current Biology*, **3**, 261–266.

Lamar, R. T., Larsen, M. J. & Kirk, T. K. (1990). Sensitivity to and degradation of pentachlorophenol by *Phanerochaete* spp. *Applied and Environmental Microbiology*, **56**, 3519–3526.

Liss, S. N., Bicho, P. A. & Saddler, J. N. (1997). Microbiology and biodegradation of resin acids in pulp mill effluents: a minireview. *Canadian Journal of Microbiology*, **75**, 599–611.

Loos, M. A. (1975). Phenoxyalkanoic acids. In *Herbicides: Chemistry, Degradation, and Mode of Action*, eds. P. C. Kearney & D. D. Kaufman, Vol. 1, pp. 1–128. New York: Marcel Dekker.

Majcherczyk, A., Zeddel, A. & Huttermann, A. (1994). Biodegradation of TNT (2,4,6-trinitrotoluene) in contaminated soil by white-rot fungi. In *Applied Biotechnology for Site Remediation*, eds. R. E. Hinchee, D. B. Anderson, F. B. Metting & G. D. Sayler, pp. 365–370. Boca Raton, FL: CRC Press.

Mansur, M., Suarez, T., Fernandez-Larrea, J. B., Brizuele, M. A. & Gonzalez, A. L. (1997). Identification of a laccase gene family in the new lignin-degrading basidiomycete CECT 20197. *Applied and Environmental Microbiology*, **60**, 2637–2646.

Mayfield, M. B., Kishi, K., Alic, M. & Gold, M. H. (1994). Homologous expression of recombinant manganese peroxidase in *Phanerochaete chrysosporium. Applied and Environmental Microbiology*, **60**, 4303–4309.

Michel, F. C. Jr, Dass, S. B., Grulke, E. A. & Reddy, C. A. (1991). Role of manganese peroxidase and lignin peroxidase of *Phanerochaete chrysosporium* in the decolorization of kraft bleach plant effluent. *Applied and Environmental Microbiology*, **57**, 2368–2375.

Michels, J. & Gottschalk, G. (1994). Inhibition of the lignin peroxidase of *Phanerochaete chrysosporium* by hydroxylamino-dinitro-toluene. *Applied and Environmental Microbiology*, **60**, 187–194.

Mileski, G. J., Bumpus, J. A., Jurek, M. A. & Aust, S. D. (1988). Biodegradation of pentachlorophenol by the white rot fungus *Phanerochaete chrysosporium. Applied and Environmental Microbiology*, **54**, 2885–2889.

Mougin, C., Laugero, C., Asther, M., Dubroca, J., Frasse, P. & Asther, M. (1994). Biotransformation of the herbicide atrazine by the white-rot fungus *Phanerochaete chrysosporium. Applied and Environmental Microbiology*, **60**, 705–708.

Mougin, C., Pericaud, C., Malosse, C., Laugero, C. & Asther, M. (1996). Biotransformation of the insecticide lindane by the white rot basidiomycete *Phanerochaete chrysosporium. Pesticide Science*, **47**, 51–59.

Murado, M. A., Tejedor, M. C. & Baluja, G. (1976). Interactions between polychlorinated biphenyls (PCBs) and soil microfungi: effects of Aroclor-1254 and other PCBs on *Aspergillus flavus* cultures. *Bulletin of Environmental Contamination and Toxicology*, **15**, 768–774.

Novotny, C., Vyas, B. R., Erbanova, P., Kubatova, & Sasek, A. V. (1997). Removal of PCBs by various white rot fungi in liquid cultures. *Folia Microbiology*, **42**, 136–140.

Okeke, B. C., Paterson, A., Smith, J. E. & Watson-Craik, I. A. (1993). Aerobic metabolism of pentachlorophenol by spent sawdust culture of 'shiitake' mushroom (*Lentinus edodes*) in soil. *Biotechnology Letters*, **15**, 1077–1080.

Okeke, B. C., Paterson, A., Smith, J. E. & Watson-Craik, I. A. (1994). The relationship between phenol oxidase activity, soluble protein and ergosterol with growth of *Lentinus* species in oak sawdust logs. *Applied Microbiology and Biotechnology*, **41**, 28–31.

Okeke, B. C., Paterson, A., Smith, J. E. & Watson-Craik, I. A. (1997). Comparative biotransformation of pentachlorophenol by solid substrate cultures of *Lentinula edodes. Applied Microbiology and Biotechnology*, **48**, 563–569.

Pickard, M. A., Roman, R., Tinoco, R. & Vázuez-Duhalt, R. (1999). Polycyclic aromatic hydrocarbon metabolism by white rot fungi and oxidation by *Coriolopsis gallica* UAMH 8260 laccase. *Applied and Environmental Microbiology*, **65**, 3805–3809.

Radtke, C., Cook, W. S. & Anderson, A. (1994). Factors affecting antagonism of the growth of *Phanerochaete chrysosporium* by bacteria isolated from soil. *Applied Microbiology and Biotechnology*, **41**, 274–280.

Raghukumar, C., D'Souza, T. M. & Reddy, C. A. (1999). Lignin-modifying enzymes of *Flavodon flavus*, a basidiomycete isolated from a coastal marine environment. *Applied and Environmental Microbiology*, **65**, 2103–2111.

Reddy, C. A. (1993). An overview of the recent advances on the physiology and molecular biology of lignin peroxidases of *Phanerochaete chrysosporium*. *Journal of Biotechnology*, **30**, 91–107.

Reddy, C. A. (1995). The potential for white-rot fungi in the treatment of pollutants. *Current Opinion in Biotechnology*, **6**, 320–328.

Reddy, C. A. & D'Souza, T. M. (1994). Physiology and molecular biology of the lignin peroxidases of *Phanerochaete chrysosporium. FEMS Microbiological Reviews*, **13**,137–152.

Reddy, G. V. B. & Gold, M. H. (1999). A two-component tetrachlorohydroquinone reductive dehalogenase system from the lignin-degrading basidiomycete *Phanerochaete chrysosporium. Biochemical and Biophysical Research Communications*, **257**, 901–905.

Reddy, G. V. B. & Gold, M. H. (2000). Degradation of pentachlorophenol by *Phanerochaete chrysosporium*: intermediates and reactions involved. *Microbiology*, **146**, 405–413.

Reddy, G. V. B., Joshi, D. K. & Gold, M. H. (1997). Degradation of chlorophenoxyacetic acids by the lignin-degrading fungus *Dichomitus squalens. Microbiology*, **143**, 2353–2360.

Ricotta, A., Unz, R. F. & Bollag, J. (1996). Role of a laccase in the degradation of pentachlorophenol. *Bulletin of Environmental Contamination and Toxicology*, **57**, 560–567.

Rieble, S., Joshi, D. K. & Gold, M. H. (1994). Aromatic nitroreductase from the basidiomycete *Phanerochaete chrysosporium. Biochemical and Biophysical Research Communications*, **205**, 298–304.

Robinson, G. K. & Lenn, M. J. (1994). The bioremediation of polychlorinated biphenyl (PCB's): pro and perspectives. *Genetic Engineering Review*, **12**, 139–188.

Rodriguez, E., Pickard, M. A. & Vázquez-Duhalt, R. (1999). Industrial dye decolorization by laccases from ligninolytic fungi. *Current Microbiology*, **38**, 27–32.

Roy-Arcand, L. & Archibald, F. S. (1991). Direct dechlorination of chlorophenolic compounds by laccases from *Trametes* (*Coriolus*) *versicolor. Enzyme and Microbial Technology*, **13**, 194–203.

Ruckenstein, E. & Wang, X. (1994). Production of lignin peroxidase by *Phanerochaete chrysosporium* immobilized on porous poly(styrene-divinylbenzene) carrier and its application to the degrading of 2-chlorophenol. *Biotechnology and Bioengineering*, **44**, 79–86.

Ryan, T. P. & Bumpus, J. A. (1989). Biodegradation of 2,4,5-trichlorophenoxyacetic acid in liquid culture and in soil by the white rot fungus *Phanerochaete chrysosporium. Applied Microbiology and Biotechnology*, **31**, 302–307.

Scheibner, K. & Hofrichter, M. (1998). Conversion of aminonitrotoluenes by fungal manganese peroxidase. *Journal of Basic Microbiology*, **38**, 51–59.

Scheibner, K., Hofrichter, M. & Fritsche, W. (1997). Mineralization of 2-amino-4,6-dinitrotoluene by manganese peroxidase of the white-rot fungus *Nematoloma frowardii. Biotechnology Letters* **19**, 835–839.

Spadaro, J. T. & Renganathan, V. (1994). Peroxidase-catalyzed oxidation of azo dyes: mechanisms of Disperse Yellow 3 degradation. *Archives of Biochemistry and Biophysics*, **312**, 301–307.

Spadaro, J. T., Gold, M. H. & Renganathan, V. (1992). Degradation of azo dyes by the lignin-degrading fungus *Phanerochaete chrysosporium. Applied and Environmental Microbiology*, **58**, 2397–2401.

Sutherland, J. B. (1992). Detoxification of polycyclic aromatic hydrocarbons by fungi. *Journal of Industrial Microbiology*, **9**, 53–62.

Sutherland, J. B., Selby, A. L., Freeman, J. P., Evans, F. E. & Cerniglia, C. E. (1991). Metabolism of *Phanerochaete chrysosporium. Applied and Environmental Microbiology*, **57**, 3310–3316.

Takada, S., Nakamura, M., Matsueda, T., Kondo, R. & Sakai, K. (1996). Degradation of polychlorinated dibenzo-*p*-dioxins and polychlorinated dibenzofurans by the white rot fungus *Phanerochaete sordida* YK-624. *Applied and Environmental Microbiology*, **62**, 4323–4328.

Thomas, D. R., Carlswell, K. & Georgiou, G. (1992). Mineralization of biphenyl and PCBs by the white rot fungus *Phanerochaete chrysosporium. Biotechnology and Bioengineering*, **40**, 1395–1402.

Thurston, C. F. (1994). The structure and function of fungal laccase. *Microbiology*, **140**, 19–26.

Valli, K. & Gold, M. H. (1991). Degradation of 2,4-dichlorophenol by the lignin-degrading basidiomycete *Phanerochaete chrysosporium. Journal of Bacteriology*, **173**, 345–352.

Valli, K., Brock, B. J., Joshi, D. K. & Gold, M. H. (1992a). Degradation of 2,4-dinitrotoluene by the lignin degrading fungus *Phanerochaete chrysosporium. Applied and Environmental Microbiology*, **58**, 221–228.

Valli, K., Wariishi, H. & Gold, M. H. (1992b). Degradation of 2,7-dichlorodibenzo-p-dioxin by the lignin-degrading basidiomycete *Phanerochaete chrysosporium. Journal of Bacteriology*, **174**, 2131–2137.

Van, A. B., Hofrichter, M., Scheibner, K., Hatakka, A. I., Naveau, H. & Agathos, S. N. (1999). Transformation and mineralization of 2,4,6-trinitrotoluene (TNT) by manganese peroxidase from the white-rot basidiomycete *Phlebia radiata. Biodegradation*, **10**, 83–91.

Vyas, B. R. M. & Molitoris, H. P. (1995). Involvement of an extracellular H_2O_2-dependent lignolytic activity of the white rot fungus *Pleurotus ostreatus* in the decolorization of remazol brilliant blue R. *Applied and Environmental Microbiology*, **61**, 3919–3927.

Vyas, B. R. M., Bakowski, S., Sasek, V. & Matucha, M. (1994). Degradation of anthracene by selected white-rot fungi. *FEMS Microbiology Ecology*, **14**, 65–70.

Wang, Z., Chen, T., Gao, Y., Breuil, C. & Hiratsuka, Y. (1995). Biological degradation of resin acids in wood chips by wood-inhabiting fungi. *Applied and Environmental Microbiology*, **61**, 222–225.

Witiich, R.-M. (1998). Degradation of dioxin-like compounds by microorganisms. *Applied Microbiology and Biotechnology*, **49**, 489–499.

Yadav, J. S. & Reddy, C. A. (1993a). Mineralization of 2,4-dichlorophenoxyacetic acid (2,4-D) and mixtures of 2,4-D and 2,4,5-trichlorophenoxyacetic acid by *Phanerochaete chrysosporium. Applied and Environmental Microbiology*, **59**, 2904–2908.

Yadav, J. S. & Reddy, C. A. (1993b). Degradation of benzene, toluene, ethylbenzene, and xylenes (BTEX) by the lignin-degrading basidiomycete *Phanerochaete chrysosporium. Applied and Environmental Microbiology*, **59**, 756–762.

Yadav, J. S., Quensen, J. F. III., Tiedje, J. M. & Reddy, C. A. (1995a). Degradation of polychlorinated biphenyl (PCB) mixtures (Aroclors 1242,

1254, and 1260) by the white rot fungus *Phanerochaete chrysosporium* as evidenced by congener specific analysis. *Applied and Environmental Microbiology*, **61**, 2560–2565.

Yadav, J. S., Wallace, R. E. & Reddy, C. A. (1995b). Mineralization of mono- and dichloro benzenes and simultaneous degradation of chloro- and methyl-substituted benzenes by the white rot fungus *Phanerochaete chrysosporium*. *Applied and Environmental Microbiology*, **61**, 677–680.

Yadav, J. S., Bethea, C. & Reddy, C. A. (2000). Mineralization of trichloroethylene (TCE) by the white rot fungus *Phanerochaete chrysosporium*. *Bulletin of Environmental Contamination and Toxicology*, **65**, 28–34.

Yano, S., Nakamura, T., Uehara, T., Furuno, T. & Takashi, A. (1995). Biotransformation of terpenoids in conifers by microorganisms. IV. Absolute configuration of C-15 in 15,16-dihydroxy-8,11,13-abietatrien-8-oic acid, a metabolite from (+)-dehydroabietic acid with *Chaetomium cochliodes*. *Mokuzai Gakkaishi*, **41**, 1226–1232.

Zhang, L. H. & Jenssen, D. (1994). Studies on intrachromosomal recombination in SP5/V79 Chinese hamster cells upon exposure to different agents related to carcinogenesis. *Carcinogenesis*, **15**, 2303–2310.

Zeddel, A., Majcherczyk, A. & Hutterman, A. (1993). Degradation of polychlorinated biphenyls by white rot fungi *Pleurotus ostreatus* and *Trametes versicolor*. *Toxicological and Environmental Chemistry*, **40**, 255–266.

4

Fungal remediation of soils contaminated with persistent organic pollutants

IAN SINGLETON

Introduction

Laboratory-based studies have shown that fungi are able to degrade a wide range of organic pollutants (see other chapters) and have great potential for use as inoculants to remediate contaminated soil. However, soil is a heterogeneous environment and it is to be expected that experiments using fungal inocula to remove pollutants will show varying degrees of success. For example, soil environmental conditions such as pH, nutrient and oxygen levels may not be optimal for fungal growth or for activity of the fungal extracellular enzymes involved in pollutant transformation. In addition, results from laboratory studies on fungal transformation of persistent organic pollutants (POPs) carried out under optimal conditions in nutritionally defined liquid media are likely to be different from those obtained in the soil environment. Despite this, fungi have been shown to transform a wide variety of POPs in soil and have been used on a large scale to remediate contaminated sites (Lamar *et al.*, 1994) . This chapter will first highlight some important issues faced by researchers when using fungi for soil remediation, provide a critical review of previous work concerning fungal transformation of organic pollutants in soil, and then discuss actual field studies using fungal inocula to remediate contaminated soil. Throughout this chapter 'pollutant' refers to persistent organic pollutants only.

Bioavailability of persistent organic pollutants in soil

The bioavailability of POPs in soil is affected by the chemical nature of the pollutant and a variety of soil parameters, both chemical and physical in nature. In this chapter bioavailability refers to the acquisition and subsequent transformation and/or metabolism of an organic pollutant. The interactions of POPs with soil are complex but essentially POPs are removed from solution by sorption to soil constituents, resulting in a decrease in pollutant bioavailability. There are a variety of sorption mechanisms, including covalent bonding and electrostatic interactions (see Head, 1998) and the extent of sorption is largely dependent on the chemical nature of the pollutant and the amount of organic matter or clay in the soil. Hydrophobic organic compounds such as pentachlorophenol (PCP) generally sorb to soil organic matter (Divincenzo & Sparks, 1997; Welp & Bruemmer, 1999) and common soil organic materials such as humic and syringic acid have been found to decrease levels of PCP transformation in liquid culture by *Phanerochaete chrysosporium* (Stevens, Badkoubi & Murarka, 1996). It is important to note that pH also affects sorption of compounds such as PCP, which becomes a water-soluble salt at high pH (Banerji, Wei & Bajpai, 1993; Lafrance *et al.*, 1994). In contrast to hydrophobic pollutants, water-soluble pollutants may be rapidly transformed in soils (unless they are extremely toxic) because of their higher bioavailability. In some instances, sorption of POPs to soil may be beneficial, especially at high concentrations as their toxicity may be reduced, enabling introduced fungi to grow in the soil. Once sorbed the question remains as to how fungi access POPs for subsequent transformation. Do fungi transform sorbed pollutants or do they require pollutants to be in solution before transformation? The mechanisms involved in pollutant transformation by fungi are complex and there is good evidence for both membrane-based transformations and the involvement of extracellular enzymes (Barr & Aust, 1994). A clear understanding of how sorption of POPs to soil constituents affects these mechanisms may allow the development of more effective and long-lasting soil remediation strategies.

Long-term contaminated soil (aged contaminated soil)

Longer periods of contact of POPS with soil constituents allow more time for sorption reactions to occur and subsequent slow migration and/or diffusion of POPS into soil micropores renders pollutants unavailable for microbial transformation even by extracellular enzymes. General microbial activity is known to be affected by pore size; for example, carbon turnover rates were lower when organic substrates were located in smaller soil pores at low soil matric potential (Killham, Amato & Ladd, 1993) and nitrifying bacteria were found to be restricted to soil pores of 136 to 214 μm in size (Fair, Jamieson & Hopkins, 1994). Evidence is also available for restriction of pollutant transformation in pores less than 1 μm diameter (see Head, 1998). Overall it is clear that however efficient a fungal soil inoculant may be in transforming POPs in the laboratory, the chemical and physical restrictions encountered in the heterogenous soil environment will prevent complete pollutant transformation. It may be possible to make residual amounts of POPs available by chemical or physical means, but the extra cost involved must be balanced with the level of clean-up required and the residual toxicity of the soil after remediation.

Experimental difficulties

Complex chemical analyses using gas chromatography and/or high performance liquid chromatography (HPLC), both if possible, coupled with mass spectrometry are required to follow rates of POP transformation in soil by fungi. Both the disappearance of the parent pollutant and the appearance of breakdown products are generally required to demonstrate that transformation has occurred. When using non-sterile soils, extra complications involve pollutant transformation by other soil microbes, which can mean that other unknown breakdown products may be produced. Both sorption to soil and enzymic bonding of POPs to soil organic matter would decrease parent pollutant concentration without the appearance of transformation products. Accordingly, amounts of transformation products created are generally low in soils, making detection more difficult; for example, only trace amounts of chlorinated anisoles were found during fungal degradation of PCP (Tuomela *et al.*, 1999). Overall the chemical analysis of POP transformation in soils must be very carefully interpreted. Radiolabelled pollutants have been used to follow biodegradation and have shown that a common pollutant fate in soil is complexation to soil organic matter (sec below, Fungal transformation and complexation of POPs in soil for more details and appropriate references). One criticism of using radiolabelling techniques could be that the method involves the addition of 'fresh' pollutant, which may not represent conditions in an aged contaminated soil. Again results must be interpreted very carefully to take account of this. More recently, there has been significant interest in methods of assessing the residual toxicity of bioremediated soils in combination with chemical analysis. It is perhaps becoming accepted that a combination of chemical and toxicological methods is required to assess the risk associated with treated soils, and that complete breakdown of POPs may not always be necessary to reduce the environmental and human health risks associated with contaminated soil. This issue is discussed further below.

Fungal remediation of contaminated soil: laboratory studies

Initial demonstrations of the ability of fungi to remove POPs from soils were often carried out under sterile conditions. Fungi were inoculated into sterile soils containing freshly added pollutant and pollutant transformation monitored. These experiments showed that fungi could transform POPs in soil but were limited in that fungi were not exposed to competition from indigenous soil organisms and to the high POP concentrations often found in contaminated soils. Results obtained in sterile soil can be very different from those seen in non-sterile soil. A comparative study of PCP degradation in sterile and non-sterile soil by *Lentinula edodes* showed that PCP transformation was less and breakdown product formation higher in non-sterile soil (Okeke *et al.*, 1997). Relatively few fungal inoculants have been used in soil (Table 4.1) and species used have tended to be restricted to white rot fungi, which are known to transform POPs even though other organisms such as *Cunninghamella* sp., *Penicillium* sp. and *Aspergillus niger* are known to be capable of transforming polycyclic aromatic hydrocarbons (PAHs) (Sutherland, 1992; Launen *et al.*, 1995; Sack *et al.*, 1997). More recent work has tended to move away from exclusive use of the white rot fungus *P. chrysosporium* to use other white rot fungi that have higher transformation capababilities and are perhaps more suited to growth in soil. A potential disadvantage with *P. chrysosporium* is its high optimum growth temperature of 40°C (Lamar & Dietrich, 1990), which may limit its potential to hotter climates. Most importantly, the literature indicates varying success when using fungi to remediate contaminated soil and this variation can be attributed to a variety of factors such as unfavourable soil conditions for fungal growth, lack of POP availability and competition with indigenous microorganisms. High concentrations of POPs may also restrict fungal growth, as seen with studies on 2,4,6-trinitrotoluene (TNT) transformation in soil by *P. chrysosporium*. In this work, fungal growth was completely inhibited by small amounts of contaminated soil (equivalent to 24 ppm TNT) added to liquid culture (Spiker, Crawford & Crawford, 1992).

Effects of the soil environment on fungal growth

Given that soil is not the natural habitat for many of the fungi useful for bioremediation, soil conditions may have to be altered to encourage their growth. There is evidence that the soil environment can dramatically affect POP transformation by fungi. For example, *L. edodes* was more effective

Table 4.1. *Examples of persistent organic pollutants transformed in soil by fungal inoculants*

Soil pollutant	Fungal inoculant	References
Alachlor	*Phanerochaete chrysosporium*	McFarland *et al.*, 1996
Atrazine	*P. chrysosporium*	Hickey, Fuster & Lamar, 1994
Benomyl	*P. chrysosporium*	Ali & Wainwright, 1994
Cresol	*Rhodotolura aurantiaca*	Middelhoven, Koorevaar & Schuur, 1992
Dibenzodioxin	*Pleurotus florida*	Rosenbrock *et al.*, 1997
	P. chrysosporium	Rosenbrock *et al.*, 1997
	Dichomitus squalens	Rosenbrock *et al.*, 1997
Polycyclic aromatic hydrocarbons (PAHs)	*P. chrysosporium*	George & Neufeld, 1989
	Phanerochaete sordida	Lamar *et al.*, 1994
	Pleurotus ostreatus	Bogan *et al.*, 1999
	Trametes versicolor	Boyle *et al.*, 1998
	Trametes hirsutus	Boyle *et al.*, 1998
	Keuhneromyces mutabilis	Sack *et al.*, 1997
	Agrocybe aegerita	Sack *et al.*, 1997
Pentachlorophenol (PCP)	*P. chrysosporium*	Lamar & Dietrich, 1990
	P. sordida	Lamar *et al.*, 1994
	Lentinula edodes	Okeke *et al.*, 1996
	T. versicolor	Tuomela *et al.*, 1999
	P. ostreatus	Ruttimann-Johnson & Lamar, 1997
	Pleurotus pulmonarius	Chiu *et al.*, 1998
2,4,6-Trinitrotoluene (TNT)	*P. chrysosporium*	Spiker, Crawford & Crawford, 1992
2,4,5-Trichlorophenoxyacetic acid	*P. chrysosporium*	Ryan & Bumpus, 1989

in transforming PCP at lower soil moisture contents while *P. chrysosporium* was more effective at higher moisture levels (Okeke *et al.*, 1996). The same study found that optimal transformation by both fungi occurred at pH 4. In addition, better fungal growth could help introduced fungi to overcome competition from indigenous soil microorganisms. The effect of soil conditions should not be overlooked and could be partly responsible for the variation in remediation success seen in the literature.

As well as soil physicochemical factors, competition from indigenous soil microbes affects POP transformation by fungal inoculants. Microbes antagonistic to *P. chrysosporium* have been isolated from soil (Ali & Wainwright, 1994; Radtke, Cook & Anderson, 1994) and different fungi are known to have different abilities to compete with soil microflora. Lang,

Eller & Zadrazil (1997) found that *Pleurotus* is a better competitor in soil than several other white rot fungi and, therefore, would appear to be a more suitable inoculant. However, in a non-sterile soil, pyrene degradation using an inoculum of *Dichomitus squalens* (a white rot fungus) was greater than that achieved using a *Pleurotus* inoculum (in der Wiesche, Martens & Zadrazil, 1996). The authors suggested that even though *Pleurotus* is a better competitor in soil than *D. squalens* the latter was perhaps better able to stimulate the overall degradation capability of the soil microflora. Clearly there are complex interactions between fungal inoculant, soil type and soil microflora that are poorly understood.

Soil inoculation and amendment techniques

To aid the colonization of white rot fungi in soil, various inoculation and soil amendment strategies have been suggested. To be effective on a large scale, these methods must be robust and cheap to apply. Accordingly, fungi have been grown and added to soil on substrates such as corn cobs (e.g. Ryan & Bumpus, 1989) and specially prepared pellets, which enabled good soil colonization by several fungal species (Lestan & Lamar, 1996). Correct preparation of fungal inoculants for introduction into soil is crucial (inoculant formulation is covered in more detail in Chapter 5). Several workers have observed that soil amendment with straw improves pollutant transformation by white rot fungi. Morgan *et al.* (1993) found that straw increased the hyphal length of white rot fungi in soil and that straw generally gave the greatest initial rate of pollutant breakdown. Importantly, Rosenbrock *et al.* (1997) demonstrated that fungal inoculation gave greater increases in breakdown of dibenzo-*p*-dioxin than simple soil amendment (straw and compost). As well as increasing fungal growth, amendments could exert beneficial effects by sorbing pollutants and hence decreasing the amount of toxic pollutant available. The amount of amendment used for optimum pollutant transformation in different studies has varied widely and probably depends on soil type, fungal inoculum, amendment type and the age of contamination. Generally significant quantities of amendment are required, for example a straw to soil ratio of 1 : 4 or a ground corn cob to soil ratio of 4 : 1 (Ryan & Bumpus, 1989; Morgan *et al.*, 1993). Clearly, consideration must be given to amendment cost, the space available for remediation on site and costs associated with subsequent soil disposal if soil amendment is to be used as a remediation strategy. Ideally, the remediated soil could be used as subsequent soil conditioner if amendments are used, thus eliminating disposal costs. Finally, composting of contaminated soil with inoculation by *P. chrysosporium* was successful on a laboratory scale and would seem to create good conditions for growth of such organisms (McFarland *et al.*, 1996).

To examine the success of different inoculation and amendment methods, a variety of techniques are available to monitor fungal growth and activity in soil. These include image analysis (Morgan *et al.*, 1993), determination of extracellular enzyme activities (in der Wiesche *et al.*, 1996, Lang *et al.*, 1997) and *in situ* fungal gene expression via extraction of fungal mRNA and subsequent quantification by reverse transcriptase polymerase chain reaction (Bogan *et al.*, 1996).

Innovative soil treatments for improving fungal remediation of contaminated soil

Useful soil treatments for increasing POP transformation rates could include addition of surfactants and elements such as manganese. Surfactants have not been widely used to increase POP transformation in fungal studies despite their frequent use in soil remediation work (see reviews by Singleton, 1994; Head, 1998). However from the information available, surfactants have had a positive effect on POP transformation by fungi, and PAH transformation in both liquid culture and soil was increased by surfactants (Boyle, Wiesner, & Richardson, 1998; Kotterman *et al.*, 1998; Bogan *et al.*, 1999). Stimulation of PAH transformation by surfactants would appear to indicate that bioavailability is a major factor controlling PAH transformation. Such results using surfactants are encouraging for further work in this area.

Manganese is another potential candidate for addition to soil as it can stimulate fungal biotransformation of atrazine in liquid culture. This stimulation was explained by manganese increasing membrane permeability and stimulating activity of manganese peroxidase (Masaphy, Henis & Levanon, 1996). As far as the author is aware, there are no studies that examine the effect of manganese addition to soil on transformation of organic compounds by fungal inoculants. Normal soil solution concentrations of manganese vary between $0.1 \, \mu mol \, l^{-1}$ in aerated alkaline soils to $400 \, \mu mol \, l^{-1}$ in submerged soils (Marschner, 1988), while stimulation of phenanthrene transformation by *Pleurotus pulmonarius* in liquid media was stimulated by up to $300 \, \mu mol \, l^{-1}$ of manganese (Masaphy *et al.*, 1996). As fungi grow best in aerated soils, which probably have reduced manganese availability, it may be beneficial to add manganese in these situations (assuming added manganese stays in an available form). Alternatively a reduction in soil pH will increase its availability (Marschner, 1988) and potentially improve fungal growth (Okeke *et al.*, 1996).

In summary, fungal inoculants are able to penetrate and colonize soil but correct formulation and/or alteration of soil conditions (by amendments, changing pH and/or water contents) to improve growth of the introduced fungi and access to pollutants will enable more efficient soil colonization and subsequent transformation of POPs.

Fungal transformation and complexation of pollutants in soil

The complete breakdown of a pollutant into its constituents is generally demonstrated by addition of ^{14}C-labelled pollutant to soil and following the production of ^{14}C-labelled carbon dioxide. It is evident from the literature that complete breakdown of POPs in non-sterile soil by fungal inoculants does not occur. Degradation levels are variable but typical levels are 14% for PAHs (Martens *et al.*, 1999) and 9–29% for PCP (Ruttimann-Johnson & Lamar, 1997; Tuomela *et al.*, 1999). As mentioned above, these studies involve addition of 'fresh' organic pollutant, which is theoretically more readily accessible than aged pollutant, so presumably, and if it were possible using $\delta^{13}C$ (see Head, 1998), studies on aged soils would reveal even less decomposition by introduced fungi. Usually a variety of transformation products are found, which are subject to further transformation by other soil microbes, extracellular enzymes, chemical reactions with other soil constituents and sorption to soil surfaces. Recognition of this has led to a focus on pollutant fates in soil. Research on PCP transformation by white rot fungi in soil has found that usually only one identifiable product is detected (pentachloroanisole) but that large amounts of unidentified compounds are also produced (Lamar & Dietrich, 1990; Ruttimann-Johnson & Lamar, 1997). An experiment using soil fumigated with methyl bromide found that most (34–65%) PCP became bound to humic acid, fulvic acid and humin. Inoculation with different white rot fungi altered the amount of PCP bound to these organic soil constituents. *Pleurotus ostreatus* bound 65% of PCP to soil organic material compared with 34–46% binding by *Irpex lacteus*, *Bjerkandera adusta* and *Trametes versicolor* (Ruttimann-Johnson & Lamar, 1997). Similarly bonding of PAHs to soil organic matter by *P. ostreatus* has been observed but to a lesser extent than seen with PCP (Bogan *et al.*, 1999). It is likely that both parent compounds and transformation products will be subject to complexation with soil organic matter. The covalent binding of POPs to humic material is probably catalysed by fungal oxidative enzymes such as peroxidases and laccases (Dawel *et al.*, 1997, Gianfreda, Xu & Bollag, 1999) and it has been suggested that laccases could be applied to reduce the availability of POPs in soil. However, it must be noted that soil properties affect the catalytic abilities of laccases, and their use in the remediation of contaminated soil must take this into consideration (see Gianfreda *et al.*, 1999).

Complexation of POPs to soil organic matter is important as it reduces their bioavailability and presumably reduces the toxicity of contaminated soil. Presently, most environmental protection authorities have regulations concerning the total amount of POPs in soil, and levels recommended depend on the future use of contaminated soil. However if POPs are rendered unavailable in soil through complexation, then total amounts may not necessarily reflect the actual toxicity of the soil as measured by observed effects on test organisms. This could mean that soil remediated by fungal inoculants is of lower risk than anticipated and could be suitable for a wider variety of applications. It appears that studies on the toxicity of remediated soil would be beneficial and may be used to alter current regulations concerning the risk associated with and subsequent uses of contaminated soil.

Potential risks associated with soil remediated by fungal inocula

Given the large amounts of POPs bound to soil organic matter by fungal inoculants, the stability of pollutant–humic material complexes and the long-term release of POPs and their transformation products from humic compounds require further study. Indeed both biological and chemical mechanisms have potential to release bound contaminants from their organic complexes and are important aspects of soil remediation work. Potential biological release mechanisms include degradation flushes of organic matter following rewetting of dried organic material (Pulleman & Tietema, 1999) and increased microbial activity as a result of extra substrate addition (Eschenbach, Wienberg & Mahro, 1998). Despite the potential for biological release of POPs, available experimental evidence shows that the associated risk is low. Drying and rewetting of soil reduced the bioavailability of several organic pollutants and was actually proposed as a potential bioremediation strategy (White, Quinones & Alexander, 1998). Additionally, substrate addition or freeze/thawing did not substantially increase PAH release from organic material (Eschenbach *et al.*, 1998). However, soil organic matter is a heterogeneous mix of different materials, some fractions of which are more labile than others. If POPs are bound to the more easily degraded organic fractions then this may lead to larger releases of toxic compounds from organic matter than expected. Importantly, chemical changes to soil organic matter may increase POP availability as treatment of soil with ethylenediaminetetraacetic acid (EDTA) treatment was found to increase the amounts of PAH bound to dissolved organic matter (Eschenbach *et al.*, 1998). This would lead to increased pollutant mobility and potentially increased bioavailability.

As mentioned above, different POPs exhibit different degrees of complexation to organic matter. For example, PAHs were not bound as extensively to soil organic matter as PCP and it appears that white rot fungi produce metabolites from PAHs that can be further degraded by normal soil microflora (Andersson & Henrysson, 1996; Kotterman *et al.*, 1998). Certainly these fungi have great potential for use in remediation of PAH-contaminated soil particularly as they can also act upon the more complex PAHs, which are very resistant to microbial attack.

Toxicity of remediated soil: risk assessment

The production of easily metabolized and less toxic transformation products may not be possible for all POPs, and the potential for a wide variety of microbial and chemical transformations means that metabolites which have increased toxicity towards different soil organisms may be formed. Recent evidence of this occurring in soil is suggested by the production of compounds toxic towards *Bacillus megaterium* growth and dehydrogenase activity in PCP-contaminated soil after 6 weeks of remediation by *P. chrysosporium* (McGrath & Singleton, 2000). In this work, methanol extracts of soil were used to assess toxicity towards *B. megaterium* and the question of toxicant bioavailability remains an issue as methanol will extract transformation products complexed to organic material. It is also likely that a longer remediation time would result in further transformation of the toxicants present. However, the potential for production of toxic intermediates and the inability of chemical analysis alone to indicate synergistic toxic interactions has resulted in an increasing number of reports detailing the use of toxicity assays to determine the success of soil remediation by both biological and chemical strategies (Gunderson *et al.*, 1997; Meier *et al.*, 1997; Salanitro *et al.*, 1997; Rocheleau *et al.*, 1999). Generally, a variety of ecotoxicological tests are carried out as the bioavailability of POPs (and their transformation products) varies for different organisms. Techniques such as the Microtox test, earthworm viability, plant toxicity and genotoxicity are used, which take into account different toxicological mechanisms and the differing ability of organisms to access POPs.

Overall the complete breakdown of POPs by fungal inoculants is not necessary as long as the parent molecules and transformation products are either strongly bound to humic material (i.e. become unavailable to living organisms), or further transformed into less toxic metabolites by indigenous soil microflora. The latter would appear to be the case when using white rot fungi to degrade PAHs. The use of relevant ecotoxicological assays in combination with soil chemical analysis to ensure full remediation of contaminated soil would seem to be the most responsible way forward.

Alternative approaches to fungal remediation of contaminated soil: *ex situ* methods

If soil contaminant levels are too toxic, or other soil conditions are unfavourable for fungal growth, then an *ex situ* treatment process for remediating soil has been suggested. POPs may be removed by a solvent wash and the solvent/POP mix subsequently treated by fungi; for example, *Bjerkandera* sp. strain BOS55 has been shown to oxidize anthracene in water/solvent mixtures and PAHs in solvent extracts of polluted soil (Field *et al.*, 1995, 1996). Higher solvent concentrations (20% v/v) killed the fungus but pollutant oxidation still occurred and was presumably catalysed by fungal extracellular enzymes. Confirming this, other workers have shown that laccases are active in organic media (Gianfreda *et al.*, 1999) and potentially purified fungal extracellular enzymes may be useful for POP detoxification in solvent extracts of soil. An alternative *ex situ* approach using a surfactant-based soil washing process and separate fungal transformation of PAHs in the resulting surfactant/PAH mixture showed potential at least on a laboratory scale (May, Schroeder & Sandermann, 1997).

Molecular techniques

Molecular approaches to improve the bioremediation activity of fungal inoculants have rarely been attempted, but efforts have been made to improve production and excretion of extracellular fungal enzymes involved in POP transformation (laccases, lignin peroxidase and manganese peroxidase). Primarily these studies have been fundamental in nature but there may be potential for increasing fungal pollutant transformation abilities. Laccase and manganese peroxidase genes have been introduced into other fungi noted for their high secretion capability, e.g. *Aspergillus oryzae* (Stewart *et al.*, 1996; Berka *et al.*, 1997) and the enzymes have been excreted into liquid growth media. In a specific attempt to improve fungal remediation, a hybrid gene was constructed consisting of a bacterial gene coding for organophosphate transformation fused to a fungal promoter and used to transform the soil fungus *Gliocladium virens* (Xu, Wild & Kenerley, 1996). Whether or not such manipulated fungi would increase pollutant transformation in contaminated soil is unknown.

Use of mycorrhizal associations

If time is not an important issue and very large amounts of soil need to be remediated (e.g. on agricultural land), then mycorrhizal associations may provide an economically attractive alternative to more engineered bio-remediation methods. At least one ectomycorrhizal fungus has ligninase activity (Cairney & Burke, 1998) and it should be possible for such fungi to transform toxic organic pollutants. In support of this, transformation of atrazine, 2,4-dichlorophenoxyacetic acid, TNT and 2,4-dichlorophenol has been observed by ectomycorrhizal fungi (Donnelly, Entry & Crawford, 1993; Meharg, Cairney & Maguire, 1997a; Meharg, Dennis & Cairney, 1997b). In addition, better transformation of 2,4-dich-lorophenol was observed when one fungus (*Suillus variegatus*) was asso-ciated with the host plant. It also appears that mycorrhizas support an associated bacterial biofilm that can transform pollutants (Sarand *et al.*, 1998, 1999). More work is required to investigate the commercial potential of mycorrhizal detoxification of contaminated soil (see also Chapter 16).

Pilot-scale and field studies

The true test of any remediation technology is in field application but because of the costs involved there are relatively few reports of pilot and full-scale studies using fungi in the literature. In most cases, fungal inocula were beneficial in treating contamination under field conditions but with certain limitations. As always, the extra cost of using inocula of any kind must be balanced with the extra remediation benefit obtained. Use of fungal inocula can only be justified if they are shown to detoxify the soil more effectively than the indigenous microflora. As discussed above, some species of white rot fungi are very effective in transforming high-molecular-weight PAHs, which is a major advantage as these compounds are normally very slowly degraded. One way of potentially reducing fungal inoculum production costs is to use waste fungal mycelium from industry. For example, spent oyster mushroom substrate and spent sawdust culture of shiitake mushroom can transform PCP (Okeke *et al.*, 1993; Chiu *et al.*, 1998). However, cheap and robust fungal inocula are also easily prepared (Lestan & Lamar, 1996) and may be cost effective depending on the volume of contaminated soil to be treated.

Results from field studies are extremely valuable for indicating future research directions and for demonstrating difficulties involved in scaling up fungal inoculation technology. The work carried out by Lamar and coworkers is invaluable in this respect (Lamar & Dietrich, 1990; Davis *et al.*, 1993; Lamar, Evans & Glaser, 1993; Lamar *et al.*, 1994). These workers concentrated on the use of *P. chrysosporium*, *Phanerochaete sordida* and *Trametes hirsuta* to detoxify PCP and creosote-contaminated soil on a field scale. Of the three fungi, *P. sordida* proved to be the most effective inoculant as it had the highest transformation capacity and the ability to grow at lower temperatures. In one field study, temperatures dropped to 8°C, which posed problems for growth of *P. chrysosporium*. Generally, fungal inoculation improved transformation levels of PCP and PAHs in the trials but, significantly, PCP levels were not reduced below those required for commercial/industrial or residential soil and concentrations of the more complex five- and six-ring PAHs were not decreased. In one trial, an attempt was made to increase the fungal inoculum level, but difficulties with applying inoculum resulted in lower initial inoculum density than anticipated (Lamar *et al.*, 1994). Overall outcomes of the work indicated that methods to decrease POP levels further are required (e.g. surfactants) and that robust and reliable inoculum production and delivery techniques are needed. To overcome the problem with complex PAHs, there is considerable potential for fungal genera such as *Pleurotus*, which have excellent capacity to transform these compounds in soil (Martens *et al.*, 1999).

Finally it is important to maintain correct moisture content, soil pH and aeration to ensure fungal growth in field applications. For best results, the optimum parameters for growth and pollutant transformation should be obtained in advance for the particular inoculant to be used and the environmental conditions maintained as close to these values as is economically possible on a large scale.

Conclusions

Fungal inoculants do grow and successfully detoxify POPs in soil at both laboratory and field scale. Detoxification may occur by degradation, transformation to less harmful compounds (with subsequent transformation by indigenous microflora) or by complexation to soil organic matter. Ideally, both chemical and ecotoxicological analyses of the remediation process should be used. Future work could examine more species of fungal inoculants, different chemical contaminants, effects of surfactants on transformation and the potential of molecular manipulation of fungi to increase extracellular enzyme production, although public concern over releasing genetically manipulated microbes into the environment may limit this last technology. The long-term stability of fungal-remediated soils needs to be assessed to satisfy potential concerns over release of pollutants from humic materials, although the risk involved would appear to be low unless the soil receives subsequent chemical additions causing pollutant mobilization. The use of mycorrhiza could have potential for low-cost treatment of large quantities of surface-contaminated soil and for areas that do not require a fast clean up. Field studies have shown the potential for fungal inocula and also the limitations in effectively managing the introduction and growth of the fungi in large quantities of soil. Only work on a large scale can demonstrate the ease and viability of the inoculation methods suggested in the literature. Given the right conditions and especially for difficult pollutants such as PAHs, fungal inocula could prove to be a useful field tool for bioremediation.

References

Ali, T. A. & Wainwright, M. (1994). Growth of *Phanerochaete chrysosporium* in soil and its ability to degrade the fungicide benomyl. *Bioresource Technology*, **49**, 197–201.

Andersson, B. E. & Henrysson, T. (1996). Accumulation and degradation of dead-end metabolites during treatment of soil contaminated with polycyclic aromatic hydrocarbons with five strains of white-rot fungi. *Applied Microbiology and Biotechnology*, **46**, 647–652.

Banerji, S. K., Wei, S. M. & Bajpai, R. K. (1993). Pentachlorophenol interactions with soil. *Water, Air and Soil Pollution*, **69**, 149–163.

Barr, D. P. & Aust, S. D. (1994). Mechanisms white rot fungi use to degrade pollutants. *Environmental Science and Technology*, **28**, 78A–87A.

Berka, R. M., Schneider, P., Golightly, E. J., Brown, S. H., Madden, M., Brown, K. M., Halkier, T., Mondorf, K. & Xu, F. (1997). Characterisation of the gene encoding an extracellular laccase of *Myceliophthora thermophila* and analysis of the recombinant enzyme expressed in *Aspergillus oryzae. Applied and Environmental Microbiology*, **63**, 3151–3157.

Bogan, B. W., Schoenike, B., Lamar, R. T. & Cullen, D. (1996). Expression of *lip* genes during growth in soil and oxidation of anthracene by *Phanerochaete chrysosporium. Applied and Environmental Microbiology*, **62**, 3697–3703.

Bogan, B. W., Lamar, R. T., Burgos, W. D. & Tien, M. (1999). Extent of humification of anthracene, flouranthrene and benzo[a]pyrene by *Pleurotus ostreatus* during growth in PAH-contaminated soil. *Letters in Applied Microbiology*, **28**, 250–254.

Boyle, D., Wiesner, C. & Richardson, A. (1998). Factors affecting the degradation of polycyclic aromatic hydrocarbons in soil by white-rot fungi. *Soil Biology and Biochemistry*, **30**, 873–882.

Cairney, J. W. G. & Burke, R. M. (1998). Extracellular enzyme activities of the ericoid mycorrhizal endophyte *Hymenoscyphus ericae* (Read) Korf and Kernan: their likely roles in the decomposition of dead plant tissues in soil. *Plant and Soil*, **205**, 181–192.

Chiu, S. W., Ching, M. L., Fong, K. L. & Moore, D. (1998). Spent oyster mushroom substrate performs better than many mushroom mycelia in removing the biocide pentachlorophenol. *Mycological Research*, **102**, 1553–1562.

Davis, M. W., Glaser, J. A., Evans, J. W. & Lamar, R. T. (1993). Field evaluation of the lignin-degrading fungus *Phanerochaete sordida* to treat creosote contaminated soil. *Environmental Science and Technology*, **27**, 2572–2576.

Dawel, G., Kaestner, M., Michels, J., Poppitz, W., Guenther, W. & Fritsche, W. (1997). Structure of a laccase-mediated product of coupling of 2,4-diamino-6-nitrotoluene to guaiacol, a model for coupling of 2,4,6-trinitrotoluene metabolites to a humic organic soil matrix. *Applied and Environmental Microbiology*, **63**, 2560–2565.

Divincenzo, J. P. & Sparks, D. L. (1997). Slow sorption kinetics of PCP on soil: concentration effects. *Environmental Science and Technology*, **31**, 977–983.

Donnelly, P. K., Entry, J. A. & Crawford, D. L. (1993). Degradation of atrazine and 2,4-dichlorophenoxyacetic acid by mycorrhizal fungi at three nitrogen concentrations in vitro. *Applied and Environmental Microbiology*, **59**, 2642–2647.

Eschenbach, A., Wienberg, R. & Mahro, B. (1998). Fate and stability of non-extractable residues of (^{14}C) PAH in contaminated soils under environmental stress conditions. *Environmental Science and Technology*, **32**, 2585–2590.

Fair, R. J., Jamieson, H. M. & Hopkins, D. W. (1994). Spatial distribution of nitrifying (ammonium oxidising) bacteria in soil. *Letters in Applied Microbiology*, **18**, 162–164.

Field, J. A., Boelsman, F., Baten, H. & Rulkens, W. H. (1995). Oxidation of anthracene in water/solvent mixtures by the white rot fungus *Bjerkandera* sp. strain BOS55. *Applied Microbiology and Biotechnology*, **44**, 234–240.

Field, J. A., Baten, H., Boelsman, F. & Rulkens, W. H. (1996). Biological elimination of polycyclic aromatic hydrocarbons in solvent extracts of polluted soil by the white rot fungus, *Bjerkandera* sp. strain BOS55. *Environmental Technology*, **17**, 317–323.

George, E. J. & Neufeld, R. D. (1989). Degradation of fluorene in soil by fungus *Phanerochaete chrysosporium*. *Biotechnology and Bioengineering*, **33**, 1306–1310.

Gianfreda, L., Xu, F. & Bollag, J.-M. (1999). Laccases: a useful group of oxidoreductive enzymes. *Bioremediation Journal*, **3**, 1–25.

Gunderson, C. A., Kostuk, J. M., Gibbs, M. H. Napolitano, G. E., Wicker, L. F., Richmond, J. E. & Stewart, A. J. (1997). Multispecies toxicity assessment of compost produced in bioremediation of explosives-contaminated sediment. *Environmental Toxicology and Chemistry*, **16**, 2529–2537.

Head, I. M. (1998). Bioremediation: towards a credible technology. *Microbiology*, **144**, 599–608.

Hickey, W. J., Fuster, D. J. & Lamar, R. T. (1994). Transformation of atrazine in soil by *Phanerochaete chrysosporium*. *Soil Biology and Biochemistry*, **26**, 1665–1671.

in der Wiesche, C., Martens, R. & Zadrazil, F. (1996). Two-step degradation of pyrene by white rot fungi and soil microorganisms. *Applied Microbiology and Biotechnology*, 46, 653–659.

Killham, K., Amato, M. & Ladd, J. N. (1993). Effect of substrate location in soil and soil pore water regime on carbon turnover. *Soil Biology and Biochemistry*, **25**, 57–62.

Kotterman, M. J. J., Rietberg, H. J., Hage, A. & Field, J. A. (1998). Polycyclic aromatic hydrocarbon oxidation by the white rot fungus *Bjerkandera* sp. strain BOS55 in the presence of non ionic surfactants. *Biotechnology and Bioengineering*, **57**, 220–227.

Lafrance, P., Marineau, L., Perreault, L. & Villenueve, J. P. (1994). Effect of natural dissolved organic matter found in groundwater on soil adsorption and transport of pentachlorophenol. *Environmental Science and Technology*, **28**, 2314–2320.

Lamar, R. T. & Dietrich, D. M. (1990). *In situ* depletion of pentachlorophenol from contaminated soil by *Phanerochaete* spp. *Applied and Environmental Microbiology*, **56**, 3093–3100.

Lamar, R. T., Evans, J. W. & Glaser, J. A. (1993). Solid-phase treatment of a PCP-contaminated soil using lignin-degrading fungi. *Environmental Science and Technology*, **27**, 2566–2571.

Lamar, R. T., Davis, M. W., Dietrich, D. M. & Glaser, J. A. (1994). Treatment of a pentachlorophenol- and creosote-contaminated soil using the lignin-degrading fungus *Phanerochaete sordida*: a field demonstration. *Soil Biology and Biochemistry*, **26**, 1603–1611.

Lang, E., Eller, G. & Zadrazil, F. (1997). Lignocellulose decomposition and production of ligninolytic enzymes during interaction of white rot fungi with soil microorganisms. *Microbial Ecology*, **31**, 1–10.

Launen, L., Pinto, L., Wiebe, C., Kiehlman, E. & Moore, M. (1995). The oxidation of pyrene and benzo[a]pyrene by nonbasidiomycete soil fungi. *Canadian Journal of Microbiology*, **41**, 477–488.

Lestan, D. & Lamar, R. T. (1996). Development of fungal inocula for bioaugmentation of contaminated soils. *Applied and Environmental Microbiology*, **62**, 2045–2052.

Marschner, H. (1988). Mechanisms of manganese acquisition by roots from soils. In *Manganese in Soils and Plants*, eds. R. G. Graham, R. J. Hannam & N. C. Uren, pp. 191–204, Dordrecht: Kluwer Academic.

Martens, R., Wolter, M., Bahadir, M. & Zadrazil, F. (1999). Mineralisation of ^{14}C-labelled highly condensed polycyclic aromatic hydrocarbons in soil by *Pleurotus* sp. Florida. *Soil Biology and Biochemistry*, **31**, 1893–1899.

Masaphy, S., Henis, Y. & Levanon, D. (1996). Manganese-enhanced biotransformation of atrazine by the white rot fungus *Pleurotus pulmonarius* and its correlation with oxidation activity. *Applied and Environmental Microbiology*, **62**, 3587–3593.

May, R., Schroeder, P. & Sandermann, H. (1997). Ex-situ process for treating PAH contaminated soil with *Phanerochaete chrysosporium*. *Environmental Science and Technology*, **31**, 2626–2633.

McFarland, M. J., Salladay, D., Ash, D. & Baiden, E. (1996). Composting treatment of alachlor impacted soil with the white rot fungus: *Phanerochaete chrysosporium*. *Hazardous Waste and Hazardous Materials*, **13**, 363–373.

McGrath, R. & Singleton, I. (2000). Pentachlorophenol transformation in soil: a toxicological assessment. *Soil Biology and Biochemistry*, **32**, 1311–1314.

Meharg, A. A., Cairney, J. W. G. & Maguire, N. (1997a). Mineralisation of 2,4-dichlorophenol by ectomycorrhizal fungi in axenic culture and in symbiosis with pine. *Chemosphere*, **34**, 2495–2504.

Meharg, A. A., Dennis, G. R. & Cairney, J. W. G. (1997b). Biotransformation of 2,4,6-trinitrotoluene (TNT) by ectomycorrhizal basidiomycetes. *Chemosphere*, **35**, 513–521.

Meier, J. R., Chang, L. W., Jacobs, S., Torsella, J., Meckes, M. C. & Smith, M. K. (1997). Use of plant and earthworm bioassays to evaluate remediation of soil from a site contaminated with polychlorinated biphenyls. *Environmental Toxicology and Chemistry*, **16**, 928–938.

Middelhoven, W. J., Koorevaar, M. & Schuur, G. W. (1992). Degradation of benzene compounds by yeasts in acidic soils. *Plant and Soil*, **145**, 37–43.

Morgan, P., Lee, S. A., Lewis, S. T., Sheppard, A. N. & Watkinson, R. J. (1993). Growth and biodegradation by white-rot fungi inoculated into soil. *Soil Biology and Biochemistry*, **25**, 279–287.

Okeke, B. C., Smith, J. E., Paterson, A. & Watson-Craik, I. A. (1993). Aerobic metabolism of pentachlorophenol by spent sawdust culture of shiitake mushroom (*Lentinula edodes*) in soil. *Biotechnology Letters*, **15**, 1077–1080.

Okeke, B. C., Smith, J. E., Paterson, A. & Watson-Craik, I. A. (1996). Influence of environmental parameters on pentachlorophenol biotransformation in soil by *Lentinula edodes* and *Phanerochaete chrysosporium. Applied Microbiology and Biotechnology*, **45**, 263–266.

Okeke, B. C., Paterson, A., Smith, J. E. & Watson-Craik, I. A. (1997). Comparative biotransformation of pentachlorophenol in soils by solid substrate cultures of *Lentinula edodes. Applied Microbiology and Biotechnology*, **48**, 211–214.

Pulleman, M. & Tietema, A. (1999). Microbial C and N transformations during drying and rewetting of coniferous forest floor material. *Soil Biology and Biochemistry*, **31**, 275–285.

Radtke, C., Cook, W. S. & Anderson, A. (1994). Factors affecting antagonism of the growth of *Phanerochaete chrysosporium* by bacteria isolated from soil. *Applied Microbiology and Biotechnology*, **41**, 274–280.

Rocheleau, S., Cimpoia, R., Paquet, L., van Koppen, I., Guiot, S. R. Hawari, J., Ampleman, G., Thiboutot, S. & Sunahara, G. I. (1999). Ecotoxicological evaluation of a laboratory-scale bioslurry treating explosives-spiked soil. *Bioremediation Journal*, **3**, 233–245.

Rosenbrock, P., Martens, R., Buscot, F., Zadrazil, F. & Munch, J. C. (1997). Enhancing the mineralisation of (U-^{14}C) dibenzo-p-dioxin in three different soils by addition of organic substrate or inoculation with white rot fungi. *Applied Microbiology and Biotechnology*, **48**, 665–670.

Ruttimann-Johnson, C. & Lamar, R. T. (1997). Binding of pentachlorophenol to humic substances in soil by the action of white rot fungi. *Soil Biology and Biochemistry*, **29**, 1143–1148.

Ryan, T. P. & Bumpus, J. A. (1989). Biodegradation of 2,4,5-trichlorophenoxyacetic acid in liquid culture and in soil by the white-rot fungus *Phanerochaete chrysosporium. Applied Microbiology and*

Biotechnology, **31**, 302–307.

Sack, U., Heinze, T. M., Deck, J., Cerniglia, C. E., Cazau, M. C. & Fritsche, W. (1997). Novel metabolites in phenanthrene and pyrene transformation by *Aspergillus niger*. *Applied and Environmental Microbiology*, **63**, 2906–2909.

Salanitro, J. P., Dorn, P. B., Heusemann, M. H., Moore, K. O., Rhodes, I. A., Rice, J. L. M., Vipond, T. E., Western, H. M. & Wisniewski, H. L. (1997). Crude oil hydrocarbon bioremediation and soil ecotoxicity assessment. *Environmental Science and Technology*, **31**, 1769–1776.

Sarand, I., Timonen, S., Nurmiaho, L. E. L., Koivula, T., Haahtela, K., Romantschuk, M. & Sen, R. (1998). Microbial biofilms and catabolic plasmid harbouring degradative fluorescent pseudomonads in Scots pine mycorrhizospheres developed petroleum contaminated soil. *FEMS Microbiology Ecology*, **27**, 115–126.

Sarand, I., Timonen, S., Koivula, T., Peltola, R., Haahtela, K., Sen, R. & Romantschuk, M. (1999). Tolerance and biodegradation of *m*-toluate by Scots pine, a mycorrhizal fungus and fluorescent pseudomonads individually and under associative conditions. *Journal of Applied Microbiology*, **86**, 817–826.

Singleton, I. (1994). Microbial metabolism of xenobiotics: fundamental and applied research. *Journal of Chemical Technology and Biotechnology*, **59**, 9–23.

Spiker, J. K., Crawford, D. L. & Crawford, R. L. (1992). Influence of 2,4,6-trinitrotoluene concentration on the degradation of TNT in explosive-contaminated soils by the white rot fungus *Phanerochaete chrysosporium*. *Applied and Environmental Microbiology*, **58**, 3199–3202.

Stevens, D. K., Badkoubi, A. & Murarka, I. P. (1996). Pentachlorophenol mineralisation by *Phanerochaete chrysosporium* in liquid culture in the presence of syringic acid or humic acid. *Hazardous Waste and Hazardous Materials*, **13**, 473–484.

Stewart, P., Whitwam, R. E., Kersten, P. J., Cullen, D. & Tien, M. (1996). Efficient expression of a *Phanerochaete chrysosporium* manganese peroxidase gene in *Aspergillus oryzae*. *Applied and Environmental Microbiology*, **62**, 860–864.

Sutherland, J. B. (1992). Detoxification of polycyclic aromatic hydrocarbons by fungi. *Journal of Industrial Microbiology*, **9**, 53–62.

Tuomela, M., Lyytikainen, M., Oivanen, P. & Hatakka, A. (1999). Mineralisation and conversion of pentachlorophenol (PCP) in soil inoculated with the white-rot fungus *Trametes versicolor*. *Soil Biology and Biochemistry*, **31**, 65–74.

Welp, G. & Bruemmer, G. W. (1999). Effects of organic pollutants on soil microbial activity: the influence of sorption, solubility and speciation. *Ecotoxicology and Environmental Safety*, **43**, 83–90.

White, J. C., Quinones, R. A. & Alexander, M. (1998). Effect of wetting and drying on the bioavailability of organic compounds sequestered in soil. *Environmental Toxicology and Chemistry*, **17**, 2378–2382.

Xu, B., Wild, J. R. & Kenerley, C. M. (1996). Enhanced expression of a bacterial gene for pesticide degradation in a common soil fungus. *Journal of Fermentation and Bioengineering*, **81**, 473–481.

5

Formulation of fungi for *in situ* bioremediation

JOAN W. BENNETT, WILLIAM J. CONNICK, JR,
DONALD DAIGLE AND KENNETH WUNCH

Introduction

Fungi play a major role in environmental biotechnology. Their morphological, physiological and reproductive strategies make them especially suited for terrestrial habitats. This book is a testament to their multifaceted role in the biodegradation of natural and xenobiotic compounds and to the major progress that has been made in our ability to use them as agents for the detoxification of hazardous wastes. Nevertheless, the fact remains that most of the successful applications have been performed in laboratory bench-top experiments. Field trials have been plagued by suboptimal results. Physical parameters such as aeration, moisture, nutrient level, pH, temperature and toxic contaminant level interact with living systems in unpredictable ways. Biological parameters such as predation and competition from the resident microbial populations also contribute to the variability of outcomes for *in situ* bioremediation. The challenge is to create remediation protocols that can be effective despite these numerous uncontrolled variables.

Two major biological strategies have been employed to increase the effectiveness of microbial bioremediation in field trials. The first is the stimulation of the indigenous population, usually through the delivery of a limiting nutrient. This practice is called *biostimulation*, and successful applications include use in marine oil spills and polycyclic aromatic hydrocarbon (PAH)-contaminated soils (Atlas & Bartha, 1992; Riser-Roberts, 1998). Nitrogen and phosphorus are the most commonly added nutrients (Liebeg & Cutright, 1999).

Composting is another form of biostimulation. In this venerable practice, mixtures of straw, manure, agricultural wastes and the like are mixed with soils, thereby stimulating the growth of various uncharacterized consortia of bacteria and fungi (Fermor, 1993). The use of composting in fungal bioremediation is reviewed by Singleton in Chapter 4.

The second biological strategy is the controlled addition of specific microorganisms to the environment. This practice is called *bioaugmentation*; the introduced microorganisms augment, but do not replace, the resident population (Walter, 1997). The introduced species may or may not already be present. In some cases, bioaugmentation populations are supplemented with a nutrient; in other cases no biostimulation adjuvant is used. Whatever the approach, a general finding is that organisms that are efficient in laboratory conditions do not fare as well in the 'real world'. Under natural conditions, laboratory strains face impoverished nutritional status and variable weather conditions. They have to compete with already established indigenous communities and they may succumb to a variety of predators. Moreover, there is often a mismatch between the normal habitat of the introduced species and the ecological niche into which it is placed. Finally, when biostimulation and bioaugmentation are used simultaneously, it is a common finding that the added nutrients favour indigenous populations so much that they overgrow the introduced species.

Although scientists and engineers have little control over conditions of weather, predator populations and a myriad of other factors, bioaugmentation systems are amenable to experimental optimization. When appropriately supplemented with carbon or other nutrient sources, delivery formulations can supply a 'head start' for organisms inoculated into habitats where they might not otherwise grow. Judicious formulation with other adjuvants (e.g. buffers, humectants, selective antibiotics, surfactants, etc.) can improve the success of *in situ* bioremediation strategies. Such formulations have received relatively little attention in the overall research effort on fungal bioremediation.

In this chapter, 'formulation' will be used as an umbrella term to describe general methods for the delivery of microorganisms for application in agricultural, industrial or environmental settings. 'Encapsulation' and 'entrapment' will be used more narrowly to refer to formulation techniques in which the microbial inoculum is embedded in a specific carrier substance.

Bacterial models

Bacterial bioaugmentation has been used in a variety of commercial contexts, for example in enhanced oil recovery (Donaldson, Chilingarian & Yen, 1989). Moreover, the preparation of bacterial inocula for drain cleaners, sewerage and wastewater treatment is big business (Stephenson & Stephenson, 1992; van Limbergen, Top & Vestraete, 1998). These inocula are usually made available as freeze-dried or air-dried solids or in liquid form as stabilized suspensions. Sometimes emulsifiers or wetting agents are added (Glasner, 1979). Similarly, the addition of nitrogen-fixing bacteria to agricultural crops is a time-honoured practice that has been used on a large scale (Thompson, 1980). Results have ranged from rapid disappearance (Ladha *et al.*, 1989) to significant levels of survival for more than 10 years (Brunel *et al.*, 1988). Fluid systems, as well as granular mixtures of vermiculite, sand or peat, are commonly used to introduce bacteria in agriculture (Okon & Hadar, 1987; van Elsas & Heijnen, 1990).

The survival of the introduced agricultural strains after they have been released into the environment can be improved by encapsulation into polymer gels such as calcium alginate. Encapsulation stabilizes introduced strains, provides a protective habitat for the strain and ensures a slow release of nutrients (Jung *et al.*, 1982; Bashan, 1986; Mueller *et al.*, 1993). Moreover, encapsulation allows living inocula to be stored in a dry, uniform state prior to application. In one successful trial of rhizosphere biocontrol agents, alginate-encapsulated *Pseudomonas fluorescens* amended with skimmed milk and bentonite clay showed higher root colonization rates than unencapsulated forms (van Elsas *et al.*, 1992). General principles of encapsulation technology for use in soils have been reviewed by Trevors *et al.* (1992).

The most extensive field studies of bacterial bioremediation for hazardous waste clean-up have involved marine oil spills. An excellent review of this large, complex and often conflicting literature is given by Swannell, Lee & McDonagh (1996). The aftermath of the Exxon Valdez disaster off the coast of Alaska in 1989 yielded some of the most robust research. In conjunction with research following other oil spills, the Exxon Valdez clean-up demonstrated that biostimulation of indigenous microbial consortia through the use of nitrogen and other fertilizers is more effective than bioaugmentation through 'seeding' with exogenous organisms (Venosa, Haines & Allen, 1992; Prince, 1993). Moreover, while there is considerable evidence that bioremediation enhances the biodegradation of petroleum on contaminated shores, there is little evidence that suggests similar effectiveness at sea (Pritchard *et al.*, 1992; Bragg *et al.*, 1994; Swannell, Lee & McDonagh, 1996). One important aftermath of this research has been increased scepticism about the effectiveness of bioaugmentation strategies in general.

Nevertheless, and on a far smaller scale, bacterial bioaugmentation using both free and encapsulated inocula has been tested for the remediation of soil contamination. Several studies have shown that the initial inoculum density is an important factor (Ramadan, El-Tayeb & Alexander, 1990). For example, Comeau, Greer & Samson (1993) in a study on 2,4-dichlorophonoxyacetic acid (2,4-D) biodegradation in soil showed that the time was reduced by 1 hour for each log increase in inoculum population over 10^5 cells ml^{-1}. Perhaps the most important parameter affecting survival of introduced bacterial inocula is nutrition. For example, both indigenous and non-indigenous strains survived well in slurries from polluted sediments in the presence of adequate carbon sources; survival of introduced strains dropped sharply when carbon was limiting (Blumenroth & Wagner-Dobler, 1998).

The encapsulation of bacteria targeted for use in terrestrial bioremediation was pioneered by Stormo & Crawford (1992). In their studies, a *Flavobacterium* entrapped in microbeads composed of agar, alginate or polyurethane retained a high rate of pentachlorophenol (PCP) biodegradation. However, in another study using a *Pseudomonas* sp. in creosote-contaminated soil slurries, Weir *et al.* (1995) found no significant differences in phenanthrene degradation using free or encapsulated inocula. Bacterial encapsulation for hazardous waste clean-up has been reviewed by Levinson *et al.* (1994) and an overview of bioaugmentation for site remediation is given by Hinchee, Frendrickson & Alleman (1995).

Fungal models

The need to develop effective formulations of fungi for environmental bioremediation has its parallel in microbial pesticides. Fungi have been widely applied in agriculture as biocontrol agents for pest management. In this approach, beneficial fungi are the active ingredients of bioherbicide, bioinsecticide and biofungicide products (Burges, 1998).

The first commercialized mycoherbicides are the anti-weed products DeVine™, a *Phytophthora palmivora* preparation that is active against stranglervine (*Morrenia odorata*), and Collego™, a *Colletotrichum gloeosporioides* preparation active against northern jointvetch (*Aeschynomene virginica*). Both DeVine™ and Collego™ are formulated as aqueous spore suspensions and applied by spraying. Although effective in the field, the highly perishable nature of the DeVine™ formulation was a major commercial disadvantage (Connick, Lewis & Quimby, 1990). For this reason, many fungal biocides are now encapsulated.

Biocontrol scientists have discovered that calcium alginate is one of the best polymers for encapsulation. Spores or mycelial fragments are mixed with an alginate solution and upon polymerization the living propagules are 'trapped' in the gel matrix. The gel allows substrate diffusion, protects the inoculum during storage as well as from adverse environmental conditions after application in the field, and can easily be supplemented with nutrients. The presence of a food source encourages rapid proliferation of the inoculum when sufficient moisture becomes available. Fillers and adjuvants can be added that improve stability and allow extrusion into a variety of uniform sizes and shapes for application. The use of alginate and other entrapment techniques for the delivery of biocontrol agents has been reviewed by Walker & Connick (1983), Connick (1988) and Connick *et al.* (1990).

Adaptations of solid-state fermentation are another approach. In this method, mycelia are grown with little or no free water on solid substrates such as grain, composted ligninocellulosic waste or other mixtures of plant materials. Both mushroom growers and koji producers use solid-state fermentation for producing fungal biomass (Smith, Berry & Kristiansen 1983; Chang, Buswell & Chiu 1993).

Many fungal formulations utilize some form of solid-state fermentation. After the mycelia have ramified through the substrate, the substrate in effect becomes a matrix that encapsulates the hyphae. When a well-colonized substrate is used directly as an ingredient, it can then be dried, extruded and shaped. In one variation on this theme, biocontrol fungi are fermented on rice flour, then combined with wheat flour, kaolin and water and finally extruded into granules. No separate step is needed to harvest conidia, and there is considerable flexibility in the choice of ingredients for the fermentation, for example damaged cereal grains can be used (Daigle *et al.*, 1998). Finally, it is possible to grow mycelia in a liquid culture and then formulate them in a dough of wheat flour, filler and water. Mycelial fragments of the biocontrol species *Alternaria cassiae*, *Alternaria crassa*, *Colletotrichum truncatum*, and *Fusarium lateritium* are mixed with semolina, kaolin and water. The resultant 'dough' is kneaded and passed through a pasta machine several times, yielding a thin sheet that is air dried and crushed into granules. This mycoherbicide product contains a homogenous mixture of fungi, nutrient and filler and has been dubbed 'Pesta' (Connick, Boyette & McAlpine, 1991).

Solid-state fermentation systems are also well developed for the preparation of entomogenous fungi such as *Metarhizium flavoviride* (Bartlett & Jaronski, 1988). Moisture level is one of the most important parameters affecting viability and virulence of *M. flavoviride* formulations (Hong, Ellis

& Moore, 1997; Moore, Longewalkd & Obognon, 1997; Hong, Jenkins & Ellis, 1999). In addition, adequate oxygenation of the substrate is import-ant for high spore yield during the production phase, while slow drying of the conidia is important for stable shelf life (Bradley *et al.*, 1992; Hong, Jenkins & Ellis, 2000).

Wood rot fungi and soil

The ability of white rot fungi, especially the model species *Phanerochaete chrysosporium*, to degrade a variety of organic pollutants completely has been documented repeatedly in laboratory trials (Barr & Aust, 1994; Bennett & Faison, 1997; Paszczynski & Crawford, 2000) and elsewhere in this book. However, xenobiotic contamination is rarely considered a prob-lem in wood, the natural habitat of white rot fungi. How can growth of these fungi be encouraged in environments where they do not naturally thrive? Several groups are investigating ways in which to improve the survival of wood rot fungi in polluted soils. This research is difficult because of the well-known pitfalls associated with measuring growth in non-sterile solid substrates. The filamentous nature of fungal growth also poses problems for quantification. In studies with wood rot fungi, indirect methods are used such as the detection of ligninolytic enzymes, the re-moval of a target xenobiotic, or the evolution of carbon dioxide.

Most of the protocols for delivering wood rot fungi for soil bioremedi-ation have been adopted from mushroom growers, who have perfected the art of producing fungal spawn on lignocellulosic waste. Therefore, wood rot species have been formulated on inexpensive substrates such as corn cobs, sawdust, wood chips, peat or wheat straw, and then these mycelia-impregnated substrates are mixed with contaminated soil (Barr & Aust, 1994; Paszczynski & Crawford, 2000). In an early successful trial, for example, *P. chrysosporium*, *Chrysosporium lignorum* and *Trametes ver-sicolor* all colonized soil and degraded 3,4-dichloroaniline and benzo[a] pyrene in soils supplemented with straw, hay or wood, although levels of breakdown were extremely low (Morgan *et al.*, 1993).

Pleurotus ostreatus is effective against a variety of PAHs in liquid culture (Belzlalel, Hadar & Cerniglia, 1996) and in soil–ligninocellulose systems (Lang, Geller & Zadrazil, 1997; Martens & Zadrazil, 1998). Several studies have shown that *P. ostreatus* inoculated on straw is superior to *P. chrysosporium* and *T. versicolor* in ability to colonize soils (Martens & Zadrazil, 1998; Novotny *et al.*, 1999). Straw was also used as a successful carrier for introducing *P. ostreatus* in trials with unsterile soils doubly contaminated with PAHs and heavy metals (Baldrian *et al.*, 2000). In addition, *P. ostreatus* was effective against pyrene, benzo[*a*]anthracene and benzo[*a*]pyrene in sterile sand microcosms amended with straw (Wolter *et al.*, 1997).

Using steam-sterilized soil microcosms, Boyle (1995) demonstrated that the growth of white rot fungi was frequently limited by carbon and nitrogen. *T. versicolor* grew well when amended with alfalfa straw, bark, sawdust or sphagnum moss; bran gave lesser simulation. In contrast, in unsterile soils, nutrient supplementation often benefited indigenous fungi more than fungal inocula. Benomyl is a biocide to which many white rot fungi are relatively resistant, so when benomyl was applied at 115 ppm, the growth of white rot fungi was improved. Complete degradation of PCP was faster in soils that had been amended with both benomyl and alfalfa straw than in any of the other systems tested (Boyle, 1995).

Gramss, Voigt & Kirsche (1999) studied the capacity of 12 species of filamentous fungi to degrade PAHs in sterile and unsterile soils. Inocula were grown on a mixture of beech wood dust and wheat straw mixture supplemented with beet sugar and flour and then introduced into PAH-spiked soils. Fungi increased PAH degradation in soils rich in organic material but inhibited PAH degradation in organic-poor soils (Gramss *et al.*, 1999).

In a similar study of PAH degradation, Martens & Zadrazil (1998) screened 45 white-rot fungi and four brown rot fungi that had been precultivated on wheat straw. Of the strains tested, none of the brown rots could colonize the soil, nor could 22 of the white-rot species. In some trials, the addition of wheat straw alone improved degradation more than did inoculation with the wood-rotting fungi. In one case, the colonizing fungus actually impeded ^{14}C-pyrene degradation, indicating that addition of suitable organic amendments to soil can dramatically improve the capacity of indigenous microbial consortia to degrade PAHs (Martens & Zadrazil, 1998). These findings are similar to those obtained from studying marine oil spills: biostimulation alone is often more effective than bioaugmentation.

Curiously, despite considerable documentation that indigenous consortia are usually more effective than single exogenous species, there has been relatively little research about bioaugmentation with mixed cultures. In natural ecosystems, the ability of fungi to bioremediate is a collaborative effort with bacteria. Seigle-Murandi *et al.* (1996) have shown that among 10 natural isolates of *P. chrysosporium*, one or more bacterial species were always present. The bacteria included *Agrobacterium radiobacter*,

Burkholderia sp. and a new taxon that is a member of the rRNA superfamily IV. The specific nature of the relationship between the fungus and the bacteria has not been elucidated but it seems likely that they work together in natural habitats (Seigle-Murandi *et al.*, 1996). In future bioaugmentation research on contaminated soils, it would make sense to decrease the emphasis on single-species inocula and test formulations with mixed bacteria–fungal consortia.

Encapsulation of fungi for bioremediation

Considerable research has demonstrated that entrapment of microbial inocula into dry, nutritive matrices improves the effectiveness of bioaugmentation in natural habitats. The Forest Products Laboratory in Madison, Wisconsin and our group at Tulane University, working in collaboration with the Southern Regional Research Center of the Agriculture Research Center, USDA, in New Orleans, Louisiana, have been the most active in studying encapsulation technologies for the delivery of fungi for bioremediation. The Forest Products Laboratory has built on strategies adapted from the mushroom spawn industry, while our group has focused on adaptation of methods from biocontrol research. Interestingly, after attempts to use a number of other formulations, both groups have found that some form of alginate encapsulation gives good results. It should be noted that alginate also has improved the consistency and effectiveness of an ectomycorrhizal inoculant (Le Tacon *et al.*, 1985).

Aspen wood chips overgrown with *P. chrysosporium* and *P. sordida* have been used to test the depletion of PCP (Lamar, Glaser & Kirk 1990; Lamar, Larson & Kirk 1990). In a successful, large-scale demonstration, fungal inoculum was produced in a proprietary formulation of a nutrient-fortified mixture of grain and sawdust. Soils inoculated with a 10% mixture of hyphae and inoculum substrate caused a 64% reduction in PCP after 20 weeks, compared with 26% for amended controls and 18% for non-amended controls (Lamar *et al.*, 1994). Although these results were promising, it was calculated that it would required 25 000 tons (22 500 tonnes) (wet weight) of fungal mycelium to treat 100 000 tons (91 000 tonnes) of soil, making commercial application impractical (Lamar *et al.*, 1994). The proprietary formulation was also successful in a *P. sordida* treatment of a creosote-contaminated soil (Davis *et al.*, 1993).

The Forest Products group proceeded to develop a novel inoculum in the form of pelleted solid substrates coated with an alginate suspension of fungal propagules and then incubated until the pellets were overgrown with a dense mycelium (Lestan & Lamar, 1996; Lestan *et al.*, 1996). Fungal viability was not reduced after spray coating pellets with a suspension of alginate and conidia of *P. chrysosporium*, chlamydospores and mycelial fragments of *P. sordida*, or mycelial fragments of *Irpex lacteus, Bjerkandera adusta* and *T. versicolor*. When introduced as sporulating mycelium on pellets into unsterile soil microcosms spiked with PCP, the mycelium-coated pellets of *I. lacteus, B. adusta* and *T. versicolor* removed over 80% of the contaminant in 4 weeks. Coated pellets on which the mycelium had *not* been allowed to proliferate failed to survive in the soil; they were especially prone to competition from indigenous fungi such as *Trichoderma* and *Fusarium* spp. Handling and application did not compromise the mechanical strength of the pellets nor impact inoculum potential (Lestan and Lamar, 1996). Compromised mechanical strength during the application process had been a problem in earlier trials with uncoated formulations of *P. sordida* (Lamar *et al.*, 1994).

Lestan *et al.* (1996) have defined 'the biological potential' of fungal inoculum for use in soil remediation as 'the amount of fungal biomass produced per unit weight or volume of fungal inoculum' (on a dry weight basis). Many of the traditional carriers such as straw, corn cobs and wood chips have low inoculum potential and require vast quantities for effective bioremediation. Quantification of biological potential is subject to the usual problems associated with assaying filamentous growth in solid substrates. Lestan *et al.* (1996) used a fluorescein diacetate hydrolysing activity (FDA) assay as an indicator of the biological potential of *P. chrysosporium* and *T. versicolor* grown on pelleted substrates. Pellets composed of aspen sawdust, starch, cornmeal and calcium lignosulfonate were ground to a uniform size, dried and then coated with a sodium alginate hydrogel that contained fungal biomass. At 24 °C, non-pelleted *P. chrysosporium* gave higher FDA activity than pelleted spawn; the highest FDA activity was found associated with growth of pelleted spawn at 39 °C. However, under the conditions of these experiments, neither *P. chrysosporium* nor *T. versicolor* with high biological potential removed PCP from contaminated soil more efficiently than pellets with lower biological potential (Lestan *et al.*, 1996). Entrapment of mycelial fragments of *I. lacteus* and *T. versicolor* was also compared in alginate, agarose, carrageenan, chitosan and gelatin; alginate gave the best growth and viability (Lestan, Lestan & Lamar 1998).

In the Pesta encapsulation strategy, spores or mycelia are embedded in a matrix that uses the gluten of wheat flour as the binding agent. This method could not be adapted for *P. chrysosporium* because the wheat gluten inhibited growth of the white rot fungus (Bennett *et al.*, 1996).

In the approach used in the Tulane University–Southern Regional Research Center collaboration, spores and mycelia were formulated directly into alginate pellets in a formulation amended with corn cob grits, sawdust or a non-nutritive clay filler (Pyrax) (Loomis *et al.*, 1997). Viability was enhanced with corn cob grits. Temperature was the most important variable in the shelf life of the inocula. Propagules embedded in alginate and stored at room temperature in the absence of nutrient supplementation were largely inactive after 2 months. Addition of sawdust or corn cob grits extended the viability of alginate-embedded mycelia; nevertheless, after 9 months, only about 20% of the pellets stored at room temperature gave growth. Refrigerated pellets have remained viable for over 4 years (Loomis *et al.*, 1997 and unpublished data). When used in laboratory toxicity tests against 2,4,6-trinitrotoluene, alginate-embedded *P. chrysosporium* gave more rapid and reproducible results than tests performed with mycelial plugs (Loomis *et al.*, 1997).

Marasmiellus troyanus, a mushroom isolated from a toxic waste site, completely degrades benzo[*a*]pyrene in liquid culture (Wunch, Alworth & Bennett 1999). *M. troyanus* grows profusely but does not sporulate in the laboratory so stable formulations require encapsulation of mycelial fragments. When formulated in alginate pellets, such formulations of *M. troyanus* removed approximately 90% of benzo[*a*]pyrene from soil microcosms after 6 weeks. The addition of a fertilizer solution with nitrogen and phosphorus does not increase the level of removal (Nemergut *et al.*, 2000). Unfortunately, the cost of alginate makes it costly as a large-scale delivery system for the bioremediation of contaminated environments. An alternative method under investigation involves solid-state fermentation in which *M. troyanus* is grown on moistened broken rice in autoclavable polypropylene bags. After several weeks of growth, the hyphal substrate mixture is extruded in a twin-screw extruder and dried in a fluid bed drier at different water activities. Granules with low moisture content have given the best viability after storage at room temperature (unpublished data).

Conclusions

The study of fungal bioremediation is dominated by studies using pure cultures, controlled environments, single chemical compounds as target xenobiotics, balanced culture media and other conditions that are amenable to experimental optimization and replication. It is unrealistic to expect such conditions to exist in the field. Changes in weather, especially temperature and moisture, can have profound effects on the outcome of bioremediation efforts. Each habitat and each pollution profile is unique. Moreover, indigenous microbial communities are well-adapted entities that often outcompete introduced species. For fungal bioaugmentation to meet its promise in environmental bioremediation, the technology must become more reproducibly effective *in situ* in spite of these many uncontrolled variables.

Part of the appeal of bioaugmentation strategies is that the introduction of exogenous species and nutrients are among the few variables that *are* amenable to experimental manipulation. Furthermore, a large body of scientific data demonstrates that bioaugmentation can speed and improve xenobiotic degradation in natural habitats. It should be remembered, however, that even with strains originating from a given ecosystem and growing effectively on a targeted substrate, such effects are frequently transient. The maintenance of population densities probably requires regular reintroduction to ensure continuous treatment efficacy (Boon *et al.*, 2000).

The goal of any formulation technology is to maintain the viability of the inoculum and to supply sources of nutrition to maintain the introduced species. In general, it has been discovered that encapsulation improves the survival and effectiveness of introduced strains. However, there is no universal carrier or single 'best' formulation for the release of fungi into natural habitats because requirements for application vary. Each organism and each habitat presents a special set of circumstances. It is common for contaminated waters and soils to contain multiple contaminants of uneven distribution. Factors such as mechanical stability, inoculum potential and degradative effectiveness have to be weighed against other factors such as shelf life, ease of preparation and ease of application. The methodologies that have yielded the most consistent and effective results in field trials, such as encapsulation in alginate, are difficult to implement at present. The lack of manufacturing technology capable of producing fungal inocula in sufficient quantities to have an environmental impact is another barrier to large-scale implementation. Therefore, the translation of scientific data into cost-effective *in situ* bioremediation will require the cooperation of environmental engineers and the investment of considerable money into research and scale-up. In addition, microbiologists should overcome their preferences for working with pure cultures and devote more research to the study of mixed consortia.

Acknowledgements

Research at Tulane University on fungal formulations for bioremediation has been supported by grants from Exxon Corporation and the US Department of Defense. We thank Jason Beadle for help in manuscript preparation.

References

Atlas, R. M. & Bartha, R. (1992). Hydrocarbon biodegradation and oil spill bioremediation. *Advances in Microbial Ecology*, **12**, 287–338.

Baldrian, P, der Wiesche, C., Gabriel, J., Nerud, F. & Zadrazil, F. (2000). Influence of cadmium and mercury on activities of ligninolytic enzymes and degradation of polycyclic aromatic hydrocarbons by *Pleurotus ostreatus* in soil. *Applied and Environmental Microbiology*, **66**, 2471–2478.

Barr, D. P. & Aust, S. D. (1994). Mechanisms white rot fungi use to degrade pollutants. *Environmental Science and Technology*, **28**, 79A–87A.

Bartlett, M. C. & Jaronski, S. T. (1988). Mass production of entomogenous fungi for biological control of insects. In *Fungi in Biological Control Systems*, ed. M. N. Burge, pp. 61–85. Manchester, UK: Manchester University Press.

Bashan, Y. (1986). Alginate beads as synthetic inoculant carriers for slow release of bacteria that affect plant growth. *Applied and Environmental Microbiology*, **51**, 1089–1098.

Belzlalel, L., Hadar, Y. & Cerniglia, C. E. (1996). Mineralization of polycyclic aromatic hydrocarbons by the white rot fungus *Pleurotus ostreatus*. *Applied and Environmental Microbiology*, **62**, 292–295.

Bennett, J. W. & Faison, B. D. (1997). Use of fungi in biodegradation. In *Manual of Environmental Microbiology*, eds. C. J. Hurst, G. R. Knudson, M. J. McInerney, L. D. Stetzenback & M. V. Walter, pp. 758–765. Washington, DC: ASM Press.

Bennett, J. W., Turner, A. J., Loomis, A. K. & Connick, W. J. Jr (1996). Comparison of alginate and 'Pesta' for formulation of *Phanerochaete chrysosporium*. *Biotechnology Techniques*, **10**, 7–12.

Blumenroth, P. & Wagner-Dobler, I. (1998). Survival of inoculants in polluted sediments: effect of strain origin and carbon source competition. *Microbial Ecology*, **35**, 279–288.

Boon, N., Goris, J., de Vos, P., Verstreate, W. & Top, E. M. (2000). Bioaugmentation of activated sludge by an indigenous 3-chloroaniline-degrading *Comamonas testosteroni* strain, 12gfp. *Applied and Environmental Microbiology*, **66**, 2906–2913.

Boyle, C. D. (1995). Development of a practical method for inducing white-rot fungi to grow into and degrade organopollutants in soil. *Canadian Journal of Microbiology*, **41**, 345–353.

Bradley, C. A., Black, W. E., Kearns, R. & Wood, P. (1992). Role of production technology in mycoinsecticide development. In *Frontiers in Industrial Mycology*, ed. G. F. Leatham, pp. 160–173. London: Chapman & Hall.

Bragg, J. R., Prince, R. C., Harner, E. J. & Atlas, R. M. (1994). Effectiveness of bioremediation for the Exxon Valdez oil spill. *Nature* (London), **368**, 413–418.

Brunel, B., Cleyet-Marel, J. C., Normand, P. & Bardin, R. (1988). Stability of *Bradyrhizobium japonicum* inoculants after introduction into soil. *Applied and Environmental Microbiology*, **54**, 2636–2642.

Burges, H. D. (1998). *Formulation of Microbial Biopesticides*, Boston, MA: Kluwer Academic.

Chang, S. T., Buswell, J. A. & Chiu, S. W. (eds.) (1993). *Mushroom Biology and Mushroom Products*. Hong Kong: Chinese University Press.

Comeau, Y., Greer, C. W. & Samson R. (1993). Role of inoculum preparation and density of bioremediation of 2-4-D-contaminated soil by bioaugmentation. *Applied Microbiology and Biotechnology*, **38**, 681–687.

Connick, W. J. Jr (1988). Formulation of living biological control agents with alginate. In *ACS Symposium Series* 371, *Pesticide Formulations: Innovations and Developments*. eds. B. Cross & B. Scher, pp. 241–250. Washington, DC: American Chemical Society.

Connick. W. J. Jr, Lewis, J. A. & Quimby, P. C. Jr (1990). Formulation of biocontrol agents for use in plant pathology. In *New Directions in Biological Control: Alternatives for Suppressing Agricultural Pests and Diseases*, eds. R. R. Baker & P. E. Dunn, pp. 345–372. New York: Alan R. Liss.

Connick, W. J. Jr, Boyette, C. D. & McAlpine, J. R. (1991). Formulation of mycoherbicides using a pasta-like process. *Biological Control*, **1**, 281–287.

Daigle, D. J. Connick, W. J. Jr, Boyette, C. D., Jackson, M. A. & Dorner, J. W. (1998). Solid-state fermentation plus extrusion to make biopesticide granules. *Biotechnology Techniques*, **12**, 715–719.

Davis, M. W., Glaser, J. A., Evans, J. W. & Lamar R. T. (1993). Field evaluation of the lignin-degrading fungus *Phanerochaete sordida* to treat creosote-contaminated soil. *Environmental Science and Technology*, **27**, 2572–2576.

Donaldson, E. D., Chilingarian, G. V. & Yen, T. F. (1989). *Microbial Enhanced Oil Recovery*. Amsterdam: Elsevier.

Fermor, T. R. (1993). Applied aspects of composting and bioconversion of lignocellulosic materials: an overview. *International Biodeterioration and Biodegradation*, **3**, 87–106.

Glasner, L. L. (1979). Microorganisms for waste treatment. In *Microbial Technology*, 2nd edn, Vol. II, eds. H. J. Peppler & D. Perlman, pp. 211–22, New York: Academic Press.

Gramss, G., Voigt, K. & Kirsche, B. (1999). Degradation of polycyclic aromatic hydrocarbons with three to seven aromatic rings by higher fungi in sterile and unsterile soils. *Biodegradation*, **10**, 51–62.

Hinchee, R. E., Frendrickson, J. & Alleman, B. C. (eds.) (1995). *Bioaugmentation for Site Remediation*. Columbus, OH: Battelle Memorial Institute.

Hong, T. D., Ellis, R. H. & Moore, D. (1997). Development of a model to predict the effect of temperature and moisture on fungal spore longevity. *Annals of Botany*, **79**, 121–128.

Hong, T. D., Jenkins, N. E. & Ellis, R. H. (1999). Fluctuating temperature and longevity of conidia of *Metarhizium flavoviride* in storage. *Biocontrol Science and Technology*, **9**, 165–176.

Hong, T. D., Jenkins, N. E. & Ellis, R. H. (2000). The effects of duration of development and drying regime on the longevity of conidia of *Metarhizium flavoviride*. *Mycological Research*, **104**, 662–665.

Jung, G., Mugnier, J., Diem, H. G. & Dommergues, Y. R. (1982) Polymer-entrapped rhizobium as an inoculant for legumes. *Plant and Soil*,

65, 219–231.

Ladha, J. K., Garcia, M., Miyan, S, Padre, A. T. & Watanabe, I. (1989). Survival of *Azorhizobium calinodans* in the soil and rhizosphere of wetland rice under *Sesbania rostrata* – rice rotation. *Applied and Environmental Microbiology*, **55**, 454–460.

Lamar, R. T., Glaser, J. A. & Kirk, T. K. (1990). Fate of pentachlorophenol (PCP) in sterile soils inoculated with the white-rot basidiomycete *Phanerochaete chrysosporium*: mineralization, volatilization and depletion of PCP. *Soil Biology and Biochemistry*, **22**, 433–440.

Lamar, R. T. Larson, M. J. & Kirk, T. K. (1990). Sensitivity to and degradation of pentachlorophenol by *Phanerochaete* spp. *Applied and Environmental Microbiology*, **56**, 3519–3526.

Lamar, R. T., Davis, M. W., Dietrich, D. M. & Glaser, J. A. (1994). Treatment of a pentachlorophenol-and creosote-contaminated soil using the lignin-degrading fungus *Phanerochaete sordida*: a field demonstration. *Soil Biology and Biotechnology*, **26**, 1603–1611.

Lang, E., Geller, G. & Zadrazil, F. (1997). Lignocellulose decomposition and production of ligninolytic enzymes during interaction of white rot fungi with soil microorganisms. *Microbial Ecology*, **34**, 1–10.

Lestan, D. & Lamar, R. T. (1996). Development of fungal inocula for bioaugmentation of contaminated soils. *Applied and Environmental Microbiology*, **62**, 2045–2052.

Lestan, D., Lestan, M., Chapelle, J. A. & Lamar, R. T. (1996) Biological potential of fungal inocula for bioaugmentation of contaminated soils. *Journal of Industrial Microbiology*, **16**, 286–294.

Lestan, D., Lestan, M. & Lamar, R. T. (1998). Growth and viability of mycelial fragments of white-rot fungi on some hydrogels. *Journal of Industrial Microbiology and Biotechnology*, **20**, 244–250.

Le Tacon, F., Jung, G., Mugnier, J., Michelot, P. & Mauperin, C. (1985). Efficiency in a forest nursery of an ectomycorrhizal fungus inoculum produced in a fermentor and entrapped in polymeric gels. *Canadian Journal of Botany*, **63**, 1664–1668.

Levinson, W. E., Stormo, K. E., Tao, H.-L. & Crawford, R. L. (1994). Hazardous waste cleanup and treatment with encapsulated or entrapped microorganisms. In *Biological Degradation and Bioremediation of Toxic Chemicals*, ed. G. R. Chaudhry, pp. 455–469. Portland, OR: Dioscorides Press.

Liebeg, E. W. & Cutright, T. J. (1999). The investigation of enhanced bioremediation through the addition of macro and micro nutrients in a PAH contaminated soil. *International Biodeterioration and Biodegradation*, **44**, 55–64.

Loomis, A. K., Childress, A. M., Daigle, D. & Bennett, J. W. (1997). Alginate encapsulation of the white rot fungus *Phanerochaete chrysosporium. Current Microbiology*, **34**, 127–130.

Martens, R. & Zadrazil, F. (1998). Screening of white-rot fungi for their ability to mineralize polycyclic aromatic hydrocarbons in soil. *Folia Microbiologica*, **43**, 97–103.

Moore, D. Longewalkd, J. & Obognon, F. (1997). Effects of rehydration on the conidial viability of *Metarhizium flavoviride* mycopesticide formulations. *Biocontrol Science and Technology*, **7**, 87–94.

Morgan, P., Lee, S. A., Lewis, S. T., Sheppard, A. N. & Watkinson, R. J. (1993).

Growth and biodegradation by white-rot fungi inoculated into soil. *Soil Biology and Biochemistry*, **25**, 279–287.

Mueller, J. G., Lin, J., Lantz, S. E. & Pritchard, P. H. (1993). Recent developments in clean-up technologies. *Remediation*, **3**, 369–381.

Nemergut, D. R., Wunch, K. G., Johnson, R. M. & Bennett, J. W. (2000). Benzo[a]pyrene removal by *Marasmiellus troyanus* in soil microcosms. *Journal of Industrial Microbiology and Biotechnology*, **25**, 116–119.

Novotny, C., Erbanova, P., Sasek, V., Kubatova, A., Cajthaml, T., Lang, E., Krahl, J. & Frantisek, Z. (1999). Extracellular oxidative enzyme production and PAH removal in soil by exploratory mycelium of white rot fungi. *Biodegradation*, **10**, 159–168.

Okon, Y. & Hadar, Y. (1987). Microbial inoculants as crop-yield enhancers. *CRC Critical Reviews in Biotechnology*, **6**, 61–85.

Paszczynski, A. & Crawford, R. C. (2000). Recent advances in the use of fungi in environmental remediation and biotechnology. In *Soil Biochemistry*, Vol. 10, eds. J. M. Bollag & G. Stotzky, pp. 379–422. New York: Marcel Dekker.

Prince, R. C. (1993). Petroleum spill bioremediation in marine environments. *Critical Reviews in Microbiology*, **19**, 217–242.

Pritchard, P. H. J., Mueller, J. G., Rogers, J. C., Kremer, F. V. & Glaser, J. E. (1992). Oil spill bioremediation: experiences, lessons and results from the Exxon Valdez oil spill in Alaska. *Biodegradation*, **3**, 315–335.

Ramadan, M. W., El-Tayeb, O. M. & Alexander, M. (1990). Inoculum size as a factor limiting success of inoculation for biodegradation. *Applied and Environmental Microbiology*, **56**, 1392–1396.

Riser-Roberts, E. (1998). *Remediation of Petroleum Contaminated Soils*. Boca Raton, FL: Lewis.

Seigle-Murandi, F., Guiraud, P., Croize, J., Falsen, E. & Eriksson, K.-E. L. (1996). Bacteria are omnipresent on *Phanerochaete chrysosporium* Burdsall. *Applied and Environmental Microbiology*, **62**, 2477–2481.

Smith, J. E., Berry, D. R. & Kristiansen, B. (eds.) (1983). *The Filamentous Fungi*, Vol. IV: *Fungal Technology*. London: Edward Arnold.

Stephenson, D. & Stephenson, T. (1992). Bioaugmentation for enhancing biological wastewater treatment. *Biotechnology Advances*, **10**, 549–559.

Stormo, K. E. & Crawford, R. L. (1992). Preparation of encapsulated microbial cells for environmental applications. *Applied and Environmental Microbiology*, **58**, 727–730.

Swannell, R. P. J., Lee, K. & McDonagh, M. (1996). Field evaluations of marine oil spill bioremediation. *Microbiological Reviews*, **60**, 342–365.

Thompson, J. A. (1980). Production and quality control of legume inoculants. In *Methods of Evaluating Biological Nitrogen Fixation*, ed. F. J. Bergersen, pp. 489–533. New York: Wiley & Sons.

Trevors, J. T., van Elsas, J. D., Lee, H. & van Overbeek, L. S. (1992). Use of alginate and other carriers for encapsulation of microbial cells for use in soil. *Microbial Releases*, **1**, 61–69.

van Elsas, J. D. & Heijnen, C. E. (1990). Methods for the introduction of bacteria into soil: a review. *Biology and Fertility of Soils*, **10**, 127–133.

van Elsas, J. D., Trevors, J. T., Jain, D., Wolters, A. C., Heijnen, C. E. & van Overbeek, L. S. (1992). Survival of, and root colonization by, alginate-encapsulated *Pseudomonas fluorescens* cells following introduction into soil. *Biology and Fertility of Soils*, 14, 14–22.

van Limbergen, H. E., Top, E. M. & Vestraete, W. (1998). Bioaugmentation in activated sludge: current features and future perspectives. *Applied Microbiology and Biotechnology*, **50**, 16–23.

Venosa, A. D., Haines, J. R. & Allen, D. M. (1992). Efficacy of commercial inocula in enhancing biodegradation of weathered crude oil contaminating a Prince William Sound Beach. *Journal of Industrial Microbiology*, **10**, 1–11.

Walker, H. L. & Connick W. J. Jr (1983). Sodium alginate for production and formulation of mycoherbidicides. *Weed Science*, **31**, 333–338.

Walter, M. V. (1997). Bioaugmentation. In *Manual of Environmental Microbiology*, eds. C. J. Hurst, G. R. Knudsen, M. J. McInerney, L. D Stetzenback & M. V. Walter, pp. 753–758. Washington, DC: ASM Press.

Weir, S. C., Dupuis, S. P., Providenti, M. A., Lee, H. & Trevors, J. T. (1995). Nutrient-enhanced survival of and phenanthrene mineralization by alginate-encapsulated and free *Pseudomonas* sp. UG14Lr cells in creosote-contaminated soil slurries. *Applied Microbiology and Biotechnology*, **43**, 946–951.

Wolter, M., Zadrazil, F., Martens, R. & Bahadir, M. (1997). Degradation of eight highly condensed polycyclic aromatic hydrocarbons by *Pleurotus* sp. Florida in solid wheat straw substrate. *Applied Microbiology and Biochemistry*, **48**, 398–404.

Wunch, K. G., Alworth, W. L. & Bennett, J. W. (1999). Mineralization of benzo[*a*]pyrene by the litter rot fungus *Marasmiellus troyanus*. *Microbiological Research*, **154**, 75–79.

6

Fungal biodegradation of chlorinated monoaromatics and BTEX compounds

JOHN A. BUSWELL

Introduction

Fungal degradation of monoaromatic compounds has clear implications for bioremediation, and the role of fungi in the removal of these contaminants from the environment has been the subject of extensive study. An understanding of the mechanisms involved in the degradation of benzenoid compounds and elucidation of the catabolic pathways is also important for predicting the recalcitrance of new products in the environment. Furthermore, enzymes catalysing key steps in a catabolic pathway could be used in the design and operation of biosensors for detecting environmental pollutants.

In view of the manifold types of monoaromatic compounds that enter the environment from various sources, this chapter has been confined to coverage of chlorinated monoaromatics and the BTEX group of compounds (benzene, toluene, ethylbenzene and m-, o and p-xylenes). Moreover, since there are already many excellent reviews available, emphasis has been given to the results of research conducted since the early 1990s. The contents cover the sources and distribution of BTEX and chlorinated monoaromatic environmental contaminants, fungal transformation studies including degradation pathways and associated enzymology, and various fungal-based bioremediation strategies employed for contaminant removal.

Sources and distribution of chlorinated monoaromatic and BTEX contaminants in the environment

Monomeric aromatic compounds are widely distributed in the environment as a result of natural synthetic and degradative processes. Many natural benzenoid compounds are synthesized by plants or are generated during the breakdown of more complex aromatic biopolymers such as lignin. However, it is the aromatic substances introduced into the environment by human activities that are of greater concern because of their potential toxicity and recalcitrance. A comprehensive list of these along with the major sources and applications have been documented by Swoboda-Colberg (1995).

Enormous quantities of refined petroleum products and bulk chemicals (e.g. benzene, alkylbenzenes), together with production intermediates such as chlorobenzenes, aniline, nitrobenzene and phthalates, are used for the industrial manufacture of solvents, pesticides, plastics and various other products (Swoboda-Colberg, 1995). Significant quantities of these contaminants inevitably enter the environment during the production processes. Soil and groundwater contamination with the monoaromatic BTEX hydrocarbons is also associated with situations such as leakages from petroleum and fuel oil underground storage tanks, the manufacture of solvent-based paints, lacquers and varnishes; and the activities of manufactured gas plants.

Chlorinated aromatic compounds have been used extensively as herbicides, for example 2,4-dichlorophenoxyacetic acid (2,4-D), 2,4,5-trichlorophenoxyacetic acid (2,4,5-T), and as solvents, fumigants and intermediates in the production of dyes, for example chlorobenzenes. Pentachlorophenol (PCP) and 2,4,6-trichlorophenol (2,4,6-TCP) are two of the most common biocides used in wood-preserving preparations, and these can be detected in soils and waters near sawmills (Valo *et al.*, 1984). Many chlorinated aromatics (e.g. chlorophenols, chloroveratroles (1,2-dimethoxybenzenes), chloroguaiacols, chlorocatechols, chlorinated cymenes and hexachlorobenzene) are also produced during chlorine-based bleaching processes in pulp and paper mills (Bjørseth, Carlberg & Møller, 1979; Huynh *et al.*, 1985) and the incineration of organic materials in the presence of chloride. In addition, 2,4-dichlorophenol (2,4-DCP) and 2,4,5-trichlorophenol (2,4,5-TCP) are used as precursors in the manufacture of the herbicides 2,4-D and 2,4,5-T. Agricultural use of these herbicides and their subsequent microbial metabolism yields, in turn, the respective chlorinated phenols as catabolic intermediates (Mikesell & Boyd, 1985).

Catabolism of monoaromatic compounds by fungi: general features

There are many reports describing the breakdown of individual monocyclic benzenoid compounds by fungi. Biodegradation of aromatics occurs via both aerobic and anaerobic systems, although all fungal systems described so far appear to be aerobic. Moreover, both extracellular and intracellular enzyme systems play important roles. Although there is a great diversity of metabolic routes operative in the dissimilation of aromatic compounds to intermediates of central metabolism, a number of common features have been established. The aerobic biodegradative process proceeds in two stages: (i) preparation of the benzenoid nucleus for ring cleavage, and (ii) ring fission and further degradation to central pathway metabolites. The introduction of hydroxyl substituents is an essential element for ring cleavage if these are not already present in both the requisite number and arrangement. At least two hydroxyl substituents (either *ortho* or *para* to each other) are necessary. It may also be necessary to remove substituent side chains, or to convert alkyl, alkoxy or other groups into hydroxyl groups. Hydroxylation reactions are catalysed by monooxygenases and by enzymic components of fungal ligninolytic systems (see below). A number of hydroxylated catabolic intermediates are now known to play central roles in the catabolism of monoaromatic compounds by fungi and to serve as substrates for fungal ring-fission enzymes. These include catechol, protocatechuic acid, gentisic acid, 1,2,4-trihydroxybenzene and various alkylated derivatives.

Aromatic ring cleavage is catalysed by dioxygenases, which insert both atoms of the oxygen molecule into the ring-fission product. Even here, two major alternative reactions exist: the *ortho*-fission (intradiol cleavage) where the oxygen atoms are inserted between the *ortho*-oriented hydroxyl substituents, and *meta*-fission (extradiol cleavage) where insertion occurs across the carbon–carbon bond immediately adjacent to the *ortho*-diphenol. However, *meta*-cleavage appears to be unusual among fungi. The existence of a *meta*-type ring cleavage in a *Penicillium* sp. was indicated by the formation of pyruvate but not β-ketoadipate from protocatechuic acid (Cain, Bilton & Darrah, 1968), while Hashimoto (1970, 1973) reported the novel 1,6-ring fission of 4-methylcatechol by *Candida tropicalis*. Catechol 1,2-dioxygenase and protocatechuate 3,4-dioxygenase catalyse the *ortho*-ring cleavage of catechol and protocatechuate to *cis,cis*-muconic acid and β-carboxymuconate, respectively (Cain *et al.*, 1968). These ring-cleavage products are further converted to succinate and acetyl-CoA via well-documented pathways through the common intermediate β-ketoadipic acid (Cain *et al.*, 1968). Where alkyl substituents are present in the ring-fission substrate (e.g. 4-methylcatechol), separate dioxygenases more active against the alkylated substrate may be elaborated for the cleavage step (Powlowski, Ingebrand & Dagley, 1985). Gentisate

1,2-dioxygenase converts gentisic acid to maleylpyruvate (Middlehoven *et al.*, 1992), and 1,2,4-trihydroxybenzene 1,2-dioxygenase catalyses the intradiol cleavage of 1,2,4-trihydroxybenzene to form maleylacetate, which is then reduced to β-ketoadipate by a NADPH-requiring enzyme (Buswell & Eriksson, 1979; Rieble, Joshi & Gold, 1994). Lignin peroxidase (LiP) from the white rot basidiomycete *Phanerochaete chrysosporium*, has also been shown to cleave the benzene nucleus (Umezawa & Higuchi, 1989).

Degradation of monomeric halogenated phenols

Degradation studies

Lyr (1963), using the white rot *Trametes versicolor*, provided one of the earliest substantiated reports describing the fungal degradation of chlorinated phenols. Duncan & Deverall (1964) also showed that approximately 43% of PCP disappeared from wood chips after 12 weeks of exposure to a *Trichoderma* sp., while Cserjesi & Johnson (1972) reported the transformation of PCP by *Trichoderma virgatum* in which 10–20% of the PCP was methylated to pentachloroanisole. Since these earlier reports, the use of radiolabelled substrates has shown that numerous different fungal species are capable of at least partial breakdown of a range of chlorinated phenolic compounds. Moreover, the application of gas chromatography with mass spectrometry (GC–MS) to identify degradation intermediates has revealed much about the major steps in the biodegradative processes.

There are now many reports describing the transformation and degradation of halogenated monoaromatics by fungi and yeasts. A systematic survey of over 1000 fungi representing various taxonomic groups indicated that most of those able to degrade PCP belonged to the Dematiaceae and the Zygomycetes, whereas a relatively lower proportion of yeasts and basidiomycetes tested were capable of doing so (Seigle-Murandi, Steiman & Benoit-Guyod, 1991; Seigle-Murandi *et al.*, 1992, 1993). Some fungi are unable to use halogenated phenols as a sole source of carbon and energy but can degrade these compounds during or following growth with phenol as a co-substrate. For example, resting phenol-grown mycelia of a strain of *Penicillium frequentans* metabolized various monohalogenated phenols and 3,4-dichlorophenol (4-DCP), and 2,4-DCP was degraded in the presence of phenol (Hofrichter, Bublitz & Fritsche, 1994). Halophenols were first oxidized to the corresponding halocatechols; 4-halocatechols were further degraded via 4-carboxymethylenebut-2-en-4-olide while 3-halocatechols underwent ring cleavage to form 2-halomuconic acids.

Dichlorophenols were converted to the corresponding catechols, and 3,5-dichlorocatechol was *O*-methylated to give two isomers of dichloroguaiacol. Co-metabolic halophenol degradation has also been reported by a strain of *Penicillium simplicissimum*, isolated from a sewage plant, which metabolized 3- and 4-chlorophenol and 4-bromophenol when grown in submerged cultures in the presence of phenol (Marr *et al.*, 1996). 3-Chlorophenol was converted to chlorohydroquinone, 4-chlorocatechol, 4-chloro-1,2,3-trihydroxybenzene and 5-chloro-1,2,3-trihydroxybenzene, indicating that hydroxylation occurred at all positions, whereas only 4-chlorocatechol was detected with 4-chlorophenol. No release of chloride was observed. Whereas the chlorophenols and bromophenol did not serve as sole carbon and energy source, fungal growth did occur in the presence of either 3- or 4-fluorophenol. Both substrates were completely broken down with equimolar release of fluoride ions. Degradation of the fluorophenols was enhanced in the presence of phenol, and substrate and co-substrate disappeared simultaneously. Although unable to use monochlorophenols as a growth substrate, phenol-grown cells of *Candida maltosa* degrade 3- and 4-chlorophenol with the formation of 4-chlorocatechol and 5-chloropyrogallol (Polnisch *et al.*, 1991), respectively. 2-Chlorophenol was partially converted to 3-chlorocatechol which was also obtained in small amounts from 3-chlorophenol. Aromatic ring fission yielded *cis,cis*-chloromuconic acid, which underwent cycloisomerization to 4-carboxymethylenebut-2-en-4-olide with concomitant release of chloride.

Many of the more recent investigations concerned with the degradation of chlorinated phenols (as well as other environmental contaminants) have focused on the lignin-degrading basidiomycete *P. chrysosporium*. The key to the capacity of this white rot fungus to degrade environmental contaminants has been linked, at least in part, to its ability to degrade lignin since both functions operate during idiophase (secondary metabolism), which is triggered most effectively by nitrogen starvation (Keyser, Kirk & Zeikus, 1978). As part of its ligninolytic system, *P. chrysosporium* produces two extracellular haem peroxidases, LiP and manganese peroxidase (MnP) (Glenn *et al.*, 1983; Tien & Kirk, 1983, 1984; Glenn & Gold, 1985) and initial attack on the lignin polymer occurs via a non-specific mechanism that has been equated to 'enzymatic combustion' (Kirk & Farrell, 1987). The recognition that the heterogeneous biopolymer contained substructures resembling many primary pollutants led researchers to surmise that the enzyme systems used by the fungus to degrade the lignin polymer might be sufficiently non-specific to attack aromatic pollutants also.

Huynh *et al.* (1985) showed that *P. chrysosporium* degraded various toxic chlorinated phenols present in pulp-mill bleach effluents including 2,4,6-TCP, polychlorinated guaiacols and polychlorinated vanillins. Extensive degradation (20–50%) and formation of water-soluble metabolites of ring carbons of PCP labelled with ^{14}C ([^{14}C-ring]-PCP) by *P. chrysosporium* was reported by Mileski *et al.* (1988). However, some of the volatile ^{14}C produced from [^{14}C-ring]-PCP by this fungus is organic (Lamar, Glaser & Kirk, 1990a), possibly giving rise to overestimates of breakdown levels. Although fungal growth was prevented by PCP concentrations of $4\,mg\,l^{-1}$ when cultures were initiated by inoculation with conidia, toxic effects were not as evident when PCP was added to established fungal biomass. Under these conditions, the fungus was able to grow and degrade [^{14}C-ring]-PCP at concentrations as high as $0.5\,g\,l^{-1}$ ($1.9\,mmol\,l^{-1}$). Highest rates of degradation were observed in cultures grown under nutrient nitrogen-limited conditions, which, together with observed temporal similarities in the patterns of both [^{14}C-ring]-lignin and [^{14}C-ring]-PCP degradation, indicated that conversion was linked to the activity of the fungal lignin-degrading system. However, substantial degradation also occurred in nitrogen-sufficient cultures, suggesting the involvement of another transformation system. In separate studies, breakdown of 2,4-DCP, 2,4,5-TCP and 2,4,6-TCP by *P. chrysosporium* occurred when the fungus was grown under secondary metabolic (i.e. ligninolytic) conditions (Valli & Gold, 1991; Joshi & Gold, 1993; Reddy *et al.*, 1998). However, both fungal mycelium and extracellular protein (culture supernatant) were required for the release of chloride from 2,4,6-TCP (Armenante, Pal & Lewandowski, 1994). Previously, Lin, Wang & Hickey (1990) observed the degradation of PCP using only *P. chrysosporium* biomass and proposed a pathway involving both extracellular peroxidases and cell-associated enzymes. Degradation of PCP by *P. chrysosporium* was also observed in cultures where ammonium lignosulfonate (LS), a waste product of the papermill industry, was used as a carbon and nitrogen source (Aiken & Logan, 1996). Three days of cultivation in either a 2% LS (nitrogen-sufficient) medium or a 0.23% LS and 2% glucose (nitrogen-sufficient) medium removed 72–75% of PCP compared with 95% removal recorded in nitrogen-deficient glucose and ammonia medium. After 13 days, over 98% of the initial PCP was either broken down or transformed into organic halides. Of the original chlorine, 58% was recovered as organic (non-PCP) halide, while 40% of the remainder was released as chloride. Extensive degradation of chlorobenzene and *o*-, *m*- and *p*-dichlorobenzenes by *P. chrysosporium* has also been described (Yadav, Wallace & Reddy, 1995). Monochlorobenzene was degraded most effectively, followed by *m*-, *o*- and *p*-dichlorobenzene in that order. Total degradation was maximal when the fungus was cultured in a rich medium (malt extract) when enzymes of the ligninolytic system were not synthesized. Interestingly, the fungus degraded both chlorobenzene and toluene simultaneously when the two compounds were presented as a mixture.

Biotransformation of PCP, including dechlorination and partial degradation, was observed in sterilized and unsterilized soils inoculated with *Lentinula edodes* (Okeke *et al.*, 1996, 1997).

More recently, wheat straw cultures of the brown rot fungi *Gloeophyllum striatum* and *G. trabeum* were reported to degrade 2,4-DCP and PCP. Up to 54% and 27%, respectively, of radiolabel was liberated as $^{14}CO_2$ from [^{14}C-ring]-labelled substrates within 6 weeks (Fahr *et al.*, 1999). *T. versicolor* grown under identical conditions released up to 42% and 43% as $^{14}CO_2$, respectively, from the two chlorinated phenols. Although high levels of laccase, MnP and manganese-independent peroxidase were expressed in cultures of the white rot fungus, no such activities were detected in straw or liquid cultures of the *Gloeophyllum* spp. Furthermore, *G. striatum* degraded both chlorophenols most efficiently under conditions where co-metabolites were lacking, i.e. on a defined mineral medium lacking sources of carbon, nitrogen and phosphate.

Biodegradation of chlorinated phenols by *P. chrysosporium* has also been studied in various bioreactor systems. The degradation rates for 2,4,6-TCP and 2,4,5-TCP by immobilized fungus in packed bed reactors was two orders of magnitude greater than in shake flasks (Pal, Lewandowski & Armenante, 1995). Degradation rates were affected by the concentrations of carbon and nitrogen sources, pH and fluid sheer stress. 4-Chlorophenol and 2,4-DCP degradation has also been studied using wood chip reactors seeded with *P. chrysosporium* (Yum & Pierce, 1998).

Degradative pathways and associated enzymology

Evidence supporting involvement of the fungal ligninolytic system in chlorophenol degradation is provided by observations that these compounds are also substrates for isolated LiP, MnP and laccase. Crude laccase preparations were shown to oxidize chlorophenols to unspecified products, with associated release of inorganic chloride (Lyr, 1963). However, more recent studies with *T. versicolor* suggest that laccase does not play an integral role in the degradation of 2-chlorophenol and PCP (Ricotta, Unz & Bollag, 1996; Grey, Hofer & Schlosser, 1998).

LiPs from *P. chrysosporium* have been shown to catalyse the

Fig. 6.1. Mechanism proposed for the lignin peroxidase-catalysed oxidation of 2,4,6-trichlorophenol to 2,6-dichloro-1,4-benzoquinone. (Reprinted from Hammel, 1992, by courtesy of Marcel Dekker, Inc.)

peroxidative 4-dechlorination of polychlorinated chlorophenols to the corresponding 1,4-benzoquinones (Hammel & Tardone, 1988). Thus, 2,4,6-TCP and PCP are oxidized quantitatively to 2,6-dichloro-1,4-benzoquinone and tetrachloro-1,4-benzoquinone (TCBQ), respectively with concomitant release of inorganic chloride. A reaction mechanism involving two sequential one-electron oxidations of the chlorophenol to yield a cyclohexadienone cation, followed by nucleophilic attack by water and elimination of chloride, is thought to be involved (Fig. 6.1) (Hammel & Tardone, 1988; Mileski *et al.*, 1988; Lin *et al.*, 1990). MnP and laccase were subsequently shown to catalyse these oxidative dechlorination reactions even more efficiently that LiP (Roy-Arcand & Archibald, 1991).

Detailed studies by Gold and coworkers involving the characterization of both fungal metabolites and of oxidation products generated by purified LiP and MnP have established the metabolic pathways for the degradation of several polychlorinated phenols by *P. chrysosporium*. Several key extracellular and cell-associated reactions are associated with the degradative processes: peroxidative dechlorination, quinone reduction, methylation and reductive dechlorination (Valli & Gold, 1991; Joshi & Gold, 1993; Reddy *et al.*, 1998; Reddy & Gold, 2000). Initial oxidation of 2,4-DCP by either LiP or MnP led to the removal of the 4-chlorine atom to yield a *p*-quinone, which was then reduced to the corresponding hydroquinone (Fig. 6.2) (Valli & Gold, 1991). Methylation of the latter generated an intermediate, 2-chloro-1,4-dimethoxybenzene, which again served as a substrate for the LiP-catalysed oxidative dechlorination. A subsequent cycle of oxidative dechlorination and reduction of the ensuing quinone ultimately yielded 1,2,4,5-tetrahydroxybenzene, which then underwent aromatic ring cleavage and further oxidation to malonic acid (Valli & Gold, 1991) (Fig. 6.2). Peroxidative dechlorination catalysed by either LiP or MnP was also the first step in the degradation pathway for 2,4,5-TCP by *P. chrysosporium* (Joshi & Gold, 1993). The reaction product, 2,5-dichloro-1,4-benzoquinone, was then converted to the corresponding hydroquinone. Although this reduction step can proceed nonenzymically, it occurred more rapidly in the presence of fungal cells and may be associated with a reported intracellular quinone reductase (Buswell, Hamp & Eriksson, 1979; Constam, *et al.*, 1991; Brock, Rieble & Gold, 1995; Brock & Gold, 1996). 2,5-Dichloro-1,4-hydroquinone was then converted to 1,2,4,5-tetrahydroxybenzene by two subsequent MnP-catalysed peroxidative dechlorination/quinone reduction cycles via 5-chloro-4-hydroxy-1,2-benzoquinone (Joshi & Gold, 1993). Methylated products were also detected but did not appear to be key catabolic

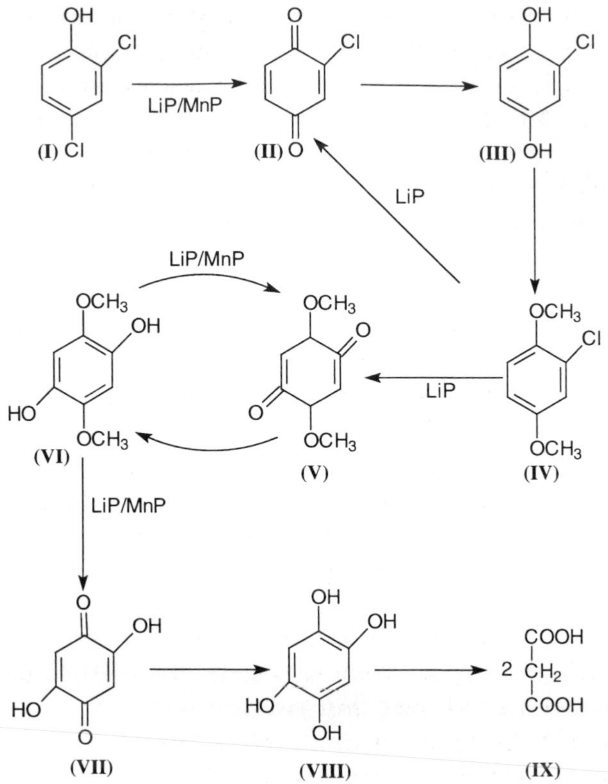

Fig. 6.2. Route proposed by Valli & Gold (1991) for 2,4-dichlorophenol degradation by *P. chrysosporium*. I, 2,4-dichlorophenol; II, 2-chloro-*p*-benzoquinone; III, 2-chloro-*p*-hydroquinone; IV, 2-chloro-1,4-dimethoxybenzene; V, 2,5-dimethoxy-*p*-benzoquinone; VI, 2,5-dimethoxy-*p*-hydroquinone; VII, 2,5-dihydroxy-*p*-benzoquinone; VIII, 1,2,4,5-tetrahydroxybenzene; IX, malonic acid; LiP, lignin peroxidase; MnP, manganese peroxidase. (Reprinted from Hammel, 1992, by courtesy of Marcel Dekker, Inc.)

intermediates in the degradation of 2,4,5-TCP. In yet another catabolic variation, 2,4,6-TCP was degraded by *P. chrysosporium* using pathways involving reductive dechlorination (Reddy *et al.*, 1998). Attack on the substrate was initiated by LiP- or MnP-mediated peroxidative dechlorination to yield the *p*-quinone. This was reduced to 2,6-dichloro-1,4-dihydroxybenzene, which underwent intracellular reductive dechlorination to 2-chloro-1,4-dihydroxybenzene. Further conversion of this intermediate proceeded either by a second reductive dechlorination to form 1,4-hydroquinone, which was then hydroxylated to 1,2,4-trihydroxybenzene, or via hydroxylation to 5-chloro-1,2,4-trihydroxybenzene, which was then reductively dechlorinated to the common key catabolite 1,2,4-trihydroxybenzene. In a more recent study of PCP degradation by *P. chrysosporium*, Reddy & Gold (2000) reported that, following initial removal of the 4-chlorine of PCP by peroxidative dechlorination and reduction of the resultant TCBQ to form 2,3,5,6-tetrachloro-1,4-dihydroxybenzene (TCDHB), the remaining chlorine substituents of this intermediate were removed by successive reductive dechlorination steps (Fig. 6.3). A two-component enzyme system that reductively dechlorinated TCDHB to trichlorohydroquinone has been identified in cell-free extracts of *P. chrysosporium* (Reddy & Gold, 1999). This system comprised of a membrane-bound component, which, in the presence of reduced glutathione (GSH), converted TCDHB to the glutathionyl conjugate, and a soluble component, which converted the conjugate to trichlorohydroquinone in the presence of GSH, cysteine or dithiothreitol. Interestingly, the reductive dechlorination steps proceeded under both nitrogen-deficient and nitrogen-sufficient conditions, indicating that the ligninolytic system of the fungus was not involved. In parallel and cross-linking pathways, TCDHB was sequentially reductively dechlorinated to hydroquinone, which was then hydroxylated to 1,2,4-trihydroxybenzene (Fig. 6.3). Alternatively, TCBQ was converted to trichlorotrihydroxybenzene by a non-enzymic 1,4-addition of H_2O (Joshi & Gold, 1994) followed by three successive reductive dechlorinations to form the key ring-fission substrate 1,2,4-trihydroxybenzene (Fig. 6.3). In addition, cross-linking between the two pathways can occur through hydroxylation of the trichloro-, dichloro- and monochlorodihydroxybenzene intermediates emanating from TCDHB (Fig. 6.3). The basidiomycete *Mycena avenacea* also metabolized PCP to 2,3,5,6-tetrachloro-*p*-benzoquinone, which was then reduced to 2,3,5,6-tetrachloro-*p*-hydroquinone (Kremer, Sterner & Heidrun, 1992). Subsequent dechlorination of this intermediate yielded 3,5,6-trichloro-2-hydroxy-*p*-benzoquinone.

Chlorophenoxyacetic acids

Biodegradation of 2,4,5-T by *P. chrysosporium* in liquid culture and in soil proceeded through chlorinated phenolic intermediates, the further transformation of which was catalysed by enzymes of the ligninolytic system (Ryan & Bumpus, 1989). However, Yadav & Reddy (1993) reported the degradation of 2,4-D and mixtures of 2,4-D and 2,4,5-T in high nitrogen and malt extract media by both wild-type and a peroxidase-negative

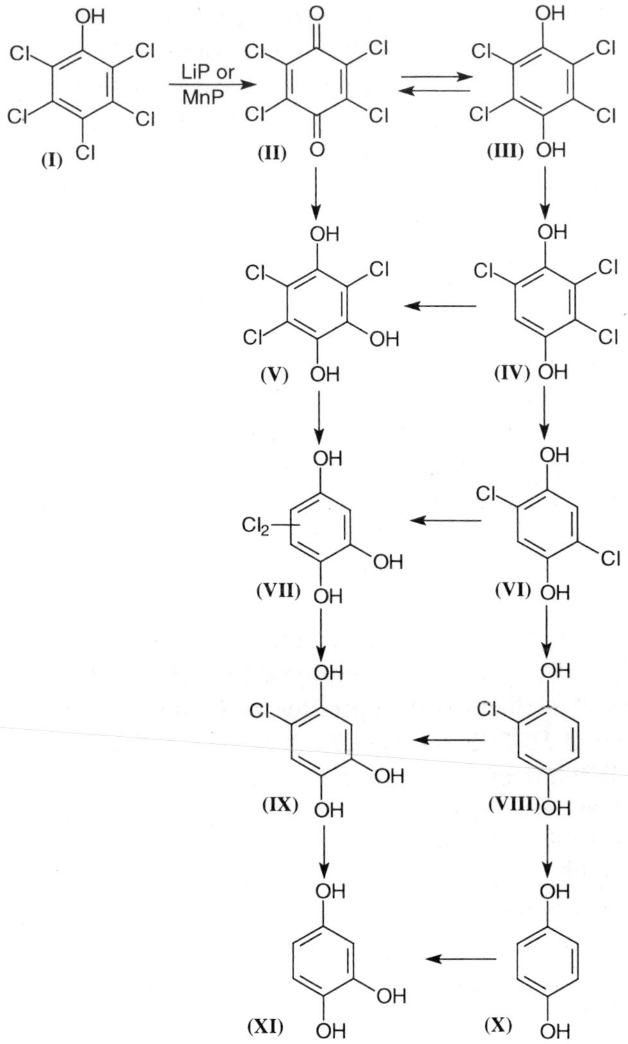

Fig. 6.3. Proposed pathways for the degradation of pentachlorophenol by *P. chrysosporium*. I, pentachlorophenol; II, tetrachlorobenzoquinone; III, 2,3,5,6-tetrachloro-1,4-dihydroxybenzene; IV, trichlorodihydroxybenzene; V, 3,5,6-trichlorotrihydroxybenzene; VI, 2,5-dichlorodihydroxybenzene; VII, dichlorotrihydroxybenzene; VIII, 2-chloro-1,4-dihydroxybenzene; IX, chlorotrihydroxybenzene; X, *p*-hydroquinone; XI, trihydroxybenzene; LiP, lignin peroxidase; MnP manganese peroxidase. (Reproduced from Reddy & Gold, 2000, with permission of the publishers.)

mutant of *P. chrysosporium*. Mass balance analysis of [^{14}C-ring]-2,4-D in malt extract cultures revealed 82.7% recovery of the radioactivity of which 38.6% was released as $^{14}CO_2$ and 27, 11.2 and 5.9% were present in the aqueous, methylene chloride and mycelial fractions, respectively. A relatively higher rate of breakdown was observed when 2,4-D and 2,4,5-T were presented as a mixture. Degradation under conditions that suppressed the synthesis of LiP and MnP, and effective degradation of 2,4-D by a *per* mutant lacking the capacity to produce these peroxidases, supported earlier observations (Yadav & Reddy, 1992) indicating that these components of the ligninolytic system were not required for degradation. Degradation of 2,4-D by mycorrhizal and free-living fungi has also been reported (Donnelly, Entry & Crawford, 1993).

In a separate study using ^{14}C-ring and ^{14}C-labelled side chains in chorophenoxyacetic acids, cleavage of the side chain and subsequent metabolism of the side-chain intermediate by both *P. chrysosporium* and *Dichomitus squalens* was catalysed by a mechanism independent of the lignin degradation system (Reddy, Joshi & Gold, 1997). Both [^{14}C-ring]- and [^{14}C-side chain]-2,4,5-T are broken down more efficiently by *D. squalens*: under the experimental conditions, 22% of the side chain label and 13% of the ring label were converted to $^{14}CO_2$ by *P. chrysosporium* compared with 65% and 32%, respectively by *D. squalens*. 4-Chlorophenol, 2,4-DCP and 2,4,5-TCP were identified by MS and HPLC as intermediates in the degradation of 4-chorophenoxyacetic acid, 2,4-D and 2,4,5-T, respectively, by *D. squalens*. Each chlorophenol intermediate was xylosylated to the chlorophenolxyloside, which was hydrolysed by an intracellular β-xylosidase back to the chlorophenol. However, in this study, further degradation of the chlorophenol intermediate occurred primarily during secondary metabolism. The answer to the apparent disparity of these results with those of Yadav & Reddy (1992, 1993) may be found in the consistent biphasic pattern associated with 2,4-D degradation observed in cultures of *P. chrysosporium* grown in malt-extract medium (Yadav & Reddy, 1993). Two peaks of activity were evident, the first after 6 days and the other after about 30 days incubation, and this second peak may be a consequence of nutrient starvation and subsequent induction of the ligninolytic system. A similar biphasic response has also been observed during PCP degradation by *P. chrysosporium*; this is the result of initial transient accumulation of pentachloroanisole and subsequent degradation at a later stage (Lamar, Larsen & Kirk, 1990b).

Degradation of BTEX organopollutants

P. chrysosporium degrades all the BTEX components either individually or as a composite mixture (Yadav & Reddy,1993). All the components of the BTEX mixture were simultaneously degraded and, except for toluene, degradation occurred to similar extents irrespective of whether the component formed part of a mixture or was tested individually. About half of the total degradation recorded for benzene and toluene involved conversion of substrate carbon to carbon dioxide. Total disappearance values for the BTEX compounds in fungal culture, as determined by GC analyses, indicated that rapid degradation occurred when the fungus was grown under non-ligninolytic conditions in a malt-extract medium in which LiP and MnP synthesis was suppressed. The lack of involvement of extracellular peroxidases in the degradative process is further supported by the comparable levels of degradation of BTEX compounds observed with the wild-type and the *per* mutant, which lacks the ability to produce LiP and MnP, and association of the degradative activity with mycelial pellets. Increased levels of degradation were recorded at 25°C compared with 37°C and under oxygenated conditions.

Polymerization of monoaromatic environmental contaminants and binding to humic substances

The binding of environmental contaminants, and their degradation products, with organic substances in the soil has long been recognized and clearly has important implications for bioremediation (Mathur & Morley, 1975; Bollag, Sjobald & Minard, 1977). From more recent studies, it is evident that contaminants are converted into soil-bound transformation products as a result of these interactions with soil humic substances (Lamar *et al.*, 1990a; Bollag 1992). Oxidative coupling of chlorophenols with naturally occurring humic acid precursors and mediated by extracellular oxidoreductases was determined using ^{14}C-labelled chemicals and by measuring the uptake of radioactivity by humic material (Bollag, 1992; Bollag, Myers & Minard, 1992). Using model systems, Bollag *et al.* (1992) were able to demonstrate the formation of covalent linkages between chlorinated phenols/carboxylic acids and fulvic and humic acids in the presence of fungal phenol oxidases. They were also able to isolate and identify cross-coupling products and to elucidate the site and type of binding. Different chlorophenols are polymerized to dimers, trimers and tetramers by the action of laccase and the extent of polymerization is dependent on the level of chlorine substitution (Dec & Bollag, 1990). Polymerization is also accompanied by dehalogenation (Dec & Bollag, 1994). High-molecular-weight polymeric material was also formed in reaction mixtures containing concentrated culture fluid from ligninolytic cultures of *P. chrysosporium*, PCP, a humic acid precursor (ferulic acid), H_2O_2 and a detergent (Ruttimann-Johnson & Lamar, 1996). Pure MnP, LiP and laccase also catalysed the polymerization reaction(s). More recently, the removal of PCP from solution using purified laccase from *T. (Coriolus) versicolor* was described (Ullah, Bedford & Evans, 2000). The products were primarily acid-stable polymers of molecular weight $\sim 80\,000$.

Binding of chlorophenols and other environmental contaminants may decrease their availability for interaction with biota and inhibit their movement via leaching. Thus, complexation of chorophenols into humus may be an environmentally beneficial phenomenon. However, the use of oxidative coupling for soil decontamination raises concerns about the ultimate fate of chlorophenols and the potential for forming dimers (e.g. polychlorinated dibenzo-*p*-dioxins, dibenzofurans and diphenylethers) that are more toxic than the parent compounds (Minard, Liu & Bollag, 1981; Svenson, Kjeller & Rappe, 1989; Oberg *et al.*, 1990).

Methylation in the biodegradation of monoaromatic environmental contaminants

The role of methylation in the degradation of certain aromatic xenobiotics is intriguing. Eriksson *et al.* (1984) demonstrated methylation of the 4-hydroxyl group of syringic acid by *P. chrysosporium* and several other white rot and soft rot fungi. These workers suggested that methylation could serve as a detoxification route for phenolics in some species. It may also play an important role in the degradation of some environmental contaminants. For example, methylation of the 2-chloro-1,4-dihydro-quinone intermediate formed during 2,4-D degradation by *P. chrysos-porium* appears to serve as a mechanism for regenerating a substrate susceptible to attack by LiP and MnP (Valli & Gold, 1991).

Harper and coworkers have demonstrated the existence of two independent mechanisms capable of methylating substituted phenols in *P. chrysos-porium*, one involving chloromethane as the methyl donor and the other linked to *S*-adenosylmethionine (Harper *et al.*, 1990; Coulter, Hamilton & Harper, 1993a; Coulter *et al.*, 1993b; Jeffers, McRoberts & Harper, 1997). Coulter *et al.* (1993a) purified an *S*-adenosylmethionine: 2,4-disubstituted phenol-*O*-methyltransferase from *P. chrysosporium* that catalysed the 4-*O*-methylation of acetovanillone. Substrate specificity studies showed that 3-methoxy- and 3,5-dimethoxy-substituted 4-hydroxybenzaldehydes, 4-benzoic acids and 4-acetophenones were the preferred substrates for the enzyme. Substituents in both the 2 and 4 positions relative to the hydroxyl group appeared to be essential for significant enzyme attack of a substrate. Provided that certain steric criteria were satisfied, the nature of the substituent was not critical. Hence, xenobiotic compounds such as 2,4-DCP and 2,4-dibromophenol were methylated almost as readily as acetovanillone. More recently, a highly specific phenolic 3-*O*-methyltransferase has been identified and purified from the same fungus (Jeffers *et al.*, 1997). However, a direct role for these methylating systems in the degradation or detoxification of monoaromatic xenobiotics has yet to be established.

Application of fungi to the bioremediation of chlorinated monoaromatics

Although much information is available on the ability of fungi to degrade monoaromatic environmental contaminants under laboratory conditions and on the various degradative pathways employed, far less is known about the effectiveness of fungi *in situ* and hence their potential for bioremediation. Feasibility studies conducted to determine the potential of white rot fungi to remediate contaminated field soils showed that *P. chrysosporium* depleted 2,4,5-T (Ryan & Bumpus, 1989) and PCP (Lamar *et al.*, 1990a) from contaminated soil samples. In the latter study, depletion was mainly through conversion of PCP to non-volatile products, the nature of which (i.e. whether they were bound to the soil particles or freely extractable) was dependent upon the soil type. Spent sawdust cultures of the shiitake mushroom *L. edodes*, when added to sterilized soil containing $200 \, \text{mg kg}^{-1}$ PCP, were also found to reduce the level of the contaminant by between 44.4 and 60.5% (Okeke *et al.*, 1993). Addition of H_2O_2 markedly enhanced PCP metabolism and GC–MS analysis showed that pentachloroanisole was a metabolic product.

In a field study to determine the ability of *P. chrysosporium* and *Phanerochaete sordida* to deplete PCP from sterilized soil contaminated with a commercial wood preservative, inoculation of soil containing 250–$400 \, \mu\text{g PCP g}^{-1}$ soil resulted in an overall depletion of 88–91% of PCP in the soil within 6.5 weeks (Lamar & Dietrich, 1990). A small proportion of this depletion (9–14%) resulted from the methylation of PCP to pentachloroanisole, which is also gradually degraded by *P. chrysosporium*

(Lamar *et al.*, 1990b). The percentage of PCA formed depended upon the fungus and on the type of soil. Since laboratory-scale soil-based studies indicated there was little degradation/volatilization of PCP, it appeared that most of the PCP was converted to non-extractable soil-bound products. These results compare favourably with those of bio-remediation studies in which mixtures of chlorophenol-degrading bacteria were used as the inoculum (Valo & Salkinoja-Salonen, 1986). Moreover, the depletion levels were achieved even though soil conditions (e.g. temperature) were suboptimal for fungal growth. It is also reported that taxonomically closely related white rot basidiomycetes exhibited significant differences in their sensitivity to PCP and ability to degrade PCP in an aqueous medium (Lamar *et al.*, 1990b).

Other studies have shown that *P. chrysosporium* degrades various xenobiotics in soil under non-sterile conditions, indicating that the fungus is able to compete under 'natural' conditions (Fernando, Aust & Bumpus, 1989). Interestingly, both *P. chrysosporium* and phenazine-producing pseudomonads could be isolated from contaminated agricultural soils even though the bacteria strongly inhibited fungal growth (Radtke, Cook & Anderson, 1994). Furthermore, conditions that most favoured the production of the fungal peroxidases LiP and MnP enhanced the growth of the antagonistic pseudomonads.

Conclusions

It is clear that fungi are able to degrade environmental contaminants both in pure culture and in sterilized soils. In several fungi, for example *P. chrysosporium*, this biodegradative ability is closely associated with the ligninolytic system of the fungus. Bioremediation potential can be increased by optimization of environmental parameters coupled with the selection of fungal strains with desirable characteristics for bioremediation: rapid growth and survival rates at high concentrations of contaminant and superior biodegradative ability. However, it remains uncertain to what extent fungi have a role in bioremediation *in situ* where they are components of a more complex microflora, which may contain antagonistic components. More studies are now required to establish the effects of these other microfloral components on fungal survival and biodegradative ability in order to understand more clearly the ecological role of fungi in the breakdown of organopollutants in the natural environment.

References

Aiken, B. S. & Logan, B. E. (1996). Degradation of pentachlorophenol by the white rot fungus *Phanerochaete chrysosporium* grown in ammonium lignosulphonate media. *Biodegradation*, **7**, 175–182.

Armenante, P. M., Pal, N. & Lewandowski, G. (1994). Role of mycelium and extracellular protein in the biodegradation of 2,4,6-trichlorophenol by *Phanerochaete chrysosporium*. *Applied and Environmental Microbiology*, **60**, 1711–1718.

Bjørseth, A., Carlberg, G. E. & Møller, M. (1979). Determination of halogenated organic compounds and mutagenicity testing of spent bleach liquors. *Science of the Total Environment*, **11**, 197–211.

Bollag, J. M. (1992). Decontaminating soil with enzymes. *Environmental Science and Technology*, **26**, 1876–1881.

Bollag, J. M., Sjobald, R. D. & Minard, R. D. (1977). Polymerization of phenolic intermediates of pesticides by a fungal enzyme. *Experientia*, **33B** 1564–1566.

Bollag, J. M., Myers, C. J. & Minard, R. D. (1992). Biological and chemical interactions of pesticides with soil organic matter. *Science of the Total Environment*, **123/124**, 205–217.

Brock, B. J. & Gold, M. H. (1996). 1,4-Benzoquinone reductase from the basidiomycete *Phanerochaete chrysosporium*: spectral and kinetic analysis. *Archives of Biochemistry and Biophysics*, **331**, 31–40.

Brock, B. J., Rieble, S. & Gold, M. H. (1995). Purification and characterization of a 1,4-benzoquinone reductase from the basidiomycete *Phanerochaete chrysosporium*. *Applied and Environmental Microbiology*, **61**, 3076–3081.

Buswell, J. A. & Eriksson, K.-E. (1979). Aromatic ring cleavage by the white rot fungus *Sporotrichum pulverulentum*. *FEBS Letters*, **104**, 258–260.

Buswell, J. A., Hamp, S. & Eriksson, K.-E. (1979). Intracellular quinone reduction in *Sporotrichum pulverulentum* by a NAD(P)H:quinone oxidoreductase. *FEBS Letters*, **108**, 229–232.

Cain, R. B., Bilton, R. F. & Darrah, J. A. (1968). The metabolism of aromatic acids by microorganisms. Metabolic pathways in the fungi. *Biochemical Journal*, **108**, 797–828.

Constam, D., Muheim, A., Zimmermann, W. & Fiechter, A. (1991). Purification and partial characterization of an intracellular NADH:quinone oxidoreductase from *Phanerochaete chrysosporium*. *Journal of General Microbiology*, **137**, 2209–2214.

Coulter, C., Hamilton, J. T. G. & Harper, D. B. (1993a). Evidence for the existence of independent chloromethane- and *S*-adenosylmethionine-utilizing systems for methylation in *Phanerochaete chrysosporium*. *Applied and Environmental Microbiology*, **59**, 1461–1466.

Coulter, C., Kennedy, J. T., McRoberts, W. C. & Harper, D. B. (1993b). Purification and properties of an *S*-adenosylmethionine:2,4-disubstituted phenol *O*-methyltransferase from *Phanerochaete chrysosporium*. *Applied and Environmental Microbiology*, **59**, 706–711.

Cserjesi, A. J. & Johnson, E. L. (1972). Methylation of pentachlorophenol by *Trichoderma virgatum*. *Canadian Journal of Microbiology*, **18**, 45–49.

Dec, J. & Bollag, J. M. (1990). Detoxification of substituted phenols by oxidoreductive enzymes through polymerization reactions. *Archives of Environmental Contamination and Toxicology*, **19**, 543–550.

Dec, J. & Bollag, J. M. (1994). Dehalogenation of chlorinated phenols during oxidative coupling. *Environmental Science and Technology*, **28**, 484–490.

Donnelly, P. K., Entry, J. A. & Crawford, D. L. (1993). Degradation of atrazine and 2,4-dichlorophenoxyacetic acid by mycorrhizal fungi at three nitrogen concentrations *in vitro*. *Applied and Environmental Microbiology*, **59**, 2642–2647.

Duncan, C. G. & Deverall, F. J. (1964). Degradation of wood preservatives by fungi. *Applied Microbiology*, **12**, 57–62.

Eriksson, K. E., Gupta, J. K., Nishida, A. & Rao, M. (1984). Syringic acid metabolism by some white rot, soft rot and brown rot fungi. *Journal of General Microbiology*, **130**, 2457–2464.

Fahr, K., Wetzstein, H. G., Grey, R. & Schlosser, D. (1999). Degradation of 2,4-dichlorophenol and pentachlorophenol by two brown rot fungi. *FEMS Microbiology Letters*, **175**, 127–132.

Fernando, T., Aust, S. D. & Bumpus, J. A. (1989). Effects of culture parameters on DDT [1,1,1-trichloro-2,2-bis(4-chlorophenyl)ethane] biodegradation by *Phanerochaete chrysosporium*. *Chemosphere*, **19**, 1387–1398.

Glenn, J. K. & Gold, M. H. (1985). Purification and characterization of an extracellular Mn(III)-dependent peroxidase from the lignin-degrading basidiomycete *Phanerochaete chrysosporium*. *Archives of Biochemistry and Biophysics*, **242**, 329–341.

Glenn, J. K., Morgan, M. A., Mayfield, M. B., Kuwahara, M. & Gold, M. H. (1983). An extracellular H_2O_2-requiring enzyme preparation involved in lignin biodegradation by the white rot basidiomycete *Phanerochaete chrysosporium*. *Biochemical and Biophysical Research Communications*, **114**, 1077–1083.

Grey, R., Hofer, C. & Schlosser, D. (1998). Degradation of 2-chlorophenol and the formation of 2-chloro-1,4-benzoquinone by mycelia and cell-free crude culture liquids of *Trametes versicolor* in relation to extracellular laccase activity. *Journal of Basic Microbiology*, **38**, 371–382.

Hammel, K. E. (1992). Oxidation of aromatic pollutants by lignin-degrading fungi and their extracellular peroxidases. In *Metal Ions in Biological Systems, Vol. 28, Degradation of Environmental Pollutants by Microorganisms and Their Metalloenzymes*, eds. H. Sigel & A. Sigel, pp. 41–60. New York: Marcel Dekker.

Hammel, K. E. & Tardone, P. J. (1988). The oxidative 4-dechlorination of polychlorinated phenols is catalysed by extracellular fungal lignin peroxidases. *Biochemistry*, **27**, 6563–6568.

Harper, D. B., Buswell, J. A., Kennedy, J. T. & Hamilton, J. T. G. (1990). Chloromethane, methyl donor in veratryl alcohol biosynthesis in *Phanerochaete chrysosporium* and other lignin-degrading fungi. *Applied and Environmental Microbiology*, **56**, 3450–3457.

Hashimoto, K. (1970). Oxidation of phenols by yeast I. A new oxidation product from *p*-cresol by an isolated strain of yeast. *Journal of General and Applied Microbiology*, **16**, 1–13.

Hashimoto, K. (1973). Oxidation of phenols by yeast II. Oxidation of phenols by *Candida tropicalis*. *Journal of General and Applied Microbiology*, **19**, 171–187.

Hofrichter, M., Bublitz, F. & Fritsche, W. (1994). Unspecific degradation of halogenated phenols by the soil fungus *Penicillium frequentans* Bi 7/2. *Journal of Basic Microbiology*, **34**, 163–172.

Huynh, V.-B., Chang, H.-M., Joyce, T. W. & Kirk, T. K. (1985). Dechlorination of chloro-organics by a white-rot fungus. *Technical Association of the Pulp and Paper Industry Journal*, **68**, 98–102.

Jeffers, M. R., McRoberts, W. C. & Harper, D. B. (1997). Identification of a phenolic 3-*O*-methyltransferase in the lignin-degrading fungus *Phanerochaete chrysosporium*. *Microbiology*, **143**, 1975–1981.

Joshi, D. K. & Gold, M. H. (1993). Degradation of 2,4,5-trichlorophenol by the lignin-degrading basidiomycete *Phanerochaete chrysosporium*. *Applied and Environmental Microbiology*, **59**, 1779–1785.

Joshi, D. K. & Gold, M. H. (1994). Oxidation of dibenzo-*p*-dioxin by lignin peroxidase from the basidiomycete *Phanerochaete chrysosporium*. *Biochemistry*, **33**, 10969–10976.

Keyser, P., Kirk, T. K. & Zeikus, J. G. (1978). Ligninolytic enzyme system of *Phanerochaete chrysosporium*: synthesized in the absence of lignin in response to nitrogen starvation. *Journal of Bacteriology*, **135**, 790–797.

Kirk, T. K. & Farrell, R. L. (1987). Enzymatic 'combustion': the microbial degradation of lignin. *Annual Reviews of Microbiology*, **41**, 465–505.

Kremer, S., Sterner, O. & Heidrun, A. (1992). Degradation of pentachlorophenol by *Mycena avenacea* TA 8480 – identification of initial dechlorinated metabolites. *Zeitschrift für Naturforschung,* **47C**, 561–566.

Lamar, R. T. & Dietrich, D. M. (1990). *In situ* depletion of pentachlorophenol from contaminated soil by *Phanerochaete* spp. *Applied and Environmental Microbiology*, **56**, 3093–3100.

Lamar, R. T., Glaser, J. A. & Kirk, T. K. (1990a). Fate of pentachlorophenol (PCP) in sterile soils inoculated with *Phanerochaete chrysosporium*: mineralization, volatilization and depletion of PCP. *Soil Biology and Biochemistry*, **22**, 433–440.

Lamar, R. T., Larsen, M. J. & Kirk, T. K. (1990b). Sensitivity to and degradation of pentachlorophenol by *Phanerochaete* spp. *Applied and Environmental Microbiology*, **56**, 3519–3526.

Lin, J.-E., Wang, H. Y. & Hickey, R. F. (1990). Degradation kinetics of pentachlorophenol degradation by *Phanerochaete chrysosporium*. *Biotechnology and Bioengineering*, **35**, 1125–1134.

Lyr, H. (1963). Enzymatische detoxifikation chlorierter phenole. *Phytopathologische Zeitschrift*, **47**, 73–83.

Marr, J., Kremer, S., Sterner, O. & Anke, H. (1996). Transformation and mineralization of halophenols by *Penicillium simplicissimum* SK9117. *Biodegradation*, **7**, 165–171.

Mathur, S. P. & Morley, H. V. (1975). A biodegradation approach for investigating pesticide incorporation into soil humus. *Soil Science*, **119**, 238–240.

Middlehoven, W. J., Coenen, A., Kraakman, B. & Sollewijn Gelpke, M. D. (1992). Degradation of some phenols and hydroxybenzoates by the imperfect ascomycetous yeasts *Candida parapsilosis* and *Arxula adeninivorans*: evidence for an operative gentisate pathway. *Antonie van Leeuwenhoek*, **62**, 181–187.

Mikesell, M. D. & Boyd, S. A. (1985). Reductive dechlorination of the pesticides 2,4-D, 2,4,5-T and pentachlorophenol in anaerobic sludges. *Journal of Environmental Quality*, **14**, 337–340.

Mileski, G. J., Bumpus, J. A., Jurek, M. A. & Aust, S. D. (1988). Biodegradation of pentachlorophenol by the white rot fungus *Phanerochaete chrysosporium*. *Applied and Environmental Microbiology*, **54**, 2885–2889.

Minard, R. D., Liu, S.-Y. & Bollag, J. M. (1981). Oligomers and quinones from 2,4-dichlorophenol. *Journal of Agricultural and Food Chemistry*, **29**, 250–252.
Oberg, L. G., Glas, B., Swanson, S. E. & Rappe, K. G. (1990). Peroxidase-catalyzed oxidation of chlorophenols to polychlorinated dibenzo-*p*-dioxins and dibenzofurans. *Archives of Environmental Contamination and Toxicology*, **19**, 930–938.
Okeke, B. C., Smith, J. E., Paterson, A. & Watson-Craik, I. A. (1993). Aerobic metabolism of pentachorophenol by spent sawdust culture of 'shiitake' mushroom (*Lentinus edodes*) in soil. *Biotechnology Letters*, **15**, 1077–1080.
Okeke, B. C., Smith, J. E., Paterson, A. & Watson-Craik, I. A. (1996). Influence of environmental parameters on pentachlorophenol biotransformation in soil by *Lentinula edodes* and *Phanerochaete chrysosporium*. *Applied Microbiology and Biotechnology*, **45**, 263–266.
Okeke, B. C., Smith, J. E., Paterson, A. & Watson-Craik, I. A. (1997). Comparative biotransformation of pentachlorophenol in soils by solid substrate cultures of *Lentinula edodes*. *Applied Microbiology and Biotechnology*, **48**, 563–569.
Pal, N., Lewandowski, G. & Armenante, P. M. (1995). Process optimization and modeling of trichlorophenol degradation by *Phanerochaete chrysosporium*. *Biotechnology and Bioengineering*, **46**, 599–609.
Polnisch, E., Kneifel, H., Franzke, H. & Hofmann, K. H. (1991). Degradation and dehalogenation of monochlorophenols by the phenol-assimilating yeast *Candida maltosa*. *Biodegradation*, **92**, 193–199.
Powlowski, J. B., Ingebrand, J. & Dagley, S. (1985). Enzymology of the *beta*-ketoadipate pathway in *Trichosporon cutaneum*. *Journal of Bacteriology*, **163**, 1136–1141.
Radtke, C., Cook, W. S. & Anderson, A. (1994). Factors affecting antagonism of the growth of *Phanerochaete chrysosporium* by bacteria isolated from soils. *Applied Microbiology and Biotechnology*, **41**, 274–280.
Reddy, G. V. B. & Gold, M. H. (1999). A two-component tetrachlorohydroquinone reductive dehalogenase system from the lignin-degrading basidiomycete *Phanerochaete chrysosporium*. *Biochemical and Biophysical Research Communications*, **257**, 901–905.
Reddy, G. V. & Gold, M. H. (2000). Degradation of pentachlorophenol by *Phanerochaete chrysosporium*: intermediates and reactions involved. *Microbiology*, **146**, 405–413.
Reddy, G. V. B., Joshi, D. K. & Gold, M. H. (1997). Degradation of chlorophenoxyacetic acids by the lignin-degrading fungus *Dichomitus squalens*. *Microbiology*, **143**, 2353–2360.
Reddy, G. V. P., Sollewijn Gelpke, M. D. & Gold, M. H. (1998). Degradation of 2,4,6-trichlorophenol by *Phanerochaete chrysosporium*: involvement of reductive dechlorination. *Journal of Bacteriology*, **180,** 5159–5164.
Ricotta, A., Unz, R. F. & Bollag, J. M. (1996). Role of a laccase in the degradation of pentachlorophenol. *Bulletin of Environmental Contamination and Toxicology*, **57**, 560–567.
Rieble, S., Joshi, D. K. & Gold, M. H. (1994). Purification and characterization of a 1,2,4-trihydroxybenzene 1,2-dioxygenase from the basidiomycete *Phanerochaete chrysosporium*. *Journal of Bacteriology*, **176**, 4838–4844.
Roy-Arcand, L. & Archibald, F. S. (1991). Direct dechlorination of chlorophenolic compounds by laccases from *Trametes* (*Coriolus*) *versicolor*. *Enzyme and Microbial Technology*, **13**, 194–203.

Ruttimann-Johnson, C. & Lamar, R. T. (1996). Polymerization of pentachlorophenol and ferulic acid by fungal extracellular lignin-degrading enzymes. *Applied and Environmental Microbiology*, **62**, 3890–3893.

Ryan, T. P. & Bumpus, J. A. (1989). Biodegradation of 2,4,5-trichlorophenoxyacetic acid in liquid culture and in soil by the white rot fungus *Phanerochaete chrysosporium. Applied Microbiology and Biotechnology*, **31**, 302–307.

Seigle-Murandi, F., Steiman, R. & Benoit-Guyod, J. L. (1991). Biodegradation potential of some micromycetes for pentachlorophenol. *Exotoxicology and Environmental Safety*, **21**, 290–300.

Seigle-Murandi, F., Steiman, R., Benoit-Guyod, J. L., Muntalif, B. & Sage, L. (1992). Relationship between the biodegradative capability of soil micromycetes for pentachlorophenol and pentachloronitrobenzene. *Science of the Total Environment*, **123/124**, 291–298.

Seigle-Murandi, F., Steiman, R., Benoit-Guyod, J. L. & Guiraud, P. (1993). Fungal degradation of pentachlorophenol by micromycetes. *Journal of Biotechnology*, **30**, 27–35.

Svenson, A., Kjeller, L.-O. & Rappe, C. (1989). Enzyme-mediated formation of 2,3,7,8-tetrasubstituted chlorinated dibenzodioxins and dibenzofurans. *Environmental Science and Technology*, **23**, 900–902.

Swoboda-Colberg, N. G. (1995). Chemical contamination of the environment: sources, types, and fate of synthetic organic chemicals. In *Microbial Transformation and Degradation of Toxic Organic Chemicals,* eds. L. Y. Young & C. E. Cerniglia, pp. 27–74. New York: Wiley-Liss.

Tien, M. & Kirk, T. K. (1983). Lignin-degrading enzyme from the hymenomycete *Phanerochaete chrysosporium* Burds. *Science*, **221**, 551–663.

Tien, M. & Kirk, T. K. (1984). Lignin-degrading enzyme from *Phanerochaete chrysosporium:* purification, characterization and catalytic properties of a unique H_2O_2-requiring oxygenase. *Proceedings of the National Academy of Sciences USA*, **81**, 2280–2284.

Ullah, M. A., Bedford, C. T. & Evans, C. S. (2000). Reactions of pentachlorophenol with laccase from *Coriolus versicolor. Applied Microbiology and Biotechnology*, **53**, 230–234.

Umezawa, T. & Higuchi, T. (1989). Cleavages of the aromatic ring and beta-O-4 bond of synthetic lignin (DHP) by lignin peroxidase. *FEBS Letters*, **242**, 325–329.

Valli, K. & Gold, M. H. (1991). Degradation of 2,4-dichlorophenol by the lignin-degrading fungus *Phanerochaete chrysosporium. Journal of Bacteriology*, **173**, 345–352.

Valo, R. & Salkinoja-Salonen, M. (1986). Microbial transformation of polychlorinated phenoxy phenols. *Journal of General Microbiology*, **32**, 505–517.

Valo, R., Kitunen, V., Salkinoja-Salonen, M. & Raisenan, S. (1984). Chorinated phenols as contaminants of soil and water in the vicinity of two Finnish sawmills. *Chemosphere*, **13**, 835–844.

Yadav, J. S. & Reddy, C. A. (1992). Non-involvement of lignin peroxidases and manganese peroxidases in 2,4,5-trichlorophenoxyacetic acid degradation by *Phanerochaete chrysosporium. Biotechnology Letters*, **14**, 1089–1092.

Yadav, J. S. & Reddy, C. A. (1993). Mineralization of 2,4-dichlorophenoxyacetic acid (2,4-D) and mixtures of 2,4-D and 2,4,5-trichlorophenoxyacetic acid by *Phanerochaete chrysosporium. Applied and Environmental Microbiology*, **59**,

2904–2908.

Yadav, J. S., Wallace, R. E. & Reddy, C. A. (1995). Mineralization of mono- and dichlorobenzenes and simultaneous degradation of chloro- and methyl-substituted benzenes by the white rot fungus *Phanerochaete chrysosporium. Applied and Environmental Microbiology*, **61**, 677–680.

Yum, K. J. & Pierce, J. J. (1998). Biodegradation kinetics of chlorophenols in immobilized-cell reactors using white-rot fungus on wood chips. *Water Environment Research*, **70**, 205–213.

7

Bioremediation of polycyclic aromatic hydrocarbons by ligninolytic and non-ligninolytic fungi

CARL E. CERNIGLIA
AND JOHN B. SUTHERLAND

Introduction

There is considerable interest in the application of biological systems to remediate polycyclic aromatic hydrocarbon (PAH) contamination in the environment (Chen *et al.*, 1999). Recent research on the bioremediation of environmentally relevant chemicals has centred on four important aspects: first, the characterization of the biodegradation processes useful for the treatment of xenobiotic compounds in soil; second, the development of technical protocols for increasing the degradation rates and substrate ranges of enzymes from microorganisms; third, the design and engineering of bioreactor systems and biotreatment strategies to optimize biodegradation processes; and fourth, development of information on the ecological and human health risks associated with exposure to the chemicals (Mueller, Cerniglia & Pritchard, 1996).

Low-molecular-weight PAHs, such as naphthalene, acenaphthene, acenaphthylene, fluorene, anthracene, and phenanthrene (Fig. 7.1), are transformed rapidly by many bacteria and fungi (Pothuluri *et al.*, 1992a,b, 1993; Sutherland *et al.*, 1995; Eriksson, Dalhammar & Borg-Karlson, 2000). High-molecular-weight PAHs, however, are more recalcitrant in the environment and resist both chemical and microbial degradation (Atlas & Cerniglia, 1995; Ahn, Sanseverino & Sayler, 1999; Kanaly & Harayama, 2000). Benzo[*a*]pyrene, one of the most recalcitrant PAHs in soil, adsorbs to the soil matrix and thus is physically unavailable to degradative bacteria and fungi (Banks, Lee & Schwab, 1999). The formation of non-extractable bound residues is a significant sink of PAHs in soils (Richnow *et al.*, 2000). PAH metabolites also can be incorporated into soil organic matter to form bound residues. Once bound into the soil organic matrix, the bioavailability of a PAH metabolite is decreased, thus reducing the hazardous potential (Boopathy, 2000).

The use of fungi to remediate PAH-contaminated environments has received widespread attention since their potential to degrade PAHs has been extensively demonstrated under laboratory conditions. In this chapter, an overview will be given of our current knowledge of the capabilities of fungi to bioremediate PAH-contaminated soil. For further information related to this subject, the reader is referred to other reviews (Cerniglia, 1984, 1992, 1993, 1997; Cerniglia, Sutherland & Crow, 1992; Sutherland, 1992; Müncnerová & Augustin, 1994; Aust, 1995; Hammel, 1995a,b; Mueller *et al.*, 1996).

Importance of polycyclic aromatic hydrocarbons

PAHs are non-polar, hydrophobic organic compounds with two or more fused benzene rings; some PAHs also have five-membered rings (Fig. 7.1). Naphthalene, also known as tar camphor, is the simplest PAH, from which numerous industrial chemical derivatives are manufactured. Naphthalene is less persistent than other PAHs in the environment and will not be considered in detail here since the focus of this review is on the more recalcitrant high-molecular-weight PAHs. Pyrolysis of organic materials is mainly responsible for the widespread occurrence of PAHs in air, sediments, water and food (Baek *et al.*, 1991). PAHs also are major constituents of crude oil, creosote and coal tar and contaminate the environment via many routes, including improper disposal of wastes from the combustion of fossil fuels, coal gasification and liquefaction, incineration of industrial wastes, wood treatment processes and accidental spillage of petroleum hydrocarbons (Harvey, 1992). They are also found in grilled and smoked foods (Lijinsky, 1991). In the atmosphere, PAHs come mainly from diesel and gasoline engine exhausts, coal-fired power plants, tobacco smoke, forest fires and farm debris fires (Finlayson-Pitts & Pitts, 1997). They are thought to be produced as an offshoot of soot formation during combustion (Siegmann, Scherrer & Siegmann, 1999).

Although PAHs are not highly water soluble, they are widespread pollutants in freshwater and seawater, particularly in estuaries and coastal waters with pollution from petroleum, coal or other heavy industries (Maldonado, Bayona & Bodineau, 1999; Mitra *et al.*, 1999; Ngabe, Bidleman & Scott, 2000). Another major source of PAHs in water is the creosote residue from wood-preserving industries (Davis *et al.*, 1993; Kennes & Lema, 1994). PAHs are common in estuarine sediments but are unevenly distributed; if they are trapped in woody debris, they may have no significant effect on aquatic organisms (Mitra *et al.*, 1999). Combustion products

138   *C. E. Cerniglia and J. B. Sutherland*

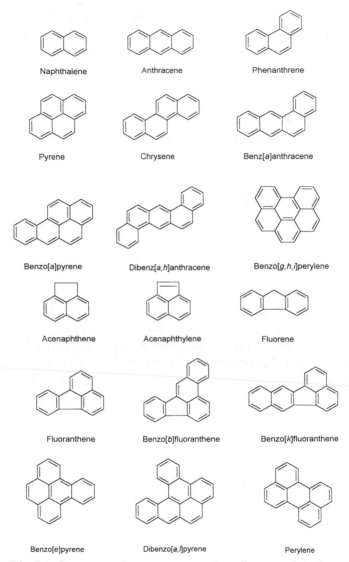

Fig. 7.1. Structures of some common polycyclic aromatic hydrocarbons.

are the major sources of PAHs in stormwater runoff from urbanized areas (Ngabe *et al.*, 2000).

Soils receive PAHs mainly from atmospheric deposition. The predominant PAHs in soils of the temperate zones include the benzofluoranthenes, chrysene, fluoranthene, and pyrene (Fig. 7.1) (Wilcke, 2000). In tropical soils, the predominant PAHs are more apt to include naphthalene, phenanthrene, and perylene (Fig. 7.1). PAHs are accumulated by earthworms and other invertebrates, but they are not taken up appreciably by most plants (Wilcke, 2000). As a result of natural and anthropogenic processes, background levels of total PAH of $1-5\,mg\,kg^{-1}$ soil now are common in urban areas and can be 10 to 20 times higher in industrial areas (Wilson & Jones, 1993). Because of the toxicity and frequent occurrence of PAHs in the environment, the US Environmental Protection Agency (EPA) has included selected PAHs among its priority pollutants list (Keith & Telliard, 1979).

Physical and chemical properties

All pure PAHs are solid, crystalline substances at room temperature. The vapour pressures are low, except for PAHs with two and three aromatic rings; as a result, they do not have a tendency to volatilize. Because of their large resonance energy, they are thermodynamically stable but are easily photooxidized at various positions on the aromatic rings. PAHs have low water solubility, which decreases as the number of condensed aromatic rings increases. This is a significant factor contributing to their persistence in the environment (Aitken *et al.*, 1998). Because they have a high photoelectric charging, they are abundant in the smallest fractions of particulate matter in the atmosphere (Siegmann *et al.*, 1999) and in freshwater and marine sediments (Tuvikene, 1995). The PAHs in moist soil may either dissolve or partition into the aqueous phase, based on the aqueous solubility and the octanol–water partition coefficient (K_{ow}). The ionization potentials of PAHs have been correlated with the degradation of these compounds by white rot fungi (Hammel, Kalyanaraman & Kirk, 1986). Some physical properties of PAHs are listed in Table 7.1.

Toxicity of PAHs

A 1761 report by physician John Hill, recognizing the link between excessive use of tobacco snuff and nasal cancer, began over two centuries of research on PAH carcinogenesis (see Cerniglia, 1984). In 1775, Percival Pott related chimney sweeps' exposure to soot with scrotal skin cancer; in 1915, Yamigiwa and Ichikawa reported that tumours formed on the ears of rabbits after repeated applications of coal tar and from 1930 to 1955, Kennaway, Hieger, Cook and Hewett established that the carcinogenic fraction of coal tar contained PAHs (see Cerniglia, 1984). In the 1970s, Miller & Miller (1985) showed that many chemicals require metabolic

Table 7.1. *Properties of selected polycyclic aromatic hydrocarbons (PAHs)*

	Molecular mass (Da)	Vapour Pressure (Pa)	Log octanol–water partition coefficient	Solubility (mg l^{-1})	Ionization potential (eV)
Naphthalene	128.18	12.0	3.58	30	8.13
Acenaphthene	154.20	4.02	3.92	3.6	7.86
Acenaphthylene	155.20	3.87	3.90	3.88	8.22
Fluorene	166.23	0.13	4.18	2.0	7.89
Phenanthrene	178.24	0.0161	4.46–4.63	1–2	7.91
Anthracene	178.24	0.001	4.45	0.015	7.43
Fluoranthene	202.26	0.001	5.22	0.25	7.95
Pyrene	202.26	0.0006	5.88–6.7	0.12–0.18	7.44
Benz[a]anthracene	228.30	20.0×10^{-5}	5.9	0.01	7.6
Chrysene	228.30	6.08×10^{-7}	5.01–7.10	0.0015–0.004	7.59
Benzo[a]pyrene	252.32	7.0×10^{-7}	5.78–6.5	0.001–0.006	7.7

From: Mackay, Shiu & Ma, 1992; Majcherczyk, Johannes & Hüttermann, 1998.

activation to express toxicity. It is now well established that PAHs must be metabolically activated by mammalian microsomal enzymes to elicit their latent mutagenic, genotoxic and carcinogenic properties.

Harvey (1996) and Harvey *et al.* (1999) recently reviewed mechanisms of PAH carcinogenesis (Fig. 7.2) and indicated that there are at least four mechanisms. (i) The dihydrodiol epoxide mechanism involves metabolic activation of the PAH by microsomal cytochrome P450 enzymes to give reactive epoxide and diol-epoxide intermediates; these form covalent adducts with DNA, perhaps resulting in mutations that lead to tumourigenesis. (ii) The radical-cation mechanism involves one-electron oxidation to generate radical-cation intermediates, which may attack DNA, resulting in depurination. (iii) The quinone mechanism involves enzymic dehydrogenation of dihydrodiol metabolites to yield quinone intermediates; these may either combine directly with DNA or enter into a redox cycle with oxygen to generate reactive oxygen species capable of attacking DNA. (iv) The benzylic oxidation mechanism entails formation of benzylic alcohols, which are converted by sulfotransferase enzymes to reactive sulfate esters and these may attack DNA. The most significant mechanism of carcinogenesis by PAHs is the diol-epoxide pathway. Since fungi in many ways mimic mammalian metabolism and have the potential to form reactive intermediates, such as dihydrodiol epoxides or quinones, the metabolic profiles of fungi with PAHs must be determined to see if potentially toxic intermediates are formed during the bioremediation of PAH-contaminated sites (Cerniglia, 1997).

Some PAHs, but not all, are acutely toxic, mutagenic, or carcinogenic. For instance, the combination of anthracene and solar ultraviolet radiation is acutely toxic and immunosuppressive to fish (Tuvikene, 1995). Benzo[*a*]pyrene, benz[*a*]anthracene, chrysene and several other PAHs, after metabolic activation by liver enzymes, induce mutations in bacteria (Harvey, 1992). Some PAH metabolites bind to DNA, RNA and proteins; the resulting adducts may cause damage directly to cells and also have teratogenic or carcinogenic effects (Harvey, 1992). Exposure to high concentrations of PAHs in the workplace has been associated with lung and bladder cancers among industrial workers (Mastrangelo, Fadda & Marzia, 1996). Although some PAHs show weak estrogenic or antiestrogenic activity (Santodonato, 1997), these effects are overshadowed by the carcinogenic properties of PAHs.

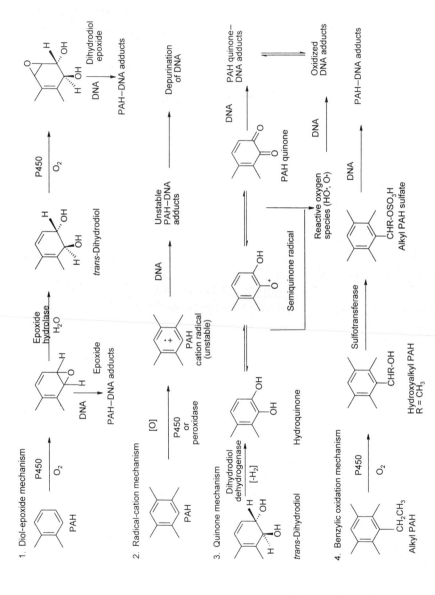

Fig. 7.2. Mechanisms of polycyclic aromatic hydrocarbons (PAHs) carcinogenesis. (After Harvey, 1996.)

Microbial degradation of PAHs

Bacteria and fungi metabolize a wide variety of PAHs, converting them either completely to carbon dioxide or to various microbial metabolites that are considered dead-end products and do not result in the production of carbon dioxide (Cerniglia, 1992, 1993). The elucidation of microbial degradation pathways is necessary to determine the extent of degradation and whether the metabolites that are formed are toxic or biologically inactive. Depending upon the enzymic repertoires of the microorganisms, different mechanisms can be used to metabolize PAHs (Fig. 7.3) (Gibson, 1982). Bacterial degradation of PAHs generally proceeds via the action of multicomponent dioxygenases to form *cis*-dihydrodiols. These compounds are subsequently dehydrogenated to form dihydroxy-PAHs, which may be substrates for ring-fission enzymes (Sutherland *et al.*, 1995). Many bacteria are capable of complete degradation of PAHs to form carbon dioxide. Recent findings also indicate that PAHs can be metabolized by monooxygenases in bacteria to form *trans*-dihydrodiols, although this activity is generally lower than the dioxygenase activity in the same organism (Heitkamp *et al.*, 1988).

Non-ligninolytic fungi metabolize PAHs by cytochrome P450 monooxygenase and epoxide hydrolase-catalysed reactions to form *trans*-dihydrodiols. These reactions are highly regio- and stereoselective. Other metabolites formed include phenols, quinones and conjugates (Sutherland *et al.*, 1995). The types of metabolite isolated are similar to those formed by mammals. Ligninolytic fungi degrade PAHs by non-specific radical oxidation, catalysed by extracellular ligninolytic enzymes, that leads primarily to PAH quinones. Some ligninolytic fungi can further metabolize PAH quinones by cleaving the aromatic rings, with subsequent breakdown of the PAH to carbon dioxide (Hammel, 1995a). Since PAHs are relatively insoluble in water and bind strongly to organic matter in sediments and soils, they may not be accessible to microbial degradation and they persist in anoxic environments (Atlas & Cerniglia, 1995). There have been recent reports demonstrating the anaerobic degradation of PAHs when nitrate, sulfate or ferric iron serves as the terminal electron acceptor (Mihelcic & Luthy, 1988; Coates, Anderson & Lovley, 1996; Zhang & Young, 1997; Meckenstock *et al.*, 2000; Rockne *et al.*, 2000).

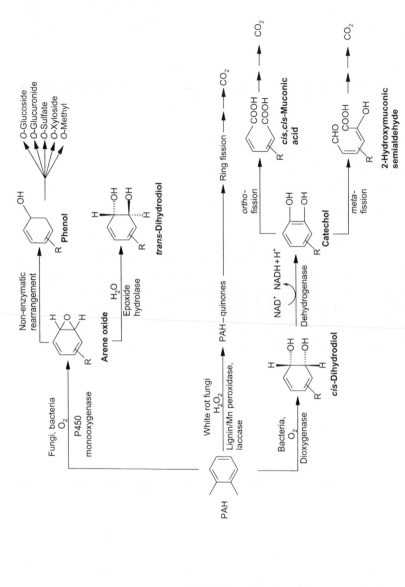

Fig. 7.3. Initial reactions in the degradation of polycyclic aromatic hydrocarbons (PAHs) by bacteria and fungi. (After Cerniglia, 1993.)

Reasons for using fungi

Yeasts and filamentous fungi are found in aquatic sediments, terrestrial habitats, acidic oil seeps and water surfaces and could be significant in natural biodegradation of PAHs (Cerniglia *et al.*, 1992; MacGillivray & Shiaris, 1993; Atlas & Cerniglia, 1995). Fungi also have advantages over bacteria since fungal hyphae can penetrate contaminated soil to reach the PAHs (Novotny *et al.*, 1999; April, Foght & Currah, 2000). Lignin-degrading fungi, which colonize wood and other lignocellulosic materials, are abundant in nature and have received considerable attention for their bioremediation potential since the enzymes that are involved in lignin breakdown can also degrade a wide range of pollutants (Bumpus *et al.*, 1985; Aust, 1995). The enzymes of lignin degradation are extracellular and have a broad substrate specificity, which makes them attractive candidates for environmental clean-up. The relative contribution of fungi to the total biotransformation of phenanthrene in coastal marine sediments has been estimated at only about 3% (MacGillivray & Shiaris, 1994). Nevertheless, many terrestrial fungi co-metabolize PAHs during growth on other substrates (Martens & Zadrazil, 1998; Márquez-Rocha, Hernández-Rodríguez & Vázquez-Duhalt, 2000) and *Rhodotorula glutinis* has even been reported to degrade phenanthrene in pure culture (Romero *et al.*, 1998). Applications of fungal technology for the remediation of PAH-contaminated soils will be discussed later in this chapter.

Bioremediation

Background on remediation technologies

Industrial and military sites that have been abandoned are usually found to be contaminated with a host of toxic substances originating from petroleum, coal and chemical residues. The clean-up of these pollutants in soil, water and refuse is often mandated by government environmental agencies (Mueller *et al.*, 1996; Löser *et al.*, 1999; Boopathy, 2000).

Many of the remediation technologies currently being used for contaminated soil and water involve not only physical and chemical treatments but also bioremediation of pollutants by microbial activity (Mueller *et al.*, 1993, 1996; Cutright & Lee, 1994; Atlas & Cerniglia, 1995; Allard & Neilson, 1997; Cha *et al.*, 1999; Straube *et al.*, 1999). Methods for the bioremediation of PAHs in soil and water have been investigated in many different laboratory, pilot-scale and full-scale tests. Bioremediation, generally with indigenous microorganisms from the polluted site, has been used for oil spills from supertankers as well as leaking underground fuel tanks (Atlas & Cerniglia, 1995). The complex mixtures of PAHs typically found in contaminated sites can generally be bioremediated if the geological, engineering, chemical and microbiological aspects of the problem can be coordinated (Allard & Neilson, 1997).

Land farming, which means spreading contaminated material over a field and tilling the soil, is sometimes useful for bioremediation (Mueller *et al.*, 1996). Spilled oil and wood-preserving wastes have been bioremediated by land-farming treatments (Haught *et al.*, 1995; Margesin & Schinner, 1999). Adding compost to contaminated soil enhances bioremediation because of the structure of the organic compost matrix (Kästner & Mahro, 1996). Compost enhances the oxidation of aromatic contaminants in soil to ketones and quinones, which eventually disappear (Wischmann & Steinhart, 1997). The ketone and quinone metabolites produced in creosote-contaminated soil can be monitored by solid-phase extraction and high-performance liquid chromatography (Meyer, Cartellieri & Steinhart, 1999). Bioreactors, which are enclosed containers for biological treatment of relatively small amounts of waste, have been used to treat soil and other materials contaminated with petroleum residues (McFarland *et al.*, 1992; Déziel, Comeau & Villemur, 1999). Even in cold environments, microorganisms are generally able to degrade at least a portion of the hydrocarbons in an oil spill after the addition of inorganic nutrients and oxygen (Margesin & Schinner, 1999). An example of the use of large-scale bioremediation was the partial clean-up of the Alaskan shoreline of Prince William Sound after the Exxon Valdez oil spill of 1989 (Atlas & Cerniglia, 1995; Margesin & Schinner, 1999; Boopathy, 2000).

Factors affecting bioremediation

Microorganisms with the ability to degrade aliphatic petroleum hydrocarbons appear to be ubiquitous in contaminated sites and so inoculation is normally unnecessary. Unfortunately, this is not the case for high-molecular-weight PAHs. Since bioremediation generally is attempted in unsterile soil, it is important to determine whether artificially introduced fungi will survive and remain active in the presence of indigenous soil bacteria. Therefore, it is critical when evaluating biotreatment options to have detailed information on the diversity of soil microflora and characteristics of the fungal inoculum to be used. The so-called intrinsic bioremediation of groundwater requires only time for natural attenuation by microbial processes (Chapelle, 1999). Usually, however, the bioremediation of PAHs requires aeration and the addition of nitrogen and phosphorus fertilizer for maximum efficiency of degradation (Atlas & Cerniglia, 1995; Mueller *et al.*, 1996; Chapelle, 1999). Other important environmental factors include redox potential, pH, soil characteristics and temperature. In alpine and polar regions, biodegradation of PAHs is slower and does not reach completion (Margesin & Schinner, 1999).

The rate of bioremediation in soils and sediments depends on the bioavailability of the contaminants; it is much slower if there are problems in mass transfer of the PAH molecules to microorganisms (Boopathy, 2000). Contaminants that are sorbed to clay minerals or organic matter are unavailable to microorganisms (Atlas & Cerniglia, 1995; Zhang, Bouwer & Ball, 1998). Even the bioremediation of groundwater in aquifers is impossible if the PAHs are trapped in spaces between soil particles that are too small for microorganisms to enter (Zhang *et al.*, 1998). Extracellular enzymes produced by fungi can catalyse oxidative coupling reactions between PAHs and soil humic material and may be an important factor in detoxifying a PAH-contaminated site (Qiu & McFarland, 1991). Bogan *et al.* (1999) showed humification of PAHs and bound residue formation after application of ligninolytic fungi to contaminated soil.

During the bioremediation of two-phase liquid systems in bioreactors, contaminants are broken down in the interface between the organic and aqueous phases (Déziel *et al.*, 1999). High concentrations of synthetic surfactants may increase the proportion of PAHs accessible to microorganisms in the aqueous phase, thus enhancing microbial degradation (Volkering, Breure & Rulkens, 1998; Bogan & Lamar, 1999; Pinto & Moore, 2000). However, it should be noted that some surfactants, especially ionic surfactants, may be toxic to microorganisms. Non-ionic surfactants, such as Tween-80 (polyoxyethylene sorbitan monooleate), may serve as growth substrates (Volkering *et al.*, 1998; Pinto & Moore, 2000). Microbially produced surfactants may prove to be more useful than synthetic surfactants for the bioremediation of oil spills in soil and water (Barkay *et al.*, 1999).

Bioremediation by bacteria, fungi, and plants

Bacteria with the ability to degrade one or more PAHs are widespread in the environment, including sites used for coal-tar disposal (Ghiorse *et al.*, 1995), and they are found even in groundwater (Chapelle, 1999). Individual bacterial strains may be able to degrade several PAHs in the laboratory but be unable to degrade all of the components of PAH mixtures in contaminated sites (Korda *et al.*, 1997). For instance, a strain of *Sphingomonas paucimobilis* degraded fluoranthene, chrysene, pyrene, benz[*a*] anthracene, benzo[*a*]pyrene, benzo[*b*]fluoranthene, and dibenz[*a,h*]anthracene, but not dibenzo[*a,l*]pyrene (Ye *et al.*, 1996); several bacteria from contaminated soils completely degraded phenanthrene, chrysene, benz[*a*] anthracene and benzo[*a*]pyrene, but not pyrene (Aitken *et al.*, 1998). Genetically engineered microorganisms have been developed with the ability to degrade a greater variety of pollutants (Chen *et al.*, 1999). However, because of the unknown risks from the release of these organisms into the environment, some of the recombinant bacterial strains intended for bioremediation have been programmed to 'commit suicide' after the depletion of the substrate (Garbisu & Alkorta, 1999).

Many fungi are able to transform PAHs and could be significant in bioremediation, although the rates of PAH biotransformation by pure cultures of fungi are lower than those for bacteria (Cerniglia, 1997). Mixed cultures of *Penicillium janthinellum* with bacteria have been shown to degrade pyrene, chrysene, benz[*a*]anthracene, benzo[*a*]pyrene and dibenz[*a,h*]anthracene (Boonchan, Britz & Stanley, 2000). Pyrene, chrysene, and benzo[*a*]pyrene have been extracted from soil with high concentrations of Tween-80 for later bioremediation by *Penicillium* spp. (Pinto & Moore, 2000). Several cultures of fungi have been cited in patents for use in the bioremediation of petroleum hydrocarbons, not necessarily including PAHs (Korda *et al.*, 1997). No commercial applications of fungi for the bioremediation of individual PAHs or mixtures of PAHs have been established, although there appears to be potential for the use of *Cunninghamella elegans*, *Rhodotorula glutinis* and white rot fungi for this purpose (Hüttermann *et al.*, 1989; Cutright, 1995; Andersson & Henrysson, 1996; Cerniglia, 1997; Romero *et al.*, 1998; Yateem *et al.*, 1998; Gramss, Voigt & Kirsche, 1999a; Andersson *et al.*, 2000).

Several investigators have demonstrated in laboratory and field experiments the potential of the white rot fungus *Phanerochaete chrysosporium* for use in the bioremediation of soil contaminated with PAHs (Haemmerli *et al.*, 1986; Hammel *et al.*, 1986, 1992; Sanglard, Leisola & Fiechter, 1986; Bumpus, 1989; Hammel, Green & Gai, 1991; Sutherland *et al.*, 1991; Dhawale, Dhawale & Dean-Ross, 1992; Kennes & Lema, 1994; Barclay, Farquhar & Legge, 1995; Bogan & Lamar, 1995; Haught *et al.*, 1995; McFarland & Qiu, 1995; Liao & Tseng, 1996; Liao *et al.*, 1997; May, Schröder & Sandermann, 1997). In a series of experiments, from laboratory bench-scale to full-scale field demonstrations, Haught *et al.* (1995) demonstrated the potential of *P. chrysosporium* and *Phanerochaete sordida* to degrade PAHs. Removal of high-molecular-weight PAHs (five rings and above) to low levels was difficult using white rot fungi. A pilot scale reactor system was developed that combined extraction of PAH-contaminated soil with a physically separate fungal bioreactor containing *P. chrysosporium* (May *et al.*, 1997). The extraction of high-molecular-weight PAHs from the soil made them bioavailable to the fungus, which led to high degradation rates.

In another study, *P. sordida* transformed PAHs with three and four rings in creosote-contaminated soil, but five- and six-ring PAHs were not degraded (Davis *et al.*, 1993). Some species of fungi are able to grow throughout the soil mass, which may or may not be an advantage for PAH degradation. Martens & Zadrazil (1998) screened a variety of wood-rotting fungi for their ability to degrade PAHs in a bioreactor containing straw and soil. A higher degradation rate (40–58% of the applied [^{14}C]-PAH as ^{14}CO$_2$) was observed in microcosms containing fungal strains that did not colonize the soil than in those inoculated with the soil-colonizing fungi. An explanation for the difference was that the indigenous soil bacteria were stimulated by compounds produced during the lysis of straw by non-colonizing fungi, which provided carbon sources to enhance bacterial growth and PAH degradation. Bogan & Lamar (1999) reported the ability of *Pleurotus ostreatus* to degrade 80% of the total PAHs in soil in 35 days.

Novotny *et al.* (1999) compared the abilities of *P. ostreatus*, *P. chrysosporium* and *Trametes versicolor* to degrade PAHs and produce ligninolytic enzymes in soil. They found that colonization of sterilized soil from straw-grown inocula and degradation of anthracene, phenanthrene and pyrene were greatest with *P. ostreatus*. The production of manganese peroxidase and laccase in soil was similar in *P. ostreatus* and *T. versicolor* but extremely low for *P. chrysosporium*. In aged soil contaminated with creosote, *P. ostreatus* degraded approximately 40% of the benzo[*a*]pyrene present after 12 weeks of incubation (Eggen & Majcherczyk, 1998; Eggen & Sveum, 1999). However, degradation was only about 1% when spent mushroom compost containing *P. ostreatus* was supplemented with fish oil and used for a soil contaminated with creosote. After 7 weeks, approximately 89% of the three-ring PAHs, 87% of the four-ring PAHs, and 48% of the five-ring PAHs had been degraded (Eggen, 1999). Removal of 86% of the priority PAHs was reported. However, the use of ligninolytic fungi for remediation of PAH-contaminated soil has not always given promising results. When *Bjerkandera* sp. strain BOS55 and *P. ostreatus* were inoculated into PAH-containing soil, the level of PAH removal was similar to those observed in abiotic controls (Kotterman, 1999). Interestingly, both fungal strains had been able to degrade PAHs extensively in pure-culture experiments in liquid nutrient media. Similar findings were reported by Harmsen & Heersche (1999), who used commercial mushroom-production wastes as a substrate for *P. ostreatus* grown in the presence of PAH-contaminated sediments, but found that it did not increase the rate of PAH degradation.

The use of plants to concentrate and metabolize toxic compounds in soil and water, called phytoremediation, has been shown to be feasible in many contaminated sites (Schnoor *et al.*, 1995; Salt, Smith & Raskin, 1998; Macek, Macková & Kás, 2000). For instance, the growth of sod-forming prairie grasses stimulates the bioremediation of benz[*a*]anthracene, chrysene, benzo[*a*]pyrene and dibenz[*a,h*]anthracene in soil (Aprill & Sims, 1990; Banks *et al.*, 1999). Phytoremediation with aquatic plants also has been investigated for the clean-up of oil spills along shorelines (Lee & de Mora, 1999). Since microorganisms are more abundant in the rhizosphere of plants than elsewhere in the soil, they are available to metabolize organic compounds at the root surface (Macek *et al.*, 2000). Ectomycorrhizal fungi, including strains of *Amanita, Leccinum* and *Suillus*, are able to degrade phenanthrene, pyrene, chrysene, and benzo[*a*]pyrene (Braun-Lullemann, Hüttermann & Majcherczyk, 1999). The arbuscular mycorrhizal fungus *Glomus mosseae* enhances the survival and growth of ryegrass in soil containing PAHs (Leyval & Binet, 1998). Ascomycetes growing on smooth cordgrass in a polluted saltmarsh ecosystem are as resistant to toxic compounds as the plants themselves (Newell, Wall & Maruya, 2000), although there is no evidence that these fungi can degrade PAHs.

PAH metabolism by non-ligninolytic fungi

Organisms

The vast majority of fungi grow on substrates other than wood and do not produce extracellular lignin peroxidases. Some of these fungi (Table 7.2) have been found to metabolize PAHs (Cerniglia, 1992, 1993, 1997; Cerniglia, Sutherland & Crow, 1992; Sutherland, 1992; Müncnerová & Augustin, 1994; Pothuluri & Cerniglia, 1994; Sutherland *et al.*, 1995). *C. elegans* and most other non-ligninolytic fungi are unable to use PAHs as sources of carbon or energy, but they may co-metabolize them while growing on other substrates. Although this process does not enhance fungal growth, it may result in a reduction of the toxic, mutagenic or carcinogenic properties

Table 7.2. *Non-ligninolytic fungi that metabolize polycyclic aromatic hydrocarbons*

Class	Genus	
Zygomycetes	*Cunninghamella blakesleeana* *C. echinulata* *C. elegans* *Mortierella ramanniana* (*Mucor ramannianus*)	*M. verrucosa* *Mucor racemosus* *Rhizopus arrhizus* *Syncephalastrum racemosum*
Ascomycetes	*Cryphonectria parasitica* *Dichotomomyces cejpii* *Morchella* spp.	*Neurospora crassa* *Saccharomyces cerevisiae* *Sporormiella australis*
Blastomycetes	*Candida krusei* *C. maltosa* *C. tropicalis* *Cryptococcus albidus*	*Rhodotorula glutinis* *R. minuta* *Trichosporon penicillatum*
Hyphomycetes	*Aspergillus niger* *A. ochraceus* *A. terreus* *A. versicolor* *Beauveria alba* *Botrytis cinerea* *Chrysosporium pannorum* *Cladosporium herbarum* *Curvularia lunata* *C. tuberculata* *Cylindrocladium destructans* *C. simplex* *Doratomyces stemonitis* *Drechslera spicifera* *Embellisia annulata* *Fusarium subglutinans* *Gliocladium virens*	*Monosporium olivaceum* *Penicillium chrysogenum* *P. janthinellum* *P. notatum* *P. simplicissimum* *Pestalotia palmarum* *Phialophora alba* *Phialophora hoffmannii* (*Lecythophora hoffmannii*) *Rhizoctonia solani* *Scopulariopsis brumptii* *Scytalidium lignicola* *Sporothrix cyanescens* *Trichoderma harzianum* *T. viride* *Verticillium lecanii*
Coelomycetes	*Cicinnobolus cesatii* *Colletotrichum dematium* *Coniothyrium fuckelii*	*C. sporulosum* *Phoma herbarum*

From: Woods & Wiseman, 1979; Cerniglia, Freeman & Mitchum, 1982; Pothuluri *et al.*, 1990, 1996; Sutherland *et al.*, 1992, 1995; MacGillivray & Shiaris, 1993; Müncnerová & Augustin, 1994; Wunder *et al.*, 1994; Casillas *et al.*, 1996; Krivobok *et al.*, 1998; Romero *et al.*, 1998; Gramss *et al.*, 1999b; Lisowska & Dlugonski, 1999; Pinto & Moore, 2000; Ravelet *et al.*, 2000.

of the PAHs (Cerniglia, White & Heflich, 1985; Pothuluri *et al.*, 1992b; Sutherland, 1992; Rudd *et al.*, 1996). In contrast to the inability of most fungi to grow on PAHs, a strain of the yeast *R. glutinis*, obtained from a polluted stream below an oil refinery, has been reported to grow exponentially on phenanthrene as a carbon and energy source (Romero *et al.*, 1998).

Pathways

Many non-ligninolytic fungi oxidize PAHs to water-soluble products (Colombo, Cabello & Arambarri, 1996), the first step being the epoxidation of one of the aromatic rings in a cytochrome P450 monooxygenase reaction to form a transient arene oxide (Fig. 7.3) (Sutherland, 1992). The arene oxide is immediately hydrated by an epoxide hydrolase to form a *trans*-dihydrodiol, and subsequent non-enzymic rearrangement may also produce a phenol (Sutherland *et al.*, 1995). *C. elegans* and *Cunninghamella echinulata* have genes for cytochrome P450 monoxygenase (Wang *et al.*, 2000) and oxidoreductase (Yadav & Loper, 2000). The oxidoreductase is induced by *n*-tetradecane (Yadav & Loper, 2000); the inducibility of the gene for the cytochrome P450 monooxygenase, the enzyme that actually binds to PAHs, is still unknown. Non-ligninolytic fungi also produce ketones and quinones from some PAHs (Pothuluri *et al.*, 1992a, 1993; Sutherland, 1992; Garon, Krivobok & Seigle-Murandi, 2000), but the mechanisms by which they are formed are not fully understood.

Some fungi metabolize the *trans*-dihydrodiols and phenols of PAHs further by sulfation, methylation or conjugation with glucose, xylose or glucuronic acid (Fig. 7.3) (Cerniglia, Freeman & Mitchum, 1982; Pothuluri *et al.*, 1990, 1996; Sutherland *et al.*, 1992; Müncnerová & Augustin, 1994; Wunder *et al.*, 1994; Casillas *et al.*, 1996). Conjugates are more water soluble than the other typical PAH metabolites. Selected examples of the metabolism of PAHs by non-ligninolytic fungi are described below.

Acenaphthene is metabolized by *C. elegans* to 6-hydroxyacenaphthenone, 1,2-acenaphthenedione, *trans*-1,2-dihydroxyacenaphthene, 1,5-dihydroxyacenaphthene, 1-acenaphthenol, 1-acenaphthenone and *cis*-1,2-dihydroxyacenaphthene (Fig. 7.4) (Pothuluri *et al.*, 1992a).

Fluorene is metabolized by *C. elegans* to 9-fluorenol, 9-fluorenone and 2-hydroxy-9-fluorenone (Fig. 7.5) (Pothuluri *et al.*, 1993). Various other fungi also oxidize fluorene (Sack & Günther, 1993; Garon *et al.*, 2000).

Anthracene is oxidized by *C. elegans* to an anthracene *trans*-1,2-dihyd-

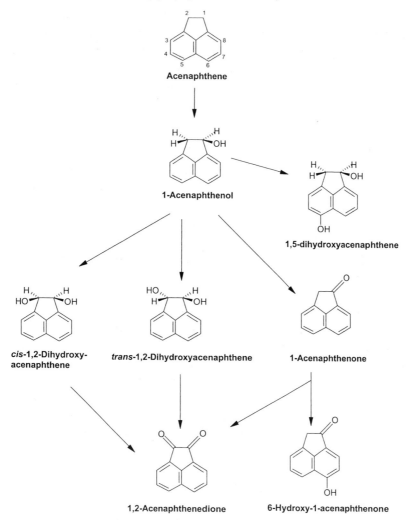

Fig. 7.4. Metabolism of acenaphthene by *Cunninghamella elegans*. (After Pothuluri *et al.*, 1992a.)

rodiol enantiomer and 1-anthryl sulfate (Fig. 7.6) (Cerniglia, 1982; Cerniglia & Yang, 1984). *Rhizoctonia solani* oxidizes it first to both enantiomers of anthracene *trans*-1,2-dihydrodiol and then further to three xyloside conjugates (Sutherland *et al.*, 1992). Many other fungi, including *A. niger, Cryphonectria parasitica, Rhizopus arrhizus, C. echinulata* and *Cladosporium herbarum*, have also been reported to metabolize anthracene (Yogambal & Karegoudar, 1997; Krivobok *et al.*, 1998; Lisowska & Dlugonski, 1999).

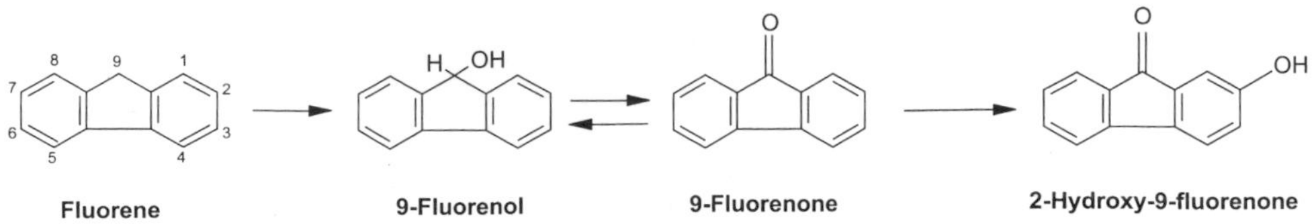

Fluorene   9-Fluorenol   9-Fluorenone   2-Hydroxy-9-fluorenone

Fig. 7.5. Metabolism of fluorene by *Cunninghamella elegans*. (After Pothuluri *et al.*, 1993.)

Phenanthrene is transformed by *C. elegans* to two enantiomers each of phenanthrene *trans*-1,2-dihydrodiol and *trans*-9,10-dihydrodiol, and by *Syncephalastrum racemosum* to two enantiomers of phenanthrene *trans*-3,4-dihydrodiol (Fig. 7.7) (Cerniglia & Yang, 1984; Sutherland *et al.*, 1991, 1993; Casillas *et al.*, 1996). Several yeasts from coastal sediments, including *Trichosporon penicillatum*, and various other fungi also transform phenanthrene (MacGillivray & Shiaris, 1993; Sack & Günther, 1993; Lisowska & Dlugonski, 1999). *A. niger, S. racemosum* and *C. elegans* may produce not only *trans*-dihydrodiols from phenanthrene but also sulfate, glucuronide and glucoside conjugates (Cerniglia *et al.*, 1989; Casillas *et al.*, 1996). *A. niger* metabolizes phenanthrene to 1-methoxyphenanthrene; the minor metabolites are 1- and 2-phenanthrol (Sack *et al.*, 1997a).

Fluoranthene is metabolized by *C. elegans* to fluoranthene *trans*-2,3-dihydrodiol, 8- and 9-hydroxyfluoranthene *trans*-2,3-dihydrodiols, 3-fluoranthene β-glucopyranoside, and 3-(8-hydroxyfluoranthene)-β-glucopyranoside (Fig. 7.8) (Pothuluri *et al.*, 1990). These metabolites have been shown to be less mutagenic to bacteria than fluoranthene (Pothuluri *et al.*, 1992b). Several other strains of fungi, including *Cryptococcus albidus, Aspergillus terreus, Cicinnobolus cesatii*, and *Penicillium* sp., also metabolize fluoranthene (Sack & Günther, 1993; Salicis *et al.*, 1999).

Benz[a]anthracene is transformed by *C. elegans* (Fig. 7.9) (Cerniglia, Dodge & Gibson, 1980a; Cerniglia, Gibson & Dodge, 1994) and by several yeasts from coastal sediments, including *Candida krusei* and *Rhodotorula minuta* (MacGillivray & Shiaris, 1993).

Chrysene is biotransformed by *C. elegans, S. racemosum* and *Penicillium* spp. (Pothuluri *et al.*, 1995; Kiehlmann, Pinto & Moore, 1996). In cultures of *C. elegans*, 2-hydroxychrysene, 2,8-dihydroxychrysene, 2-hydroxychrysene sulfate and 8-hydroxy-2-*O*-chrysene sulfate have been identified (Fig. 7.10) (Pothuluri *et al.*, 1995); in *P. janthinellum*, chrysene *trans*-1,2-dihydrodiol has been identified (Kiehlmann *et al.*, 1996).

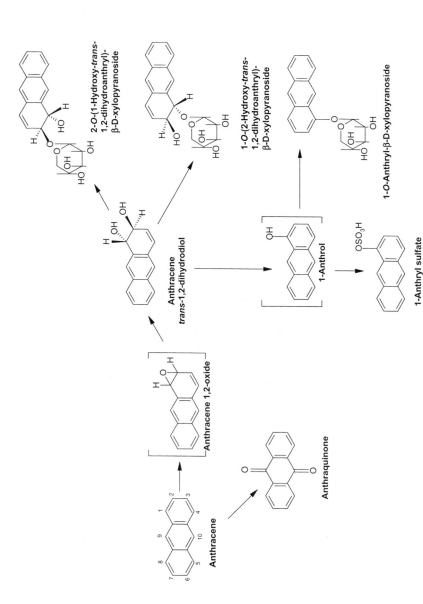

Fig. 7.6. Metabolites produced from anthracene by *Cunninghamella elegans* and *Rhizoctonia solani*. The sulfate conjugate is produced by *C. elegans* and the xyloside conjugates by *R. solani*. (After Cerniglia, 1982; Sutherland *et al.*, 1992.)

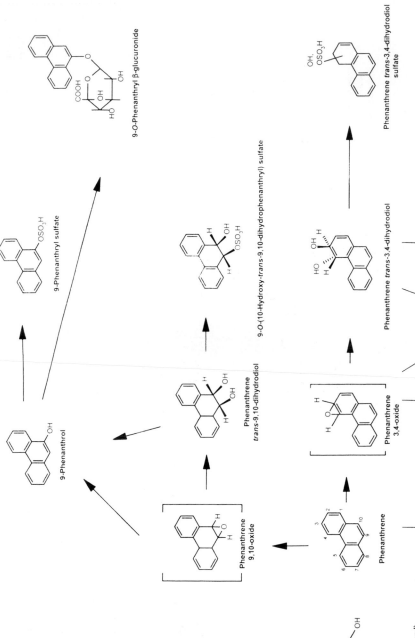

9-O-Phenanthryl β-glucuronide

9-Phenanthryl sulfate

9-O-(10-Hydroxy-trans-9,10-dihydrophenanthryl) sulfate

Phenanthrene trans-3,4-dihydrodiol sulfate

9-Phenanthrol

Phenanthrene trans-9,10-dihydrodiol

Phenanthrene trans-3,4-dihydrodiol

Phenanthrene 9,10-oxide

Phenanthrene 3,4-oxide

Phenanthrene

1-O-(2-Hydroxy-phenanthryl) β-glucopyranoside

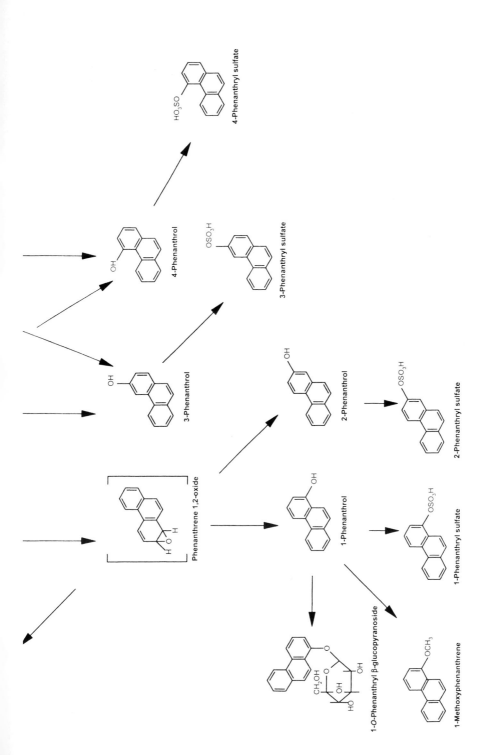

Fig. 7.7. Metabolites produced from phenanthrene by *Aspergillus niger*, *Cunninghamella elegans* and *Syncephalastrum racemosum*. Not all of the metabolites are produced by all three of the fungi. (After Casillas *et al.*, 1996; Sack *et al.*, 1997a.)

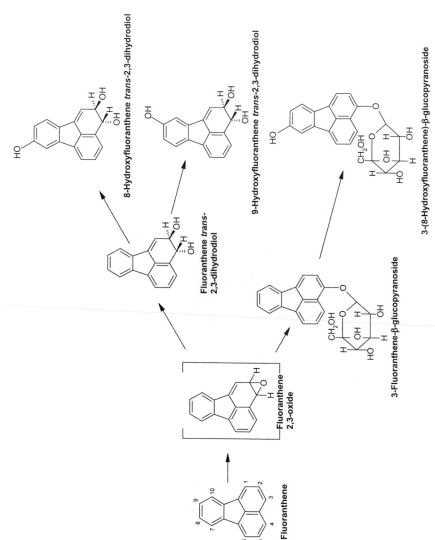

Fig. 7.8. Metabolism of fluoranthene by *Cunninghamella elegans*. (Modified after Pothuluri *et al.*, 1990.)

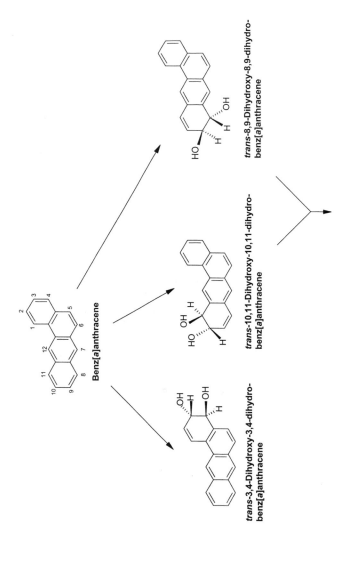

Fig. 7.9. Metabolism of benz[*a*]anthracene by *Cunninghamella elegans*. (After Cerniglia, 1984.)

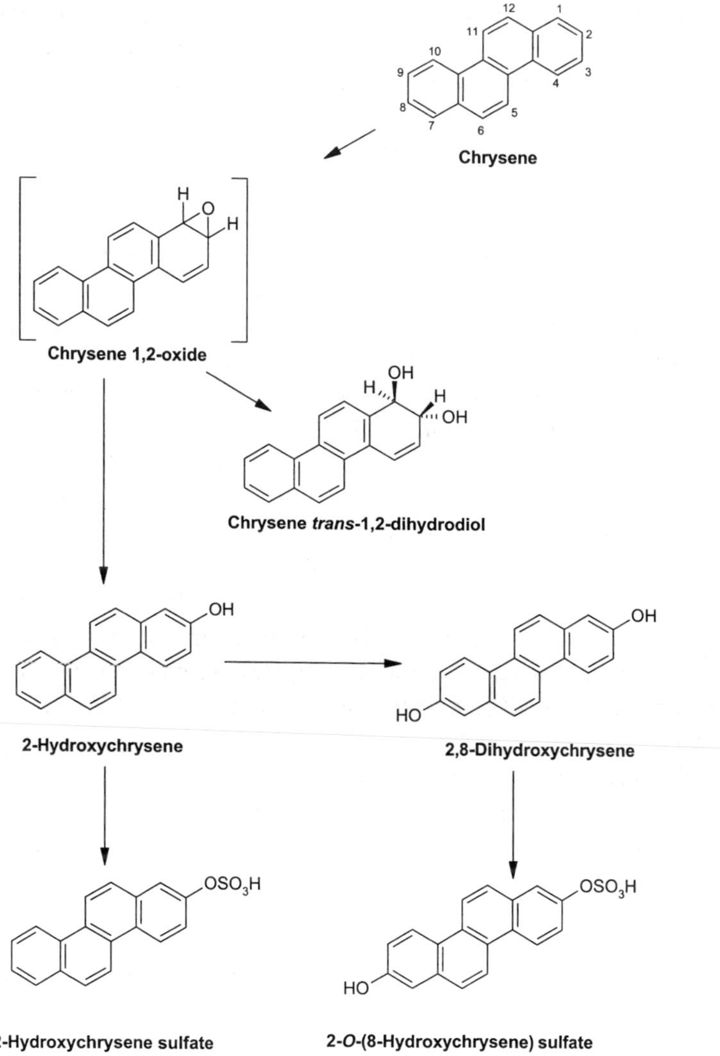

Fig. 7.10. Metabolism of chrysene by *Cunninghamella elegans*. (After Pothuluri *et al.*, 1995.)

Pyrene is oxidized by *A. niger, C. elegans, P. janthinellum, S. racemosum* and other fungi to 1-pyrenol, 1,6- and 1,8-pyrenequinones and 1,6- and 1,8-dihydroxypyrene (Fig. 7.11) (Cerniglia *et al.*, 1986; Wunder *et al.*, 1994; Launen *et al.*, 1995; in der Wiesche, Martens & Zadrazil, 1996). *A. niger* also produces 1-pyrenyl sulfate, 1-hydroxy-8-pyrenyl sulfate and 1-methoxypyrene (Wunder *et al.*, 1994, 1997; Sack *et al.*, 1997a). In *P. janthinellum*, glucose inhibits the oxidation of pyrene (Launen, Pinto & Moore, 1999). Other fungi from aquatic sediments and contaminated soils, including strains of *Mucor racemosus, Phialophora alba, Coniothyrium fuckelii*, and *Penicillium* sp., have also been shown to degrade pyrene (Sack & Günther, 1993; Ravelet *et al.*, 2000).

Perylene is metabolized by several fungi, including *Morchella* spp., *Botrytis cinerea, Scytalidium lignicola, Trichoderma* sp. and others, to unknown products (Gramss *et al.*, 1999a).

Benzo[*a*]pyrene, which is highly carcinogenic if it is metabolized by liver enzymes, is hydroxylated by the cytochrome P450-containing aryl hydrocarbon hydroxylase from *Saccharomyces cerevisiae* (Wiseman & Woods, 1979; Woods & Wiseman, 1979). *C. elegans* oxidizes benzo[*a*]pyrene to *trans*-dihydrodiols, diol epoxides, phenols and quinones (Fig. 7.12) (Cerniglia & Gibson, 1979; 1980a,b; Cerniglia, Mahaffey & Gibson, 1980b). *Aspergillus ochraceus* produces *trans*-dihydrodiols and quinones (Dutta *et al.*, 1983; Ghosh *et al.*, 1983; Wunch, Feibelman & Bennett, 1997). *P. janthinellum, S. racemosum* and other fungi also oxidize benzo[*a*] pyrene (Launen *et al.*, 1995). Microsomes from the yeasts *Kluyveromyces marxianus* and *S. cerevisiae* show a type I binding spectrum with benzo[*a*] pyrene (Engler *et al.*, 2000), indicating that it is a substrate for cytochrome P450.

Benzo[*e*]pyrene, a non-carcinogenic benzopyrene isomer, is metabolized by strains of *C. elegans* to 3-benzo[*e*]pyrenyl sulfate, 1-hydroxy-3-benzo[*e*] pyrenyl sulfate and benzo[*e*]pyrene 3-*O*-β-glucopyranoside (Fig. 7.13) (Pothuluri *et al.*, 1996).

Metabolism of polycyclic aromatic hydrocarbons by ligninolytic fungi

A group of lignin-degrading fungi in the class Basidiomycetes have received considerable attention for their ability to degrade a wide variety of structurally diverse chemical pollutants (Morgan, Lewis & Watkinson, 1991; Shah *et al.*, 1992; Morgan *et al.*, 1993; Field *et al.*, 1993; Aust, 1995; Hammel, 1995a,b; Kotterman, 1999). *P. chrysosporium, T. versicolor, P. ostreatus, Bjerkandera adusta, Bjerkandera* sp. strain B0S55 and *Nematoloma frowardii* have been the most extensively studied. Lignin peroxidase (LiP), manganese-dependent peroxidase (MnP), manganese-independent peroxidase and laccase are the enzymes that have been implicated in the oxidation of PAHs. Since these enzymes are extracellular,

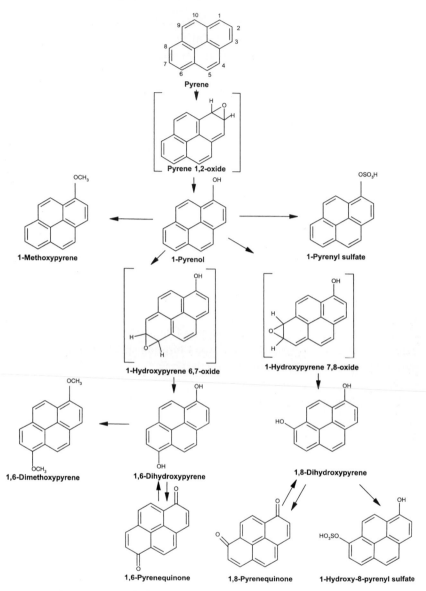

Fig. 7.11. Metabolism of pyrene by *Cunninghamella elegans* and *Aspergillus niger*. (After Cerniglia *et al.*, 1986; Wunder *et al.*, 1994; Sack *et al.*, 1997a.)

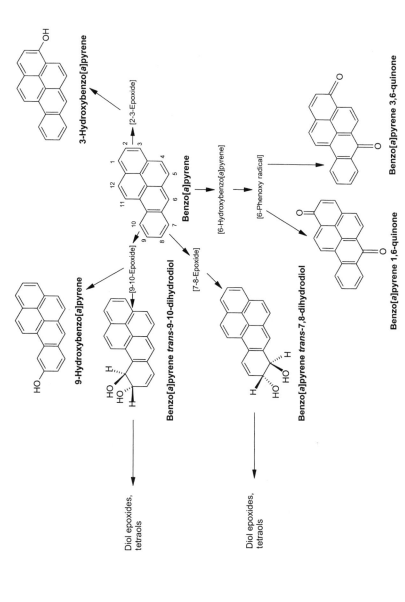

Fig. 7.12. Metabolism of benzo[a]pyrene by *Cunninghamella elegans*. (After Cerniglia & Gibson, 1979; Cerniglia *et al.*, 1992.)

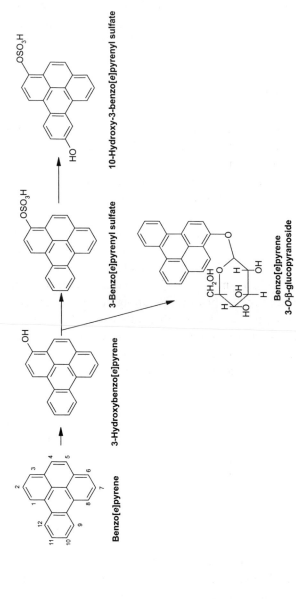

Fig. 7.13. Metabolism of benzo[e]pyrene by *Cunninghamella elegans*. (Modified after Pothuluri *et al.*, 1996.)

have low substrate specificity and may diffuse into the soil matrix where the PAHs are entrapped, investigators have attempted to use ligninolytic fungi for the degradation of recalcitrant PAHs (Brodkorb & Legge, 1992; Bogan & Lamar, 1996; Lang *et al.*, 1996; Sack & Fritsche, 1997; Boyle, Wiesner & Richardson, 1998; Novotny *et al.*, 1999; Pickard *et al.*, 1999).

Phanerochaete chrysosporium

Bumpus *et al.* (1985) published a seminal paper demonstrating the potential of the white rot fungus *P. chrysosporium* to degrade chemically diverse environmental pollutants including benzo[*a*]pyrene. These experiments were performed in liquid media with low concentrations of nitrogen. LiPs are induced during secondary metabolism under nutrient-deficient cultural conditions (Bogan *et al.*, 1996a). *P. chrysosporium* grows optimally at 39°C, a higher temperature than most white rot fungi. This fungus has considerable promise in bioremediation technologies, since the LiP and MnP, coupled with a system producing H_2O_2, generate Fenton-type radicals that oxidatively attack not only lignin but also other aromatic compounds (Aust, 1995; Hammel, 1995b). Another important feature of *P. chrysosporium*, compared with most non-ligninolytic fungi, is the ability to cleave aromatic rings and eventually break down PAHs (Hammel *et al.*, 1991). LiP ionizes aromatic compounds to form aryl cation radicals, which undergo further oxidation to form quinones. For example, the major pathway for anthracene degradation is via 9,10-anthraquinone, with subsequent ring cleavage to phthalic acid and finally carbon dioxide (Fig. 7.14) (Hammel *et al.*, 1991). Purified forms of LiP and MnP have also been shown to oxidize anthracene, pyrene, fluorene and benzo[*a*]pyrene to the corresponding quinones (Fig. 7.14) (Bogan, Lamar & Hammel, 1996b; Haemmerli *et al.*, 1986; Hammel *et al.*, 1986, 1991). Some PAHs, up to six aromatic rings, are oxidized by manganese-dependent lipid peroxidation reactions, both *in vitro* and *in vivo* (Bogan & Lamar, 1995; Bogan *et al.*, 1996b,c). Cytochrome P450 has also been purified and implicated in the hydroxylation of benzo[*a*]pyrene by *P. chrysosporium* (Masaphy *et al.*, 1996).

Since phenanthrene has an ionization potential of 8.19 Ev, it is not considered to be a LiP substrate (Hammel *et al.*, 1986). However, Hammel *et al.* (1992) later demonstrated that *P. chrysosporium* metabolizes phenanthrene initially to phenanthrene 9,10-quinone and then to a ring cleavage product, 2,2′-diphenic acid, under ligninolytic conditions (Fig. 7.14). Lipid peroxidation by the MnP of *P. chrysosporium* has also been implicated in

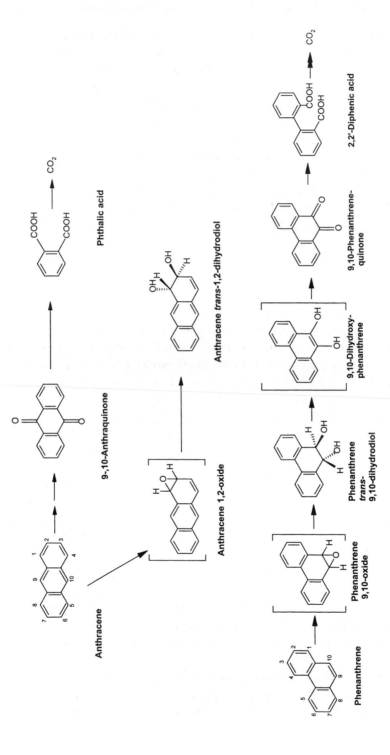

Fig. 7.14. Metabolism of anthracene and phenanthrene by ligninolytic fungi.

phenanthrene oxidation (Moen & Hammel, 1994). Degradation of phenanthrene by *P. chrysosporium* under non-ligninolytic conditions, in liquid media with high nitrogen, results in a different metabolic profile from ligninolytic incubations. *P. chrysosporium* metabolizes phenanthrene to phenanthrene *trans*-3,4- and *trans*-9,10-dihydrodiols, 3-, 4- and 9-phenanthrols and a glucoside conjugate of 9-phenanthrol (Sutherland *et al.*, 1991). Based on the metabolites formed from phenanthrene in the culture broth of *P. chrysosporium*, it was speculated that the initial steps in the oxidation of phenanthrene under non-ligninolytic conditions are catalysed by cytochrome P450 and epoxide hydrolase. Tatarko & Bumpus (1993) observed that LiP H8 oxidizes 9-phenanthrol to 9,10-phenanthrenequinone. Therefore, depending upon the growth conditions, *P. chrysosporium* has multiple enzymic pathways for the metabolism of phenanthrene.

Trametes versicolor

T. versicolor has been used in biodegradation studies because of its strong extracellular laccase production; the laccase of *T. versicolor* oxidizes most of the 16 PAHs listed by the US EPA as priority pollutant chemicals. PAH oxidation by laccase from this and other species is enhanced by the addition of mediator compounds, such as 2,2'-azino-bis(3-ethylbenzothiazoline-6-sulfonic acid) and 1-hydroxybenzotriazole (Böhmer, Messner & Srebotnik, 1998; Majcherczyk, Johannes & Hüttermann, 1998; Majcherczyk & Johannes, 2000). Benzo[*a*]pyrene and perylene are partially converted to polymeric products. Small amounts of quinones and ketones are the main oxidation products from anthracene (9,10-anthraquinone), benzo[*a*]pyrene (benzo[*a*]pyrene 1,6-, 3,6- and 6,12-quinones) and fluorene (9-fluorenone) (Collins *et al.*, 1996; Johannes, Majcherczyk & Hüttermann, 1996; Majcherczyk *et al.*, 1998). The laccase of *T. versicolor* in combination with 1-hydroxybenzotriazole oxidizes acenaphthene and acenaphthylene to a variety of compounds; the primary metabolites are 1,2-acenaphthenedione and 1,8-naphthalic acid (Johannes, Majcherczyk & Hüttermann, 1998). The role of natural mediators, including phenols and aromatic amines, in the degradation of PAHs by laccase is now beginning to be unveiled (Johannes & Majcherczyk, 2000).

When *T. versicolor* is cultured in media with high nitrogen and manganese, it can completely oxidize fluorene (Collins & Dobson, 1996). Unlike purified LiP, MnP of *T. versicolor* is capable of oxidizing phenanthrene and fluorene, which have ionization potentials greater than 7.55 Ev (Table 7.1). Sack *et al.* (1997b) showed that *T. versicolor* can degrade pyrene and phenanthrene in liquid and straw cultures. The major sites of enzymic attack are the K-regions of pyrene and phenanthrene, forming pyrene *trans*-4,5-dihydrodiol and phenanthrene *trans*-9,10-dihydrodiol, respectively (Figs. 7.14 and 7.15). The results also suggest that *T. versicolor* produces intracellular enzymes, such as cytochrome P450 and epoxide hydrolase, that can attack PAHs (Sack *et al.*, 1997b).

The above studies indicate that *T. versicolor* has both intracellular and extracellular enzymes that are important in the degradation of PAHs. The capability of laccase and natural mediator systems to degrade PAHs *in vitro* may have applications in the detoxification of these environmentally persistent pollutants.

Pleurotus ostreatus

The edible oyster mushroom, *P. ostreatus*, has been shown to be an efficient PAH degrader (Bezalel, Hadar & Cerniglia, 1996a, 1998; Wolter *et al.*, 1997). Although many of the experiments have been done in nutrient-rich liquid media, the fungus has potential for practical application in the decontamination of soils polluted with PAHs (Eschenback *et al.*, 1995; Baldrian *et al.*, 2000) since mycelia on solid substrates, such as straw and spent mushroom waste, can be used as economical sources of inoculum (Eggen, 1999; Novotny *et al.*, 1999). Furthermore, *P. ostreatus* grows well in soil and can compete with indigenous soil bacteria (Martens & Zadrazil, 1998; Martens *et al.*, 1999). In contrast to *P. chrysosporium, P. ostreatus* does not produce LiP activity (Hatakka, 1990), but its abilities to degrade lignin and PAHs may be linked with laccase and MnP activities. Bezalel *et al.* (1996a) demonstrated that *P. ostreatus* can degrade phenanthrene, anthracene, fluorene, pyrene and benzo[*a*]pyrene (Figs. 7.14 and 7.15). Further studies on the isolation and identification of polar and organic soluble metabolites suggest that the metabolism of *P. ostreatus* is similar to that of non-ligninolytic fungi and indicate that intracellular cytochrome P450 monooxygenase and epoxide hydrolase are responsible for the initial enzymic attack (Bezalel *et al.*, 1996b,c; Bezalel, Hadar & Cerniglia, 1997). There appears to be no direct correlation between the activities of laccase and MnP and the oxidation of PAHs, suggesting that these extracellular enzymes may be involved in the later reactions of PAH degradation (Bezalel *et al.*, 1996c; Schutzendubel *et al.*, 1999).

Phenanthrene is metabolized to phenanthrene *trans*-9,10-dihydrodiol

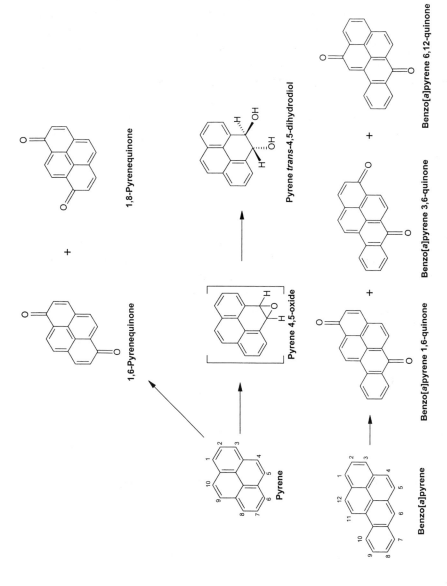

Fig. 7.15. Metabolism of pyrene and benzo[*a*]pyrene by ligninolytic fungi.

and 2,2'-diphenic acid under non-ligninolytic conditions by *P. ostreatus* (Fig. 7.14) (Bezalel *et al.*, 1996b). Oxygen-18 incorporation experiments indicate that *P. ostreatus* initially oxidizes phenanthrene stereoselectively, using cytochrome P450 and epoxide hydrolase to form a dihydrodiol, predominantly as the (9*R*,10*R*) enantiomer (Bezalel *et al.*, 1996b). This is similar to the major enantiomer produced by the non-ligninolytic fungus *C. elegans* but is different from the major enantiomer (9*S*,10*S*) produced by *P. chrysosporium* (Sutherland *et al.*, 1991, 1993). As in phenanthrene metabolism, pyrene is also metabolized by *P. ostreatus* in the K-region (4,5-positions) to form pyrene *trans*-4,5-dihydrodiol, with the predominant enantiomer in the (*R, R*)-configuration (Bezalel *et al.*, 1996c).

P. ostreatus initially metabolizes anthracene at two different sites on the molecule. One enzymic attack is at the 1,2-bond, to form anthracene *trans*-1,2-dihydrodiol. The major enantiomer of this *trans*-dihydrodiol is in the (*S,S*)-configuration, which is similar to that formed by *C. elegans*. The oxidation of anthracene to 9,10-anthraquinone by *P. ostreatus* has been demonstrated in several laboratories (Fig. 7.14) (Vyas *et al.*, 1994; Bezalel *et al.*, 1996c). The involvement of extracellular enzymes also has been implicated in 9,10-anthraquinone formation by *P. chrysoporium, T. versicolor, Bjerkandera* sp. and *Coriolopsis polyzona* (Hammel *et al.*, 1991; Field *et al.*, 1992; Vyas *et al.*, 1994).

Based on the above studies, *P. ostreatus* has multiple degradation pathways for the metabolism of PAHs. Cytochrome P450 monoxygenase has been demonstrated to be responsible for the initial attack on several PAHs (Bezalel *et al.*, 1997). However, free radical attack by extracellular oxidative enzymes also occurs to form quinones. In a related species, *Pleurotus pulmonarius*, cytochrome P450 was purified and implicated in benzo[*a*]pyrene hydroxylation (Masaphy *et al.*, 1995; Masaphy, Krinfeld & Levanon, 1998; Masaphy, Lamb & Kelly, 1999). Since many of the experiments were performed in nutrient-rich liquid media in flasks, however, the relationship between the intracellular and extracellular enzymes of *P. ostreatus* in PAH degradation remains an open question.

Other fungi

Other ligninolytic fungi have been screened for PAH degradation and compared with known white rot fungi that are PAH degraders. The white rot fungus *Bjerkandera* sp. strain BOS55 can oxidize a variety of PAHs including benzo[*a*]pyrene (Fig. 7.15) (Field *et al.*, 1992, 1995; Kotterman *et al.*, 1994; Kotterman, Wasseveld & Field, 1996; Grotenhuis *et al.*, 1998). This fungus produces LiP, MnP and manganese-independent peroxidase, which play major roles in PAH metabolism (Field *et al.*, 1996b). *Bjerkandera* sp. strain BOS55 was able to metabolize three- and four-ring PAHs in solvent extracts of polluted soil (Field *et al.*, 1996a). The high-molecular-weight PAHs (five and six aromatic rings) were poorly degraded, mainly because of the high concentration of the water-miscible solvent acetone, which solubilizes PAHs but is toxic to the fungal cells (Field *et al.*, 1996a). A subsequent study showed that non-ionic surfactants like Tween 80 increase the bioavailability of PAHs. This increased the solubility of the PAHs and, therefore enhanced by up to fivefold the oxidation of anthracene, pyrene and benzo[*a*]pyrene by the ligninolytic enzymes (Kotterman *et al.*, 1998a).

Further studies with this fungus also suggest that it has potential application in remediation of polluted waste sites (Kotterman, Vis & Field, 1998b). *Bjerkandera* sp. BOS55 and indigenous bacteria from activated sludge or forest soil were combined and evaluated for benzo[*a*]pyrene degradation. There was a stimulation in the breakdown of benzo[*a*]pyrene as well as a significant decrease in its mutagenicity compared with the fungal culture without the soil microflora.

A crude preparation of MnP from the South American white rot fungus *N. frowardii* oxidized mixtures of eight different PAHs, as well as five individual PAHs, including anthracene, phenanthrene, pyrene, fluoranthene and benzo[*a*]pyrene (Sack, Hofrichter & Fritsche, 1997c,d). Activity of the MnP was stimulated by the addition of glutathione. The values for PAH degradation ranged between 2.5 and 7.3%.

Sack *et al.* (1997b) screened several wood-decaying fungi, including *T. versicolor, Kuehneromyces mutabilis, Flammulina velutipes, Laetiporus sulphureus* and *Agrocybe aegerita*, for phenanthrene and pyrene degradation. Many of the fungi screened were able to break down the PAHs in liquid and straw cultures. Phenanthrene *trans*-1,2-, 3,4- and 9,10-dihydrodiols and the corresponding phenolic isomers were identified from cultures incubated with phenanthrene (Fig. 7.14). Pyrene *trans*-4,5-dihydrodiol and 1-hydroxypyrene were isolated from cultures incubated with pyrene (Fig. 7.15) (Sack *et al.*, 1997b).

Purified laccase from *Pycnoporus cinnabarinus* oxidizes benzo[*a*]pyrene to the corresponding 1,6-, 3,6- and 6,12-quinones (Fig. 7.15) (Rama *et al.*, 1998) as in incubations with purified ligninases of *P. chrysosporium* (Haemmerli *et al.*, 1986). The reaction, which was conducted in a bench-scale reactor, also required the exogenous mediator 2,2′-azino-bis(3-ethyl-benzothiazoline-6-sulfonic acid).

In a report on six white rot fungal strains isolated from soils in Korea, Song (1999) reported that *Irpex lacteus* had broken down 15.6% of pyrene added in liquid media after 4 weeks of incubation. One strain of *T. versicolor*, as well as *P. chrysosporium* and *P. ostreatus*, had degraded lesser amounts. *Marasmiellus troyanus*, which was isolated from a toxic waste site, degraded more benzo[*a*]pyrene than did *P. chrysosporium* (Wunch, Alworth & Bennett, 1999). After 15 days of incubation, 8.1% of the benzo[*a*]pyrene was degraded with *M. troyanus* compared with 1.1% for *P. chrysosporium*. Pyrene is metabolized by different strains of *Crinipellis stipitaria*, a member of the Agaricales that has not been shown to degrade lignin, to 1-hydroxypyrene, 1,6- and 1,8-dihydroxypyrene, 1,6- and 1,8-pyrenequinone, pyrene *trans*-4,5-dihydrodiol, 1-pyrenylsulfate, 6,8-dihydroxy-3-methylisocoumarin, 6-hydroxypyrene 1-sulfate and pyrene 1,6-disulfate (Lambert *et al.*, 1994; Lange *et al.*, 1994, 1995, 1996).

Conclusions

Several of the papers mentioned in this review demonstrate the potential of non-ligninolytic and ligninolytic fungi to remediate PAH-contaminated soils. Since most non-ligninolytic fungi do not cleave the aromatic rings of PAHs, much of the research since the late 1980s has focused on the use of white-rot fungi for the biodegradation of PAHs. The extracellular enzymes (including LiPs, MnPs and laccases) of these ligninolytic fungi metabolize PAHs to quinones, cleave aromatic rings and, to a limited extent, produce carbon dioxide. Numerous experiments have shown that white-rot fungi can transform individual PAHs and complex mixtures, not only in laboratory cultures but also in soil spiked with PAHs. The resulting quinones, free-radical intermediates and carboxylic acids may bind to the organic and inorganic components of soil to reduce the bioavailability of the PAHs. Therefore, the biotransformation process may be characterized as a sequestration that can lead to eventual detoxification. However, the mechanisms of these oxidative and ring-cleavage enzymes, including both intracellular and extracellular processes, and the humification process in soil still need intensive research.

Since many different treatment techniques have been proposed for the remediation of contaminated sites, methods are usually selected based on feasibility, effectiveness, time requirements and cost. The use of fungi for bioremediation has more often been found effective in the laboratory than in the field. The recent report of the utilization of phenanthrene by a yeast, if confirmed and extended to other PAHs, will be of interest for field studies. Future success in bioremediation will require a greater under-standing of the capabilities of the selected fungi and their interactions with abiotic and biotic factors in the soil environment. For example, the intro-duced fungi must be capable of surviving and competing with indigenous soil bacteria. This is critical because bacteria could either inhibit the growth of the fungal inoculum or, in combination with fungi, enhance the total degradation of the PAHs. Research in several laboratories has shown that ligninolytic fungi have the ability to colonize straw, wood chips and mushroom compost for use in PAH-contaminated soil. This approach could be effective as a remediation strategy.

Before fungi can be used for the bioremediation of PAH-contaminated sites, the environmental and nutritional factors that influence biodegrada-tion rates must be considered. A successful method will also have to be reliable enough to meet government regulatory requirements. Methods for the detoxification of PAH residues in the environment must be as rapid, cost-effective and environmentally safe as possible.

Acknowledgements

The authors would like to thank Danny Tucker, Sandra Malone and Pat Fleischer for illustrations and clerical assistance.

References

Ahn, Y., Sanseverino, J. & Sayler, G. S. (1999). Analyses of polycyclic aromatic hydrocarbon-degrading bacteria isolated from contaminated soil. *Biodegradation*, **10**, 149–157.

Aitken, M. D., Stringfellow, W. T., Nagel, R. D., Kazunga, C. & Chen, S.-H. (1998). Characteristics of phenanthrene-degrading bacteria isolated from soils contaminated with polycyclic aromatic hydrocarbons. *Canadian Journal of Microbiology*, **44**, 743–752.

Allard, A.-S. & Neilson, A. H. (1997). Bioremediation of organic waste sites: a critical review of microbiological aspects. *International Biodeterioration and Biodegradation*, **39**, 253–285.

Andersson, B. E. & Henrysson, T. (1996). Accumulation and degradation of dead-end metabolites during treatment of soil contaminated with polycyclic aromatic hydrocarbons with five strains of white-rot fungi. *Applied Microbiology and Biotechnology*, **46**, 647–652.

Andersson, B. E., Welinder, L., Olsson, P. A., Olsson, S. & Henrysson, T. (2000). Growth of inoculated white-rot fungi and their interactions with the bacterial community in soil contaminated with polycyclic aromatic hydrocarbons, as measured by phospholipid fatty acids. *Bioresource Technology*, **73**, 29–36.

April, T. M., Foght, J. M. & Currah, R. S. (2000). Hydrocarbon-degrading filamentous fungi isolated from flare pit soils in northern and western Canada. *Canadian Journal of Microbiology*, **46**, 38–49.

Aprill, W. & Sims, R. C. (1990). Evaluation of the use of prairie grasses for stimulating polycyclic aromatic hydrocarbon treatment in soil. *Chemosphere*, **20**, 253–265.

Atlas, R. M. & Cerniglia, C. E. (1995). Bioremediation of petroleum pollutants. *BioScience*, **45**, 332–339.

Aust, S. D. (1995). Mechanisms of degradation by white rot fungi. *Environmental Health Perspectives*, **103**, 59–61.

Baek, S. O., Field, R. A., Goldstone, M. E., Kirk, P. W., Lester J. N. & Perry, R. (1991). A review of atmospheric polycyclic aromatic hydrocarbons: sources, fate, and behavior. *Water, Air and Soil Pollution*, **60**, 279–300.

Baldrian, P., in der Wiesche, C., Gabriel, J., Nerud, F. & Zadrazil, F. (2000). Influence of cadmium and mercury on activities of ligninolytic enzymes and degradation of polycyclic aromatic hydrocarbons by *Pleurotus ostreatus* in soil. *Applied and Environmental Microbiology*, **66**, 2471–2478.

Banks, M. K., Lee, E. & Schwab, A. P. (1999). Evaluation of dissipation mechanisms for benzo[a]pyrene in the rhizosphere of tall fescue. *Journal of Environmental Quality*, **28**, 294–298.

Barclay, C. D., Farquhar, G. F. & Legge, R. L. (1995). Biodegradation and sorption of polyaromatic hydrocarbons by *Phanerochaete chrysosporium*. *Applied Microbiology and Biotechnology*, **42**, 958–963.

Barkay, T., Navon-Venezia, S., Ron, E. Z. & Rosenberg, E. (1999). Enhancement of solubilization and biodegradation of polyaromatic hydrocarbons by the bioemulsifier Alasan. *Applied and Environmental Microbiology*, **65**, 2697–2702.

Bezalel, L., Hadar, Y. & Cerniglia, C. E. (1996a). Mineralization of polycyclic aromatic hydrocarbons by the white-rot fungus *Pleurotus ostreatus*. *Applied and Environmental Microbiology*, **62**, 292–295.

Bezalel, L., Hadar, Y., Fu, P. P., Freeman, J. P. & Cerniglia, C. E. (1996b). Metabolism of phenanthrene by the white-rot fungus *Pleurotus ostreatus*. *Applied and Environmental Microbiology*, **62**, 2547–2553.

Bezalel, L., Hadar, Y., Fu, P. P., Freeman, J. P. & Cerniglia, C. E. (1996c). Initial oxidation products in the metabolism of pyrene, anthracene, fluorene and dibenzothiophene by the white-rot fungus *Pleurotus ostreatus*. *Applied and Environmental Microbiology*, **62**, 2554–2559.

Bezalel, L., Hadar, Y. & Cerniglia, C. E. (1997). Enzymatic mechanisms involved in phenanthrene degradation by the white rot fungus *Pleurotus ostreatus*. *Applied and Environmental Microbiology*, **63**, 2495–2501.

Bezalel, L., Hadar, Y. & Cerniglia, C. E. (1998). Degradation of polycyclic aromatic hydrocarbons by the white rot fungus *Pleurotus ostreatus*. In *Advances in Biotechnology*, ed. A. Pandey, pp. 405–421. New Delhi: Educational Publishers and Distributors.

Bogan, B. W. & Lamar, R. T. (1995). One-electron oxidation in the degradation of creosote polycyclic aromatic hydrocarbons by *Phanerochaete chrysosporium*. *Applied and Environmental Microbiology*, **61**, 2631–2635.

Bogan, B. W. & Lamar, R. T. (1996). Polycyclic aromatic hydrocarbon-degrading capabilities of *Phanerochaete laevis* HHB-1625 and its extracellular ligninolytic enzymes. *Applied and Environmental Microbiology*, **62**, 1597–1603.

Bogan, B. W. & Lamar, R. T. (1999). Surfactant enhancement of white-rot fungal

PAH soil remediation. In *The Fifth International In Situ and On-Site Bioremediation Symposium, Phytoremediation and Innovative Strategies for Specialized Remedial Applications*, San Diego, California, 19–22 April, 1999, eds. A. Leason & B. C. Allman, pp. 81–86. Columbus OH: Battelle Press.

Bogan, B. W., Schoenike, B., Lamar, R. T. & Cullen, D. (1996a). Expression of *lip* genes during growth in soil and oxidation of anthracene by *Phanerochaete chrysosporium. Applied and Environmental Microbiology*, **62**, 3697–3703.

Bogan, B. W., Lamar, R. T. & Hammel, K. E. (1996b). Fluorene oxidation *in vivo* by *Phanerochaete chrysosporium* and *in vitro* during manganese peroxidase-dependent lipid peroxidation. *Applied and Environmental Microbiology*, **62**, 1788–1792.

Bogan, B. W., Schoenike, B., Lamar, R. T. & Cullen, D. (1996c). Manganese peroxidase mRNA and enzyme activity levels during bioremediation of polycyclic aromatic hydrocarbon-contaminated soil with *Phanerochaete chrysosporium. Applied and Environmental Microbiology*, **62**, 2381–2386.

Bogan, B. W., Lamar, R. T., Burgos, W. D. & Tien, M. (1999). Extent of humification of anthracene, fluoranthene, and benzo[*a*]pyrene by *Pleurotus ostreatus* during growth in PAH-contaminated soils. *Letters in Applied Microbiology*, **28**, 250–254.

Böhmer, S., Messner, K. & Srebotnik, E. (1998). Oxidation of phenanthrene by a fungal laccase in the presence of 1-hydroxybenzotriazole and unsaturated lipids. *Biochemical and Biophysical Research Communications*, **244**, 233–238.

Boonchan, S., Britz, M. L. & Stanley, G. A. (2000). Degradation and mineralization of high-molecular-weight polycyclic aromatic hydrocarbons by defined fungal-bacterial cocultures. *Applied and Environmental Microbiology*, **66**, 1007–1019.

Boopathy, R. (2000). Factors limiting bioremediation technologies. *Bioresource Technology*, **74**, 63–67.

Boyle, D., Wiesner, C. & Richardson, A. (1998). Factors affecting the degradation of polyaromatic hydrocarbons in soil by white-rot fungi. *Soil Biology and Biochemistry*, **30**, 873–882.

Braun-Lullemann, A., Hüttermann, A. & Majcherczyk, A. (1999). Screening of ectomycorrhizal fungi for degradation of polycyclic aromatic hydrocarbons. *Applied Microbiology and Biotechnology*, **53**, 127–132.

Brodkorb, T. S. & Legge, R. L. (1992). Enhanced biodegradation of phenanthrene in oil tar-contaminated soils supplemented with *Phanerochaete chrysosporium. Applied and Environmental Microbiology*, **58**, 3117–3121.

Bumpus, J. A. (1989). Biodegradation of polycyclic aromatic hydrocarbons by *Phanerochaete chrysosporium. Applied and Environmental Microbiology*, **55**, 154–158.

Bumpus, J. A., Tien, M., Wright, D. & Aust, S. D. (1985). Oxidation of persistent environmental pollutants by a white rot fungus. *Science*, **228**, 1434–1436.

Casillas, R. P., Crow, S. A. Jr, Heinze, T. M., Deck, J. & Cerniglia, C. E. (1996). Initial oxidative and subsequent conjugative metabolites produced during the metabolism of phenanthrene by fungi. *Journal of Industrial Microbiology*, **16**, 205–215.

Cerniglia, C. E. (1982). Initial reactions in the oxidation of anthracene by *Cunninghamella elegans. Journal of General Microbiology*, **128**, 2055–2061.

Cerniglia, C. E. (1984). Microbial metabolism of polycyclic aromatic hydrocarbons. *Advances in Applied Microbiology*, **30**, 31–71.

Cerniglia, C. E. (1992). Biodegradation of polycyclic aromatic hydrocarbons. *Biodegradation*, **3**, 351–368.

Cerniglia, C. E. (1993). Biodegradation of polycyclic aromatic hydrocarbons. *Current Opinion in Biotechnology*, **4**, 331–338.

Cerniglia, C. E. (1997). Fungal metabolism of polycyclic aromatic hydrocarbons: past, present and future applications in bioremediation. *Journal of Industrial Microbiology and Biotechnology*, **19**, 324–333.

Cerniglia, C. E. & Gibson, D. T. (1979). Oxidation of benzo[*a*]pyrene by the filamentous fungus *Cunninghamella elegans*. *Journal of Biological Chemistry*, **254**, 12174–12180.

Cerniglia, C. E. & Gibson, D. T. (1980a). Fungal oxidation of benzo[*a*]pyrene and (±)-*trans*-7,8-dihydroxy-7,8-dihydrobenzo[*a*]pyrene: Evidence for the formation of a benzo[*a*]pyrene 7,8-diol-9,10-epoxide. *Journal of Biological Chemistry*, **255**, 5159–5163.

Cerniglia, C. E. & Gibson, D. T. (1980b). Fungal oxidation of (±)-9,10-dihydroxy- 9,10-dihydrobenzo[*a*]pyrene: formation of diastereomeric benzo[*a*]pyrene 9,10-diol 7,8-epoxides. *Proceedings of the National Academy of Sciences USA*, **77**, 4554–4558.

Cerniglia, C. E., & Yang, S. K. (1984). Stereoselective metabolism of anthracene and phenanthrene by the fungus *Cunninghamella elegans*. *Applied and Environmental Microbiology*, **47**, 119–124.

Cerniglia, C. E., Dodge, R. H. & Gibson, D. T. (1980a). Studies on the fungal oxidation of polycyclic aromatic hydrocarbons. *Botanica Marina*, **23**, 121–124.

Cerniglia, C. E., Mahaffey, W. & Gibson, D. T. (1980b). Fungal oxidation of benzo[a]pyrene: formation of (−)-*trans*-7,8-dihydroxy-7,8-dihydrobenzo[*a*]pyrene by *Cunninghamella elegans*. *Biochemical and Biophysical Research Communications*, **94**, 226–232.

Cerniglia, C. E., Freeman, J. P. & Mitchum, R. K. (1982). Glucuronide and sulfate conjugation in the fungal metabolism of aromatic hydrocarbons. *Applied and Environmental Microbiology*, **43**, 1070–1075.

Cerniglia, C. E., White, G. L. & Heflich, R. H. (1985). Fungal metabolism and detoxification of polycyclic aromatic hydrocarbons. *Archives of Microbiology*, **143**, 105–110.

Cerniglia, C. E., Kelly, D. W., Freeman, J. P. & Miller, D. W. (1986). Microbial metabolism of pyrene. *Chemico-Biological Interactions*, **57**, 203–216.

Cerniglia, C. E., Campbell, W. L., Freeman, J. P. & Evans, F. E. (1989). Identification of a novel metabolite in phenanthrene metabolism by the fungus *Cunninghamella elegans*. *Applied and Environmental Microbiology*, **55**, 2275–2279.

Cerniglia, C. E., Sutherland, J. B. & Crow, S. A. (1992). Fungal metabolism of aromatic hydrocarbons. In *Microbial Degradation of Natural Products*, ed. G. Winkelmann, pp. 193–217. Weinheim: VCH Press.

Cerniglia, C. E., Gibson, D. T. & Dodge, R. H. (1994). Metabolism of benz[*a*]anthracene by the filamentous fungus *Cunninghamella elegans*. *Applied and Environmental Microbiology*, **60**, 3931–3938.

Cha, D. K., Chiu, P. C., Kim, S. D. & Chang, J. S. (1999). Treatment technologies. *Water Environment Research*, **71**, 870–885.

Chapelle, F. H. (1999). Bioremediation of petroleum hydrocarbon-contaminated ground water: the perspectives of history and hydrology. *Ground Water*, **37**, 122–132.

Chen, W., Bruhlmann, F., Richins, R. D. & Mulchandani, A. (1999). Engineering of improved microbes and enzymes for bioremediation. *Current Opinion in Biotechnology*, **10**, 137–141.

Coates, J. D., Anderson, R. T. & Lovley, D. R. (1996). Oxidation of polycyclic aromatic hydrocarbons under sulfate-reducing conditions. *Applied and Environmental Microbiology*, **62**, 1099–1101.

Collins, P. J. & Dobson, A. D. W. (1996). Oxidation of fluorene and phenanthrene by Mn(II) dependent peroxidase activity in whole cultures of *Trametes* (*Coriolus*) *versicolor*. *Biotechnology Letters*, **18**, 801–804.

Collins, P. J., Kotterman, M. J. J., Field, J. A. & Dobson, A. D. W. (1996). Oxidation of anthracene and benzo[*a*]pyrene by laccases from *Trametes versicolor*. *Applied and Environmental Microbiology*, **62**, 4563–4567.

Colombo, J. C., Cabello, M. & Arambarri, A. M. (1996). Biodegradation of aliphatic and aromatic hydrocarbons by natural soil microflora and pure cultures of imperfect and lignolytic fungi. *Environmental Pollution*, **94**, 355–362.

Cutright, T. J. (1995). Polycyclic aromatic hydrocarbon biodegradation and kinetics using *Cunninghamella echinulata* var. *elegans*. *International Biodeterioration and Biodegradation*, **35**, 397–408.

Cutright, T. J. & Lee, S. (1994). Microorganisms and metabolic pathways for remediation of PAH contaminated soil. *Fresenius Environmental Bulletin*, **3**, 413–421.

Davis, M. W., Glaser, J. A., Evans, J. W. & Lamar, R. T. (1993). Field evaluation of the lignin-degrading fungus *Phanerochaete sordida* to treat creosote-contaminated soil. *Environmental Science and Technology*, **27**, 2572–2576.

Déziel, E., Comeau, Y. & Villemur, R. (1999). Two-liquid-phase bioreactors for enhanced degradation of hydrophobic/toxic compounds. *Biodegradation*, **10**, 219–233.

Dhawale, S. W., Dhawale, S. S., & Dean-Ross, D. (1992). Degradation of phenanthrene by *Phanerochaete chrysosporium* occurs under ligninolytic as well as non-ligninolytic conditions. *Applied and Environmental Microbiology*, **58**, 3000–3006.

Dutta, D., Ghosh, D. K., Mishra, A. K. & Samanta, T. B. (1983). Induction of benzo[*a*]pyrene hydroxylase in *Aspergillus ochraceus* TS: evidence of multiple forms of cytochrome P-450. *Biochemical and Biophysical Research Communications*, **115**, 692–699.

Eggen, T. (1999). Application of fungal substrate from commercial mushroom production – *Pleurotus ostreatus* – for bioremediation of creosote contaminated soil. *International Biodeterioration and Biodegradation*, **44**, 117–126.

Eggen, T. & Majcherczyk, A. (1998). Removal of polycyclic aromatic hydrocarbons (PAH) in contaminated soil by white rot fungus *Pleurotus ostreatus*. *International Biodeterioration and Biodegradation*, **41**, 111–117.

Eggen, T. & Sveum, P. (1999). Decontamination of aged creosote polluted soil: the influence of temperature, white rot fungus *Pleurotus ostreatus*, and pretreatment. *International Biodeterioration and Biodegradation*, **43**, 125–133.

Engler, K. H., Kelley, S. L., Coker, R. D. & Evans, I. H. (2000). Toxin-binding properties of cytochrome P450 in *Saccharomyces cerevisiae* and *Kluyveromyces marxianus*. *Biotechnology Letters*, **22**, 3–8.

Eriksson, M., Dalhammar, D. & Borg-Karlson, A.-K. (2000). Biological

178   *C. E. Cerniglia and J. B. Sutherland*

degradation of selected hydrocarbons in an old PAH/creosote contaminated soil from a gas work site. *Applied Microbiology and Biotechnology*, **53**, 619–626.

Eschenback, A., Kästner, M., Wienberg, R. & Mahro, B. (1995). Microbial PAH degradation in soil material from a contaminated site – mass balance experiments with *Pleurotus ostreatus* and different ^{14}C PAH. In *Contaminated Soil '95*, eds. W. J. van der Bring, R. Bosman, & F. Arendt, pp. 377–378. Dordrecht, the Netherlands: Kluwer Academic.

Field, J. A., de Jong, E., Costa, G. F. & de Bont, J. A. M. (1992). Biodegradation of polycyclic aromatic hydrocarbons by new isolates of white rot fungi. *Applied and Environmental Microbiology*, **58**, 2219–2226.

Field, J. A., de Jong, E., Costa, G. F. & de Bont, J. A. M. (1993). Screening for ligninolytic fungi applicable to the biodegradation of xenobiotics. *Trends in Biotechnology*, **11**, 44–49.

Field, J. A., Boelsma, F., Baten, H. & Rulkens, W. H. (1995). Oxidation of anthracene in water/solvent mixtures by the white-rot fungus, *Bjerkandera* sp. strain BOS55. *Applied Microbiology and Biotechnology*, **44**, 234–240.

Field, J. A., Baten, H., Boelsma, F. & Rulkens, W. H. (1996a). Biological elimination of polycyclic aromatic hydrocarbons in solvent extracts of polluted soil by the white rot fungus, *Bjerkandera* sp. strain BOS55. *Environmental Technology*, **17**, 317–323.

Field, J. A., Vledder, R. H., van Zelst, J. G. & Rulkens, W. H. (1996b). The tolerance of lignin peroxidase and manganese-dependent peroxidase to miscible solvents and the *in vitro* oxidation of anthracene in solvent: water mixtures. *Enzyme and Microbial Technology*, **18**, 300–308.

Finlayson-Pitts, B. J. & Pitts, J. N. (1997). Tropospheric air pollution: ozone, airborne toxics, polycyclic aromatic hydrocarbons, and particles. *Science*, **276**, 1045–1052.

Garbisu, C. & Alkorta, I. (1999). Utilization of genetically engineered microorganisms (GEMs) for bioremediation. *Journal of Chemical Technology and Biotechnology*, **74**, 599–606.

Garon, D., Krivobok, S. & Seigle-Murandi, F. (2000). Fungal degradation of fluorene. *Chemosphere*, **40**, 91–97.

Ghiorse, W. C., Herrick, J. B., Sandoli, R. L. & Madsen, E. L. (1995). Natural selection of PAH-degrading bacterial guilds at coal-tar disposal sites. *Environmental Health Perspectives*, **103** (Suppl. 5), 107–111.

Ghosh, D. K., Dutta, D., Samanta, T. B. & Mishra, A. K. (1983). Microsomal benzo[a]pyrene hydroxylase in *Aspergillus ochraceus* TS: assay and characterization of the enzyme system. *Biochemical and Biophysical Research Communications*, **113**, 497–505.

Gibson, D. T. (1982). Microbial degradation of hydrocarbons. *Toxicological and Environmental Chemistry*, **5**, 237–250.

Gramss, G., Voigt, K.-D. & Kirsche, B. (1999a). Degradation of polycyclic aromatic hydrocarbons with three to seven aromatic rings by higher fungi in sterile and unsterile soils. *Biodegradation*, **10**, 51–62.

Gramss, G., Kirsche, B., Voigt, K.-D., Günther, T. & Fritsche, W. (1999b). Conversion rates of five polycyclic aromatic hydrocarbons in liquid cultures of fifty-eight fungi and the concomitant production of oxidative enzymes. *Mycological Research*, **103**, 1009–1018.

Grotenhuis, T., Field, J., Wasseveld, R. & Rulkens, W. (1998). Biodegradation of polyaromatic hydrocarbons (PAH) in polluted soil by the white rot fungus

Bjerkandera. Journal of Chemical Technology and Biotechnology, **71**, 359–360.

Haemmerli, S. D., Leisola, M. S. A., Sanglard, D. & Fiechter, A. (1986). Oxidation of benzo[*a*]pyrene by extracellular ligninases of *Phanerochaete chrysosporium. Journal of Biological Chemistry*, **261**, 6900–6903.

Hammel, K. E. (1995a). Mechanisms for polycyclic aromatic hydrocarbon degradation by ligninolytic fungi. *Environmental Health Perspectives*, **103**, 41–43.

Hammel, K. E. (1995b). Organopollutant degradation by ligninolytic fungi. In *Microbial Transformation and Degradation of Toxic Chemicals*, eds. L. Y. Young & C. E. Cerniglia, pp. 331–336. New York: Wiley-Liss.

Hammel, K. E., Kalyanaraman, B. & Kirk, T. K. (1986). Oxidation of polycyclic aromatic hydrocarbons and dibenzo[*p*]dioxins by *Phanerochaete chrysosporium* ligninase. *Journal of Biological Chemistry*, **261**, 16948–16952.

Hammel, K. E., Green, B. & Gai, W. Z. (1991). Ring fission of anthracene by a eukaryote. *Proceedings of the National Academy of Science USA*, **88**, 10605–10608.

Hammel, K. E., Gai, W. Z., Green, B. & Moen, M. A. (1992). Oxidative degradation of phenanthrene by the ligninolytic fungus *Phanerochaete chrysosporium. Applied and Environmental Microbiology*, **58**, 1832–1838.

Harmsen, J. & Heersche, J. (1999). Use of residual substrate from mushroom farms to stimulate biodegradation of poorly available PAH. In *The Fifth International In Situ and On-Site Bioremediation Symposium, Phytoremediation and Innovative Strategies for Specialized Remedial Applications*, San Diego, California, 19–22 April, 1999, eds. A. Leason & B. C. Allman, pp. 87–92. Columbus, OH: Battelle Press.

Harvey, R. G. (1992). *Polycyclic Aromatic Hydrocarbons: Chemistry and Carcinogenicity.* Cambridge, UK: Cambridge University Press.

Harvey, R. G. (1996). Mechanisms of carcinogenesis of polycyclic aromatic hydrocarbons. *Polycyclic Aromatic Compounds*, **9**, 1–23.

Harvey, R. G., Penning, T. M., Jarabak, J. & Zhang, F. J. (1999). Role of quinone metabolites in PAH carcinogenesis. *Polycyclic Aromatic Compounds*, **16**,13–20.

Hatakka, A. (1990). Lignin-modifying enzymes from selected white-rot fungi: production and role in lignin degradation. *FEMS Microbiology Reviews*, **13**, 125–135.

Haught, R. C., Neogy, R., Vonderhaar, S. S., Krishnan, E. R., Safferman, S. I. & Ryan, J. (1995). Land treatment alternatives for bioremediating wood preserving wastes. *Hazardous Waste and Hazardous Materials*, **12**, 329–344.

Heitkamp, M. A., Freeman, J. P., Miller, D. W. & Cerniglia, C. E. (1988). Pyrene degradation by a *Mycobacterium* sp.: identification of ring oxidation and ring fission products. *Applied and Environmental Microbiology*, **54**, 2556–2565.

Hüttermann, A., Loske, D., Majcherczyk, A., Zadrazil, F., Waldinger, P. & Lorson, H. (1989). Reclamation of PAH-contaminated soils with active fungus-straw-substrata. In *Recycling International 3*, ed. K. J. Thome-Kozmiensky, pp. 2192–2199. Berlin: EF-Verlag.

in der Wiesche, C., Martens, R. & Zadrazil, F. (1996). Two-step degradation of pyrene by white-rot fungi and soil microorganisms. *Applied Microbiology and Biotechnology*, **46**, 653–659.

Johannes, C. & Majcherczyk, A. (2000). Natural mediators in the oxidation of polycyclic aromatic hydrocarbons by laccase mediator systems. *Applied and Environmental Microbiology*, **66**, 524–528.

Johannes, C., Majcherczyk, A. & Hüttermann, A. (1996). Degradation of anthracene by laccase of *Trametes versicolor* in the presence of different mediator compounds. *Applied Microbiology and Biotechnology*, **46**, 313–317.

Johannes, C., Majcherczyk, A. & Hüttermann, A. (1998). Oxidation of acenaphthene and acenaphthylene by laccase of *Trametes versicolor* in a laccase-mediator system. *Journal of Biotechnology*, **61**, 151–156.

Kanaly, R. A. & Harayama, S. (2000). Biodegradation of high-molecular-weight polycyclic aromatic hydrocarbons by bacteria. *Journal of Bacteriology*, **182**, 2059–2067.

Kästner, M. & Mahro, B. (1996). Microbial degradation of polycyclic aromatic hydrocarbons in soils affected by the organic matrix of compost. *Applied Microbiology and Biotechnology*, **44**, 668–675.

Keith, L. H. & Telliard, W. A. (1979). Priority pollutants I – a perspective view. *Environmental Science and Technology*, **13**, 416–423.

Kennes, C. & Lema, J. M. (1994). Degradation of major compounds of creosotes (PAH and phenols) by *Phanerochaete chrysosporium*. *Biotechnology Letters*, **16**, 759–764.

Kiehlmann, E., Pinto, L. & Moore, M. (1996). The transformation of chrysene to *trans*-1,2-dihydrochrysene by filamentous fungi. *Canadian Journal of Microbiology*, **42**, 604–608.

Korda, A., Santas, P., Tenente, A. & Santas, R. (1997). Petroleum hydrocarbon bioremediation: sampling and analytical techniques, in situ treatments and commercial microorganisms currently used. *Applied Microbiology and Biotechnology*, **48**, 677–686.

Kotterman, M. J. J. (1999). Development of white rot fungal technology for PAH degradation. In *The Fifth International In Situ and On-Site Bioremediation Symposium, Phytoremediation and Innovative Strategies for Specialized Remedial Applications*, San Diego, California, 19–22 April, 1999, eds. A. Leason & B. C. Allman, pp. 69–74. Columbus, OH: Battelle Press.

Kotterman, M. J. J., Heessels, E., de Jong, E. & Field, J. A. (1994). The physiology of anthracene biodegradation by the white rot fungus *Bjerkandera* sp. strain BOS55. *Applied Microbiology and Biotechnology*, **42**, 179–186.

Kotterman M. J. J., Wasseveld, R. A. & Field, J. A. (1996). Hydrogen peroxide production as a limiting factor in xenobiotic compound oxidation by nitrogen-sufficient cultures of *Bjerkandera* sp. strain BOS55 overproducing peroxidases. *Applied and Environmental Microbiology*, **62**, 880–885.

Kotterman, M. J. J., Rietberg, H.-J., Hage, A. & Field, J. A. (1998a). Polycyclic aromatic hydrocarbon oxidation by the white-rot fungus *Bjerkandera* sp. strain BOS55 in the presence of nonionic surfactants. *Biotechnology and Bioengineering*, **57**, 220–227.

Kotterman, M. J. J., Vis, E. H. & Field, J. A. (1998b). Successive mineralization and detoxification of benzo[*a*]pyrene by the white rot fungus *Bjerkandera* sp. strain BOS55 and indigenous microflora. *Applied and Environmental Microbiology*, **64**, 2853–2858.

Krivobok, S., Miriouchkine, E., Seigle-Murandi, F. & Benoit-Guyod, J.-L. (1998). Biodegradation of anthracene by soil fungi. *Chemosphere*, **37**, 523–530.

Lambert, M., Kremer, S., Sterner, O. & Anke, H. (1994). Metabolism of pyrene by the basidiomycete *Crinipellis stipitaria* and identification of pyrenequinones and their hydroxylated precursors in strain JK375. *Applied and Environmental Microbiology*, **60**, 3597–3601.

Lang, E., Nerud, F., Novotna, E., Zadrazil, F. & Martens, R. (1996). Production of ligninolytic exoenzymes and ^{14}C-pyrene mineralization by *Pleurotus* sp. in lignocellulose substrate. *Folia Microbiologica*, **41**, 489–493.

Lange, B., Kremer, S., Sterner, O. & Anke, H. (1994). Pyrene metabolism in *Crinipellis stipitaria*: Identification of *trans*-4,5-dihydro-4,5-dihydroxypyrene and 1-pyrenylsulfate in strain JK364. *Applied and Environmental Microbiology*, **60**, 3602–3606.

Lange, B., Kremer, S., Sterner, O. & Anke, H. (1995). Induction of secondary metabolism by environmental pollutants: metabolization of pyrene and formation of 6,8-dihydroxy-3-methylisocoumarin by *Crinipellis stipitaria* JK 364. *Zeitschrift für Naturforschung*, **50**, 806–812.

Lange, B., Kremer, S., Sterner, O. & Anke, H. (1996). Metabolism of pyrene by basidiomycetous fungi of the genera *Crinipellis, Marasmius*, and *Marasmiellus*. *Canadian Journal of Microbiology*, **42**, 1179–1183.

Launen, L. A., Pinto, L. J., Wiebe, C., Kiehlmann, E. & Moore, M. M. (1995). The oxidation of pyrene and benzo[*a*]pyrene by nonbasidiomycete soil fungi. *Canadian Journal of Microbiology*, **41**, 477–488.

Launen, L. A., Pinto, L. J. & Moore, M. M. (1999). Optimization of pyrene oxidation by *Penicillium janthinellum* using response-surface methodology. *Applied Microbiology and Biotechnology*, **51**, 510–515.

Lee, K. & de Mora, S. (1999). In situ bioremediation strategies for oiled shoreline environments. *Environmental Technology*, **20**, 783–794.

Leyval, C. & Binet P. (1998). Effect of polyaromatic hydrocarbons in soil on arbuscular mycorrhizal plants. *Journal of Environmental Quality*, **27**, 402–407.

Liao, W. L. & Tseng, D. H. (1996). Biotreatment of naphthalene by PAH-acclimated pure culture with white-rot fungus *Phanerochaete chrysosporium*. *Water Science and Technology*, **34**, 73–79.

Liao, W. L., Tseng, D. H., Tsai, Y. C. & Chang, S. C. (1997). Microbial removal of polycyclic aromatic hydrocarbons by *Phanerochaete chrysosporium*. *Water Science and Technology*, **35**, 255–264.

Lijinsky, W. (1991). The formation and occurrence of polynuclear aromatic hydrocarbons associated with food. *Mutation Research*, **259**, 251–261.

Lisowska, K. & Dlugonski, J. (1999). Removal of anthracene and phenanthrene by filamentous fungi capable of cortexolone 11-hydroxylation. *Journal of Basic Microbiology*, **39**, 117–125.

Löser, C., Ulbricht, H., Hoffmann, P. & Seidel, H. (1999). Composting of wood containing polycyclic aromatic hydrocarbons (PAHs). *Compost Science and Utilization*, **7**, 16–32.

Macek, T., Macková, M. & Kás, J. (2000). Exploitation of plants for the removal of organics in environmental remediation. *Biotechnology Advances*, **18**, 23–24.

MacGillivray A. R. & Shiaris, M. P. (1993). Biotransformation of polycyclic aromatic hydrocarbons by yeasts isolated from coastal sediments. *Applied and Environmental Microbiology*, **59**, 1613–1618.

MacGillivray, A. R. & Shiaris, M. P. (1994). Relative role of eukaryotic and prokaryotic microorganisms in phenanthrene transformation in coastal sediments. *Applied and Environmental Microbiology*, **60**, 1154–1159.

Mackay, D., Shiu, W. Y. and Ma, K. C. (1992). *Illustrated Handbook of Physical-Chemical Properties and Environmental Fate of Organic Chemicals.* Boca Raton, FL: Lewis.

Majcherczyk, A. & Johannes, C. (2000). Radical mediated indirect oxidation of a PEG-coupled polycyclic aromatic hydrocarbon (PAH) model compound by fungal laccase. *Biochimica et Biophysica Acta*, **1471**, 157–162.

Majcherczyk, A., Johannes, C. & Hüttermann, A. (1998). Oxidation of polycyclic aromatic hydrocarbons (PAH) by laccase of *Trametes versicolor*. *Enzyme and Microbial Technology*, **22**, 335–341.

Maldonado, C., Bayona, J. M. & Bodineau, L. (1999). Sources, distribution, and water column processes of aliphatic and polycyclic aromatic hydrocarbons in the northwestern Black Sea water. *Environmental Science and Technology*, **33**, 2693–2702.

Margesin, R. & Schinner, F. (1999). Biological decontamination of oil spills in cold environments. *Journal of Chemical Technology and Biotechnology*, **74**, 381–389.

Márquez-Rocha, F. J., Hernández-Rodríguez, V. Z. & Vázquez-Duhalt, R. (2000). Biodegradation of soil-adsorbed polycyclic aromatic hydrocarbons by the white rot fungus *Pleurotus ostreatus*. *Biotechnology Letters*, **22**, 469–472.

Martens, R. & Zadrazil, F. (1998). Screening of white-rot fungi for their ability to mineralize polycyclic aromatic hydrocarbons in soil. *Folia Microbiologica*, **43**:97–103.

Martens, R., Wolter, M., Bahadir, M. & Zadrazil, F. (1999). Mineralization of ^{14}C-labelled highly-condensed polycyclic aromatic hydrocarbons in soils by *Pleurotus* sp. Florida. *Soil Biology and Biochemistry*, **31**, 1893–1899.

Masaphy, S., Levanon, D., Henis, Y., Venkateswarlu, K. & Kelly, S. L. (1995). Microsomal and cytosolic cytochrome P450 mediated benzo[*a*]pyrene hydroxylation in *Pleurotus pulmonarius*. *Biotechnology Letters*, **17**, 969–974.

Masaphy, S., Levanon, D., Henis, Y., Venkateswarlu, K. & Kelly, S. L. (1996). Evidence for cytochrome P-450 and P-450-mediated benzo(a)pyrene hydroxylation in the white rot fungus *Phanerochaete chrysosporium*. *FEMS Microbiology Letters*, **135**, 51–55.

Masaphy, S., Krinfeld, B. & Levanon, D. (1998). Induction of linoleic acid supported benzo[*a*]pyrene hydroxylase activity by manganese in the white rot fungus *Pleurotus pulmonarius*. *Chemosphere*, **36**, 2933–2940.

Masaphy, S., Lamb, D. C. & Kelly, S. L. (1999). Purification and characterization of a benzo[*a*]pyrene hydroxylase from *Pleurotus pulmonarius*. *Biochemical and Biophysical Research Communications*, **266**, 326–329.

Mastrangelo, G., Fadda, E. & Marzia, V. (1996). Polycyclic aromatic hydrocarbons and cancer in man. *Environmental Health Perspectives*, **104**, 1166–1170.

May, R., Schröder, P. & Sandermann, H. Jr (1997). Ex-situ process for treating PAH-contaminated soil with *Phanerochaete chrysosporium*. *Environmental Science and Technology*, **31**, 2626–2633.

McFarland, M. J. & Qiu, X. J. (1995). Removal of benzo[*a*]pyrene in soil composting systems amended with the white rot fungus *Phanerochaete chrysosporium*. *Journal of Hazardous Materials*, **42**, 61–70.

McFarland, M. J., Qiu, X. J., Sims, J. L., Randolph, M. E. & Sims, R. C. (1992). Remediation of petroleum impacted soils in fungal compost bioreactors. *Water Science and Technology*, **25**, 197–206.

Meckenstock, R. U., Annweiler, E., Michaelis, W., Richnow, H. H. & Schink, B. (2000). Anaerobic naphthalene degradation by a sulfate-reducing enrichment culture. *Applied and Environmental Microbiology*, **66**, 2743–2747.

Meyer, S., Cartellieri, S. & Steinhart, H. (1999). Simultaneous determination of PAHs, hetero-PAHs (N, S, O), and their degradation products in creosote-contaminated soils. Method development, validation, and application to hazardous waste sites. *Analytical Chemistry*, **71**, 4023–4029.

Milhelcic, J. R. & Luthy, R. G. (1988). Degradation of polycyclic aromatic hydrocarbon compounds under various redox conditions in soil-water systems. *Applied and Environmental Microbiology*, **54**, 1182–1187.

Miller, E. C. & Miller, J. A. (1985). Some historical perspectives on the metabolism of xenobiotic chemicals to reactive electrophiles. In *Bioactivation of Foreign Compounds*, ed. M. W. Anders, pp. 3–28. Orlando, FL: Academic Press.

Mitra, S., Dickhut, R. M., Kuehl, S. A. & Kimbrough, K. L. (1999). Polycyclic aromatic hydrocarbon (PAH) source, sediment deposition patterns, and particle geochemistry as factors influencing PAH distribution coefficients in sediments of the Elizabeth River, VA, USA. *Marine Chemistry*, **66**, 113–127.

Moen, M. A. & Hammel, K. E. (1994). Lipid peroxidation by the manganese peroxidase of *Phanerochaete chrysosporium* is the basis for phenanthrene oxidation by the intact fungus. *Applied and Environmental Microbiology*, **60**, 1956–1961.

Morgan, P., Lewis, S. T. & Watkinson, R. J. (1991). Comparison of abilities of white-rot fungi to mineralize selected xenobiotic compounds. *Applied Microbiology and Biotechnology*, **34**, 693–696.

Morgan, P., Lee, S. A., Lewis, S. T., Sheppard, A. N. & Watkinson, R. J. (1993). Growth and biodegradation by white rot fungi inoculated into soil. *Soil Biology and Biochemistry*, **25**, 279–287.

Mueller, J. G., Lantz, S. E., Ross, D., Colvin, R. J., Middaugh, D. P. & Pritchard, P. H. (1993). Strategy using bioreactors and specially selected microorganisms for bioremediation of groundwater contaminated with creosote and pentachlorophenol. *Environmental Science and Technology*, **27**: 691–698.

Mueller, J. G., Cerniglia, C. E. & Pritchard, P. H. (1996). Bioremediation of environments contaminated by polycyclic aromatic hydrocarbons. In *Bioremediation: Principles and Applications*, eds. R. L. Crawford & D. L. Crawford, pp. 125–194. Cambridge, UK: Cambridge University Press.

Müncnerová, D. & Augustin, J. (1994). Fungal metabolism and detoxification of polycyclic aromatic hydrocarbons: a review. *Bioresource Technology*, **48**, 97–106.

Newell, S. Y., Wall, V. D. & Maruya, K. A. (2000). Fungal biomass in saltmarsh grass blades at two contaminated sites. *Archives of Environmental Contamination and Toxicology*, **38**, 268–273.

Ngabe, B., Bidleman, T. F. & Scott, G. I. (2000). Polycyclic aromatic hydrocarbons in storm runoff from urban and coastal South Carolina. *Science of the Total Environment*, **255**, 1–9.

Novotny, C., Erbanova, P., Sasek, V., Kubatova, A., Cajthaml, T., Lang, E., Krahl, J. & Zadrazil, F. (1999). Extracellular oxidative enzyme production and PAH removal in soil by exploratory mycelium of white rot fungi. *Biodegradation*, **10**, 159–168.

Pickard, M. A., Roman, R., Tinoco, R. & Vázquez-Duhalt, R. (1999). Polycyclic aromatic hydrocarbon metabolism by white rot fungi and oxidation by *Coriolopsis gallica* UAMH 8260 laccase. *Applied and Environmental Microbiology*, **65**, 3805–3809.

Pinto, L. J. & Moore, M. M. (2000). Release of polycyclic aromatic hydrocarbons from contaminated soils by surfactant and remediation of this effluent by *Penicillium* spp. *Environmental Toxicology and Chemistry*, **19**, 1741–1748.

Pothuluri, J. V. & Cerniglia, C. E. (1994). Microbial metabolism of polycyclic aromatic hydrocarbons. In *Biological Degradation and Bioremediation of Toxic Chemicals*, ed. G. R. Chaudhry, pp. 92–124. Portland, OR: Dioscorides Press.

Pothuluri, J. V., Freeman, J. P., Evans, F. E. & Cerniglia, C. E. (1990). Fungal transformation of fluoranthene. *Applied and Environmental Microbiology*, **56**, 2974–2983.

Pothuluri, J. V., Freeman, J. P., Evans, F. E. & Cerniglia, C. E. (1992a). Fungal metabolism of acenaphthene by *Cunninghamella elegans*. *Applied and Environmental Microbiology*, **58**, 3654–3659.

Pothuluri, J. V., Heflich, R. H., Fu, P. P. & Cerniglia, C. E. (1992b). Fungal metabolism and detoxification of fluoranthene. *Applied and Environmental Microbiology*, **58**, 937–941.

Pothuluri, J. V., Freeman, J. P., Evans, F. E. & Cerniglia, C. E. (1993). Biotransformation of fluorene by the fungus *Cunninghamella elegans*. *Applied and Environmental Microbiology*, **59**, 1977–1980.

Pothuluri, J. V., Selby, A., Evans, F. E., Freeman, J. P. & Cerniglia, C. E. (1995). Transformation of chrysene and other polycyclic aromatic hydrocarbon mixtures by the fungus *Cunninghamella elegans*. *Canadian Journal of Botany*, **73**, 1025–1033.

Pothuluri, J. V., Evans, F. E., Heinze, T. M. & Cerniglia, C. E. (1996). Formation of sulfate and glucoside conjugates of benzo[*e*]pyrene by *Cunninghamella elegans*. *Applied Microbiology and Biotechnology*, **45**, 677–683.

Qiu, X. J. & McFarland, M. J. (1991). Bound residue formation in PAH contaminated soil composting using *Phanerochaete chrysosporium*. *Hazardous Waste and Hazardous Materials*, **8**, 115–126.

Rama, R., Mougin, C., Boyer, F.-D., Kollmann, A., Malosse, C. & Sigoillot, J.-C. (1998). Biotransformation of benzo[*a*]pyrene in bench scale reactor using laccase of *Pycnoporus cinnabarinus*. *Biotechnology Letters*, **20**, 1101–1104.

Ravelet, C., Krivobok, S., Sage, L. & Steiman, R. (2000). Biodegradation of pyrene by sediment fungi. *Chemosphere*, **40**, 557–563.

Richnow, H. H., Annweiler, E., Koning, M., Lüth, J.-C., Stegmann, R., Garms, C., Francke, W. & Michaelis, W. (2000). Tracing the transformation of labelled [1-^{13}C]phenanthrene in a soil bioreactor. *Environmental Pollution*, **108**, 91–101.

Rockne, K. J., Chee-Sanford, J. C., Sanford, R. A., Hedlund, B. P., Staley, J. T. & Strand, S. E. (2000). Anaerobic naphthalene degradation by microbial pure cultures under nitrate-reducing conditions. *Applied and Environmental Microbiology*, **66**, 1595–1601.

Romero, M. C., Cazau, M. C., Giorgieri, S. & Arambarri, A. M. (1998). Phenanthrene degradation by microorganisms isolated from a contaminated stream. *Environmental Pollution*, **101**, 355–359.

Rudd, L. E., Perry J. J., Houk, V. S., Williams, R. W. & Claxton, L. D. (1996). Changes in mutagenicity during crude oil degradation by fungi.

Biodegradation, **7**, 335–343.

Sack, U. & Fritsche, W. (1997). Enhancement of pyrene mineralization in soil by wood-decaying fungi. *FEMS Microbiology Ecology*, **22**, 77–83.

Sack, U. & Günther, T. (1993). Metabolism of PAH by fungi and correlation with extracellular enzymatic activities. *Journal of Basic Microbiology*, **33**, 269–277.

Sack, U., Heinze, T. M., Deck, J., Cerniglia, C. E., Cazau, M. C. & Fritsche, W. (1997a). Novel metabolites in phenanthrene and pyrene transformation by *Aspergillus niger*. *Applied and Environmental Microbiology*, **63**, 2906–2909.

Sack, U., Heinze, T. M., Deck, J., Cerniglia, C. E., Martens, R., Zadrazil, F. & Fritsche, W. (1997b). Comparison of phenanthrene and pyrene degradation by different wood-decaying fungi. *Applied and Environmental Microbiology*, **63**, 3919–3925.

Sack, U., Hofrichter, M. & Fritsche, W. (1997c). Degradation of polycyclic aromatic hydrocarbons by manganese peroxidase of *Nematoloma frowardii*. *FEMS Microbiology Letters*, **152**, 227–234.

Sack, U., Hofrichter, M. & Fritsche, W. (1997d). Degradation of phenanthrene and pyrene by *Nematoloma frowardii*. *Journal of Basic Microbiology*, **37**, 287–293.

Salicis, F., Krivobok, S., Jack, M. & Benoit-Guyod, J. L. (1999). Biodegradation of fluoranthene by soil fungi. *Chemosphere*, **38**, 3031–3039.

Salt, D. E., Smith, R. D. & Raskin, I. (1998). Phytoremediation. *Annual Review of Plant Physiology and Molecular Biology*, **49**, 643–668.

Sanglard, D., Leisola, M. S. A. & Fiechter, A. (1986). Role of extracellular ligninase in biodegradation of benzo[a]pyrene by *Phanerochaete chrysosporium*. *Enzyme and Microbial Technology*, **8**, 209–212.

Santodonato, J. (1997). Review of the estrogenic and antiestrogenic activity of polycyclic aromatic hydrocarbons: relationship to carcinogenicity. *Chemosphere*, **34**, 835–848.

Schnoor, J. L., Licht, L. A., McCutcheon, S. C., Wolfe, N. L. & Carreira, L. H. (1995). Phytoremediation of organic and nutrient contaminants. *Environmental Science and Technology*, **29**, 318A–323A.

Schutzendubel, A., Majcherczyk, A., Johannes, C. & Hüttermann, A. (1999). Degradation of fluorene, anthracene, phenanthrene, fluoranthene, and pyrene lacks connection to the production of extracellular enzymes by *Pleurotus ostreatus* and *Bjerkandera adusta*. *International Biodeterioration and Biodegradation*, **43**, 93–100.

Shah, M. M., Barr, D. P., Chung, N. & Aust, S. D. (1992). Use of white rot fungi in the degradation of environmental chemicals. *Toxicology Letters*, **64–65**, 493–501.

Siegmann, K., Scherrer, L. & Siegmann, H. C. (1999). Physical and chemical properties of airborne nanoscale particles and how to measure the impact on human health. *Journal of Molecular Structure (Theochem)*, **458**, 191–201.

Song, H. G. (1999). Comparison of pyrene biodegradation by white rot fungi. *World Journal of Microbiology and Biotechnology*, **15**, 669–672.

Straube, W. L., Jones-Meehan, J., Pritchard, P. H. & Jones, W. R. (1999). Bench-scale optimization of bioaugmentation strategies for treatment of soils contaminated with high molecular weight polyaromatic hydrocarbons. *Resources Conservation and Recycling*, **27**, 27–37.

Sutherland, J. B. (1992). Detoxification of polycyclic aromatic hydrocarbons by fungi. *Journal of Industrial Microbiology*, **9**, 53–62.

Sutherland, J. B., Selby, A. L., Freeman, J. P., Evans, F. E. & Cerniglia, C. E. (1991). Metabolism of phenanthrene by *Phanerochaete chrysosporium*. *Applied and Environmental Microbiology*, **57**, 3310–3316.

Sutherland, J. B., Selby, A. L., Freeman, J. P., Fu, P. P., Miller, D. W. & Cerniglia, C. E. (1992). Formation of xyloside conjugates from anthracene *trans*-1,2-dihydrodiol and 1-anthrol by *Rhizoctonia solani*. *Mycological Research*, **96**, 509–517.

Sutherland, J. B., Fu, P. P, Yang, S. K., von Tungeln, L. S. & Cerniglia, C. E. (1993). Enantiomeric composition of the *trans*-dihydrodiols produced from phenanthrene by *Phanerochaete chrysosporium* and *Cunninghamella elegans*. *Applied and Environmental Microbiology*, **59**, 2145–2149.

Sutherland, J. B., Rafii, F., Khan, A. A. & Cerniglia, C. E. (1995). Mechanisms of polycyclic aromatic hydrocarbon degradation. In *Microbial Transformation and Degradation of Toxic Organic Chemicals*, eds. L. Y. Young & C. E. Cerniglia, pp. 269–300. New York: Wiley-Liss.

Tatarko, M. & Bumpus, J. A. (1993). Biodegradation of phenanthrene by *Phanerochaete chrysosporium*: on the role of lignin peroxidase. *Letters in Applied Microbiology*, **17**, 20–24.

Tuvikene, A. (1995). Responses of fish to polycyclic aromatic hydrocarbons (PAHs). *Annales Zoologici Fennici*, **32**, 295–309.

Volkering, F., Breure, A. M. & Rulkens, W. H. (1998). Microbiological aspects of surfactant use for biological soil remediation. *Biodegradation*, **8**, 401–417.

Vyas, B. R. M., Bakowski, S., Sasek, V. & Matucha, M. (1994). Degradation of anthracene by selected white rot fungi. *FEMS Microbiology Ecology*, **14**, 65–70.

Wang, R.-F., Cao, W.-W., Khan, A. A. & Cerniglia, C. E. (2000). Cloning, sequencing, and expression in *Escherichia coli* of a cytochrome P450 gene from *Cunninghamella elegans*. *FEMS Microbiology Letters*, **188**, 55–61.

Wilcke, W. (2000). Polycyclic aromatic hydrocarbons (PAHs) in soil – a review. *Journal of Plant Nutrition and Soil Science*, **163**, 229–248.

Wilson, S. C. & Jones, K. C. (1993). Bioremediation of soil contaminated with polynuclear aromatic hydrocarbons (PAHs): a review. *Environmental Pollution*, **81**, 229–249.

Wischmann, H. & Steinhart, H. (1997). The formation of PAH oxidation products in soils and soil/compost mixtures. *Chemosphere*, **35**, 1681–1698.

Wiseman, A. & Woods, L. F. J. (1979). Benzo[*a*]pyrene metabolites formed by the action of yeast cytochrome P-450/P-448. *Journal of Chemical Technology and Biotechnology*, **29**, 320–324.

Wolter, W., Zadrazil, F., Martens, R. & Bahadir, M. (1997). Degradation of eight highly condensed polycyclic aromatic hydrocarbons by *Pleurotus* sp. Florida in solid wheat straw substrate. *Applied Microbiology and Biotechnology*, **48**, 398–404.

Woods, L. F. J. & Wiseman, A. (1979). Metabolism of benzo[*a*]pyrene by the cytochrome P-450/P-448 of *Saccharomyces cerevisiae*. *Biochemical Society Transactions*, **7**, 124–127.

Wunch, K. G., Feibelman, T. & Bennett, J. W. (1997). Screening for fungi capable of removing benzo[*a*]pyrene in culture. *Applied Microbiology and Biotechnology*, **47**, 620–624.

Wunch, K. G., Alworth, W. L. & Bennett, J. W. (1999). Mineralization of benzo[*a*]pyrene by *Marasmiellus troyanus*, a mushroom isolated from a toxic waste site. *Microbiology Research*, **154**, 75–79.

Wunder, T., Kremer, S., Sterner, O. & Anke, H. (1994). Metabolism of the polycyclic aromatic hydrocarbon pyrene by *Aspergillus niger* SK 9317. *Applied Microbiology and Biotechnology*, **42**, 636–641.

Wunder, T., Marr, J., Kremer, S., Sterner, O. & Anke, H. (1997). 1-Methoxypyrene and 1,6-dimethoxypyrene: two novel metabolites in fungal metabolism of polycyclic aromatic hydrocarbons. *Archives of Microbiology*, **167**, 310–316.

Yadav, J. S. & Loper, J. C. (2000). Cloning and characterization of the cytochrome P450 oxidoreductase gene from the zygomycete fungus *Cunninghamella*. *Biochemical and Biophysical Research Communications*, **268**, 345–353.

Yateem, A., Balba, M. T., Al-Awadhi, N. & El-Nawawy, A. S. (1998). White rot fungi and their role in remediating oil-contaminated soil. *Environment International*, **24**, 181–187.

Ye, D., Siddiqi, M. A., MacCubbin, A. E., Kumar, S. & Sikka, H. C. (1996). Degradation of polynuclear aromatic hydrocarbons by *Sphingomonas paucimobilis*. *Environmental Science and Technology*, **30**, 136–142.

Yogambal, R. K. & Karegoudar, T. B. (1997). Metabolism of polycyclic aromatic hydrocarbons by *Aspergillus niger*. *Indian Journal of Experimental Biology*, **35**, 1021–1023.

Zhang, W.-X., Bouwer, E. J. & Ball, W. P. (1998). Bioavailability of hydrophobic organic contaminants: effects and implications of sorption-related mass transfer on bioremediation. *Ground Water Monitoring and Remediation*, **18**, 126–138.

Zhang, W. & Young, L. Y. (1997). Carboxylation as an initial reaction in the anaerobic metabolism of naphthalene and phenanthrene by sulfidogenic consortia. *Applied and Environmental Microbiology*, **63**, 4759–4764.

8

Pesticide degradation

SARAH E. MALONEY

Introduction

Although responsible for saving and improving the quality of human life, pesticides have exerted a significant detrimental effect on the environment and have caused serious health problems, resulting in severe criticism of their use (Hayes, 1986). There is often a fundamental conflict between the need for a sustained level of biological activity of a pesticide in the environment and the requirement that the chemical should be degraded to non-toxic and ecologically safe products (Hill, 1978; Casida & Quistad, 1998). The era of modern synthetic pesticides largely dates from 1939 when the insecticidal properties of 1,1,1-trichloro-2,2-bis(p-chlorophenyl)ethane (DDT) were discovered (Tessier, 1982). Unlike naturally occurring organic compounds, which are readily degraded upon introduction into the environment, some pesticides such as DDT are extremely resistant to biodegradation by native microflora (Rochkind-Dubinsky, Sayler & Blackburn, 1987a). In most cases, the persistence can be explained by the chemical structure and by the degree of water solubility. In addition, some of these pesticides tend to accumulate in organisms at different trophic levels of the food chain. Chlorinated organic pesticides are one of the major groups of toxic chemicals responsible for environmental contamination and an important potential risk to human health (Kullman & Matsumura, 1996).

The most common pesticides are herbicides, insecticides and fungicides, where herbicides account for nearly 50% of all the pesticides used in developed countries and insecticides account for 75% of all pesticides used in developing countries. There are currently over 2500 pesticides in use (Anon, 1998). The trend of insecticide evolution since the 1960s has been towards increased specificity and less persistence in order to minimize adverse effects on non-target species (Casida & Quistad, 1998). However, even the synthetic pyrethroid insecticides, which were developed to replace many of the more environmentally persistent and highly toxic organochlorine, organophosphorus and methylcarbamate insecticides, have under certain conditions caused problems in the environment as a result of their adverse effects on non-target species/organisms (Woodhead, 1983; Solomon, 1986; Zabel, Seager & Oakley, 1988). The extensive use of herbicides such as the chlorophenoxyalkanoates and triazines, and other toxic chlorinated hydrocarbons used in pesticides such as pentachlorophenol (PCP) and the polychlorinated biphenyls (PCBs) has also led to the contamination of many terrestrial and aquatic ecosystems (Gilbertson, 1989; Paszczynski & Crawford, 1995; McAllister, Lee & Trevors, 1996).

Much attention has been directed towards the use of microorganisms for bioremediation of organopollutants, including pesticides (Bollag, 1974; Fewson, 1988; Shelton *et al.*, 1996). Microbially mediated decomposition, often through co-metabolism, is the major, and sometimes the only, mechanism for the permanent removal or modification of pesticides in soils (Horvarth, 1972). Most work has been carried out on insecticide decomposition, with less attention on herbicides and even less for fungicides. In contrast to fungi, bacteria have been extensively studied and exploited for use in the degradation of pesticides (see reviews by Wallnofer & Engelhardt, 1973; Cork & Krueger, 1991; Moorman, 1994). This is primarily because of their ease of culture, more rapid growth rates and amenity to genetic manipulation (Kumar, Mukerjii & Lal, 1996). Increasing numbers of studies, however, clearly demonstrate that fungi, and in particular white rot fungi, are able to degrade a large number of pollutants, including pesticides (Aust, 1990, 1993; Kirk, Lamar & Glaser, 1992; Barr & Aust, 1994a; Paszczynski & Crawford, 1995). The enormous structural diversity of the pollutants degraded by these fungi (Fig. 8.1), has increased the interest in their use for bioremediation (Barr & Aust, 1994b; Arisoy, 1998). The principal biochemical reactions in the primarily co-metabolic degradation of pesticides by fungi include oxidation, reduction, hydroxylation, aromatic ring cleavage, hydrolysis, dehalogenation, methylation and demethylation, dehydrogenation, ether cleavage, condensation and conjugate formation.

White rot fungi are characterized by their ability to degrade lignin in wood, a structurally complex, naturally occurring and environmentally persistent heteropolymer, the most abundant renewable organic material on earth (Kirk *et al.*, 1992). Lignin contains numerous substructures that are also found in common pollutants, for example phenolics and biphenyls (Bumpus *et al.*, 1985; Lamar, Larsen & Kirk, 1990; Kirk *et al.*, 1992). This unique ability to degrade lignin and a large range of pollutants depends upon the production and secretion of a group of highly potent, non-specific, non-stereoselective, extracellular enzymes, which form part of the

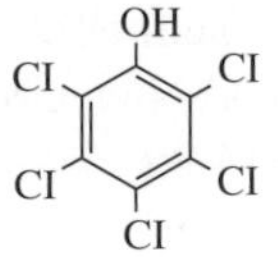

Pentachlorophenol 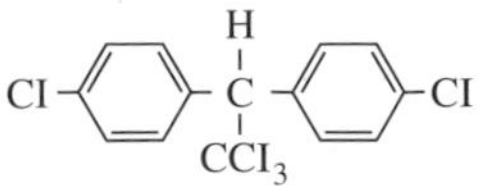   1,1,1-Trichloro-2,2-bis(4-chlorophenyl)ethane (DDT)

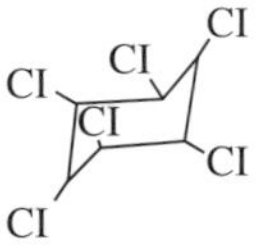

Lindane
(γ-isomer of 1,2,3,4,5,6-hexachlorocyclohexane)   Parathion

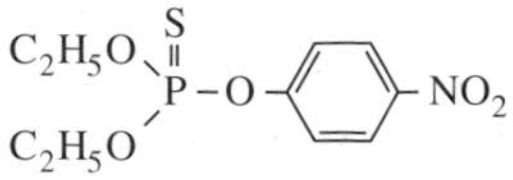

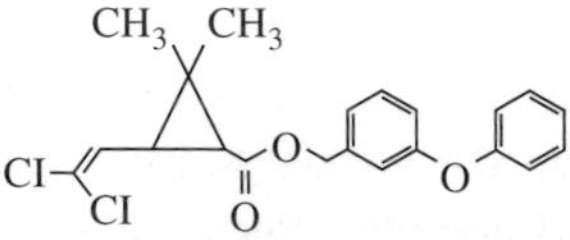

Permethrin 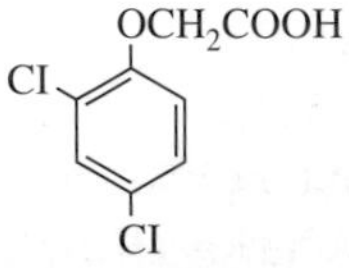   2,4-Dichlorophenoxyacetate (2,4-D)

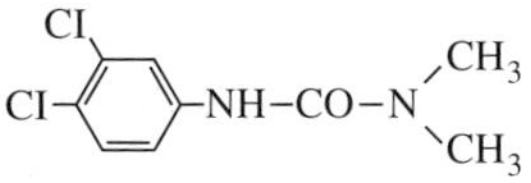

Diuron 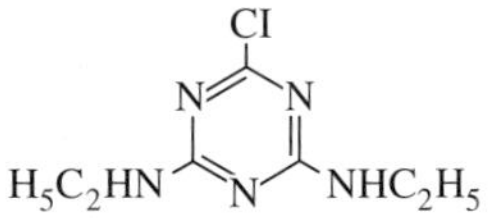   Simazine

Fig. 8.1. Examples of pesticides degraded by fungi. (Adapted from Maloney, 1991.)

lignin-degradation system (Kirk & Chang, 1981; Evans, 1987; Fernando & Aust, 1994; Aust, 1995; Bennett & Faison, 1997; Arisoy, 1998). The major components of the lignin-degrading enzyme system include lignin peroxidases (LiPs), manganese-dependent peroxidases (MnPs) and an H_2O_2-generating system. Other important components include glucose oxidases, glyoxal oxidases, lactases, reductases, methylases, veratryl alcohol, oxalate, quinones, quinone reductases and laccases, the last being widely distributed copper-containing enzymes (Barr & Aust, 1994b). The most widely studied wood-rotting fungus for xenobiotic biodegradation is *Phanerochaete chrysosporium* and PCP is one of the most well-studied xenobiotics, in terms of fungal degradation (Higson, 1991; Barr & Aust, 1994a; McAllister *et al.*, 1996).

Pentachlorophenol

Thousands of tons of PCP and chlorinated phenols are produced annually (Reineke, 1984; McAllister *et al.*, 1996). Their widespread use in agriculture and industry as fungicides, insecticides, herbicides and disinfectants, together with their acute toxicity, has led to the contamination of both terrestrial and aquatic ecosystems. As a consequence, they are listed as priority pollutants and their use is now restricted (Rao, 1978; Keith & Teillard, 1979; Crosby, 1981; Anon, 1998). A variety of microorganisms have been shown to degrade PCP, primarily by a pathway involving dechlorination, hydroxylation and methylation (Stanlake & Finn, 1982; Rochkind-Dubinsky *et al.*, 1987b; McAllister *et al.*, 1996). Degradation of PCP at $600\,mg\,kg^{-1}$ soil was demonstrated in soil systems containing bacteria (Middledorp, Briglia & Salkinoja-Salonen, 1990). Fixed-film bioreactors, consisting of a mixed microbial consortium as a biofilm on softwood bark, were used successfully by Salkinoja-Salonen *et al.* (1983) to remove PCP from water. Biodegradation in the environment, however, can often be slow, since PCP may form part of a complex mixture of different chemicals and occur at concentrations that would be quite toxic to the indigenous population. In addition, degradation may be taking place under less favourable, anaerobic conditions (Engelhardt *et al.*, 1986).

Certain fungi, particularly white rot fungi, seem to be more successful than other microorganisms because they are often able to tolerate and/or detoxify some of these pesticides within mixtures of chemicals and at relatively toxic concentrations (Cserjesi & Johnson, 1972; Mileski *et al.*, 1988; Lamar *et al.*, 1990; Kirk *et al.*, 1992). The biodegradation potential of some fungi for PCP has shown that they can be tolerant to concentrations as high as $500–1000\,mg\,l^{-1}$ (Seigle-Murandi, Steiman & Benoit-Guyod, 1991). In general, however, fungi are not efficient at degrading PCP in liquid culture or soil systems (Mileski *et al.*, 1988; Lamar *et al.*, 1990; Lamar & Dietrich, 1992). Unlike bacteria, fungi do not normally use PCP as a source of carbon and energy but degrade PCP through a fortuitous/co-metabolic reaction using wood components like lignocellulose as their primary carbon source. *Aspergillus, Penicillium, Fusarium* and *Paecilomyces* spp. are all able to degrade PCP but do not use it for growth (Rochkind-Dubinsky *et al.*, 1987b; Seigle-Murandi *et al.*, 1991). Of the white rot fungi, *P. chrysosporium* and *Phanerochaete sordida* have received particular attention in terms of PCP degradation/detoxification. *Trametes* and *Phellinus* spp. have also shown potential as PCP degraders (Alleman, Logan & Gilbertson, 1992), while Roy-Arcand and Archibald (1991) demonstrated dechlorination of chlorophenols by the laccases of *Trichoderma versicolor*. Mileski *et al.* (1988) used purified ligninase from *P. chrysosporium* to convert PCP into 2,3,5,6-tetrachloro-2,5-cyclohexadiene-1,4-dione (THCD). A dual degradation pathway was proposed in which PCP is converted to TCHD, which is then degraded by mycelial enzymes, and PCP breakdown occurs without an initial peroxidative step (Lin, Wang & Hickey, 1990). The multistep degradation pathway for PCP has been further elucidated by Reddy and Gold (2000). They characterized the fungal metabolites and oxidation products generated by purified LiP and MnP from *P. chrysosporium*. Tetrachlorobenzoquinone was shown to be degraded by two parallel pathways with several cross-pathway steps, involving a combination of intracellular reductive dechlorination reactions and extracellular hydroxylation reactions. As a result, PCP was shown to be completely dechlorinated. A silicone membrane biofilm reactor has also been used to study fungal LiP production and PCP degradation. PCP disappeared in the bioreactor at a rate of $10.5 \, \text{mg} \, \text{l}^{-1}$ per day, five times faster than in flasks (Venkatadri *et al.*, 1992). White rot fungi grown as a mycelial mat have been shown to metabolize PCP at concentrations as high as $500 \, \text{mg} \, \text{l}^{-1}$ ($1.9 \, \text{mmol} \, \text{l}^{-1}$), which are the concentrations present in some wood treatment plant effluents and contaminated soils (Mileski *et al.*, 1988; McAllister *et al.*, 1996).

Phanerochaete sp. has been shown to detoxify PCP by methylation using its lignin-degrading system (McBain *et al.*, 1995). The primary transformation product is pentachloroanisole, which although less toxic than the parent compound is more lipophilic and may bioaccumulate (McAllister *et al.*, 1996). Pentachloroanisole can become coupled to humic materials via co-polymerization, reactions catalysed by fungal phenol oxidases such as laccases and peroxidases (McAllister *et al.*, 1996). The resulting 'polymers' have been regarded as non-toxic and relatively resistant to microbial degradation in soils (Cserjesi & Johnson, 1972). It has been suggested, however, that the nature, toxicity and stability of these soil-bound products need to be further investigated (Lichtenstein, 1980; Kirk *et al.*, 1992).

Several field trials of selected strains of the white rot fungi *P. chrysos-*

porium and *P. sordida* in different soils have shown extensive and rapid conversion of PCP in mixtures of creosote to pentachloroanisole and other non-extractable soil-bound products ($> 80\%$ conversion of PCP within 6 weeks) (Lamar & Dietrich, 1990; Kirk *et al.*, 1992; Lamar *et al.*, 1994). The inoculum used in these studies consisted of a cheap and abundant solid substrate, such as wood chips or grain–sawdust mixtures supporting fungal hyphae, which was mixed into the contaminated soil with periodic tilling. Wood chips are used as vehicles for fungal inoculation and can serve as an additional carbon source for fungi (Lamar *et al.*, 1990; Lamar & Evans, 1993). Other studies have suggested that *Trametes* spp. may be a better candidate for direct inoculation into contaminated sites than *Phanerochaete* spp. because they are more resistant to PCP and do not accumulate toxic intermediates (Alleman *et al.*, 1992; McAllister *et al.*, 1996). In addition, the greater the growth of the fungi, the more tolerant the white rot fungi are to PCP (Mileski *et al.*, 1988; Lamar *et al.*, 1990; Alleman *et al.*, 1992). One of the barriers to successful implementation of fungal bioaugmentation is the development of an inexpensive high-quality fungal inoculum with uniformly high biological potential (Lestan *et al.*, 1996). Recent advances in delivery and application to maintain and optimize inoculum potential have led to the development of pelleted and powdered fungal inocula for more effective bioremediation of soils (Illman, 1993; Loomis *et al.*, 1997; Walter, 1997).

Although methylation is the dominant degradative process in fungi, the enzymes responsible have not been isolated or characterized. More studies are required to elucidate PCP degradation pathways, particularly the enzyme systems of *Trametes* spp. which do not accumulate the intermediate pentachloroanisole. More recently, a solid substrate monoculture of *Lentinula edodes* was used to achieve rapid rates of PCP degradation in soil, during which both PCP and pentachloroanisole were shown to be degraded. With mixed microflora, however, the rates of degradation were slower and pentachloroanisole was not degraded (Okeke *et al.*, 1997). *L. edodes*, in contrast to *Phanerochaete* spp., appears to remain active at the lower temperatures that are typical of temperate soils of central and northern Europe (Okeke *et al.*, 1996).

Other chlorophenols that are considered priority pollutants include 2,4,5-trichlorophenol, which is used as a precursor for the synthesis of the herbicide 2,4,5-trichlorophenoxyacetic acid (2,4,5-T), and 2,4-dichlorophenol, a breakdown product of the herbicide 2,4-dichlorophenoxyacetic acid (2,4-D). Hammel and Tardone (1988) reported that LiPs catalysed oxidative 4-dechlorination of several polychlorinated phenols, including 2,4-di-, 2,4,5-tri, 2,4,6-trichlorophenol, and PCP, producing *p*-benzoquinones. Meharg, Cairney and Maguire (1997) demonstrated degradation of 2,4-dichlorophenol by ectomycorrhizal fungi in axenic culture and in symbiosis with pine. Joshi and Gold (1993) showed that under secondary metabolite conditions *P. chrysosporium* rapidly degraded 2,4,5-trichlorophenol. The multistep pathway involved cycles of peroxidase-catalysed oxidative dechlorination reactions followed by quinone reduction reactions to yield the key non-chlorinated intermediate 1,2,4,5-tetrahydroxybenzene, which was presumed to be ring cleaved with subsequent degradation to carbon dioxide (Valli & Gold, 1991). Since 2,4,5-trichlorophenol is broken down very rapidly, the methylation reactions observed with other substrates such as PCP apparently do not predominate in this case. This unique fungal pathway contrasts with several common bacterial pathways as all three chlorine atoms were thought to be removed before ring cleavage occurred. In bacteria, phenolic groups are introduced by aromatic ring hydroxylation, and ring cleavage of the corresponding chlorocatechol produced by bacteria could generate toxic acylhalide intermediates (Reineke & Knackmuss, 1988).

Organochlorines

Between the 1940s and 1970s, organochlorine compounds were the most widely used agricultural pesticides (de Schrijver & de Mot, 1999). It was found, however, that the long-term persistence of these toxic insecticides had created both environmental and health problems, with a tendency towards bioaccumulation (Johnson & Kennedy, 1973; Anon, 1979; van de Waerdt, 1983; Spynu, 1989). As a consequence, the use of several organochlorine insecticides, including DDT, lindane (γ-isomer of 1,2,3,4,5,6-hexachlorocylohexane), chlordane, heptachlor, endosulfan (1,4,5,6,7,7-hexachloro-5-norbornene-2,3-dimethanol cyclic sulfite), aldrin and dieldrin, as well as the organochlorine herbicides dalapon and 2,4,5-T, was limited or banned in technologically advanced countries (Anon, 1998). However, many are still used in developing countries where the disadvantages from their extensive use is considered to be outweighed by their benefits, for example in malaria control (Lal & Saxena, 1982; van de Waerdt, 1983; Anon, 1999; Taverne, 1999).

DDT

There have been many studies on the bacterial degradation of DDT (see

Lal & Saxena, 1982). The main degradation route in bacteria under anaerobic conditions involves successive reductive dechlorination of the trichloromethyl group to 1,1-dichloro-2,2-bis(4-chloropheny)ethane (DDD), which then undergoes further dechlorination followed by oxidation of the carboxylic acid, decarboxylation and hydroxylation prior to further degradation that includes ring cleavage (Johnsen, 1976). These co-metabolic reactions require consortia of bacteria in order to achieve complete degradation of DDT (de Schrijver & de Mot, 1999). Under aerobic conditions, dehydrochlorination occurs (Kumar *et al.*, 1996).

Several early studies have reported degradation of DDT by fungi including *Aspergillus, Fusarium* and *Trichoderma* spp. (Matsumura & Boush, 1968). Later studies showed that the pathway for fungal degradation of DDT was clearly different from that of the major pathway proposed for bacterial degradation of DDT (Subba-Rao & Alexander, 1985; Bumpus, Powers & Sun, 1993; Paszczynski & Crawford, 1995). Although DDD was detected, fungal enzymes also catalysed the oxidation of the benzylic carbon of DDT by hydroxylation to form 2,2,2-trichloro-1,1-bis(4-chlorophenyl)ethanol (dicofol) a tertiary alcohol, which makes the compound more liable to bond cleavage and subsequent breakdown via the dechlorinated intermediate 2,2-dichloro-1,1-bis(4-chlorophenyl)ethanol (FW-152) (Fernando & Aust, 1994). Several white rot fungi including *P. chrysosporium, Pleurotus ostreatus, Phellinus weivii*, and *T. versicolor* have been shown to degrade DDT, but *P. chrysosporium* was shown to be the most efficient (Bumpus & Aust, 1987). Experiments evaluating the ability of white rot fungi and in particular *P. chrysosporium* to degrade DDT by the ligninase system showed that DDT was degraded under nitrogen-deficient conditions (Bumpus and Aust, 1987). It was concluded that lignin-degrading enzymes were involved, although later studies failed to elucidate either reductive-dechlorination products or aromatic ring-cleavage products (Kohler *et al.*, 1988). It is now thought that reductive dechlorination of DDT occurs in non-ligninolytic cultures of white rot fungi (Hammel, 1989; Juhasz & Naidu, 1999). Fernando, Aust & Bumpus (1989) investigated the effect of culture parameters on the degradation of DDT by *P. chrysosporium*. Optimal carbon sources proved to be cellulose and starch, providing a constant source of glucose both for DDT degradation and natural degradation of lignocellulose. In later studies, Fernando and Aust (1994) developed a soil system that contained ground corn cobs where a 4:1 ratio of corn cobs to silt loam soil with a 40% moisture content proved to be optimal in supporting fungal growth and sustaining DDT degradation over a 90-day period.

Other organochlorine insecticides, including aldrin, dieldrin, heptach-lor, methoxychlor, chlordane, lindane, mirex, isodrin and endrin, have been used extensively for controlling mites and other insects (Anon, 1998). Their rates of disappearance in the environment range on average from 3 (aldrin) to 8 (dieldrin) years. These insecticides have a multiring alicyclic structure containing alkenyl (olefinic) bonds, and biological conversion of these organochlorines, via dechlorination, dehydrochlorination, oxidation and/or isomerization, are common mechanisms of degradation in the environment (Lal & Saxena, 1982).

Lindane

Commercial formulations of lindane contain a mixture of isomers includ-ing the γ- and β-isomers. Lindane metabolism, including microbial metab-olism, has been reviewed by Macholz and Kujawa (1985). Microbial degradation occurs more readily under anaerobic than aerobic conditions. Anaerobic biodegradation by reductive dechlorination is well established (Jagnow, Haider & Ellwardt, 1977; Mohn & Tiedje, 1992). Under strictly anaerobic conditions, the likely metabolites, which are less toxic than the parent compound, are the γ-isomers of tetrachlorocyclohexene and mono-chlorobenzene. Under aerobic conditions, pentachlorocyclohexene may be the major degradation product detected (Ohisa & Yamaguchi, 1978). Isomerization of lindane from the γ- to the α-isomer has also been observed (Lal & Saxena, 1982). There have been few studies that have shown fungal degradation of lindane. Most studies demonstrate degradation of other cyclodienes such as endrin and aldrin, by fungi such as *Trichoderma viride* (Patil, Matsumura & Bousch, 1970). More recently, Singh and Kuhad (1999) compared the ability of *Trametes hirsutus* to degrade lindane in liquid culture with that of *P. chrysosporium*. They showed that *T. hirsutus* degraded lindane faster than *P. chrysosporium*, although the degradation pathway appeared to be the same in both cultures. Two metabolites were detected, the dechlorinated tetrachlorocyclohexane and oxidized tetrach-lorocyclohexanol, which confirmed the findings of Mougin *et al.* (1996). Further studies are required to identify and characterize the enzyme sys-tems involved in white rot degradation of lindane.

Aldrin, dieldrin and heptachlor

Reductive dechlorination of other cyclodienes by mixed populations of anaerobes has been demonstrated (Maule, Plyte & Quirk, 1987). Microbial degradation by fungi such as *Trichoderma, Fusarium* and *Penicillium* spp. often leads to the formation of an epoxide ring structure in these molecules, which tend to be more stable than the parent compound (Tu, Miles & Harris, 1968).

Dieldrin is more persistent than aldrin in soil and, of the few microorganisms able to metabolize it, *T. viride* has been shown to hydrolyse dieldrin to *trans*-aldrinol and other solvent- and water-soluble metabolites (Matsumura, Bousch & Tai, 1968). Degradation of dieldrin has also been demonstrated by *Trichoderma koningii*, which was able to ring cleave dieldrin (Bixby, Bousch & Matsumura, 1971). Morgan, Lewis and Watkinson (1991) showed some degradation of dieldrin by *T. versicolor*, *Chrysosporium lignorum* and *P. chrysosporium*, although no attempt was made to identify the water-soluble metabolites. Heptachlor has been shown to be metabolized by soil microorganisms including fungi, where a range of metabolites were produced by oxidation and hydrolysis, including heptachlor epoxide and 1-hydroxychlordene (Miles, Tu & Harris, 1969).

The ability of *P. chrysosporium* to degrade some of these organochlorine insecticides has been investigated in both liquid culture and soil–corn cob matrices (Kennedy, Aust & Bumpus, 1990). Of the six insecticides investigated, dieldrin, aldrin, heptachlor, chlordane, lindane and mirex, only lindane and chlordane underwent significant biodegradation (22.8% and 14.9%, respectively, in the soil–corn cob matrices over a 60-day period). Although degradation was shown to occur under nitrogen-limited conditions, the mechanisms involved and transformation products detected have not yet been fully elucidated.

Endosulfan

Endosulfan has been extensively used throughout the world as a broad-spectrum pesticide and is the only cyclodiene insecticide still registered for use in the USA. As a consequence, endosulfan contamination is common throughout the environment (Kullman & Matsumura, 1996). In soil, endosulfan can be degraded by a wide variety of microorganisms including soil fungi (El-Zorgani & Omer, 1974; Martens, 1976). Degradation rates, however, are usually low and the primary transformation products are endosulfan sulfate and endosulfandiol. These oxidative metabolites can be equally as toxic and persistent as the parent compound. *P. chrysosporium* has been shown to utilize both oxidative and hydrolytic pathways in the relatively rapid metabolism of this pesticide. Two distinct pathways exist, depending on whether metabolism takes place under ligninolytic (nutrient-deficient) or non-ligninolytic (nutrient-rich) conditions (Kullman & Matsumura, 1996). The transformation products include endosulfan sulfate, endosulfan diol, endosulfan hydroxyether and another metabolite tentatively identified as endosulfan dialdehyde. Sudhakar and Dikshit (1999) described the use of low-cost adsorbents, including the macrofungus *Sojar caju*, for adsorption of endosulfan from water. Although the fungal biomass did not prove to be as efficient as wood charcoal, the fungus was still able to remove over 80% of a $10\,mg\,l^{-1}$ initial concentration of endosulfan after 24 hours.

Organophosphates

Organophosphates are broad-spectrum insecticides that have been in use since the 1960s and which are relatively non-persistent compared with the organochlorines. Their half-lives in soils are measured in weeks or months, rather than years (Tessier, 1982). However, several are highly toxic to mammals (e.g. parathion) (MacRae, 1989). The group contains many well-known insecticides such as malathion, parathion, methyl parathion, fenitrothion and diazinon. They are metabolized by many different microorganisms, particularly members of the *Pseudomonas, Arthrobacter, Streptomyces* and *Thiobacillus* genera and by fungi in the *Trichoderma* genus (Matsumura & Benezet, 1978). These insecticides are primarily detoxified and degraded by hydrolysis and oxidation (Mulla, Mian & Kawecki, 1981). Studies on fungal degradation of organophosphorus insecticides have involved primarily liquid cultures (Omar, 1998). An example is the enzymic hydrolytic cleavage of fenitrothion and fenitrooxon by *T. viride* (Baarschers & Heitland, 1986). The primary transformation product, 3-methyl-4-nitrophenol, was thought to be liable to further transformation by co-metabolism (Baarschers & Heitland, 1986). This fungus had already been shown to degrade diazinon and parathion (Kiigemagi *et al.*, 1958).

Malathion

Malathion is degraded by carboxyesterase and strong activity of this enzyme has been detected in several fungi (*Aspergillus, Penicillium* and *Rhizoctonia* spp.; Mostafa *et al.*, 1972). Omar (1998) isolated 13 fungal species, including *Aspergillus, Fusarium, Penicillium* and *Trichoderma* spp., from soil that were able to degrade several organophosphorus insecticides including malathion. Hasan (1999) also demonstrated fungal utilization and degradation of several organophosphate pesticides including malathion. Several *Aspergillus* species including *A. terreus*, *A. flavus* and *A. sydowii* showed the greatest potential for utilizing organophosphorus pesticides as phosphorus and carbon sources, indicative of both carboxyesterase and phosphatase activity.

Parathion

Parathion-bound residues have been detected in the environment, particularly under anaerobic conditions (Lichtenstein, Liang & Koeppe, 1983). Opinions are still divided on whether bound pesticide residues such as that of parathion pose an environmental problem (Stott *et al.*, 1983; MacRae, 1989). Results from a study by Omar (1998) indicate that several fungal species have the potential to degrade some components of these pesticides.

Chlorpyrifos

Chlorpyrifos has a broad range of insecticidal activity but is also highly toxic to mammals (Anon, 1991). This insecticide, although quite inhibitory to bacterial populations, has been shown to stimulate fungal growth (Al-Mihanna, Salama & Abdalla, 1998). A mixed population of plant pathogenic fungi (*Fusarium* sp., *Rhizoctonia solani*, *Cladosporium cladosporiodes, Cephalosporium* sp., *T. viride, Alternaria alternata* and *Cladorrhinum brunnescens*) exhibited biodegradation of chlorpyrifos in liquid culture (Al-Mihanna *et al.*, 1998). The mixed population was shown to be more efficient for biodegradation of this insecticide than pure cultures of these fungi.

Carbamates

Carbamates are broad-spectrum pesticides that replaced many organochlorine and organophosphorus insecticides because of lower toxicity and recalcitrance (Rajagopal *et al.* 1984). They find wide applications, not only as insecticides but also as herbicides and fungicides (Machemer & Pickel, 1994 a,b).

Carbaryl and carbofuran

The most widely used carbamate is carbaryl (1-naphthyl-*N*-methylcarbamate), although its use is now restricted in the UK (Anon, 1998). Barik (1984) has reviewed the fate of carbaryl in the environment. Microbial degradation has been demonstrated where hydroxylation and hydrolytic cleavage of carbaryl were the main degradative processes. Bollag (1974) showed that a number of soil fungi were able to degrade carbaryl, producing a variety of different metabolites including 1-naphthol. The biodegradation of the insecticide carbofuran (2,3-dihydro-2,2-dimethyl-7-benzofuranyl-N-methyl carbamate) has been demonstrated by pure cultures of *Aspergillus niger* and *Fusarium graminearum* (Salama, 1998). Carbofuran was metabolized by oxidation and hydrolysis, followed by conjugation of the metabolites (Sjoblad & Bollag, 1981).

Pyrethroids

The photostable synthetic pyrethroid insecticides are an economically important group of insecticides (Leahey, 1985; Miller, 1988). They constitute about 25% of the total pesticides used worldwide (Johri, Saxena & Lal, 1997). In soils, pyrethroids appear to be relatively non-persistent (half-life 5–90 days), particularly in mineral soils that are well-drained and fertile (Chapman *et al.*, 1981). Increased persistence of the pyrethroids was observed in less-fertile soils and under anaerobic conditions (Roberts, 1981). Pyrethroids are highly toxic to a number of non-target organisms, such as terrestrial or aquatic insects, crustacea and fish (Jolly *et al.*, 1977–8). Consequently, under certain circumstances, pyrethroids pose problems in the environment because of such effects (Solomon, 1986; Zabel *et al.*, 1988). Ester cleavage appears to be the first major step in pyrethroid transformation, generating a number of non-insecticidal polar products that readily leach into the aqueous phase (Elliott, Janes & Potter, 1978; Leahey, 1985; Sakata, Mikami & Yamada, 1992a,b).

Soil culture enrichments were set up in the presence of radiolabelled permethrin by Kaufman *et al.* (1977) to isolate pure cultures of soil microorganisms capable of its metabolism. These authors also used aqueous soil suspensions of *Fusarium oxysporum* Schlect (known to possess esterase–amidase activity) but were unable to detect radiolabelled CO_2 or transformation products. Ohkawa *et al.* (1978) and Khan *et al.* (1988) demonstrated degradation of fenvalerate and deltamethrin, respectively, in microbial cultures isolated from soil. The bacterial isolates appeared to be more active than the fungi, although no details were given on the fungal species involved. The degradation of residual pesticides including fenvalerate and permethrin, commonly used in viniculture, was also studied in the yeast *Saccharomyces cerevisiae* (Fatichenti *et al.*, 1984). The primary non-insecticidal transformation products detected were 3-phenoxybenzyl alcohol and dihalovinyl acid moieties, which were the same metabolites detected in bacterial cultures (Maloney, Maule & Smith, 1988). The ecological significance on non-target species of these pesticides and metabolites has not been well investigated. A study by Stratton and Corke (1982a,b), found that 3-phenoxybenzyl alcohol and its acid were inhibitory towards the growth of fungi, algae and cyanobacteria. The median effective concentration of permethrin ranged from 60 to 100 ppm for fungal growth. Tu (1980) investigated the effects of permethrin, cypermethrin, deltamethrin and fenvalerate on soil microorganisms at concentrations of 0.5 and $5\,mg\,kg^{-1}$. At these doses, nitrification and microbial respiration were increased, suggesting the microbial degradation of these compounds. High concentrations of pyrethroids were said to exert a significant effect on microorganisms including fungi, but no further details were given (Johri *et al.*, 1997). Further work is required on microbial degradation, in particular fungal degradation, of pyrethroids, together with toxic effects under environmental conditions.

Herbicides

There is growing concern over the potential for contamination of surface water and groundwater by herbicides (Shelton *et al.*, 1996). Over 150 chemicals are used as herbicides throughout the world, of which only a small proportion have partially characterized metabolic pathways. In addition, relatively few soil-applied herbicides have been shown to be degraded by pure cultures of microorganisms. This is probably because of the wide variety of structural groups of most herbicides, which require different catabolic enzyme systems not usually found within a single organism. Fungal species (e.g. *Fusarium* and *Aspergillus* spp.) have, however, been shown to oxidize herbicide carbon–carbon and carbon–chloride bonds non-specifically, while growing with other carbon co-substrates (Kaufman & Blake, 1973).

Chlorinated phenoxyacetates

2,4-D, 2,4,5-T and 2-methyl-4-chlorophenoxyacetate (MCPA) have been widely used for the selective control of broad-leaved weeds and as defoliants since the 1940s. Concern over possible mutagenicity of these herbicides has resulted in their restricted use and increased interest in their biochemistry and metabolism. As a consequence, many studies have been carried out on their degradation in the environment (Rochkind-Dubinsky *et al.*, 1987c). In general, these herbicides appear to degrade quite rapidly in most ecosystems, although persistence/recalcitrance is a potential problem at high concentrations or in less fertile soils, especially for 2,4,5-T (MacRae, 1989). Most of the available results are from the bacterial degradation of 2,4-D (Rosenberg & Alexander, 1980; Yadav & Reddy, 1993). Ether bond cleavage, ester hydrolysis and hydroxylation are the major mechanisms used in the bacterial co-metabolism of the chlorophenoxyacetate herbicides, generating chlorophenols and chlorocatechols (de Schrijver & de Mot, 1999). Kilbane *et al.* (1983) were able to demonstrate the removal of over 90% of the more persistent 2,4,5-T, present in soil at $1\,mg\,g^{-1}$ soil, by a pure bacterial culture, even after repeated applications of up to $20\,mg\,g^{-1}$ soil. Generally, adaptation by enrichment is required, after which increased degradation rates can be achieved. The bulk, however, of the herbicides do tend to remain as chlorinated anilines, which become strongly bound to soil humic substances (Sjoblad & Bollag, 1981).

Several fungi, including *Aspergillus* and *Penicillium* spp. have been shown to degrade 2,4-D in soils, by ester hydrolysis and ring hydroxylation (Faulkner & Woodcock, 1964). Mycorrhizal fungi were shown to degrade 2,4-D through incorporation of herbicide carbon into tissue and not by decomposition (Donnelly, Entry & Crawford 1993). The biodegradation of the more persistent 2,4,5-T by *P. chrysosporium* has been demonstrated in both nutrient nitrogen-limited liquid cultures and contaminated soil supplemented by a corn cob mixture, although the mechanism of degradation has not been fully elucidated (Ryan & Bumpus, 1989). Yadav and Reddy (1993) have demonstrated biodegradation of 2,4-D and mixtures of 2,4-D and 2,4,5-T in liquid culture by *P. chrysosporium*. They showed that a relatively higher rate of breakdown of mixtures of 2,4-D and 2,4,5-T was observed than when they were tested alone. The use of an organism with a broad degradative ability is of importance when considering bioremediation of contaminated environments. Although bacteria may degrade these herbicides more rapidly than fungi, there are few bacteria that are able to metabolize mixtures of these herbicides (Haughland *et al.*, 1990).

MCPA has been shown to be metabolized by hydroxylation to 4-chloro-5-hydroxy-2-methylphenoxyacetic acid by several fungi, including *Aspergillus, Fusarium* and *Penicillium* spp. (Cripps & Roberts, 1978). Although bioremediation of MCPA has not yet been demonstrated, laboratory studies have shown that consortia of bacteria and fungi can break down MCPA under aerobic conditions (Duah-Yentumi & Kuwatsuka, 1982;

MacRae, 1989). Under anaerobic conditions, however, toxic concentrations of a metabolite 5-chloro-*o*-cresol accumulated, which slowed down MCPA catabolism (MacRae, 1989).

Phenylamides

The majority of herbicides used in agriculture belong to this class of herbicide. They comprise the acetanilides, the phenylureas and the phenylcarbamates. In many cases these herbicides are degraded to substituted anilines (Cripps & Roberts, 1978).

Acetanilides

Examples of acetanilides include propanil, alachlor, metolachlor and propachlor, which have all been widely used for pre-emergence weed control and are relatively common contaminants in groundwater aquifers (Stamper & Tuovinen, 1998). Biodegradation, often involving co-metabolism, is the single most important factor in controlling the disappearance of these herbicides in both aerobic and anaerobic environments, where low degradation rates have been reported under anaerobic conditions (MacRae, 1989; Stamper & Tuovinen, 1998). Detoxification involves primarily conjugation through glutathione-*S*-transferase between the tripeptide glutathione and the chloroacetamide moiety of the herbicide (Field & Thurman, 1996). These conjugates are subsequently degraded by microbial hydrolytic cleavage and oxidation. A number of soil fungi including *Fusarium* and *Penicillium* spp. and *Aureobasidium pullulans* can hydrolyse these herbicides with concomitant formation of the corresponding aniline. These herbicides are used as a sole source of carbon and energy (Bartha & Pramer, 1970). An acylamidase cleaves these compounds, forming the chlorinated aniline and an aliphatic moiety, which is used for growth (Cripps & Roberts, 1978).

Alachlor Tiedje and Hagedorn (1975) isolated a soil fungus, *Chaetomium globosum*, that was able to use alachlor as its sole source of carbon and energy. In contrast, Smith and Phillips (1975) showed that *Rhizoctonia solani* could only degrade alachlor under co-metabolic conditions. More recently, Ferry *et al.* (1994) studied alachlor breakdown by the white-rot fungi *Ceriporiopsis subvermispora, Phlebia tremellosa* and *P. chrysosporium* and a brown-rot fungus *Fomitopsis pinicola*. The low extent of degradation (~10%), coupled with the detection of partial degradation products suggested a co-metabolic transformation of alachlor.

Metolachlor and propachlor   Co-metabolism of the herbicide metolachlor has been reported by pure cultures of both bacteria and fungi, including white rot fungi (Liu, Freyer & Bollag, 1991). The cultures were able to dechlorinate metolachlor to a varying extent, but complete degradation did not occur. Sorption and partitioning in soils readily occurs, which limits their bioavailability. Fungi, therefore, have an advantage over bacteria, because their mycelial growth maximizes both physical/mechanical and enzymic contact with the environment (Bennett & Faison, 1997). Kaufman and Blake (1973) showed that propachlor could be degraded by pure cultures of *Aspergillus ustus, Fusarium solani, F. oxysporum, Penicillium* sp. and *T. viride.*

Phenylureas

The phenylureas were first developed after World War II as herbicides. Today there are about 25 phenylurea herbicides being marketed as pre- or postemergence herbicides for the control of annual grasses and broad-leaved weeds, for example in cereals. Typical examples include monuron, linuron, diuron, neburon and chlorbromuron (Cripps & Roberts, 1978). They can persist in soils on average between 4 and 18 months and are quite phytotoxic, being specific inhibitors of photosynthesis. The urea herbicides do, however, undergo microbial degradation (Murrey, Rieck & Lynd, 1969; Ross & Tweedy, 1973; Mudd, Greaves & Wright, 1983; Deping, Ruiwei & Wei, 1991). Maier-Bode and Hartel (1981) have reviewed the fate of linuron and monuron in soil. The major mechanism of detoxification involves successive dealkylation or N-dealkoxylation and sometimes N-methoxylation of the side chain, where the principal metabolite is 3,4-dichloroaniline (Rochkind-Dubinsky *et al.*, 1987c). A more recent study has shown that long-term application of phenylurea herbicides affects both the structure and metabolic potential of soil microbial communities, in particular bacterial isolates (Fantroussi *et al.*, 1999). This contrasts with an earlier study that suggested that phenylurea herbicides exerted little effect on the soil microbial population (Roslycky, 1977).

Few studies have dealt with the degradation of phenylurea herbicides by fungi (Weinberger & Bollag, 1972; Kaufman & Blake, 1973). Despite this, many fungi, including *Aspergillus* and *Penicillium* spp., have been shown to use these herbicides as either sole nitrogen or sole carbon sources: *Aspergillus nidulans* was found to be the most effective isolate for degradation of linuron (Schroeder, 1970). Tillmanns, Wallnoefer & Engelhardt (1978) demonstrated the oxidative dealkylation of five phenylurea herbicides, including diuron, by the fungus *Cunninghamella echinulata*. Vroumsia *et al.* (1996), carried out a comparative study to determine which soil fungus was the most efficient in the degradation of the more recently developed phenylurea herbicides, chlortoluron and isoproturon, together with one of the more phytotoxic and older examples of a phenylurea herbicide, diuron. *Rhizoctonia solani* proved to be the most efficient, with degradation rates, in liquid culture, of over 70% for all three phenylurea herbicides, where isoproturon was the most easily degraded. Weinberger and Bollag (1972) also showed that this fungus had a broad substrate specificity for phenylureas. Khadrani *et al.* (1999) showed that many strains of micromycetes were able to degrade the three phenylurea herbicides chlortoluron, isoproturon and diuron. Isoproturon was the most readily degraded and the basidiomycetes *Bjerkandera adusta* and *Oxysporus* sp. proved to be the most effective degraders in this study.

Fungi, as well as bacteria, can, under certain conditions such as high concentrations, produce peroxidases that can polymerize the chloroanilines. A laccase of the white rot fungus *T. versicolor* polymerized 4-chloroaniline to produce oligomers ranging in size from dimers to pentamers (Hoff, Liu & Bollag, 1985). The action of these extracellular laccases in the oxidative coupling of toxic intermediates of pesticide decomposition may be a detoxification reaction (MacRae, 1989).

3,4-Dichloroaniline is a product of the biodegradation of a number of herbicides, such as the phenylamide herbicides and it has generally been found to be resistant to degradation, partly because of its polymerization and binding to humic acid in the soil (Sjoblad & Bollag, 1981). Single bacterial strains, *Pseudomonas* sp. and the white rot fungus *P. chrysosporium* have, however, been shown to degrade breakdown products of the phenylamide herbicides such as 3,4-dichloroaniline (MacRae, 1989).

Phenylcarbamates

Widely used phenylcarbamate herbicides include barban, propham and swep, which are effective at low levels of application and are not generally persistent in most soils (Cripps & Roberts, 1978). Phenylcarbamates are readily detoxified by hydrolysis of the ester linkage, producing aniline and substituted anilines (Kaufman, 1967). The formation of anilines results in the loss of herbicidal activity. Aniline is readily metabolized, possibly through catechol. Several species of fungi have been reported to grow on these herbicides, including *Fusarium* sp. (Kaufman & Blake, 1973).

s-*Triazines*

s-Triazines are heterocyclic nitrogen derivatives with herbicidal properties that were first discovered in the early 1950s (Cripps & Roberts, 1978). The most commonly used triazines are the 2-chloro-s-triazine herbicides atrazine (2-chloro-4-(ethylamino)-6-(isopropylamino)-s-triazine) and simazine. They are widely used, in very large quantities, for weed control in field crops such as corn or sorghum (de Schrijver & de Mot, 1999). They are also used as a non-selective herbicide for vegetation control in non-crop land. The triazines are moderately persistent in soils (1–2 years), and their extensive use has led to the contamination of terrestrial and aquatic ecosystems, leading to restrictions on their use in agriculture (Anon, 1998). Cook and Hutter (1981) have reviewed the biodegradation of triazine herbicides. Hydrolytic and oxidative degradation of these herbicides, in particular atrazine, in soil has been well documented (Kaufman & Kearney, 1970; Cook, 1987). Pure cultures, however, often fail to degrade s-triazines beyond the point of ring cleavage. Degradation is only initiated with the formation of a deaminated metabolite. Several early studies showed that pure cultures of fungi, including *Aspergillus, Fusarium, Penicillium, Rhizopus and Trichoderma* spp. were able to use these herbicides as sole sources of carbon or nitrogen (Kaufman, Kearney & Sheets, 1965). The most common degradative mechanisms in fungi appeared to be dealkylation and deamination, and in some cases also hydrolytic dechlorination (Kaufman & Kearney, 1970; Cripps & Roberts, 1978). Degradation of atrazine through incorporation of herbicide carbon into fungal tissue has been demonstrated in nine mycorrhizal fungi (Donnelly *et al.*, 1993). Biotransformation of atrazine in liquid cultures by *P. chrysosporium* produced hydroxylated and/or N-dealkylated metabolites (Mougin *et al.*, 1994). In addition, both *Streptomyces* spp. and *P. chrysosporium* have been reported to transform this herbicide in contaminated soils (Shelton *et al.*, 1996; Newcombe & Crowley, 1999). Although the mechanism of transformation was not fully elucidated for *P. chrysosporium*, it was thought that approximately 30% was immobilized by production of a bound residue (Hickey, Fuster & Lamar, 1994).

Fungicides

Fungicides are used as both protectants and eradicants to treat soilborne, seedborne, and airborne plant diseases caused by fungi (Newman, 1978). There are over 100 compounds used as fungicides, including organomercurials, dithiocarbamates, organophosphorus compounds, aromatic compounds (including the relatively persistent quintozene, with a half-life in months to a year), heterocyclic compounds and the aliphatic compounds such as chloropicrin and allyl alcohol (Woodcock, 1978). Early studies showed that these fungicides could be detoxified/degraded at various rates by a number of different soil microorganisms, including fungi (see Kaars Sijpesteijn, Dekhuijzen & Vonk, 1977). For example, *T. viride* can use allyl alcohol as an energy source and *A. niger* was shown to demethylate triadimefon and inactivate semesan (Munnecke & Solberg, 1958; Woodcock, 1978). Other studies have shown that *A. niger* (van Tiegh) can metabolize several 1,2,4-triazolylmethane fungicides and *Botrytis cinerea, T. versicolor, Cladosporium cucumerinum* and *Fusarium culmorum* can stereospecifically reduce the enantiomeric triadimefon to the less fungitoxic diastereoisomeric triadimenol (Deas & Clifford, 1982; Deas, Clark & Carter, 1984). More recent studies have shown that *P. chrysosporium* aided biodegradation and remobilization of the fungicide anilazine in humic soil fractions (Liebich, Buranel & Fuhr, 1999).

Polychlorinated biphenyls

There are 209 different congeners of PCBs. They are extremely insoluble, chemically unreactive and heat-stable compounds that can be found in pesticides and were extensively used in industrial applications until the mid-1970s (Hutzinger, Safe & Zitko, 1974). Concerns about their acute and chronic toxicity and their potential carcinogenic role and bioaccumulation in the food chain led to a ban on the use of PCBs in many developed countries (Safe, 1989; Kimbrough & Jensen, 1989). However, their prevalence and recalcitrance has resulted in worldwide contamination and magnification within the food chain (Gilbertson, 1989; Van-Oostdam *et al.*, 1999). As a result, extensive efforts have been made to isolate microorganisms able to degrade a broad range of PCB congeners at acceptably high rates for treatment applications (Abramowicz, 1990).

PCBs do not serve as sole carbon sources but are generally co-metabolized by biphenyl-degrading microorganisms such as *Pseudomonas, Achromobacter* and *Acinetobacter* spp. (Furukawa, Tomizuka & Kamibayashi, 1983; Bedard *et al.*, 1986). Reductive dechlorination of PCBs by bacteria has also been observed under anaerobic conditions (Quensen, Tiedje & Boyd, 1988). Higher chlorinated congeners (over five chlorine atoms per molecule) tend to be more resistant to microbial degradation (Furukawa, 1982). Most of these studies have been carried out using commercial mixtures such as Aroclor and both pure and mixed microbial cultures. The main mechanisms for degradation indicate hydroxylation, ring cleavage and degradation of the non-chlorinated ring of the molecule.

The formation of chlorinated benzoic acids from chlorinated biphenyls is the most common route of PCB degradation (Furukawa, 1982), and biodegradation generally requires a mixed culture of bacteria (Golyshin *et al.*, 1999).

Many fungi have also been tested for their ability to degrade PCBs (Beaudette, 1998). *Rhizopus japonicus* has been shown to convert 4-chlorophenyl to 4-chloro-4'-hydroxybiphenyl and 4,4'-dichlorobiphenyl to an unidentified hydroxylated metabolite (Wallnofer *et al.*, 1973). *C. echinulata* Thaxter metabolized 2,5-dichloro-4'-isopropylbiphenyl by oxidation of the isopropyl group to form 2,5-dichloro-4'-biphenylcarboxylic acid and by hydroxylation of the chlorine-substituted phenyl group (Tulp, Tillmanns & Hutzinger, 1977). *A. niger* has been shown to degrade some congeners found in Clophen A30 (Dmochewitz & Ballschmiter, 1988). Various white rot fungi, including *T. versicolor* and *P. chrysosporium*, have been shown to degrade small amounts of the recalcitrant commercial PCB mixtures Aroclor 1254 and Delor 106 in liquid culture (Eaton, 1985; Novotny *et al.*, 1997). Low concentrations of the surfactant Triton X-100 appeared to increase the bioavailability of the PCB congener 2,4',5-trichlorobiphenyl for oxidation by *T. versicolor* (Beaudette *et al.*, 2000). Like bacterial systems, the fungi preferentially degrade lesser chlorinated biphenyls and biphenyl. Bumpus *et al.* (1987) demonstrated degradation of Aroclor 1242 in nitrogen-deficient cultures of *P. chrysosporium*. Yadav *et al.* (1995) showed degradation of mixtures of PCBs (Aroclors 1242, 1254 and 1260) by *P. chrysosporium*. Sasek *et al.* (1993) demonstrated substantial PCB degradation when a mixed population of white rot fungi were used together with methylotrophic and hydrocarbon-utilizing yeasts. It was suggested by Thomas, Carswell & Georgiou, (1992) that the fungus, as a mycelial mat, could be used as a biological filter to treat low concentrations of PCBs or moderately chlorinated biphenyls. Field trials, however, have not proved to be so effective as the PCB-degrading organisms introduced at contaminated sites do not compete well with indigenous populations (Unterman *et al.*, 1988). Donnelly and Fletcher (1995) have proposed that ectomycorrhizal fungi compete better with existing populations of organisms and have the potential to degrade chlorinated organic compounds including several pesticides (Donnelly *et al.*, 1993). In their study, they looked at the ability of 21 different fungi to metabolize 19 different congeners with varying chlorine content and substitution patterns. They found that 14 of the 21 ectomycorrhizal fungi tested degraded PCBs. Both the number of PCB congeners metabolized and the extent of metabolism varied among the fungi, where less chlorinated congeners were again preferentially degraded. Analysis of the data showed no correlation between taxonomically related species and metabolism of structurally similar PCB congeners. Several of the fungi that metabolized PCBs had previously been shown to degrade the herbicides 2,4-D and atrazine, demonstrating their broad specificity (Donnelly *et al.*, 1993). The degradative ability of these ectomycorrhizal fungi was as effective, and in some cases more effective, as that of white rot fungi, including *P. chrysosporium*. Beaudette (1998) looked at the ability of 12 white rot fungi to biodegrade six PCB congeners. Four of the fungi, including strains of *B. adusta, P. ostreatus* and *T. versicolor* were found to be more effective in biodegradation than *P. chrysosporium*. However, Kubatova *et al.* (1998) showed that a *Pseudomonas* sp. was more efficient in the degradation of a dichlorobiphenyl (3,3'-dichlorobiphenyl) than the white-rot fungus *P. ostreatus* under the experimental conditions used. The mechanism of PCB biodegradation has not been definitely determined for any of these fungi. It is thought that lignin-degrading enzymes other than ligninases and MnPs may be involved in the oxidation of these PCBs (Thomas *et al.*, 1992). The transformation pathways of 4-fluorobiphenyl were investigated with several mycorrhizal fungi, including the ectomycorrhizal *Tylospora fibrilosa* (Green *et al.*, 1999). Two major metabolites were identified, 4-fluorobiphen-4'-ol and 4-fluorobiphen-3'-ol, which suggested that *meta* cleavage of the less halogenated ring did not take place as would be expected (Higson, 1992; Dietrich, Hickey & Lamar, 1995). Although complete degradation was not achieved, these fungi were able to initiate biotransformation.

Polychlorinated dibenzo-*p*-dioxins and polychlorinated dibenzofurans

Polychlorinated dibenzo-*p*-dioxins (PCDDs) and polychlorinated dibenzofurans (PCDFs) like the PCBs are hazardous compounds that can bioaccumulate. They are unintentionally generated during combustion of domestic and industrial waste and can also be formed in the process of producing chlorine-containing herbicides (Safe, 1990). The release into the ecosphere from anthropogenic sources has created a strong demand for legislation and executive activities (Wittich, 1998). Degradation rates of PCDDs and PCDFs in the environment have been shown to be extremely low (Matsumura, Quensen & Tsushimoto, 1983). Slow microbial degradation of these compounds does occur but is limited to those compounds with four or fewer chlorines. Degradation of a mixture of 10 various

Table 8.1. *Advantages of white rot fungi over bacterial systems for bioremediation*

- Uses inexpensive and abundant lignocellulosic material as a nutrient source, which cannot be used by other microorganisms and, therefore, gives the white rot fungi a competitive advantage over other microorganisms
- Tolerant of relatively high concentrations of pollutants through extracellular degradation of pollutants by a potent oxidizing system
- Both physical/mechanical and enzymic contact with the environment is maximized by mycelial growth, unlike single cell microorganisms
- Able to survive in the presence of a number of different xenobiotics that may be acutely toxic to other microorganisms
- Degrades a mixture of chemicals by a non-specific free-radical-based mechanism
- No preconditioning or enrichment required for a particular pollutant, unlike many bacterial systems
- Can tolerate a wide range of environmental conditions, in terms of oxygen levels, temperature range, pH range and moisture levels
- Rate of degradation or biotransformation of a pollutant is proportional to the concentration of the pollutant and so the solubility of pollutant is not important, in contrast to most bacterial systems

chlorinated PCDDs and PCDFs has been demonstrated in liquid cultures of both *P. sordida* and *P. chrysosporium*, where various metabolites including chlorocatechols were detected (Takada *et al.*, 1996). Both these fungi showed no clear structural dependence for degradation of PCDDs and PCDFs, suggesting that the degradative process may be a free-radical process, involving peroxidases, showing little specificity, which is typical of white rot fungi (Wittich, 1998). *Cunninghamella* sp. has been shown to oxidize and hydroxylate the unchlorinated dibenzo-*p*-dioxin and dibenzofuran, a process that was thought to involve P450 monooxygenases. Fortuitous transformations by multiple hydroxylations of dibenzofuran are widespread amongst fungi, including yeasts (Hammer & Schauer, 1997). Oxygen-radical-mediated polymerization of 2-hydroxydibenzofuran by laccases of the white rot fungi *T. versicolor* and *Pycnoporus cinnabarinus* has also been reported (Jonas *et al.*, 1998).

Conclusions

Fungi constitute an important part of the microbial community in the complex and dynamic soil system, yet only a limited number of studies have been carried out in terms of pesticide degradation. The majority of degradation studies have involved bacteria, despite the fact that successful application of bacterial strains to practical problems of soil and water contamination still requires solutions to problems of survival and establishment of the biodegradative strain in the soil, water or wastewater environment. Fungi may show much promise, in particular members of the white rot fungi, where there are clear advantages in terms of bioremediation *in vivo* (Kirk & Chang, 1981; Aust, 1990; Bennett & Faison, 1997) (Table 8.1).

Genetic manipulation may offer another means of engineering microorganisms to deal with a pollutant, including pesticides, that may be present in a contaminated site. Molecular aspects of pesticide degradation have received little attention despite the fact that catabolic genes responsible for the degradation of several pesticides have been identified, isolated and cloned into various other organisms including fungi (Kumar *et al.*, 1996). One example is *Gliocladium virens*, where an effective genetic transformation system has been developed for introduction of foreign genes (Thomas & Kenerly, 1989). Since this fungus also shows a lack of sensitivity towards several organophosphates and strains of *G. virens* are ecologically competent and excellent biocontrol agents, this fungus offers a unique potential for the bioremediation of contaminated soil (Papaizas, 1985).

Optimization of degradative gene expression in other bioremedially useful organisms such as *P. chrysosporium* holds promise for pesticide degradation. However, despite the encouraging results shown in some areas of fungal bioremediation, more work is required to develop reliable application methods and to understand the more complex degradation mechanisms and ecology of fungi. Fungi show great versatility but have not as yet shown as convincing utility as bacteria within the bioremediation industry.

References

Abramowicz, D. A. (1990). Aerobic and anaerobic biodegradation of PCBs: a review. *Critical Reviews in Biotechnology*, **10**, 241–251.

Alleman, B. C., Logan, B. E. & Gilbertson, R. L. (1992). Toxicity of pentachlorophenol to six species of white rot fungi as a function of chemical dose. *Applied and Environmental Microbiology*, **58**, 4048–4050.

Al-Mihanna, A. A., Salama, A. K. & Abdalla, M. Y. (1998). Biodegradation of chlorpyrifos by either single or combined cultures of some soilborne plant pathogenic fungi. *Journal of Environmental Science and Health*, **B33**, 693–704.

Anon (1979). DDT and its derivatives. *Environmental Health Criteria 9* Geneva: World Health Organization.

Anon (1991). *The Pesticide Manual*, 9th edn, pp. 166–167. London: HMSO for

212   S. E. Maloney

The British Crop Protection Council.

Anon (1998). *Pesticides 1998*. London: HMSO for the Pesticides Safety Directorate & Health and Safety Executive.

Anon (1999). DDT lives on. *New Scientist*, 18 September, p. 5.

Arisoy, M. (1998). Biodegradation of chlorinated organic compounds by white rot fungi. *Bulletin of Environmental Contamination and Toxicology*, **60**, 872–876.

Aust, S. D. (1990). Degradation of environmental pollutants by *Phanerochaete chrysosporium*. *Microbial Ecology*, **20**, 197–209.

Aust, S. D. (1993). The fungus among us: use of white rot fungi to biodegrade environmental pollutants. *Environmental Health Perspectives*, **101**, 232–233.

Aust, S. D. (1995). Mechanisms of degradation by white rot fungi. *Environmental Health Perspectives*, **103**, 59–61.

Baarschers, W. H. & Heitland, H. S. (1986). Biodegradation of fenitrothion and fenitrooxon by the fungus *Trichoderma viride*. *Journal of Agricultural and Food Chemistry*, **34**, 707–709.

Barik, S. (1984). Metabolism of insecticides by microorganisms. In Insecticide Microbiology, ed. R. Lal, pp. 87–128. Berlin: Springer-Verlag.

Barr, D. P. & Aust, S. D. (1994a). Pollutant degradation by white rot fungi. *Reviews in Environmental Contamination and Toxicology*, **138**, 49–72.

Barr, D. P. & Aust, S. D. (1994b). Critical Review. Mechanisms white rot fungi use to degrade pollutants. *Environmental Science and Technology*, **28**, 78–87.

Bartha, R. & Pramer, D. (1970). Metabolism of acylanilide herbicides. *Advances in Applied Microbiology*, **13**, 317–341.

Beaudette L. (1998). Comparison of GC and mineralization experiment for measuring loss of PCB congeners in cultures of white rot fungi. *Applied and Environmental Microbiology*, **64**, 2020–2025.

Beaudette, L. A., Ward, O. P., Pickard, M. A. & Fedorak, P. M. (2000). Low surfactant concentration increases fungal mineralization of a polychlorinated biphenyl congener but has no effect on overall metabolism. *Letters in Applied Microbiology*, **30**, 155–160.

Bedard, D. L., Unterman, R., Bopp, L. H., Brennan, M. J., Haberl, M. L. & Johnson, C. (1986). Rapid assay for screening and characterizing microorganisms for the ability to degrade polychlorinated biphenyls. *Applied and Environmental Microbiology*, **51**, 761–768.

Bennett, J. W. & Faison, B. D. (1997). Use of fungi in bioremediation. In *Manual of Environmental Microbiology*, ed. C. J. Hurst, pp. 758–765. Washington, DC: American Society for Microbiology.

Bixby, M. W., Bousch, G. M. & Matsumura, F. (1971). Degradation of dieldrin to carbon dioxide by soil fungus *Trichoderma koningii*. *Bulletin of Environmental Contamination and Toxicology*, **64**, 491–494.

Bollag, J. M. (1974). Microbial transformation of pesticides. *Advances in Applied Microbiology*, **18**, 75–130.

Bumpus, J. A. & Aust, S. D. (1987). Biodegradation of DDT [1,1,1-trichloro-2,2-bis (4-chlorophenyl)ethane] by the white rot fungus *Phanerochaete chrysosporium*. *Applied and Environmental Microbiology*, **53**, 2001–2008.

Bumpus, J. A., Tien, M., Wright, D. & Aust, S. D. (1985). Oxidation of persistent environmental pollutants by a white rot fungus. *Science*, **228**, 1434–1436.

Bumpus, J. A., Fernando, T., Mileski, G. J. & Aust, S. D. (1987). Biodegradation of organopollutants by *Phanerochaete chrysosporium:* practical considerations. In *Proceedings of the 13th* Annual Research Symposium on Treatment of Hazardous Waste, pp. 401–411. Cincinatti, OH: US Environmental Protection Agency.

Bumpus, J. A., Powers, R. H. & Sun, T. (1993). Biodegradation of DDE (1,1-dichloro-2,2-bis(4-chlorophenyl)ethane by *Phanerochaete chrysosporium. Mycologia,* **97**, 95–98.

Casida, J. E. & Quistad, G. B. (1998). Golden age of insecticide research: past, present, or future? *Annual Review in Entomology,* **43**, 1–16.

Chapman, R. A., Tu, C. M., Harris, C. R. & Cole, C. (1981). Persistence of five pyrethroid insecticides in sterile and natural, mineral and organic soil. *Bulletin of Environmental Contamination and Toxicology,* **26**, 513–519.

Cook, A. M. (1987). Biodegradation of *s*-triazine xenobiotics. *FEMS Microbiology Reviews,* **46**, 93–116.

Cook, A. M. & Hutter, R. (1981). Degradation of *s*-triazines: a critical view of biodegradation. In *Microbial Degradation of Xenobiotics and Recalcitrant Compounds,* eds. T. Leisinger, R. Hutter, A. M. Cook & J. Nuesch, pp. 237–249. London: Academic Press.

Cork, D. J. & Krueger, J. P. (1991). Microbial transformation of herbicides and pesticides. *Advances in Applied Microbiology,* **36**, 1–11.

Cripps, R. E. & Roberts, T. R. (1978). Microbial degradation of herbicides. In *Pesticide Microbiology,* eds. I. R. Hill & S. J. L. Wright, pp. 669–730. London: Academic Press.

Crosby, D. G. (1981). Environmental chemistry of pentachlorophenol. *Pure and Applied Chemistry,* **53**, 1051–1080.

Cserjesi, A. J. & Johnson, E. L. (1972). Methylation of pentachlorophenol by *Trichoderma virgatum. Canadian Journal of Microbiology,* **18**, 45–49.

Deas, A. H. B. & Clifford, D. R. (1982). Metabolism of the 1,2,4-triazolylmethane fungicides, triadimefon, triadimenol and diclobutrazol, by *Aspergillus niger* (van Tiegh). *Pesticide Biochemistry and Physiology,* **17**, 120–133.

Deas, A. H. B., Clark, T. & Carter, G. A. (1984), The enantiomeric composition of triadimenol produced during metabolism of triadimefon by fungi. Part II. Differences between fungal species. *Pesticide Science,* **15**, 71–77.

Deping, L., Ruiwei, X. & Wei, J. (1991). The degradation, residue and vertical distribution of chlortoluron in soils. *Turang* (Nanjing), **23**, 307–310.

de Schrijver, A. & de Mot, R. (1999). Degradation of pesticides by Actinomycetes. *Critical Reviews in Microbiology,* **25**, 85–119.

Dietrich, D., Hickey, W. J. & Lamar, R. (1995). Degradation of 4,4′-dichlorobiphenyl, 3,3′,4,4′-tetrachlorobiphenyl and 2,2′,4,4′,5,5′-hexachlorobiphenyl by the white rot fungus *Phanerochaete chrysosporium. Applied and Environmental Microbiology,* **61**, 3904–3909.

Dmochewitz, S. & Ballschmiter, K. (1988). Microbial transformation of technical mixtures of polychlorinated biphenyls (PCB) by the fungus *Aspergillus niger. Chemosphere,* **17**, 111–121.

Donnelly, P. K. & Fletcher, J. S. (1995). PCB metabolism by ectomycorrhizal fungi. *Bulletin of Environmental Contamination and Toxicology,* **54**, 507–513.

Donnelly, P. K., Entry, J. A. & Crawford, D. L. (1993). Degradation of atrazine and 2,4-dichlorophenoxyacetic acid by mycorrhizal fungi at three nitrogen concentrations *in vitro. Applied and Environmental Microbiology,* **59**, 2642–2647.

Duah-Yentumi, S. & Kuwatsuka, S. (1982). Microbial degradation of benthiocarb, MCPA and 2,4-D herbicides in perfused soils amended with organic matter and chemical fertilizers. *Soil Science and Plant Nutrition*, **28**, 19–26.

Eaton, D. C. (1985). Mineralization of polychlorinated biphenyls by *Phanerochaete chrysosoporium*; a ligninolytic fungus. *Enzyme and Microbial Technology*, **7**, 194–196.

Elliott, M., Janes, N. F. & Potter, C. (1978). The future of pyrethroids in insect control. *Annual Review of Entomology*, **23**, 443–469.

El-Zorgani, G. A. & Omer, M. E. H. (1974). Metabolism of endosulfan isomer in *Aspergillus niger*. *Bulletin of Environmental Contamination and Toxicology*, **12**, 182–185.

Engelhardt, G., Wallnofer, P. R., Muecke, W. & Renner, G. (1986). Transformation of pentachlorophenol under environmental conditions. *Toxicology and Environmental Chemistry*, **11**, 233–252.

Evans, C. S. (1987). Lignin degradation. *Process Biochemistry*, **22**, 102–104.

Fantroussi, S., Verschuere, L., Verstraete, W. & Top, E. M. (1999). Effect of phenylurea herbicides on soil microbial communities estimated by analysis of 16S rRNA gene fingerprints and community-level physiological profiles. *Applied and Environmental Microbiology*, **65**, 982–988.

Fatichenti, F., Farris, G. A., Dieana, P., Cabras, P., Meloni, M. & Pirisi, F. M. (1984). The effect of *Saccharomyces cerevisiae* on concentration of dicarboxylmide and acylanilide fungicides, and pyrethroid insecticides during fermentation. *Applied Microbiology and Biotechnology*, **20**, 419–421.

Faulkner, J. K. & Woodcock, D. (1964). Metabolism of 2,4-dichlorophenoxyacetic acid by *Aspergillus niger* van Tiegh. *Nature*, **203**, 865–866.

Fernando, T. & Aust, S. D. (1994). Biodegradation of toxic chemicals by white rot fungi. In *Biological Degradation and Biochemistry of Toxic Chemicals*, ed. G. R. Chaudhry, pp. 386–402. London: Chapman & Hall.

Fernando, T., Aust, S. D. & Bumpus, J. A. (1989). Effects of culture parameters on DDT biodegradation by *Phanerochaete chrysosoporium*. *Chemosphere*, **19**, 1387–1398.

Ferry, M. L., Koskinen, W. C., Blanchette, R. A. & Burns, T. A. (1994). Mineralization of alachlor by lignin-degrading fungi. *Canadian Journal of Microbiology*, **40**, 795–800.

Fewson, C. A. (1988). Biodegradation of xenobiotics and other persistent compounds: the causes of recalcitrance. *Trends in Biotechnology*, **6**, 148–153.

Field, J. A. & Thurman, E. M. (1996). Glutathione conjugation and contaminant transformation. *Environmental Science and Technology*, **30**, 1413–1416.

Furukawa, K. (1982). Microbial degradation of polychlorinated biphenyls (PCBs). In *Biodegradation and Detoxification of Environmental Pollutants*, ed. A. M. Chakrabarty, pp. 33–57. Boca Raton, FL: CRC Press.

Furukawa, K., Tomizuka, N. & Kamibayashi, A. (1983). Metabolic breakdown of kaneclors (polychlorobiphenyls) and their products by *Acinetobacter* sp. *Applied and Environmental Microbiology*, **46**, 140–145.

Gilbertson, M. (1989). Effects on fish and wildlife populations. In 2nd edn, Vol. 4, *Halogenated Biphenyls, Terphenyls, Naphthalenes, Dibenzodioxins and Related Products. Topics in Environmental Health*, eds. R. D. Kimbrough & A. A. Jensen, pp. 103–127. Amsterdam: Elsevier.

Golyshin, P. M., Fredrickson, H. L., Giuliano, L., Rothmel, R., Timmis, K. N. & Yakimov, M. M. (1999). Effects of novel biosurfactants on biodegradation of polychlorinated biphenyls by pure and mixed bacterial cultures. *New*

Microbiology, **22**, 257–267.

Green, N. A., Meharg, A. A., Till, C., Troke, J. & Nicholson, J. K. (1999). Degradation of 4- fluorobiphenyl by mycorrhizal fungi as determined by ^{19}F NMR spectroscopy and ^{14}C radiolabelling analysis. *Applied and Environmental Microbiology,* **65**, 4021–4027.

Hammel, K. E. (1989). Organopollutant degradation by ligninolytic fungi. *Enzyme Microbiology and Technology,* **11**, 776–777.

Hammel, K. E. & Tardone, P. J. (1988). The oxidative 4-dechlorination of polychlorinated phenols is catalyzed by extracellular fungal lignin peroxidases. *Biochemistry,* **27**, 6563–6568.

Hammer, E. & Schauer, F. (1997). Fungal hydroxylation of dibenzofuran. *Mycological Research,* **101**, 433–436.

Hasan, H. A. (1999). Fungal utilization of organophosphate pesticides and their degradation by *Aspergillus flavus* and *A. sydowii* in soil. *Folia-Microbiologica Prague,* **44**, 77–84.

Haughland, R. A., Schlemm, D. J., Lyons, R. P. & Sferra, P. R. (1990). Degradation of the chlorinated phenoxyacetate herbicides 2,4-dichlorophenoxyacetic acid and 2,4,5- trichlorophenoxyacetic acid by pure and mixed bacterial cultures. *Applied and Environmental Microbiology,* **56**, 1357–1362.

Hayes, W. J. (1986). *Pesticide Studies in Man.* Baltimore: Williams & Williams.

Hickey, W. J., Fuster, D. J. & Lamar, R. T. (1994). Transformation of atrazine in soil by *Phanerochaete chrysosporium. Soil Biology and Biochemistry,* **26**, 1665–1671.

Higson, F. K. (1991). Degradation of xenobiotics by white-rot fungi. *Reviews of Environmental Contamination and Toxicology,* **122**, 111–152.

Higson, F. K. (1992). Microbial degradation of biphenyl and its derivatives. *Advances in Applied Microbiology,* **37**, 135–164.

Hill, I. R. (1978). Microbial transformation of pesticides. In *Pesticide Microbiology,* eds. I. R. Hill & S. J. L. Wright, pp. 137–202, London: Academic Press.

Hoff, T., Liu, S.-Y. & Bollag, J. M. (1985). Transformation of halogen-, alkyl-, and alkoxy- substituted anilines by a laccase of *Trametes versicolor. Applied and Environmental Microbiology,* **49**, 1040–1045.

Horvarth, R. S. (1972). Microbial co-metabolism and the degradation of organic compounds in nature. *Bacteriological Reviews,* **36**, 146–155.

Hutzinger, O., Safe, S. & Zitko, V. (1974). Commercial PCB preparations: properties and composition. In *The Chemistry of PCB's,* ed. O. Hutzinger, pp. 7–39. Cleveland, OH: CRC Press.

Illman, D. L. (1993). Hazardous waste treatment using fungus enters market place. *Science and Technology,* **2**, 26–29.

Jagnow, G., Haider, K. & Ellwardt, P. (1977). Anaerobic dechlorination and degradation of hexachlorocyclohexane isomers by anaerobic and facultative anaerobic bacteria. *Archives of Microbiology,* **115**, 285–292.

Johnsen, R. E. (1976). DDT metabolism in microbial systems. *Residue Review,* **61**, 1–28.

Johnson, B. T. & Kennedy, J. O. (1973). Biomagnification of *p,p'*-DDT and methoxychlor by bacteria. *Applied and Environmental Microbiology,* **26**, 66–71.

Johri, A. K., Saxena, D. M. & Lal, R. (1997). Interaction of synthetic pyrethroids with microorganisms: a review. *Microbios,* **89**, 151–156.

Jolly, A. L., Graves, J. B., Avault, J. W. & Koonce, K. L. (1977–8). Effects of a new insecticide on aquatic animals. *Louisiana Agriculture,* **21,** 3–4.

Jonas, U., Hammer, E., Schauer, F. & Bollag, J.-M. (1998). Transformation of 2-hydroxydibenzofuran by laccases of the white rot fungi *Trametes versicolor* and *Pycnoporus cinnabarinus* and characterization of oligomerization products. *Biodegradation,* **9,** 60–65.

Joshi, D. K. & Gold, M. H. (1993). Degradation of 2,4,5-trichlorophenol by the lignin-degrading basidiomycete *Phanerochaete chrysosporium. Applied and Environmental Microbiology,* **59,** 1779–1785.

Juhasz, A. L. & Naidu, R. (1999). Apparent degradation of 1,1,1-trichloro-2,2-bis-(*p*- chlorophenyl)ethane (DDT) by a *Cladosporium* sp. *Biotechnology Letters,* **21,** 991–995.

Kaars Sijpesteijn, A., Dekhuijzen, H. M. & Vonk, J. W. (1977). Biological conversion of fungicides in plants and microorganisms. In *Antifungal Compounds,* Vol. 2, eds. M. R. Siegel & H. D. Sisler, pp. 91–147. New York: Marcel Dekker.

Kaufman, D. D. (1967). Degradation of carbamate herbicides. *Journal of Agricultural and Food Chemistry,* **15,** 582–586.

Kaufman, D. D. & Blake, J. (1973). Microbial degradation of several acetamide, acylanilide, carbamate, toluidine, and urea pesticides. *Soil Biology and Biochemistry,* **5,** 2797–2803.

Kaufman, D. D. & Kearney, P. C. (1970). Microbial degradation of *s*-triazine herbicides. *Residue Reviews,* **32,** 235–265.

Kaufman, D. D., Kearney, P. C. & Sheets, T. J. (1965). Microbial degradation of simazine. *Journal of Agriculture and Food Chemistry,* **13,** 238–242.

Kaufman, D. D., Haynes, S. C., Jordan, E. G. & Kayser, A. J. (1977). Permethrin degradation in soil and microbial cultures. In *American Chemical Society Symposium Series,* No. 42, *Synthetic Pyrethroids,* ed. M. Elliott, pp. 147–161. Washington, DC: American Chemical Society.

Keith, L. H. & Teillard, W. A. (1979). Priority pollutants I – a perspective view. *Environmental Science and Technology,* **13,** 416–423.

Kennedy, D. W., Aust, S. D. & Bumpus, J. A. (1990). Comparative biodegradation of alkyl halide insecticides by the white rot fungus *Phanerochaete chrysosporium* (BKM-F-1767). *Applied and Environmental Microbiology,* **56,** 2347–2353.

Khadrani, A., Seigle-Murandi, F., Steiman, R. & Vroumsia, T. (1999). Degradation of three phenylurea herbicides (chlorotoluron, isoproturon and diuron) by micromycetes isolated from soil. *Chemosphere,* **38,** 3041–3050.

Khan, S. U., Behki, R. M., Tapping, R. I. & Akhtar, M. H. (1988). Deltamethrin residues in organic soils under laboratory conditions and its degradation by a bacterial strain. *Journal of Agricultural and Food Chemistry,* **36,** 636–638.

Kiigemagi, U., Morrison, H. E., Roberts, J. E. & Bollen, W. B. (1958). Biological and chemical studies on the decline of soil insecticides. *Journal of Economic Entomology,* **51,** 198–204.

Kilbane, J. J., Chatterjee, D. K. & Chakrabarty, A. M. (1983). Detoxification of 2,4,5- trichlorophenoxyacetic acid from contaminated soil by *Pseudomonas cepacia. Applied and Environmental Microbiology,* **45,** 1697–1700.

Kimbrough, R. D. & Jensen, A. A. (1989). In *Topics in Environmental Health,* 2nd edn, Vol. 4, *Halogenated Biphenyls, Terphenyls, Naphthalenes, Dibenzodioxins and Related Products,* eds. R. D. Kimbrough & A. A. Jensen. Amsterdam: Elsevier.

Kirk, T. K. & Chang, H. M. (1981). Potential applications of bio-ligninolytic systems. *Enzyme Microbiology and Technology*, **3**, 189–196.

Kirk, T. K., Lamar, R. T. & Glaser, J. A. (1992). The potential of white rot fungi in bioremediation. In *Biotechnology and Environmental Science: Molecular Approaches*, ed. S. Mongkolsuk, pp. 131–138. New York: Plenum Press.

Kohler, A., Jager, A., Willeshansen, H. & Graf, H. (1988). Extracellular ligninase of *Phanerochaete chrysosporium* Burdsall has no role in degradation of DDT. *Applied Microbiology and Biotechnology*, **29**, 618–620.

Kubatova, A., Matucha, M., Erbanova, P., Novotny, C., Vlasakova, V. & Sasek, V. (1998). Investigation into PCB biodegradation using uniformly ^{14}C-labelled dichlorobiphenyl. *Isotopes Environmental Health Studies*, **34**, 325–334.

Kullman, S. W. & Matsumura, F. (1996). Metabolic pathways utilized by *Phanerochaete chrysosporium* for degradation of the cyclodiene pesticide endosulfan. *Applied and Environmental Microbiology*, **62**, 593–600.

Kumar, S., Mukerjii, K. G. & Lal, R. (1996). Molecular aspects of pesticide degradation by microorganisms. *Critical Reviews in Microbiology*, **22**, 1–26.

Lal, R. & Saxena, D. M. (1982). Accumulation, metabolism and effect of organochlorine insecticides on microorganisms. *Microbiological Reviews*, **46**, 95–127.

Lamar, R. T. & Dietrich, D. M. (1990). *In situ* depletion of pentachlorophenol from contaminated soil by *Phanerochaete* sp. *Applied and Environmental Microbiology*, **56**, 3093–3100.

Lamar, R. T. & Dietrich, D. M. (1992). Use of lignin-degrading fungi in the disposal of pentachlorophenol-treated wood. *Journal of Industrial Microbiology*, **9**, 181–191.

Lamar, R. T. & Evans, J. W. (1993). Solid-phase treatment of a pentachlorophenol- contaminated soil using lignin-degrading fungi. *Environmental Science and Technology*, **27**, 2566–2571.

Lamar, R. T., Larsen, M. J. & Kirk, T. K. (1990). Sensitivity to and degradation of pentachlorophenol by *Phanerochaete* sp. *Applied and Environmental Microbiology*, **56**, 3519–3526.

Lamar, R. T., Davis, M. W., Dietrich, D. M. & Glaser, J. A. (1994). Treatment of a pentachlorophenol- and creosote-contaminated soil using the lignin-degrading fungus *Phanerochaete sordida*: a field demonstration. *Soil Biology and Biochemistry*, **26**, 1603–1611.

Leahey, J. P. (1985). *The Pyrethroid Insecticides*. London: Taylor and Francis.

Lestan, D., Lestan, M., Chapelle, J. A. & Lamar, R. T. (1996). Biological potential of fungal inocula for bioaugmentation of contaminated soils. *Journal of Industrial Microbiology*, **16**, 286–294.

Lichtenstein, E. P. (1980). 'Bound' residues in soils and transfer of soil residues in crops. *Residue Reviews*, **76**, 147–153.

Lichtenstein, E. P., Liang, T. T. & Koeppe, M. (1983). Effects of soil mixing and flooding on the fate of metabolism of ^{14}C-fonofos and ^{14}C-parathion in open and closed agricultural microcosms. *Journal of Economics and Entomology*, **76**, 233–238.

Liebich, J., Buranel, P. & Fuhr, F. (1999). Microbial release and degradation of nonextractable anilazine residues. *Journal of Agricultural and Food Chemistry*, **47**, 3905–3910.

Lin, J. E., Wang, H. Y. & Hickey, R. F. (1990). Degradation kinetics of pentachlorophenol by *Phanerochaete chrysosporium*. *Biotechnology and*

Bioengineering, **35**, 1125–1134.

Liu, S. Y., Freyer, A. J. & Bollag, J. M. (1991). Microbial dechlorination of the herbicide metolachlor. *Journal of Agricultural and Food Chemistry,* **39**, 631–634.

Loomis, A. K., Childress, A. M., Daigle, D. & Bennett, J. W. (1997). Alginate encapsulation of the white rot fungus *Phanerochaete chrysosporium. Current Microbiology,* **34**, 127–130.

Machemer, L. H. & Pickel, M. (1994a). Carbamate insecticides. *Toxicology,* **91**, 29–32.

Machemer, L. H. & Pickel, M. (1994b). Carbamate herbicides and fungicides, *Toxicology,* **91**, 105–108.

Macholz, R. M. & Kujawa, M. (1985). Recent state of lindane metabolism Part III. *Residue Reviews,* **94**, 119–149.

MacRae, I. C. (1989). Microbial metabolism of pesticides and structurally related compounds. In *Reviews in Environmental Contamination and Toxicology,* ed. G. W. Ware, pp. 1–87. Berlin: Springer-Verlag.

Maier-Bode, H. & Hartel, K. (1981). Linuron and monolinuron. *Residue Reviews,* **77**, 1–352.

Maloney, S. E. (1991). Microbial transformation of synthetic pyrethroid insecticides. PhD Thesis, University of London.

Maloney, S. E., Maule, A. & Smith, A. R. W. (1988). Microbial transformation of the pyrethroid insecticides: permethrin, deltamethrin, Fastac, fenvalerate and fluvalinate. *Applied and Environmental Microbiology,* **54**, 2874–2876.

Martens, M. (1976). Degradation of [8,9-^{14}C] endosulfan by soil microorganisms. *Applied and Environmental Microbiology,* **6**, 853–858.

Matsumura, F. & Benezct, H. J. (1978). Microbial degradation of insectides. In *Pesticide Microbiology,* eds. I. R. Hill & S. J. L. Wright, pp. 623–667. London: Academic Press.

Matsumura, F. & Bousch, G. M. (1968). Degradation of insecticides by a soil fungus *Trichoderma viride. Journal of Economics and Entomology,* **61**, 610–612.

Matsumura, F., Bousch, G. M. & Tai, A. (1968). Breakdown of dieldrin in the soil by a microorganism. *Nature* (London), **219**, 965–967.

Matsumura, F., Quensen, J. & Tsushimoto, G. (1983). Microbial degradation of TCDD in a model ecosystem. *Environmental Science and Research,* **26**, 191–219.

Maule, A, Plyte, S. & Quirk, A. V. (1987). Dehalogenation of organochlorine insecticides by mixed anaerobic microbial populations. *Pesticide Biochemistry and Physiology,* **27**, 229–236.

McAllister, K. A., Lee, H. & Trevors, J. T. (1996). Microbial degradation of pentachlorophenol. *Biodegradation,* **7**, 1–40.

McBain, A., Cui, F., Herbert, L. & Ruddick, J. N. R. (1995). The microbial degradation of chlorophenolic preservatives in spent, pressure-treated timber. *Biodegradation,* **6**, 47–55.

Meharg, A. A., Cairney, J. W. G. & Maguire, N. (1997). Mineralization of 2,4-dichlorophenol by ectomycorrhizal fungi in axenic culture and in symbiosis with pine. *Chemosphere,* **34**, 2495–2504.

Middledorp, P. J. M., Briglia, M. & Salkinoja-Salonen, M. S. (1990). Biodegradation of pentachlorophenol in natural soil by inoculated *Rhodococcus chlorophenolicus. Microbial Ecology,* **20**, 123–139.

Miles, J. R. W., Tu, C. M. & Harris, C. R. (1969). Metabolism of heptachlor and its degradation products by soil microorganisms. *Journal of Economics and*

Entomology, **62**, 1334–1337.

Mileski, G. J., Bumpus, J. A., Jurek, M. A. & Aust, S. D. (1988). Biodegradation of pentachlorophenol by the white rot fungus *Phanerochaete chrysosporium. Applied and Environmental Microbiology,* **54**, 2885–2889.

Miller, T. A. (1988). Mechanisms of resistance to pyrethroid insecticides. *Parasitology Today,* **4**, S8–S10.

Mohn, W. W. & Tiedje, J. M. (1992). Microbial reductive dehalogenation. *Microbiological Reviews,* **56**, 482–507.

Moorman, T. B. (1994). Pesticide degradation by soil microorganisms: environmental, ecological and management effects. *Advances in Soil Science,* **33**, 121–125.

Morgan, P., Lewis, S. T. & Watkinson, R. J. (1991). Comparison of abilities of white-rot fungi to mineralize selected xenobiotic compounds. *Applied Microbiology and Biotechnology,* **34**, 693–696.

Mostafa, I. Y., Bahig, M. R. E., Fakhr, I. M. I. & Adam, Y. (1972). Metabolism of organophosphorus insecticides. XIV Malathion breakdown by soil fungi. *Journal for Nature Research* (German), **27b**, 1115–1116.

Mougin, C., Laugero, C., Asther, M., Dubroca, J., Frasse & Asther M. (1994). Biotransformation of the herbicide atrazine by the white rot fungus *Phanerochaete chrysosporium. Applied and Environmental Microbiology,* **60**, 705–708.

Mougin, C., Pericaud, C., Malosse, C., Laugero, C. & Asther, M. (1996). Biotransformation of the insecticide lindane by the white rot basidiomycetes *Phanerochaete chrysosporium. Pesticide Science,* **47**, 51–59.

Mudd, P. J., Greaves, M. P. & Wright, S. J. L. (1983). The persistence and metabolism of isoproturon in soil. *Weed Research,* **23**, 239–246.

Mulla, M. S., Mian, L. S. & Kawecki, J. A. (1981). Distribution, transport, and fate of the insecticides malathion and parathion in the environment. *Residue Reviews,* **81**, 1–159.

Munnecke, D. E. & Solberg, R. A. (1958). Inactivation of semesan in soil by fungi. *Phytopathology,* **48**, 396–400.

Murrey, D. S., Rieck, W. L. & Lynd, J. Q. (1969). Microbial degradation of five substituted urea herbicides. *Weed Science,* **17**, 52–56.

Newcombe, D. A. & Crowley, D. E. (1999). Bioremediation of atrazine-contaminated soil by repeated applications of atrazine-degrading bacteria. *Applied Microbiology and Biotechnology,* **51**, 877–882.

Newman, J. F. (1978). Pesticides. In *Pesticide Microbiology*, eds. I. R. Hill & S. J. L. Wright, pp. 1–16. London: Academic Press.

Novotny, C., Vyas, B. R. M., Erbanova, P., Kubatova, A. & Sasek, V. (1997). Removal of PCBs by various white rot fungi in liquid cultures. *Folia Microbiologica,* **42**, 136–140.

Ohisa, N. & Yamaguchi, M. (1978). Gamma-BHC degradation accompanied by growth of *Clostridium rectum* isolated from paddy field soil. *Agricultural Biology and Chemistry,* **42**, 1819–1823.

Ohkawa, H., Nambu, K., Inui, H. & Miyamoto, J. (1978). Metabolic fate of fenvalerate (sumicidin) in soil and by soil microorganisms. *Journal of Pesticide Science,* **3**, 129–141.

Okeke, B. C., Paterson, A., Smith, J. E. & Watson-Craik, I. A. (1996). Influence of environmental parameters on pentachlorophenol biotransformation in soil by *Lentinula edodes* and *Phanerochaete chrysosporium. Applied Microbiology and Biotechnology,* **45**, 263–266.

220   S. E. Maloney

Okeke, B. C., Paterson, A., Smith, J. E. & Watson-Craik, I. A. (1997). Comparative biotransformation of pentachlorophenol in soils by solid substrate cultures of *Lentinula edodes*. *Applied Microbiology and Biotechnology*, **48**, 563–569.

Omar, S. A. (1998). Availability of phosphorus and sulfur of insecticide origin by fungi. *Biodegradation*, **9**, 327–336.

Papaizas, G. S. (1985). *Trichoderma* and *Gliocladium* biology, ecology and potential for biocontrol. *Annual Review in Phytopathology*, **23**, 23–25.

Paszczynski, A. & Crawford, R. L. (1995). Potential for bioremediation of xenobiotic compounds by the white rot fungus *Phanerochaete chrysosporium*. *Biotechnology Progress*, **11**, 368–379.

Patil, K. C., Matsumura, F. & Bousch, G. M. (1970). Degradation of endrin, aldrin and DDT by soil microorganisms. *Applied Microbiology*, **19**, 879–881.

Quensen, J. F., Tiedje, J. M. & Boyd, S. A. (1988). Reductive dechlorination of polychlorinated biphenyls by anaerobic microorganisms from sediments. *Science*, **242**, 752–754.

Rajagopal, B. S., Brahmaprakash, G. P., Reddy, B. R., Singh, U. D. & Sethunathan, N. (1984). Effect and persistence of selected carbamate pesticides in soil. *Residue Reviews*, **93**, 1–199.

Rao, K. R. (1978). *Pentachlorophenol: Chemistry, Pharmacology and Environmental Toxicology*. New York: Plenum Press.

Reddy, G. V. B. & Gold, M. H. (2000). Degradation of pentachlorophenol by *Phanerochaete chrysosporium:* intermediates and reactions involved. *Microbiology*, **146**, 405–413.

Reineke, W. (1984). Microbial degradation of halogenated aromatic compounds. In *Microbiology Series*, Vol. 13, *Microbial Degradation of Organic Compounds*, ed. D. T. Gibson, pp. 319–361. New York: Marcel Dekker.

Reineke, W. & Knackmuss, H. J. (1988). Microbial degradation of haloaromatics. *Annual Review of Microbiology*, **42**, 263–287.

Roberts, T. R. (1981). The metabolism of the synthetic pyrethroids in plants and soils. In *Progress in Pesticide Biochemistry*, Vol. 1, eds. T. R. Roberts & D. H. Hutson, pp. 115–146. Chichester: Wiley & Sons.

Rochkind-Dubinsky, M. L., Sayler, G. S. & Blackburn, J. W. (1987a). DDT and related compounds. In *Microbiology Series, Microbial Decomposition of Chlorinated Aromatic Compounds*. Vol. 18, ed. M. L. Rochkind-Dubinsky, pp. 153–162. New York: Marcel Dekker.

Rochkind-Dubinsky, M. L., Sayler, G. S. & Blackburn, J. W. (1987b). Chlorophenols and pentachlorophenol. In *Microbiology Series, Microbial Decomposition of Chlorinated Aromatic Compounds*. Vol. 18, ed. M. L. Rochkind-Dubinsky, pp. 95–107. New York: Marcel Dekker.

Rochkind-Dubinsky, M. L., Sayler, G. S. & Blackburn, J. W. (1987c). Chlorophenoxy and chlorophenyl herbicides. In *Microbiology Series, Microbial Decomposition of Chlorinated Aromatic Compounds*. Vol. 18, ed. M. L. Rochkind-Dubinsky, pp. 108–123. New York: Marcel Dekker.

Rosenberg, A. & Alexander, M. (1980). Microbial metabolism of 2,4,5-trichlorophenoxyacetic acid in soil, soil suspensions and in axenic culture. *Journal of Agricultural and Food Chemistry*, **28**, 297–302.

Roslycky, E. B. (1977). Response of soil microbiota to selected herbicide treatments. *Canadian Journal of Microbiology*, **23**, 426–433.

Ross, J. A. & Tweedy, B. G. (1973). Degradation of four phenylurea herbicides by mixed populations of microorganisms from two soil types. *Soil Biology and Biochemistry,* **5**, 739–745.

Roy-Arcand, L. & Archibald, F. S. (1991). Direct dechlorination of chlorophenolic compounds by laccases from *Trametes (Coriolus) versicolor. Enzyme and Microbial Technology,* **13**, 194–203.

Ryan, T. P. & Bumpus, J. A. (1989). Biodegradation of 2,4,5-trichlorophenoxyacetic acid in liquid culture and in soil by the white rot fungus *Phanerochaete chrysosporium. Applied Microbiology and Biotechnology,* **31**, 302–307.

Safe, S. (1989). Polychlorinated biphenyls (PCBs): mutagenicity and carcinogenicity. *Mutation Research,* **220**, 31–47.

Safe, S. (1990). Polychlorinated biphenyls (PCBs), dibenzo-*p*-dioxins (PCDDs), dibenzofurans (PCDFs) and related compounds: environmental and mechanistic considerations which support the development of toxic equivalency factors (TEFs). *Critical Reviews in Toxicology,* **21**, 51–88.

Sakata, S., Mikami, N. & Yamada, H. (1992a). Degradation of pyrethroid optical isomers in soil. *Journal of Pesticide Science,* **17**, 169–180.

Sakata, S., Mikami, N. & Yamada, H. (1992b). Degradation of pyrethroid optical isomers by soil microorganisms. *Journal of Pesticide Science,* **17**, 181–189.

Salama, A. K. M. (1998). Metabolism of carbofuran by *Aspergillus niger* and *Fusarium graminearum. Journal of Environmental Science and Health,* **B33**, 252–266.

Salkinoja-Salonen, M. S., Hakulinen, R., Valo, R. & Apajalahti, J. (1983). Biodegradation of recalcitrant organochlorine compounds in fixed film reactors. *Water Science and Technology,* **15**, 309–319.

Sasek, V., Volfova, O., Erbanova, P., Vyas, B. R. M. & Matucha, M. (1993). Degradation of PCBs by white-rot fungi, methylotrophic and hydrocarbon utilizing yeasts and bacteria. *Biotechnology Letters,* **15**, 521–526.

Schroeder, M. (1970). The microbial decomposition of substituted urea herbicides. *Weed Science,* 81–82.

Seigle-Murandi, F., Steiman, R. & Benoit-Guyod, J. L. (1991). Biodegradation potential of some micromycetes for pentachlorophenol. *Ecotoxicology and Environmental Safety,* **21**, 290–300.

Shelton, D. R., Khader, S., Karns, J. S. & Pogell, B. M. (1996). Metabolism of twelve herbicides by *Streptomyces. Biodegradation,* **7**, 129–136.

Singh, B. K. & Kuhad, R. C. (1999). Biodegradation of lindane (γ-hexachlorocyclohexane) by the white-rot fungus *Trametes hirsutus. Letters in Applied Microbiology,* **28**, 238–241.

Sjoblad, R. D. & Bollag, J.-M. (1981). Oxidative coupling of aromatic compounds by enzymes from soil microorganisms. In *Soil Biochemistry,* eds. E. A. Paul & J. N. Ladd, pp. 113–152. New York: Marcel Dekker.

Smith, A. E. & Phillips, D. V. (1975). Degradation of alachlor by *Rhizoctonia solani. Agronomic Journal,* **67**, 347–350.

Solomon, K. R. (1986). *Pyrethroids: Their Effects on Aquatic and Terrestrial Ecosystems.* [Publication 24376]. Ontario, Canada: Associate Committee on Scientific Criteria for Environmental Quality, National Research Council Canada.

Spynu, E. I. (1989). Predicting pesticide residues to reduce crop contamination. *Reviews in Environmental Contamination and Toxicology,* **109**, 89–107.

Stamper, D. M. & Tuovinen, O. H. (1998). Biodegradation of the acetanilide herbicides alachlor, metolachlor and propachlor. *Critical Reviews in*

Microbiology, **24**, 1–22.

Stanlake, G. J. & Finn, R. K. (1982). Isolation and characterization of a pentachlorophenol-degrading bacterium. *Applied and Environmental Microbiology,* **44**, 1421–1427.

Stott, D. E., Martin, J. P., Focht, D. D. & Haider, K. (1983). Biodegradation, stabilization in humus, and incorporation into soil biomass of 2,4-D and chlorocatechol carbons. *Soil Science Society of America Journal,* **47**, 66–70.

Stratton, G. W. & Corke, C. T. (1982a). Toxicity of the insecticide permethrin and some degradation products towards algae and cyanobacteria. *Environmental Pollution,* **29**, 71–80.

Stratton, G. W. & Corke, C. T. (1982b). Comparative fungitoxicity of the insecticide permethrin and ten degradation products. *Pesticide Science,* **13**, 679–685.

Subba-Rao, R. V. & Alexander, M. (1985). Bacterial and fungal co-metabolism of 1,1,1-trichloro-2,2-bis-(4-chlorophenyl)ethane (DDT) and its breakdown products. *Applied and Environmental Microbiology,* **49**, 509–516.

Sudhakar, Y. & Dikshit, A. K. (1999). Adsorbent selection for endosulfan removal from water environment. *Journal of Environmental Science and Health,* **B34**, 97–118.

Takada, S., Nakamura, M., Matsueda, T., Kondo, R. & Saki, K. (1996). Degradation of polychlorinated dibenzo-*p*-dioxins (PCDDs) and polychlorinated dibenzofurans (PCDFs) by the white rot fungus *Phanerochaete sordida* YK-624. *Applied and Environmental Microbiology,* **62**, 4323–4328.

Taverne, J. (1999). DDT – to ban or not to ban? *Parasitology Today,* **15**, 180–181.

Tessier, J. (1982). The path to Deltamethrin. In *Deltamethrin Monograph,* ed. J. Lhoste, pp. 25–36. France: Roussel Uclaf Publications.

Thomas, M. D. & Kenerly, C. M. (1989). Transformation of *Gliocladium. Current Genetics,* **15**, 415–420.

Thomas, D. R., Carswell, K. S. & Georgiou, G. (1992). Mineralization of biphenyl and polychlorinated biphenyls by the white rot fungus *Phanerochaete chrysosporium. Biotechnology and Bioengineering,* **40**, 1395–1402.

Tiedje, J. M. & Hagedorn, M. L. (1975). Degradation of alachlor by a soil fungus, *Chaetomium globosum. Journal of Agricultural and Food Chemistry,* **23**, 77–80.

Tillmanns, G. M., Wallnoefer, P. R. & Engelhardt, G. (1978). Oxidative dealkylation of five phenylurea herbicides by the fungus *Cunninghamella echinulat. Chemosphere,* **7**, 59–64.

Tu, C. M. (1980). Influence of five pyrethroid insecticides on microbial population and activities in soil. *Microbial Ecology,* **5**, 321–327.

Tu, C. M., Miles, J. R. W. & Harris, C. R. (1968). Soil microbial degradation of aldrin. *Life Science,* **7**, 311–322.

Tulp, M., Tillmanns, G. M. & Hutzinger, O. (1977). Environmental chemistry of PCB-replacement compounds. V. The metabolism of chloroisopropylbiphenyls in fish, frogs, fungi and bacteria. *Chemosphere,* **6**, 223–230.

Unterman, R., Bedard, D. L., Brennan, M. J., Bopp, L. H., Mondello, F. J., Brooks, R. E., Mobley, D. P., McDermott, J. B., Schwartz, C. C. & Dietrich, D. K. (1988). Biological approaches for PCB degradation. In *Reducing Risks*

From *Environmental Chemicals Through Biotechnology*, ed. G. S. Omen, pp. 253–269. New York: Plenum Press.

Valli, K. & Gold, M. H. (1991). Degradation of 2,4-dichlorophenol by the lignin-degrading fungus *Phanerochaete chrysosporium. Journal of Bacteriology,* **173**, 345–352.

van de Waerdt, J. (1983). Drins and related insecticides: toxicology, use and alternatives; with special reference to Latin America and SE Asia. *Ecoscript 23.* Netherlands: PAN for the Foundation for Ecological Development Alternatives.

Van-Oostdam, J., Gilman, A., Dewailly, E., Usher, P., Wheatley, B., Kuhnlein, H., Neve, S., Walker, J., Tracy, B., Feeley, M., Jerome, V. & Kwavnick, B. (1999). Human health implications of environmental contaminants in Arctic Canada; a review. *Science of the Total Environment,* **230**, 1–82.

Venkatadri, R., Tsai, S. P., Vukanic, N. & Hein, L. B. (1992). Use of a biofilm membrane reactor for the production of lignin peroxidase and treatment of pentachlorophenol by *Phanerochaete chrysosporium. Hazardous Waste and Hazardous Matter,* **9**, 231–243.

Vroumsia, T., Steiman, R., Seigle-Murandi, F., Benoit-Guyod, J. L. & Khadrani, A. (1996). Biodegradation of three substituted phenylurea herbicides (chlorotoluron, diuron and isoproturon) by soil fungi. A comparative study. *Chemosphere,* **33**, 2045–2056.

Wallnofer, P. R. & Engelhardt, G. (1973). Microbial degradation of pesticides. *Chemosphere,* **2**, 1–9.

Wallnofer, P. R., Engelhardt, G., Safe, S. & Hutzinger, O. (1973). Microbial hydroxylation of 4-chlorobiphenyl and 4,4'-dichlorobiphenyl. *Chemosphere,* **2**, 69–72.

Walter, M. V. (1997). Bioaugmentation. In *Manual of Environmental Microbiology,* ed. C. J. Hurst, pp. 753–757. Washington, DC: American Society for Microbiology.

Weinberger, M. & Bollag, J. M. (1972). Degradation of chlorbromuron and related compounds by the fungus *Rhizoctonia solani. Applied Microbiology,* **24**, 750–754.

Wittich, R.-M. (1998). Degradation of dioxin-like compounds by microorganisms. *Applied Microbiology and Biotechnology,* **49**, 489–499.

Woodcock, D. (1978). Fungicides, fumigants and nematocides. In *Pesticide Microbiology,* eds. I. R. Hill & S. J. L. Wright, pp. 731–780. London: Academic Press.

Woodhead, D. (1983). Permethrin trials in the Meltham sewage catchment area. *Water Services,* 198–202.

Yadav, J. S. & Reddy, C. A. (1993). Mineralisation of 2,4-D and mixtures of 2,4-D and 2,4,5 T by *Phanerochaete chrysosporium. Applied and Environmental Microbiology,* **59**, 2904–2908.

Yadav, J. S., Quensen, J. F., Tiedje, J. M. & Reddy, C. A. (1995). Degradation of polychlorinated biphenyl mixtures (Aroclors 1242, 1254 and 1260) by the white rot fungus *Phanerochaete chrysosporium* as evidenced by congener-specific analysis. *Applied and Environmental Microbiology,* **61**, 2560–2565.

Zabel, T. F., Seager, J. & Oakley, S. D. (1988). *Proposed Environmental Quality Standards for List II Substances in Water, Mothproofing Agents.* [Environmental Strategy Standards and Legislation Unit TR 261 publication.] London: Water Research Council.

9

Degradation of energetic compounds by fungi

DAVID A. NEWCOMBE
AND RONALD L. CRAWFORD

Introduction

Energetic compounds have important roles in military and civilian applications, and their production represents a considerable portion of the chemical manufacturing industry. Soils and waters at a significant number of sites worldwide have become contaminated with energetic organonitro compounds as a result of manufacturing and decommissioning of ordnance (Rosenblatt *et al.*, 1991). Kaplan (1990) describes hazardous energetic organonitro compounds as a class of synthetic chemical characterized by the presence of a nitroaromatic, nitrate ester or nitramine functional group or moiety. The relative toxicity, mutagenicity and recalcitrance of these compounds in the environment has led to intensive research for innovative technologies to treat contaminated wastes, soils and waters (Kaplan, 1990, 1992; Rosenblatt *et al.*, 1991).

Technologies have been developed to reduce or remove hazardous energetic organonitro compounds from particular waste streams and from the environment in general. Physical treatment technologies include activated carbon absorption, air stripping, filtration and incineration. Chemical treatment technologies include solvent extraction, surfactant precipitation and neutralization (Kaplan, 1990). Biological treatment technologies include denitrification (Kaplan, 1990), batch and continuous fermentation systems (Funk *et al.*, 1995a,b; Razo-Flores *et al.*, 1997; Lenke *et al.*, 1998) and composting (Isbister *et al.*, 1984; Williams, Ziegenfuss & Sisk, 1992; Funk *et al.*, 1995b; Emery & Faessler, 1997; Tuomi, Coover & Stroo, 1997; Lenke *et al.*, 1998). A biological approach is often desirable because of its relatively low cost compared with chemical or physical treatment technologies and the innocuous nature of the typical by-products, carbon dioxide and water.

An important factor determining the feasibility of using a biological remediation system is the amenability of the toxic compound(s) to biological attack. Only a few naturally occurring nitroorganic compounds have been found to date (Gorontzy *et al.*, 1994), so the occurrence of degrader organisms in the environment would be expected to be rare. This hypothesis is supported by the relative longevity of hazardous energetic nitro-organic compounds in the environment. However, bacteria and fungi that can degrade these compounds have been isolated. Many studies have been performed in recent years on the characterization of organisms, identification of degradation pathways and fate of the parent organonitro compounds. These data have been extensively reviewed (Kaplan, 1990, 1992; Rosenblatt *et al.*, 1991; Higson, 1992; Gorontzy *et al.*, 1994; Lewis *et al.*, 1997). Some of the most extensive breakdown of organonitro compounds during biodegradation experiments has been observed in cultures of *Phanerochaete chrysosporium* (Fernando & Aust, 1991a,b), which is known to produce enzymes that degrade complex compounds such as lignin. The non-specific lignin-degrading enzymes produced by *P. chrysosporium* are capable of catalysing the oxidation of many xenobiotic compounds (Paszczynski & Crawford, 1995). It is the occurrence of these enzymes and the ability of fungi to grow on complex substrates that has led researchers to examine the potential of fungi to degrade hazardous organonitro compounds. This chapter reviews the research performed on this topic.

Nitroaromatics

The prevalence of the nitroaromatic compounds 2,4,6-trinitrotoluene (TNT), *N*-methyl-*N*,2,4,6-tetranitroaniline and dinitrotoluenes at contaminated sites has led researchers to investigate their degradation. Unfortunately, an increasing degree of nitro-substitution apparently renders the aromatic ring electron deficient to the point that it no longer acts as a substrate for electrophilic oxygenation mechanisms. For example, Spanggord *et al.* (1991) were able to demonstrate that 2,4-dinitrotoluene was degraded via oxygenation by a *Pseudomonas* sp. To our knowledge, no studies have demonstrated a similar bacterial degradation pathway for TNT (Lewis *et al.*, 1997; Blotevogel & Gorontzy, 2000), leading researchers to seek out fungi that could degrade nitroaromatic compounds.

2,4,6-Trinitrotoluene

Historically, TNT (Fig. 9.1) is one of the most widely used military explosives in the world (Rosenblatt *et al.*, 1991). Harter (1985) estimated

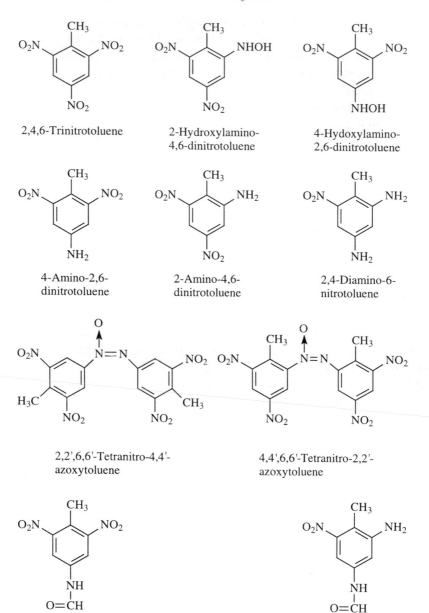

Fig 9.1. Chemical structures of 2,4,6-trinitrotoluene (TNT) and fungal transformation products.

the worldwide production of TNT to be at 10^6 kg per year. The persistence of TNT in soils contaminated during World War II and the Korean conflict reveals its relative resistance to degradation by indigenous microorganisms, which is a result, in part, of its toxicity to biological systems. The electrophilic nature of the nitro group causes TNT readily to oxidize biological reductants, causing toxicity directly or by formation of other reactive products such as nitroarene radicals (Mason & Josephy, 1985). In addition, the nitro groups draw electrons from the aromatic π bonds, effectively reducing the electron density of the conjugated aromatic system. As a result, TNT is resistant to degradation via electrophilic attack by oxygenases (Vorbeck *et al.*, 1994, 1998; Rieger & Knackmuss, 1995). In order for the aromatic ring to be cleaved, organisms must first remove or transform the nitro groups. The abundance, persistence and resistance of TNT make it one of the most intensely studied hazardous organonitro compounds with respect to bioremediation. Consequently, the largest body of work and the bulk of this chapter focus on the fungal degradation of TNT, especially by *P. chrysosporium.*

Some of the earliest published work on fungal degradation of TNT was that of Klausmeier, Osmon & Walls (1974). During their research on the effects of TNT on microorganisms, a *Rhizopus nigricans* strain was found to mediate the removal of 100 ppm TNT from a minimal medium. Parrish (1977) screened 190 fungi from 98 genera and found that 183 could transform 100 ppm TNT to reduced products. The majority of the strains reduced TNT at the 4-nitro position to 4-hydroxylamino-2,6-dinitrotoluene (4HADNT) and then to 4-amino-2,6-dinitrotoluene (4ADNT) (Fig. 9.1).

Fernando, Bumpus & Aust (1990) studied the ability of *P. chrysosporium* in a nitrogen-limiting medium to degrade TNT. Nitrogen limitation in this medium mimics lignolytic conditions and thus maximizes the expression of peroxidases by *P. chrysosporium*. Approximately 35% of the radioactivity from TNT labelled in the ring by ^{14}C added at a concentration of 1.3 ppm was trapped as $^{14}CO_2$. Of the remaining radioactivity, 25% was water-soluble material, and 16% was extracted into methylene chloride and eluted from a high performance liquid chromatography (HPLC) system as material more polar than TNT (Fernando *et al.*, 1990). When higher concentrations of TNT were used in aqueous or soil incubations, degradation and overall transformation were less extensive over the 90-day time periods studied. These data indicated that *P. chrysosporium* is capable of extensive degradation of TNT and stimulated much more work on characterizing the process and devising technology to exploit it (Tsai, 1991; Sublette, Ganapathy & Schwartz, 1992). Spiker, Crawford & Crawford (1992) found that TNT was inhibitory to spore germination of *P. chrysosporium* at concentrations greater than 5 ppm. This toxicity could be related to the activity of TNT as an oxidant, since reduction to aminodinitrotoluenes (ADNT) relieved toxicity. The well-characterized pathway of TNT degradation by *P. chrysosporium* is initiated by the reduction of a nitro group, followed by oxidation and subsequent aromatic ring cleavage (Stahl & Aust, 1993a; Michels & Gottschalk, 1994, 1995). Reductive steps are thought to be catalysed by aromatic nitroreductase activity (Stahl & Aust, 1993a,b; Michels & Gottschalk, 1995). This activity by *P. chrysosporium* was found to be membrane bound in one study (Stahl & Aust, 1993b; Reible, Joshi & Gold, 1994) and soluble in another (Michels & Gottschalk, 1995). The transient appearance of nitroso and hydroxylamino intermediates (Fig. 9.1) indicates that reduction is stepwise (Bumpus & Tatarko, 1994; Michels & Gottschalk, 1994; Reible *et al.*, 1994) and probably catalysed by a nitroreductase enzyme. Maximum evolution of $^{14}CO_2$ from cultures incubated with uniformly labelled ([^{14}C-UL]-labelled) TNT was correlated with expression of lignolytic activity, including production of various peroxidases (Hawari *et al.*, 1999). High levels of the reduced metabolite 4ADNT were found to inhibit lignin peroxidase activity (Michels & Gottschalk, 1994; Bumpus & Tatarko, 1994), causing inhibition in the cleavage of the aromatic ring (Michels & Gottschalk, 1994). 4HADNT was also found to serve as a substrate for and to be oxidized by lignin peroxidase (Michels & Gottschalk, 1994; Bumpus & Tatarko, 1994) and manganese peroxidase (Michels & Gottschalk, 1995), causing the formation of azoxytetranitrotoluenes (Fig. 9.1). Once 4HADNT had been reduced to 4ADNT, additional reactions were observed to take place before oxidation by the lignolytic systems. 4ADNT could be formylated to give 4-formamido-2,6-dinitrotoluene, which was reduced to give 2-amino-4-formamido-6-nitrotoluene (Fig. 9.1) (Donnelly *et al.*, 1997; Hawari *et al.*, 1999). This compound has been observed to slowly transform to diaminonitrotoluene (DANT), which accumulated under non-lignolytic conditions (Stahl & Aust, 1993a; Michels & Gottschalk, 1995). Under lignolytic conditions, DANT did not accumulate and was not subject to further transformations (Michels & Gottschalk, 1995). Recent research points toward the importance of manganese peroxidase in the oxidation of reduced TNT metabolites. Scheibner & Hofrichter (1998) showed that cell-free preparations of manganese peroxidase from cultures of *Nematoloma frowardii* and *Stropharia rugosoannulata* transformed TNT, 2-amino-4,6-dinitrotoluene

(2ADNT), 4ADNT and 2,6-diamino-4-nitrotoluene (2,6DANT) to unknown metabolites (Scheibner, Hofrichter & Fritsche, 1997a; Scheibner & Hofrichter, 1998). Furthermore, these authors noted that the presence of reduced thiols like glutathione or the amino acid L-cysteine considerably enhanced the rate and extent of biodegradation. In similar experiments with manganese peroxidase prepared from *Phlebia radiata*, van Aken *et al.* (1999) showed that manganese peroxidase caused more extensive oxidation of the aromatic ring with an increasing number of reduced amino groups present.

Many studies have been undertaken to screen fungi other than *P. chrysosporium* for their ability to degrade TNT. In an effort to find fungi that catalysed the initial reductive steps more rapidly than *P. chrysosporium*, and that were more tolerant to high concentrations of TNT, Bayman and Radkar (1997) studied eight fungi. Three species, *Trichoderma viride, Cladosporium resinae* and *Alternaria alternata* were less affected by 100 ppm TNT than *P. chrysosporium* during inhibition assays. Two species, *C. resinae* and *Cunninghammella echinulata* var. *elegans*, were able to catalyse the initial reductive steps more rapidly than *P. chrysosporium*. These authors did not observe any evolution of $^{14}CO_2$ or ^{14}C volatile compounds from [^{14}C-UL]-labelled TNT by the cultures tested but did not grow these strains under any other conditions. On the basis of these results, they suggested a two-step process for TNT biodegradation by fungi in which more tolerant fungi are initially used to detoxify a system, after which *P. chrysosporium* is added to enhance breakdown (Bayman & Radkar, 1997). Meharg, Dennis & Cairney (1997) tested the ability of four ectomycorrhizal basidiomycetes to transform TNT. In their study on one isolate, *Suillus variegatus*, the addition of protease inhibitors to the culture medium resulted in the ability of cell-free extracts to reduce TNT. However, the presence of mycelia enhanced the reduction of TNT over that of cell-free extracts (Meharg *et al.*, 1997). In a study of four white rot fungi, Donnelly *et al.* (1997) grew the fungi under non-lignolytic conditions and observed the transformation rates of TNT. They confirmed the conclusions of earlier studies, that all four strains were able to transform TNT under the given conditions. However, the transformation rate did not equal the detoxification rate, as shown by mutagenicity assays. Scheibner *et al.* (1997b) screened 91 fungal strains belonging to 32 genera of wood- and litter-decaying basidiomycetes. In this study, all the strains could catalyse the initial reduction reactions; however, micromycetes could accumulate higher amounts of ADNTs (Scheibner *et al.*, 1997b). A second screen was performed to select strains able to oxidize [^{14}C-ring] radiolabelled TNT. The highest ring oxidation rates were observed in wood- and litter-decaying basidiomycetes (Scheibner *et al.*, 1997b). Van Aken *et al.* (1997) subjected *P. radiata* to TNT transformation studies under lignolytic and non-lignolytic conditions. Their results were very similar to those of previous studies with white rot fungi. Finally, Samson *et al.* (1998) tested the ability of *Ceratocystis coerulescens, Lentinus lepideus* and *Trichoderma harzianum* to remove TNT from liquid culture medium. The degradation patterns were similar to those presented above.

Our laboratory has recently studied the ability of a brown rot fungus, *Gloeophyllum trabeum*, to transform TNT. Two-week old cultures were able to transform 50 ppm TNT to below detection limits within 3 days. Metabolites identified to date by HPLC and gas chromatography/mass spectrophotometery are 2 ADNT, 4 ADNT and 2,4-amino-6-nitrotoluene (D. A. Newcombe, A. Paszczynski & R. L. Crawford, unpublished data). *G. trabeum* can grow in a minimal salts, nitrogen-limiting medium with TNT concentrations $\leq 40 \, \mathrm{mg\,l}^{-1}$ (D. A. Newcombe, A. Paszczynski & R. L. Crawford, unpublished data). These data are encouraging, considering the relatively lower toxicity threshold of *P. chrysosporium* when growing in the presence of TNT (Spiker *et al.*, 1992). Work is continuing on characterization of the degradation pathway and system that may be responsible.

Dinitrotoluene

Dinitrotoluenes contaminate various sites, either as a result of the TNT-manufacturing process or as a constituent in propellants added to control burn rates and reduce hygroscopicity (Rosenblatt *et al.*, 1991). Bacteria that extensively decompose 2,4-dinitrotoluene have been isolated (Fig. 9.2), and their degradation pathways are reasonably well known (Spain, 1995a,b). It has been shown that 2,4-dinitrotoluene is readily attacked by dioxygenase enzymes (An, Gibson & Spain, 1994), so little work has been carried out on the fungal degradation of dinitrotoluenes (Suen, Haigler & Spain, 1996). However, McCormick, Cornell & Kaplan (1978) and Parrish (1977) have described 2,4-dinitrotoluene degradation by fungal cultures. Metabolites identified in a culture of *Mucosporium* sp. were 2-amino-4-nitrotoluene, 4-amino-2-nitrotoluene, 2,2'-dinitro-4,4'-azoxytoluene, 4,4'-dinitro-2,2'-azoxytoluene, 4-acetamido-2-nitrotoluene and an unidentified compound presumed to be an azoxytoluene (Fig. 9.2) (McCormick *et al.*, 1978). Valli *et al.* (1992) described the initial steps of degradation of

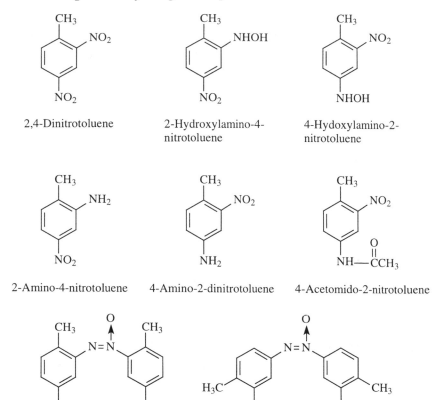

Fig 9.2. Chemical structures of 2,4-dinitrotoluene and fungal transformation products.

2,4-dinitrotoluene by *P. chrysosporium* as a reductive process yielding
2-amino-4-nitrotoluene. After reduction, the amine group was eliminated
as ammonium, whereas the other nitro group was eliminated as nitrite. In
incubations with [^{14}C-ring]-labelled TNT, 34% of the label was recovered
as $^{14}CO_2$ (Valli *et al.*, 1992). In a study in which *C. coerulescens*, *L. lepideus*
and *T. harzianum* were incubated with 2,4-dinitrotoluene and glucose,
reduced metabolites were detected (Samson *et al.*, 1998). *C. coerulescens*
and *T. harzianum* degraded dinitrotoluene more rapidly and to a greater
extent than *L. lepideus*. The authors also concluded that lignolytic conditions favoured rapid and complete degradation of 2,4-dinitrotoluene
(Samson *et al.*, 1998).

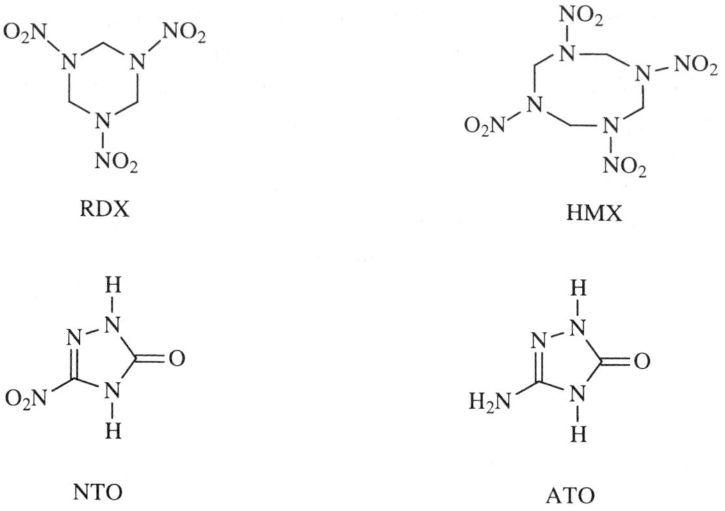

Fig 3. Chemical structures of RDX, HMX, NTO and the NTO fungal metabolite ATO (see text for full names).

Nitramines

The nitramine organonitro explosives included in this review are hexa-hydro-1,3,5-trinitro-1,3,5-triazine (RDX; British code name for Research Department or Royal Demolition Explosive), 1,3,5,7-tetranitro-1,3,5,7-tetrazocyclooctane (HMX; British code name for High Melting or His Majesty's Explosive) and 5-nitro-1,2,4-triazol-3-one (NTO) (Fig. 9.3). They have the same properties that confound the biodegradation of nitroaromatic compounds. The nitramines are fairly insoluble in aqueous solution and either the parent compound or its degradation products have been shown to be toxic to some biological systems (Le Campion, Vandais & Ouazzani, 1999). RDX and HMX are known to be degraded under anaerobic conditions by bacterial systems (Kaplan, 1990, 1992; Gorontzy *et al.*, 1994). However, the possibility of more rapid or *in situ* remediation of these compounds has sparked research into fungal systems.

RDX and HMX

RDX is prepared alone (composition A), is mixed with TNT (composition B) (Rosenblatt *et al.*, 1991) or is mixed with various plasticizers (composition C) (Urbanski, 1984). Along with TNT, it is one of the most predominant explosives used in military applications (Fernando & Aust, 1991a), and has been in use since World War II (McLellan, Hartley & Brower, 1992). RDX-laden wastewater has been successfully treated by physical, chemical and biological methods (McCormick, Cornell & Kaplan, 1981; Rosenblatt *et al.*, 1991), although contaminated soils are much more problematic to treat. These challenges led Fernando & Aust (1991a) to study RDX degradation by *P. chrysosporium*. In experiments with pure cultures of *P. chyrsosporium* grown under nitrogen-limiting conditions for 30 days, the authors recovered 67% of [^{14}C]-labelled RDX as $^{14}CO_2$, 20.2% in the soluble fraction, 4.8% in a methylene chloride extractant and 2.1% in association with the mycelium. Upon incubation of the fungus with contaminated soil amended with corn cobs and spiked with [^{14}C]-labelled RDX, they recovered 76% of the label as $^{14}CO_2$, while 4.5% was present in an acetonitrile extract and 9.7% was associated with a tightly bound fraction (Fernando & Aust, 1991a). No intermediates were identified in this study. Bayman, Ritchey & Bennett (1995) compared the tolerance to RDX of *C. resinae*, *C. echinulata* var *elegans*, *P. chrysosporium* and *Cyathus pallidus*. None of these strains of fungi exhibited significant inhibition compared with non-contaminated controls for 100 ppm RDX in radial colony growth assays. In experiments in which the fungi were grown in non-lignolytic conditions, *C. resinae* transformed the greatest amount of RDX; however, the metabolites were not identified (Bayman *et al.*, 1995).

As a by-product of RDX synthesis, HMX has been found at sites contaminated with RDX (Rosenblatt *et al.*, 1991; Gorontzy *et al.*, 1994). However, HMX is increasingly being used as a propellant and in high-grade explosives (Rosenblatt *et al.*, 1991). Very few studies, if any, have looked at HMX degradation by fungi. Since HMX is a common co-contaminant with RDX, it would be expected that the studies mentioned above involved transformations of HMX, even if it was not purposefully observed.

NTO

NTO represents one of a new generation of energetic organonitro compounds that performs to the specifications of RDX but is more stable under various conditions (Becuwe & Delclos, 1993; Le Campion *et al.*, 1999). Le Campion *et al.* (1999) isolated *Bacillus licheniformis* and a *Penicillium* sp. from an aqueous industrial waste contaminated with NTO. In addition to these isolates, the authors screened eight bacteria and 22 fungi for their ability to transform NTO. All of the microorganisms in question catalysed the transformation of NTO to 5-amino-1,2,4-triazol-3-

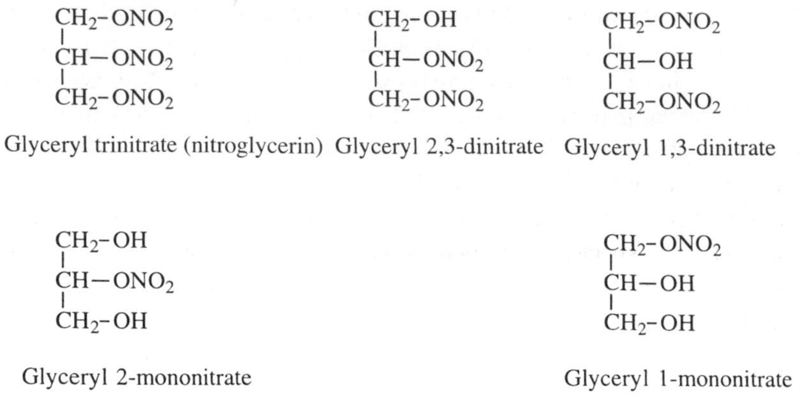

$$
\begin{array}{ccc}
\mathrm{CH_2-ONO_2} & \mathrm{CH_2-OH} & \mathrm{CH_2-ONO_2} \\
| & | & | \\
\mathrm{CH-ONO_2} & \mathrm{CH-ONO_2} & \mathrm{CH-OH} \\
| & | & | \\
\mathrm{CH_2-ONO_2} & \mathrm{CH_2-ONO_2} & \mathrm{CH_2-ONO_2}
\end{array}
$$

Glyceryl trinitrate (nitroglycerin) Glyceryl 2,3-dinitrate Glyceryl 1,3-dinitrate

$$
\begin{array}{cc}
\mathrm{CH_2-OH} & \mathrm{CH_2-ONO_2} \\
| & | \\
\mathrm{CH-ONO_2} & \mathrm{CH-OH} \\
| & | \\
\mathrm{CH_2-OH} & \mathrm{CH_2-OH}
\end{array}
$$

Glyceryl 2-mononitrate   Glyceryl 1-mononitrate

Fig 9.4. Chemical structures of glyceryl trinitrate and fungal metabolites.

one (ATO) (Fig. 9.3) via nitro group reduction (Le Campion *et al.*, 1999). *B. licheniformis* was the only microorganism that could further transform ATO. Further work will be needed to determine if *P. chrysosporium* is able to degrade ATO further.

Nitrate esters

The most famous of the nitrate ester explosives is nitroglycerin (NG), also known as glyceryl trinitrate (GTN), which in its pure form is too unstable for practical use. It is usually added to nitrocellulose (NC) (Fig. 9.4) in blasting gelatins, and it is also a component of double-base and triple-base propellants (Rosenblatt *et al.*, 1991). GTN has been the focus of fungal degradation studies and will be discussed in the following section.

Nitroglycerin

Since GTN is found not only in explosives and propellants but also in drugs used to treat angina pectoris (Ducrocq, Servy & Lenfant, 1990), it is present in many waste streams. Ducrocq, Servy & Lenfant (1989) examined *Geotrichum candidum* for its ability to transform GTN. Although they found that *G. candidum* was able to stoichiometrically transform GTN to glyceryl dinitrate (GDN) and glyceryl mononitrate (GMN) (Fig. 9.4), they hypothesized that toxicity of the compounds was the limiting factor in the transformation of GTN (Ducrocq *et al.*, 1989). These data agreed with the earlier work of Wendt, Cornell & Kaplan (1978), who observed that the microbial transformation of GTN proceeded stepwise via the dinitrate and mononitrate isomers, with each succeeding step proceeding at a slower rate. To test the ability of *P. chrysosporium* to transform GTN, Ducrocq *et al.* (1990) cultured it and five other fungi in an undefined rich medium, added GTN after about 72 hours, and observed the metabolites via HPLC. They documented the presence of 2-GMN and 1-GMN in all the cultures. However, in the *P. chrysosporium* cultures they observed a regioselectivity to the 2-GMN species (Ducrocq *et al.*, 1990). *P. chrysosporium* also exhibited the most efficient and extensive transformation of GTN of the six cultures studied. Again, these researchers observed definite toxicity thresholds for the fungi at higher concentrations of GTN and noted that toxicity increased dramatically in the absence of an exogenous carbon source (Ducrocq *et al.*, 1990). This group next examined GTN degradation by *P. chrysosporium* using electron paramagnetic resonance (EPR). When the fungus was grown in non-lignolytic medium and inoculated with GTN, the liberation of nitrite ions was observed via HPLC (Servent *et al.*, 1992). However, the absence of nitrate ions originating directly from GTN suggested that an esterase-type reaction was not involved. Furthermore, EPR analysis of mycelial samples showed the presence of nitric oxide (NO) and the appearance of heme protein–NO and non-heme protein–NO complexes, indicating that NO may be produced directly from GTN. The involvement of a glutathione transferase-like system in the evolution of nitrite from GTN was also proposed (Servent *et al.*, 1992). Zhang *et al.* (1997) isolated *Penicillium corylophilum* Dierckx from a moist double-based propellant. They postulated that since GTN is the only water-soluble component of the propellant, the fungus was using GTN as a growth substrate. Initial studies showed that *P. corylophilum* could partially transform GTN to GDN and GMN metabolites (Zhang *et al.*, 1997). In an effort to optimize culture conditions for maximal transformation, the researchers added glucose and ammonium nitrate to the growth medium. Under these culture conditions, *P. corylophilum* was able to completely transform GTN to GMN within 168 hours and degrade GMN to below detection limits within 336 hours (Zhang *et al.*, 1997). Metabolites resulting from the degradation of GMN were not discussed.

Nitrocellulose

NC is a major component of most gun propellants (Sharma *et al.*, 1995) and is often added to GTNs in double-based propellent formulations (Rosenblatt *et al.*, 1991). Kaplan (1990) suggested that a chemical treatment should precede biological treatment in bioreactors because hydrolysis of the nitro groups allows a mixed microbial culture to degrade NC. Researchers looking for alternatives to chemical pretreatment began to study the degradation of NC by fungi. Sharma *et al.* (1995) combined a cellulolytic fungus, *Sclerotium rolfsii*, with a denitrifying fungus, *Fusarium solani*, to remove NC from liquid medium. By measuring an increase in biomass and a decrease in cellulose weight, these researchers found that 31% of the added NC was transformed in 3 days. They surmised that the limited transformation was caused by exhaustion of the buffering capacity, as shown by a severe drop in pH throughout the experiment (Sharma *et al.*, 1995). Sharma *et al.* (1995) next used the *P. corylophilum* species that they had isolated from moist double-based propellant containing GTN in NC degradation experiments. *P. corylophilum* could use NC as the sole nitrogen source in a mineral salts medium amended with starch or xylan as a carbon source, but only 20% of the NC was degraded under these conditions. In an attempt to improve degradation, cultures *of P. corylophilum* and the denitrifying fungus *F. solani* were combined, but no significant enhancement was observed in the combined cultures (Zhang *et al.*, 1997). Neither NC degradation pathways nor metabolites were identified in this study.

Current and future applied research

The ability of fungi to produce extracellular enzymes and factors that can degrade complex organic compounds has sparked research on their use in decontamination of explosives-laden soils and waters. One full-scale treatment process, composting, has been used to treat a wide variety of sludges and soils contaminated with hazardous energetic organonitro compounds (Ibister *et al.*, 1984; Williams *et al.*, 1992; Tuomi *et al.*, 1997). Fungi are thought to play a vital role in the composting treatment processes (Bayman *et al.*, 1995). Energetic compounds not covered in this review, such as HMX, N-methyl-N-2,4,6-tetranitroaniline (TETRYL), nitroguanidine (NQ) and pentaerythritol tetranitrate (PETN), as well as the compounds covered here have been treated in mixed wastes via composting (Isbister *et al.*, 1984; Williams *et al.*, 1992; Tuomi *et al.*, 1997). Unfortunately, in studies using radiolabel mass balance control, many of the energetic compounds are reductively transformed and found tightly bound to the organic fractions of the composts (Kaplan & Kaplan, 1982; Isbister *et al.*, 1984; Pennington *et al.*, 1995). Dawel *et al.* (1997) have described the structure of a laccase-mediated coupling product of 2,4-diamino-6-

nitrotoluene and guaiacol, together with several trinuclear coupling products including 5-(2-amino-3-methyl-4-nitroanilino)-3,3'-dimethoxy-4,4'-diphenoquinone. These coupling reactions were suggested as a model for the coupling of TNT metabolites to humic and other organic soil fractions. Whether the energetic compounds are irreversibly bound or not is controversial. Some studies have shown a reduction in the toxicity of hazardous organonitro compounds upon treatment with composting (Isbister *et al.*, 1984). However, other studies using a mammalian system have suggested that compost-bound residues may be less tightly associated than previously thought (Palmer *et al.*, 1997). These results encourage the search for new fungi and enzymes that can degrade hazardous organonitro compounds to innocuous products.

Another example of using fungi in full-scale field applications is seen in the study by Sublette *et al.* (1992), where cultures of *P. chrysosporium* were immobilized on the discs of a rotating biological contactor in order to treat the TNT waste stream known as pink water. During treatment, *P. chrysosporium* effectively removed TNT and RDX to allowable limits from this mixed waste. Fernando & Aust (1991b) have also demonstrated the utility of *P. chrysosporium* to remove TNT in batch reactors.

Further research is needed to identify the fungal genes responsible for degradation of hazardous nitroorganic compounds and to elucidate the degradation pathways. Understanding the mechanisms of enzymes capable of catalysing the destructive reactions may lead to the development of more efficient hybrid organisms. Ultimately, these organisms could be adapted to field-scale bioremediation schemes.

References

An, D., Gibson, D. T. & Spain, J. C. (1994). Oxidative release of nitrite from 2-nitrotoluene by a three-component enzyme system from *Pseudomonas* sp. strain JS42. *Journal of Bacteriology*, **176**, 7462–7467.

Bayman, P. & Radkar, G. V. (1997). Transformation and tolerance of TNT (2,4,6-trinitrotoluene) by fungi. *International Biodeterioration and Biodegradation*, **39**, 45–53.

Bayman, P., Ritchey, S. D. & Bennett, J. W. (1995). Fungal interactions with the explosive RDX (hexahydro-1,3,5-trinitro-1,3,5-triazine). *Journal of Industrial Microbiology*, **15**, 418–423.

Becuwe, A. & Delclos, A. (1993). Low-sensitivity explosive compound for low vulnerability warheads. *Propellants and Explosives*, **18**, 1–10.

Blotevogel, K. H. & Gorontzy, T. (2000). Microbial degradation of compounds with nitro functions. In *Environmental Processes II*, Vol. 11b, 2nd edn, ed. J. Klein, pp. 274–302. Weinheim: Wiley-VCH.

Bumpus, J. A. & Tatarko, M. (1994). Biodegradation of 2,4,6-trinitrotoluene by

Phanerochaete chrysosporium: identification of initial degradation products and the discovery of a TNT metabolite that inhibit lignin peroxidases. *Current Microbiology*, **28**, 185–190.

Dawel, G., Kastner, M., Michels, J., Poppitz, W., Gunther, W. & Fritsche, W. (1997). Structure of a laccase-mediated product of coupling of 2,4-diamino-6- nitrotoluene to guaiacol, a model for coupling of 2,4,6-trinitrotoluene metabolites to a humic organic soil matrix. *Applied and Environmental Microbiology*, **63**, 2560–2565.

Donnelly, K. C., Chen, J. C., Huebner, H. J., Brown, K. W., Autenrieth, R. L. & Bonner, J. S. (1997). Utility of four strains of white rot fungi for the detoxification of 2,4,6-trinitrotoluene in liquid culture. *Environmental Toxicology and Chemistry*, **16**, 1105–1110.

Ducrocq, C., Servy, C. & Lenfant, M. (1989). Bioconversion of glyceryl trinitrate into mononitrates by *Geotrichum candidum*. *FEMS Microbiology Letters*, **53**, 219–222.

Ducrocq, C., Servy, C. & Lenfant, M. (1990). Formation of glyceryl 2-mononitrate by regioselective bioconversion of glyceryl trinitrate: efficiency of the filamentous fungus *Phanerochaete chrysosporium*. *Biotechnology and Applied Biochemistry*, **12**, 325–330.

Emery, D. D. & Faessler, P. C. (1997). First production-level bioremediation of explosives-contaminated soil in the United States. *Annals of the New York Academy of Sciences*, **829**, 326–340.

Fernando, T. & Aust, S. D. (1991a). Biodegradation of munition waste, TNT, (2,4,6-trinitrotoluene), and RDX (hexahydro-1,3,5-trinitro-1,3,5-triazine) by *Phanerochaete chrysosporium*. In *ACS Symposium Series* No. 468, *Emerging Technology in Hazardous Waste Management II*, eds. D. W. Tedder & F. G. Pohland, pp. 214–232. Washington, DC: American Chemical Society.

Fernando, T. & Aust, S. D. (1991b). Biological decontamination of water contaminated with explosives by *Phanerochaete chrysosporium*. In *Institute of Gas Technology 3rd Annual Symposium: Gas, Oil, Coal, and Environmental Biotechnology III*, pp. 193–206. Des Plaines, IL: Institute of Gas Technology.

Fernando, T., Bumpus, J. A. & Aust, S. D. (1990). Biodegradation of TNT (2,4,6-trinitroluene) by *Phanerochaete chrysosporium*. *Applied and Environmental Microbiology*, **56**, 1666–1671.

Funk, S. B., Crawford, D. L., Crawford, R. L., Mead, G. & Davis-Hoover, W. (1995a). Full-scale anaerobic bioremediation of trinitrotoluene (TNT) contaminated soil. *Applied Biochemistry and Biotechnology*, **51/52**, 625–633.

Funk, S. B., Crawford, D. L., Roberts, D. J. & Crawford, R. L. (1995b). Two-stage bioremediation of TNT contaminated soils. In *Bioremediation of Pollutants in Soil and Water*, ed. B. S. Schepart, pp. 177–189. Philadelphia, PA: American Society of Testing and Materials.

Gorontzy, T., Drzyzga, O., Kahl, M. W., Bruns-Nagel, D., Breitung, J., von Loew, E. & Blotevogel, K. H. (1994). Microbial degradation of explosives and related compounds. *Critical Reviews in Microbiology*, **20**, 265–284.

Harter, D. R. (1985). The use and importance of nitroaromatic chemicals in the chemical industry. In *Toxicity of Nitroaromatic Compounds*, ed. D. E. Rickert, pp. 1–14. New York: Hemisphere.

Hawari, J., Halasz, A., Beaudet, S., Paquet, L., Ampleman, G. & Thiboutot, S. (1999). Biotransformation of 2,4,6-trinitrotoluene with *Phanerochaete chrysosporium* in agitated cultures at pH 4.5. *Applied and Environmental Microbiology*, **65**, 2977–2986.

Higson, F. K. (1992). Microbial degradation of nitroaromatic compounds. *Advances in Applied Microbiology*, **37**, 1–19.

Isbister, J. D., Anspach, G. L., Kitchens, J. F. & Doyle, R. C. (1984). Composting for decontamination of soils containing explosives. *Microbiologica*, **7**, 47–73.

Kaplan, D. L. (1990). Biotransformation pathways of hazardous energetic organo-nitro compounds. In *Biotechnology and Biodegradation*, eds. D. Kamely, A. Chakabaity & G. S. Omlan, pp. 154–181. Houston, TX: Gulf.

Kaplan, D. L. (1992). Biological degradation of explosives and chemical agents. *Current Opinion in Biotechnology*, **3**, 253–260.

Kaplan, D. L. & Kaplan, A. M. (1982). Thermophilic biotransformations of 2,4,6-trinitrotoluene under simulated composting conditions. *Applied and Environmental Microbiology*, **44**, 757–760.

Klausmeier, R. E., Osmon, J. L. & Walls, D. R. (1974). The effect of trinitrotoluene on microorganisms. *Developments in Industrial Microbiology*, **15**, 309–317.

Le Campion, L., Vandais, A. & Ouazzani, J. (1999). Microbial remediation of NTO in aqueous industrial wastes. *FEMS Microbiology Letters*, **176**, 197–203.

Lenke, H., Warrelmann, J., Daun, G., Hund, K., Sieglen, U., Walter, U. & Knackmuss, H. (1998). Biological treatment of TNT-contaminated soil. 2. Biologically induced immobilization of the contaminants and full-scale application. *Environmental Science and Technology*, **32**, 1964–1971.

Lewis, T. A., Ederer, M. M., Crawford, R. L. & Crawford, D. L. (1997). Microbial transformation of 2,4,6-trinitrotoluene. *Journal of Industrial Microbiology and Biotechnology*, **18**, 89–96.

Mason, R. P. & Josephy, P. D. (1985). Free radical mechanism of nitroreductase. In *Toxicity of Nitroaromatic Compounds*, ed. D. E. Rickert, pp. 121–140. New York: Hemisphere.

McCormick, N. G., Cornell, J. H. & Kaplan, A. M. (1978). Identification of biotransformation products from 2,4-dinitrotoluene. *Applied and Environmental Microbiology*, **35**, 945–948.

McCormick, N. G., Cornell, J. H. & Kaplan, A. M. (1981). Biodegradation of hexahydro-1,3,5-trinitro-1,3,5-triazine (RDX). *Applied and Environmental Microbiology*, **42**, 817–823.

McLellan, W. L., Hartley, W. R. & Brower, M. E. (1992). Hexahydro-1,3,5-trinitro-1,3,5-triazine (RDX). In *Drinking Water Health Advisory: Munitions*, eds. W. C. Roberts & W. R. Hartley, pp. 133–180. Boca Raton, FL: Lewis.

Meharg, A. A., Dennis, G. R. & Cairney, J. W. G. (1997). Biotransformation of 2,4,6-trinitrotoluene (TNT) by ectromycorrhizal basidiomycetes. *Chemosphere*, **35**, 513–521.

Michels, J. & Gottschalk, G. (1994). Inhibition of the lignin peroxidase of *Phanerochaete chrysosporium* by hydroxylamino-dinitrotoluene, an early intermediate in the degradation of 2,4,6-trinitrotoluene. *Applied and Environmental Microbiology*, **60**, 187–194.

Michels, J. & Gottschalk, G. (1995). Pathway of 2,4,6-trinitrotoluene (TNT) degradation by *Phanerochaete chrysosporium*. In *Biodegradation of Nitroaromatic Compounds*, ed. J. C. Spain, pp. 135–149. New York: Plenum Press.

Palmer, W. G., Beaman, J. R., Walters, D. M. & Creasia, D. A. (1997). Bioavailability of TNT residues in composts of TNT-contaminated soil.

240   *D. A. Newcombe and R. L. Crawford*

Journal of Toxicology and Environmental Health, **51**, 97–108.

Parrish, F. W. (1977). Fungal transformation of 2,4-dinitrotoluene and 2,4,6-trinitrotoluene. *Applied and Environmental Microbiology*, **34**, 232–233.

Paszczynski, A. & Crawford, R. L. (1995). Potential for bioremediation of xenobiotic compounds by the white rot fungus *Phanerochaete chrysosporium*. *Biotechnology Progress*, **11**, 368–379.

Pennington, J. C., Hayes, C. A., Myers, K. F., Ochmean, M., Gunnison, D., Felt, D. R. & McCormick, E. G. (1995). Fate of 2,4,6-trinitrotoluene in a simulated compost system. *Chemosphere*, **30**, 429–438.

Razo-Flores, E., Matamala, A., Lettinga, G. & Field, J. A. (1997). Complete anaerobic biodegradation of nitroaromatics in continuous reactors. In *The Second In Situ and On-Site Bioremediation Symposium*, ed. B. C. Alleman & A. Leeson, pp. 31–32. Columbus, OH: Battelle Press.

Reible, S., Joshi, D. K. & Gold, M. H. (1994). Aromatic nitroreductase from the basidiomycete *Phanerochaete chrysosporium*. *Biochemical and Biophysical Research Communications*, **205**, 298–304.

Rieger, P.-G. & Knackmuss, H. (1995). Basic knowledge and perspectives on biodegradation of 2,4,6-trinitrotoluene and related nitroaromatic compounds in contaminated soil. In *Biodegradation of Nitroaromatic Compounds*, ed. J. C. Spain, pp. 1–18. New York: Plenum Press.

Rosenblatt, D. H., Burrows, E. P., Mitchell, W. R. & Parmer, D. L. (1991). Organic explosives and related compounds. In *The Handbook of Environmental Chemistry*, ed. O. Hutzinger, pp. 195–234. Berlin: Spring-Verlag.

Samson, J., Langlois, E., Lei, J., Piche, Y. & Chenevert, R. (1998). Removal of 2,4,6-trinitrotoluene and 2,4-dinitrotoluene by fungi *Ceratocystis coerulescens, Lentinus lepideus* and *Trichoderma harzianum*. *Biotechnology Letters*, **20**, 355–358.

Scheibner, K. & Hofrichter, M. (1998). Conversion of aminonitrotoluenes by fungal manganese peroxidase. *Journal of Basic Microbiology*, **38**, 51–59.

Scheibner, K., Hofrichter, M. & Fritsche, W. (1997a). Mineralization of 2-amino-4,6-dinitrotoluene by manganese peroxidase of the white rot fungus *Nematoloma frowardii*. *Biotechnology Letters*, **19**, 835–839.

Scheibner, K., Hofrichter, M., Herre, A., Michels, J. & Fritsche, W. (1997b). Screening for fungi intensively mineralizing 2,4,6-trinitrotoluene. *Applied Microbiology and Biotechnology*, **47**, 452–457.

Servent, D., Ducrocq, C., Henry, Y., Servy, C. & Lenfant, M. (1992). Multiple enzymatic pathways involved in the metabolism of glyceryl trinitrate in *Phanerochaete chrysosporium*. *Biotechnology and Applied Biochemistry*, **15**, 257–266.

Sharma, A., Sundaram, S. T., Zhang, C. & Brodman, B. W. (1995). Nitrocellulose degradation by a coculture of *Sclerotium rolfsii* and *Fusarium solani*. *Journal of Industrial Microbiology*, **15**, 1–4.

Spain, J. C. (1995a). *Biodegradation of Nitroaromatic Compounds*. New York: Plenum Press.

Spain, J. C. (1995b). Biodegradation of nitroaromatic compounds. *Annual Review of Microbiology*, **49**, 523–555.

Spanggord, R. J., Spain, J. C., Nishino, S. F. & Mortelmans, K. E. (1991). Biodegradation of 2,4-dinitrotoluene by a *Pseudomonas* sp. *Applied and Environmental Microbiology*, **57**, 3200–3205.

Spiker, J. K., Crawford, D. L. & Crawford, R. L. (1992). Influence of

2,4,6-trinitrotoluene (TNT) concentration on the degradation of TNT in explosive-contaminated soils by the white rot fungus *Phanerochaete chrysosporium. Applied and Environmental Microbiology*, **58**, 3199–3202.

Stahl, J. D. & Aust, S. D. (1993a). Metabolism and detoxification of TNT by *Phanerochaete chrysosporium. Biochemical and Biophysical Research Communications*, **192**, 477–482.

Stahl, J. D. & Aust, S. D. (1993b). Plasma membrane dependent reduction of 2,4,6-trinitrotoluene by *Phanerochaete chrysosporium. Biochemical and Biophysical Research Communications*, **192**, 471–476.

Sublette, K. L., Ganapathy, E. V. & Schwartz, S. (1992). Degradation of munitions wastes by *Phanerochaete chrysosporium. Applied Biochemistry and Biotechnology*, **34/35**, 709–723.

Suen, W. C., Haigler, B. E. & Spain, J. C. (1996). 2,4-Dinitrotoluene dioxygenase from *Burkholderia* sp. strain DNT: similarity to naphthalene dioxygenase. *Journal of Bacteriology*, **178**, 4926–4934.

Tsai, T. S. (1991). Biotreatment of red water – a hazardous stream from explosive manufacture – with a fungal system. *Hazardous Waste and Hazardous Materials*, **8**, 231–244.

Tuomi, E., Coover, M. P. & Stroo, H. F. (1997). Bioremediation using composting or anaerobic treatment for ordnance-contaminated soils. *Annals of the New York Academy of Sciences*, **829**, 160–178.

Urbanski, T. (1984). *Chemistry and Technology of Explosives*, Vol 4. Oxford: Pergamon Press.

Valli, K., Brock, B. J., Joshi, D. K. & Gold, M. H. (1992). Degradation of 2,4-dinitrotoluene by the lignin-degrading fungus *Phanerochaete chrysosporium. Applied and Environmental Microbiology*, **58**, 221–228.

van Aken, B., Skubisz, K., Naveau, H. & Agathos, S. N. (1997). Biodegradation of 2,4,6-trinitrotoluene (TNT) by the white rot basidiomycete *Phlebia radiata. Biotechnology Letters*, **19**, 813–817.

van Aken, B., Hofrichter, M., Scheibner, K., Hatakka, A. I., Naveau, H. & Agathos, S. N. (1999). Transformation and mineralization of 2,4,6-trinitrotoluene (TNT) by manganese peroxidase from the white rot basidiomycete *Phlebia radiata. Biodegradation*, **10**, 83–91.

Vorbeck, C., Lenke, H., Fischer, P. & Knackmuss, H. J. (1994). Identification of a hydride–Meisenheimer complex as a metabolite of 2,4,6-trinitrotoluene by a *Mycobacterium* strain. *Journal of Bacteriology*, **176**, 932–934.

Vorbeck, C., Lenke, H., Fischer, P., Spain, J. C. & Knackmuss, H. J. (1998). Initial reductive reactions in aerobic microbial metabolism of 2,4,6-trinitrotoluene. *Applied and Environmental Microbiology*, **64**, 246–252.

Wendt, T. M., Cornell, J. H. & Kaplan, A. M. (1978). Microbial degradation of glycerol nitrates. *Applied and Environmental Microbiology*, **36**, 693–699.

Williams, R. T., Ziegenfuss, P. S. & Sisk, W. E. (1992). Composting of explosives and propellant contaminated soils under thermophilic and mesophilic conditions. *Journal of Industrial Microbiology and Biotechnology*, **9**, 137–144.

Zhang, Y. Z., Sundaram, S. T., Sharma, A. & Brodman, B. W. (1997). Biodegradation of glyceryl trinitrate by *Penicillium corylophilum* Dierckx. *Applied and Environmental Microbiology*, **63**, 1712–1714.

10

Use of wood-rotting fungi for the decolorization of dyes and industrial effluents

JEREMY S. KNAPP, ELI J. VANTOCH-WOOD
AND FUMING ZHANG

Introduction

With increasing awareness among the general public of the problems of water pollution has come a realization among effluent dischargers that the colour in effluents represents a problem in itself. Colourless effluents are less visible, attract less attention and cause less concern than coloured effluents. This is despite the fact that often chromophores may be present in very small amounts and may pose no significant threat to the environment, other than turning a river red or purple! Having said this, there can be significant problems of toxicity associated with some chromophores (Brown & De Vito, 1993) and many coloured effluents contain damaging materials in addition to chromophores. The focus of this contribution is the removal of colour from effluents, and in particular how wood-rotting fungi can be used for this purpose. It is perhaps useful to consider briefly what alternative processes are available before examining the possible roles of fungi.

The main processes used for colour removal are physicochemical and chemical treatments (Laing, 1991; Cooper, 1995) all of which have some drawbacks. Physicochemical treatments include flocculation and coagulation, adsorption, ion exchange, ultrafiltration and reverse osmosis. These processes (apart from expense) have the problem that contaminant chemicals are not destroyed; they are simply removed from effluents and relocated elsewhere – usually disposed of to landfill or by incineration. Chemical processes mainly involve bleaching using chlorine-based chemicals, ozone or peroxides. Bleaching with chlorine can be highly effective and is relatively cheap but has the disadvantage that it produces organochlorine compounds that can be highly toxic and recalcitrant to biodegradation. Ozonolysis has many advantages but is very expensive because of the high consumption of electricity. Other processes include wet air oxidation and catalytic oxidation.

Biological processes for decolorization have been investigated to a considerable extent. However, most of the conventional aerobic biological treatment processes are of limited applicability to coloured effluents. In some cases adsorption by activated sludge has been observed (Shaul *et al.*, 1986) but this is of limited value since it involves removal rather than destruction. Biodegradation in most conventional treatment plants is limited since many (most?) chromophores are relatively recalcitrant to biodegradation. This recalcitrance is not surprising since most chromophores are complex in structure and often large in size. Furthermore, synthetic chromophores, like artificial dyes, often contain xenobiotic structural units. The fine detail of a chemical's structure can have a marked effect on its biodegradability (Alexander, 1965, 1994): the nature, number and position of substituents are important factors (Pitter & Chudoba, 1990). In chromophores, some of the important structural elements are aromatic sulfonic acid units, azo bonds and conjugated aromatic rings – all of which decrease biodegradability. In addition to chemical structure, environmental factors influence biodegradation processes and these include temperature, pH and the presence or absence of oxygen or other terminal electron acceptors (Alexander, 1965). For many chromophores, degradation has only been reported in aerobic conditions as a consequence of the obligate requirement for oxygen. However, oxygen can inhibit certain metabolic processes. Some biological reactions that modify dyes are reductive and preferentially occur in anaerobic conditions, an example being the reductive cleavage of azo dyes, which is catalysed by many bacterial species (see Chung, Stevens & Cerniglia, 1992).

Several reductive biological processes have been proposed for treatment of effluents containing azo dyes. Some involve defined bacterial strains (Haug *et al.*, 1991), others use anaerobic digestion (Carliell *et al.*, 1995) while a recently proposed process uses conventional, unacclimated activated sludge incubated under anaerobic conditions (Bromley-Challenor *et al.*, 2000). In all cases, the azo dyes are reduced to give two or more aromatic amines; some of these are degradable under aerobic conditions (Haug *et al.*, 1991) but other amines can be highly recalcitrant and some are toxic and carcinogenic (Brown & De Vito, 1993). Aromatic amines are often unstable and can be subject to spontaneous oxidation and subsequent dimerization or polymerization, often to produce a new chromophore (Knapp & Newby, 1995; Kudlich *et al.*, 1999). Several dyes (e.g. methylene blue) are subject to reversible reductive decolorization and are commonly used as redox indicators, but this process is not biodegradative. Reductive approaches to decolorization of coloured wastes may have some potential where chromophores have been shown to be reducible (like azo dyes). However, for some effluents, for example pulping and molasses-based effluents, anaerobic digestion has been shown to be ineffective in colour reduction.

White rot fungi

There has been much interest in the possible use of wood-rotting fungi as agents of biodegradation, most attention having been paid to the white rots. It is perhaps appropriate at this stage to define the term white rot fungi. White rot fungi are filamentous fungi that inhabit the wood of dead and dying trees. They are 'higher fungi', most are members of the Basidiomycota (e.g. *Phanerochaete chrysosporium* or *Trametes (Coriolus) versicolor*) but some are in the Ascomycota (e.g. *Xylaria polymorpha* or *X. hypoxylon*). *Trametes versicolor* and *Coriolus versicolor* are synonyms for the same organism and it is unclear which name is 'correct'. In this review we refer to this organism by the name used in the reports referred to. The characteristic feature of white rots is their ability to degrade lignin, a property that they share with simultaneous rot fungi and some 'litter-decomposing fungi'. It is generally considered that white rots cannot degrade cellulose without prior removal of lignin, while simultaneous rots remove both polymers at the same time. However, this distinction is probably a matter of degree and not a 'real' difference. The three-dimensional polyaromatic polymer lignin has a structural role in woody plants and also has a protective effect against fungal pathogens. It is the main barrier against enzymic degradation of cellulose and it has been shown repeatedly that degradation of lignocellulosic materials is slowed down by the presence of lignin, removal of which leads to accelerated cellulolysis. White rots are so-called because of the white appearance of the rotted wood, which is thought to be caused partly by the absence of lignin and partly oxidized aromatic lignin derivatives. In contrast, the brown rot fungi (e.g. *Serpula lacrymans* (dry rot)) remove cellulose, and hemicellulose, but leave the lignin as a brown residue, modified to some extent but largely undegraded.

   P. chrysosporium cannot grow on lignin alone and can only degrade it when provided with additional carbon and energy sources (Kirk, Connors & Zeikus, 1976). It is widely assumed that this is true for other white rot fungi. Removal of lignin by white rots is, therefore an essential pre- or corequisite for the degradation of cellulose, which yields energy and fixed carbon for the fungus. Depolymerization and degradation of hemicelluloses is less dependent on lignin removal than is degradation of cellulose. It is, therefore, likely that the process of ligninolysis is 'fuelled' to some extent by hemicellulose degradation.

The mechanism of lignin biodegradation

Because of its complex structure, the degradation of lignin is an unusual process amongst biological depolymerizations. This is not surprising since lignin is a compound of huge size and indefinite structure (Higuchi, 1980; Monties, 1994). Ligninolysis is, of course, an extracellular process. Unlike most biological polymers, lignin has no single repeating structural unit and few bonds that are, under normal conditions, subject to hydrolysis. The structure of lignin varies from plant to plant (Monties, 1994), but it is assembled by condensation of a number of different (but related) monomers that have a substituted phenyl propanoid structure. The process of biological ligninolysis remained a mystery until the mid 1980s when an enzyme that modified the structure of lignin model compounds was discovered in culture filtrates of *P. chrysosporium* (Glenn *et al.*, 1983; Tien & Kirk, 1983). The enzyme was shown to be a peroxidase and is now generally known as lignin peroxidase (LiP or ligninase). Since then, other peroxidases have been found in ligninolytic fungi that are thought to be important in degradation of lignin, lignin-model compounds and other aromatics. These include manganese-dependent peroxidase (MnP) and manganese-independent peroxidase. In addition to peroxidases, laccase is probably also involved. This copper-containing phenol oxidase enzyme (Thurston, 1994; Leontievsky *et al.*, 1997) has long been known to be associated with ligninolytic fungi although its exact role in ligninolysis remains uncertain (Youn, Hah & Kang, 1995).

The mechanism of lignin degradation by white rots has been referred to by Kirk & Farrell (1987) as 'enzymic combustion'. It is a random process, as befits such a complex material, and the details are still not entirely clear. The peroxidases are haem-containing proteins that catalyse one-electron oxidation reactions using H_2O_2 as the oxidant. Since lignin is such a large and hydrophobic compound, the chances of these enzymes forming a conventional enzyme–substrate complex with lignin are limited. In addition, ligninolytic enzymes are often cell bound rather than truly cell free. It is generally considered that the role of the peroxidases is to oxidize low-molecular-weight materials (mediators) to form powerful oxidants.

These diffuse to the lignin and bring about local oxidation by abstracting electrons. The partly modified lignin molecule is then open to spontaneous attack by water or oxygen, which causes further degradation. Fragments of lignin that are detached by such oxidations are then subject to further attack by peroxidases, either directly (by formation of enzyme–substrate complexes) or indirectly. Other enzymes that do not directly attack the intact lignin structure (e.g. laccase) may now oxidize the fragments.

The low-molecular-weight oxidants (mediators) used depend on the fungus and the enzyme. The best known examples are the oxidation of veratryl alcohol (VA) to veratryl radical cations by LiP and the oxidation of Mn^{2+} to Mn^{3+} by MnP. Trivalent manganese is very unstable and is thought to be stabilized *in vivo* by formation of chelates with organic acids, for example malonate and oxalate, derived from fungal metabolism. These low-molecular-weight mediators operate with the enzymes to bring about depolymerization. VA is a natural secondary metabolite of many white rot fungi (de Jong, Field & de Bont *et al.*, 1994) but is sometimes added to cultures to stimulate ligninolysis. Other compounds are probably also involved as mediators. Laccase, which is not a peroxidase but uses oxygen, is usually thought of as a phenol oxidase of broad specificity although it can also generate low-molecular-weight redox mediators and attack lignin using them (Youn *et al.*, 1995). Some laccases can, in the presence of suitable mediators, oxidize Mn^{2+} to Mn^{3+} – which can, in turn, oxidize other compounds (Archibald and Roy, 1992). This type of mechanism considerably extends the range of compounds and materials oxidizable by laccases. It has also been shown that certain LiPs can oxidize manganese (Khindaria, Barr & Aust, 1995) and MnP can oxidize some aromatics without the intervention of manganese ions (Heinfling *et al.*, 1998). For the peroxidases to operate there is a need for a reliable supply of H_2O_2. This is produced by the fungi through oxidase enzymes, for example glucose or cellobiose oxidase, which reduce molecular oxygen to H_2O_2. The requirement for peroxide is one of the reasons why a co-substrate is needed for lignin degradation.

Most of the original research was carried out on *P. chrysosporium* and it is clear that although other white rots may have similar processes they differ in the enzymes involved and the factors that control the process. White rot fungi can be put into groups according to the combinations of enzymes that they produce. Hatakka (1994) proposed three groups: (i) the LiP–MnP group, (ii) the MnP–laccase group, and (iii) the LiP–laccase group. However, some fungi do not fit neatly into these groups. Tuor, Winterhalter & Fiechter (1995) suggested five groupings, one of which

Table 10.1. *Groups of xenobiotic compounds that have been reported to be degraded by white rot fungi*

	Examples
Pesticides	Alachlor, 1,1,1-Trichloro-2,2-bis(4-chlorophenyl)ethane (DDT), Lindane
Chloroanilines	
Chlorinated phenols	Dichlorophenols, trichlorophenols, pentachlorophenol, chloroguaiacols
Dioxins	
Dyes	Azo dyes, triphenylmethane dyes (e.g. crystal violet), metal phthalocyanins, polymeric dyes, anthraquinones
Nitrotoluenes	Dinitrotoluenes,; trinitrotoluene (TNT)
Phenols	Phenol, *p*-cresol
Phenoxyacetic acid herbicides	2,4-Dichlorophenoxyacetic acid (2,4-D), 2,4,5-trichlorophenoxyacetic acid (2,4,5-T)
Polyaromatic hydrocarbons (PAHs)	Anthracene, benzo[*a*]pyrene, fluorene, perylene, phenanthrene, pyrene
Polychlorinated biphenyls (PCBs)	

produces LiP, MnP and laccase, another produces only LiP and another group produces no peroxidases. The complement of enzymes will influence the ability of these organisms to degrade not only their natural substrates but also xenobiotic compounds and effluents. It should not be assumed that members of the same group will behave the same in terms of their biodegradative activities.

Biodegradation of xenobiotic compounds by white rot fungi

White rot fungi have the ability to degrade many xenobiotic compounds with a wide variety of structures. The potential for using white rot fungi to treat pollutants and bioremediate contaminated land has been reviewed (see Field *et al.*, 1993; Barr & Aust, 1994; Reddy, 1995). Perhaps the most important thing to observe is the wide range of structures involved, mostly aromatic (Table 10.1). Some of the compounds, for example polychlorinated biphenyls (PCBs) and polycyclic aromatic hydrocarbons (PAHs), are hydrophobic and sparingly soluble in water – factors that often limit biodegradability. White rots are able to degrade these compounds because, generally, the degradations are catalysed by the ligninolytic system: an intrinsically non-specific system that deals in nature with a complex material that is hydrophobic and insoluble in water. However, it should not be assumed that the ligninolytic system is infallible. Kay-Shoemake & Watwood (1996) showed that there are limitations with the LiP of *P. chrysosporium*, which was incapable of degrading polyethylene glycol and *tert*-butylmethylether. It has been suggested that susceptibility to degradation by this LiP is related to the ionization potential of the putative substrate (Kay-Shoemake & Watwood, 1996; ten Have *et al.*, 1998). It appears that an ionization potential of $< 9.0 \, \text{eV}$ is required if a compound is to be oxidized by LiP (ten Have *et al.*, 1998). However, it is possible that this requirement will vary between the ligninolytic systems of different white rots, depending on the battery of enzymes that they possess. Treatment with white rot fungi or their enzymes can, sometimes, lead to complete degradation (as indicated by release of $^{14}CO_2$), or at least partial breakdown (Arjmand & Sandermann, 1985; Ryan & Bumpus, 1989). In other cases, the evidence presented (e.g. decolorization of a dye only demonstrates primary biodegradation, although more complete biodegradation may be occurring.

The mode of attack on xenobiotic chemicals is normally oxidative. While oxidation can lead to biodegradation, it can also lead to polymerization or dimerization since initial oxidative attack can produce reactive chemical species that can conjugate, to similar or dissimilar species, rather than breaking down. Most early research on biodegradation by white rot fungi employed *P. chrysosporium*. However, in recent years a much wider range of organisms has been studied both in terms of biodegradation of xenobiotics in general and coloured materials in particular (Table 10.2). Additionally, many unidentified isolates have been used since it is difficult to identify basidiomycetes from cultured mycelium, which often does not produce asexual spores or have any obvious distinguishing features. *P. chrysosporium* is undoubtedly the most commonly studied organism, with *T. versicolor* clearly 'second favourite'. *Pleurotus ostreatus*, *Bjerkandera adusta* and *Lentinus edodes* are also frequently used, with other species being studied to a much lesser extent.

General issues affecting decolorization by white rots

Most of this chapter deals with the use of white rot fungi to decolourize coloured compounds and materials that occur in industrial effluents. This research has concentrated on a small range of effluent types and materials, listed in Table 10.3.

Table 10.2. *Genera and species of white rot fungi used in biodegradation/decolorization studies*

Bjerkandera spp. and *B. adusta*
Chrysonilia sitophila
Chrysosporium lignorum
Cyathus bulleri and other *Cyathus* spp.
Dichomitus squalens
Flammulina velutipes
Funalia trogii
Ganoderma lacidum
Hericium erinaceum
Lentinus (Lentinula) edodes
Merulius tremellosus
Phanerochaete chrysosporium, P. flavido-alba
Phlebia radiata
Pleurotus ostreatus, P. eryngii, P. sajor-caju
Polyporus frondosus,
Pycnoporus cinnabarinus
Schizophyllum commune
Trametes (Coriolus) versicolor, T. hirsuta and other Trametes spp.

Problems in comparison between studies

Measurement of decolorization

It is difficult to compare much of the research reviewed here since very different methods are used to measure colour and its removal. Some studies use an overall measure of colour using colorimeters with filters while others assess colour by use of light absorption spectra (sometimes visible only and sometimes ultraviolet/visible). Colour can be reported either in absorbance (A) units or, particularly in papers referring to pulp and paper industries, Platinum-Cobalt Units (PCU). Sometimes measurement is only made at a single wavelength, often λ_{max}, but sometimes a standard wavelength is used (such as 465 nm). A method occasionally employed involves measurement at two wavelengths (Glenn & Gold, 1983). The first is usually λ_{max} and the second is in a region of the spectrum that changes little, if at all, with decolorization. A decrease in the absorbance ratio (of the first over the second wavelength) indicates decolorization. This method can be used to distinguish between degradative colour removal and adsorption. When decolorization is degradative, the absorbance ratio decreases, but when adsorption occurs the ratio tends to remain fairly constant since there are similar decreases in absorbance

Table 10.3. *List of effluents and materials considered for decolorization by white rots*

Category of effluent	Example	Types of chromophore
Chemical industry effluents	Various	Wide range including dyes
Cotton bleaching effluents	Cotton black liquor	Unknown, possibly melanoidins or similar
Dyes and textile dyeing and manufacturing effluents	Dye house effluents	Dyes, numerous different chromophores, e.g. azo, triphenylmethane, metal phthalocyanin, anthraquinone
Molasses-based effluents	Molasses spent wash from alcohol distilling	Melanoidins
Olive oil milling effluents		Polyphenolics
Paper making and pulping effluents	Kraft effluent, sulfite liquor	Depends on process; lignin fragments, lignin sulfonate, chlorolignins, chlorinated phenolics

across the whole spectrum. One drawback of this method is that it does not actually inform how much colour has been removed or the percentage change in colour. Measurement at a single wavelength can be useful and rapid but may be misleading since decolorization in one part of the visible spectrum is sometimes accompanied by increases in light absorption at other wavelengths. For example, Knapp & Newby (1999) showed that treatment of a yellow/black industrial effluent (containing azo-linked chromophores) by white rot fungi resulted in a large decrease in A_{390}. However, with some fungi, and under some conditions, there was a concomitant increase in A_{480} that resulted in the treated effluent having a red colour. Although A_{390} may have been reduced by as much as 75–85 units and A_{480} may only have increased by 3–4 units, sometimes the treated effluent actually appeared darker rather than lighter in colour, because of the greater visual perception of red colours.

Recent research in our laboratory on decolorization of blue and black dyes has shown that colour in the blue region (550 to 650 nm) is often easily removed (Zhang, 1997; E. J. Vantoch-Wood, unpublished data). However, with some fungi (under certain conditions) a residual red coloration (λ_{max} ca. 490–520 nm) remains. With decolorization of Reactive Black 5 by strains of *C. versicolor* we may get nearly 100% reduction in A_{598} but only 80% reduction in A_{520}, the colour changing from dark blue/black to light red. There is, however, considerable variation between fungal species and strains: some giving much less residual red coloration. The development of new colours clearly involves production of different chromophores, which may arise through degradation of the original chromophore to smaller entities that are still coloured. It is also possible that new chromophores may arise through spontaneous (or enzyme-catalysed) coupling of degradation products of the original chromophore. It appears that no research has been done to characterize new chromophores arising during treatment of dyes or effluents. However, it is well known that laccase and MnP can catalyse coupling of chemicals to give new chromophores. Consequently, in observing decolorization, it is essential to note changes that occur across the whole visible spectrum and not to restrict observations to one wavelength. In some cases, we do not have complete decolorization but, rather, qualitative and quantitative changes in the colour. Comparisons of decolorization of different chromophores on the basis of the change in light absorbance can be of limited value since the chromophores will have different extinction coefficients. Ideally, changes in absorbance should be related to changes in concentrations on a molar or weight basis. Unfortunately, this is not always possible since extinction coefficients may not be available, the exact identity of the chromophore may be unknown or a mixture of chromophores may be present. The best basis for comparison is probably the percentage decrease in light absorption at specified wavelengths together with the actual change in the absorbance, but this must always be taken in the context of the amount of material present. The more material present, the more bonds there are to be broken: 99% decolorization of a solution of a dye at $10 \, \text{mg} \, \text{l}^{-1}$ is less impressive than 90% decolorization when the concentration is $500 \, \text{mg} \, \text{l}^{-1}$. In many investigations, only dilute solutions of dyes or effluents have been tested; such data must be interpreted cautiously. Extrapolation to higher concentrations is generally unwise but particularly so in this case since higher concentrations are often toxic or inhibitory. It is clear, for example, that the dye Orange II is inhibitory to *P. chrysosporium* at relatively low concentrations,although not all fungi are equally susceptible (Knapp, Zhang & Tapley, 1997). For industrial effluents, even ostensibly similar effluents can vary enormously in colour and composition (both content and concentration). Effluents are often diluted prior to study and it is vital to note the extent of dilution. In practice, dilution of effluents is undesirable, it is expensive, regulatory authorities dislike it and it means that much larger reactors will be required. It is normally best to treat effluents at the highest concentration possible.

Adsorption

There are two main mechanisms for biological decolorization: adsorption to the biomass and biodegradation. Not all chromophores adsorb to particular types of biomass. However, there has been much research that demonstrates the potential of biomass as an adsorbent. Many forms of biomass have been investigated and fungal biomass has attracted much attention (e.g. Bousher, Shen & Edyvean, 1997). With some fungi, adsorption is the only decolorization mechanism, but with white rots both adsorption and degradation can occur. The relative importance of each mechanism can be difficult to assess since it is difficult to find suitable controls that contain enzymically inactive mycelium. Heat-killed mycelia have been used but their adsorptive properties can differ from those of live mycelia (E. J. Vantoch-Wood, unpublished data). Treatment with biocides (e.g. mercuric chloride or sodium azide) can kill the mycelium but do not guarantee that the enzymic complement will be inactivated. Furthermore, some biocides can complex with certain chromophores (J. S. Knapp, unpublished data). In some studies, controls have not been included to account for adsorption. However, this may not always be necessary since if colour removal is extensive then visual examination of the fungal biomass is often enough to confirm whether significant adsorption has occurred. In a study by Knapp, Newby & Reece (1995), discs of fungal mycelium were exposed to concentrated solutions of dyes with very high light absorbance (sometimes 50–100 units at λ_{max}). Decolorization was often complete leaving a white mycelium with no trace of adsorbed dye, whereas heat-killed controls and inactive fungal strains had deeply coloured mycelial mats. Many commercial dyes, and other chromophores, are negatively charged at neutral pH and strong adsorption to the fungal surface (which is also likely to be negatively charged) may be unlikely. However, with positively charged dyes like the triphenylmethanes or basic azo dyes adsorption through charge interaction is likely.

With white rot fungi, adsorption does not appear to be the principal mechanism of decolorization. It is likely that adsorption can play a part in the overall process, since prior adsorption to fungal mycelium may serve to bring chromophores into closer contact with the degradative enzymes, which are often largely associated with the cell surface (Evans *et al.*, 1991, 1994). After initial adsorption, oxidative degradation will then occur. Evidence presented by Tatarko & Bumpus (1998) and Wang & Yu (1998) showed that there is an initial rapid adsorption of dyes onto fungal mycelium followed by slower degradative decolorization of adsorbed dye. Adsorption of three dyes onto live and dead mycelium has been described using the Langmuir model (Wang & Yu, 1998), while Knapp *et al.* (1997) showed that adsorption of Orange II to heat-killed mycelium conformed to the Freundlich adsorption model.

Conditions for decolorization

Buffering and pH

Like most filamentous fungi, white rots normally show optimal growth at acidic pH values. Furthermore, growth in carbohydrate-containing media generally causes acidification of the medium, the extent of which depends on the carbon source and the type and amount of buffering present. The decreases in pH observed can be surprising; for example, Newby (1994) reported that in glucose-mineral salts medium *Piptoporus betulinus* reduced the pH to approximately 2 to 3, and sometimes to less than 2. Most fungi studied reduced pH to a lesser degree; for example, *C. versicolor* gave pH values in the range 4 to 4.4 and *P. ostreatus* gave 5.5 to 6. Since fungal growth on carbohydrates often results in production of organic acids it is likely that fungi may produce their own buffering. However, later in the growth of the culture, utilization of accumulated acids may lead to a loss of buffering and further pH changes.

Most research on white rot decolorization has involved batch cultures, sometimes agitated, but often in static conditions. Under these circumstances, cultures have only rarely been subject to pH control during growth, investigators relying on the initial poising of pH together with adequate buffering. Decolorization of an azo dye (Knapp *et al.*, 1997) and cotton bleaching effluent (Zhang, Knapp & Tapley, 1999a) has been shown to depend on initial pH. If the pH is not excessive (usually between 5.5 and 7) then the fungus is able to reduce it to 4–5.5, at which decolorization seems optimal. If pH is too high (usually > 7), then either decolorization will not occur or it will only occur when pH is reduced to the optimal region; low initial pHs can also be disadvantageous. Optimum pH will depend on the fungus, the decolorization under consideration and the type of medium, but generally appears to be in the range 4–5.5. Care must be taken to distinguish between the optimum pH for particular enzyme systems and the best pH for prolonged activity of isolated enzymes or live fungal cultures. For example, the optimum pH for LiP of *P. chrysosporium* (based on initial reaction rates) appears to be approximately 2.3–2.5. Although activity is most rapid at this pH, the enzyme is unstable and soon loses activity. In long-term use, the selected pH for enzyme reactors must be a compromise between the optimum pH for activity and stability. Interestingly, pH optima of enzymes may vary according to the substrate being converted; for example, the laccase of *Pycnoporus cinnabarinus* has different optima for guaiocol, syringaldizine and the artificial laccase substrate ABTS (2,2'-azino-bis(3-ethylbenzthiazoline 6-sulfonate) (McCarthy *et al.*, 1999). Similarly with purified isoenzymes of LiP from *P. chrysosporium*, Ollika *et al.* (1993) found that pH optima for decolorization varied according to the dye being treated and also varied between isoenzymes. Young & Yu (1997) also showed that pH/decolorization profiles varied with the dye being treated. In systems using live mycelia, the optimum pH represents a further compromise between the pH optima of several enzymes (notably laccases, peroxidases and the oxidases that produce H_2O_2) and the pH at which the mycelium, as a whole, can grow and operate effectively. The optimum pH of any decolorization process, therefore, should be assessed on a case by case basis, taking into account the organism (or enzyme used) and the materials to be treated. Precedents suggest that optimum pHs are likely to be in the range 4–5.5.

Several buffers have been used in decolorization studies. Initial studies on the LiP system of *P. chrysosporium* used 2,2'-dimethyl succinate as a buffer and consequently this has been widely used, but with little rationale. Dimethyl succinate is expensive and the cheaper 3,3'-dimethyl glutarate has been shown to be an effective alternative (Knapp *et al.*, 1997). Acetate is cheaper still and has been shown by some to be a useful buffer (Bakshi, Gupta & Sharma, 1999; Swamy & Ramsay, 1999a) although Kirk *et al.* (1978) reported that it is toxic to *P. chrysosporium*. Many buffers that are effective in the pH range 4–5.5 have been tested and there is little to suggest that any one of them is innately better than the others for all situations. Some buffers may complex with certain metal ions and alter their availability in an advantageous or disadvantageous way. For example, tartrate can complex and stabilize Mn^{3+}. Phosphate may precipitate certain metals, which may be good or bad depending on the exact nature of the process under consideration. Where salts of organic acids are used in growth media, a potential problem arises if the acid is utilized by the fungus as a carbon source, since its utilization will inevitably result in an increase in pH. 2,2'-Dimethyl succinate and 3,3'-dimethyl glutarate both contain quarternary carbon atoms and, therefore, are recalcitrant to biodegradation. This is a useful property in small-scale laboratory experiments but means that they will be of no use in large-scale applications of this technology. Given the self-buffering capacity of some white rot fungi, it is not certain that buffers are essential for decolorization processes, provided that a suitable pH can be attained. Indeed, if automatic pH control is available it should be a better option than the use of buffers. If white rot fungi are to be used for practical full-scale decolorization, omission of buffers has the advantage that additional materials are not added to the effluent that may increase the biological/chemical oxygen demand (BOD/COD) (organic acids) or cause eutrophication (phosphate).

Nutrition

Fungal nutrition has repeatedly been shown to be of enormous importance in the effectiveness of fungal decolorization systems using live mycelium. It is also crucial for systems in which separated enzymes are used, since physiology of the fungus is the key to obtaining high yields of active enzymes (Kirk & Farrell, 1987; Reddy & D'Souza, 1994). In treatment systems and research projects using pure chemicals and defined synthetic effluents, it is relatively easy to control availability of important nutrients. This is not the case for real industrial effluents that vary with time and location and are often difficult to analyse. Attention has focused on the supply of carbon and nitrogen sources, with some work on mineral nutrients and other additives.

Carbon source   In most decolorization studies, a carbon source is provided to stimulate growth and to provide a supply of oxidants by the fungus to 'fuel' the decolorization process. For example, glucose can be oxidized by fungal oxidases to give gluconic acid and H_2O_2, while certain fungal metabolites (like glyoxyal) can be similarly oxidized. The vast majority of studies use glucose as the carbon source. Alternatives that have been used include xylose, fructose, sucrose, maltose, cellobiose, glycerol and ethanol. Starch and xylan seem to be useful, but surprisingly cellulose and its derivatives are much less so. The necessity of adding a carbon source has rarely been questioned. Knapp *et al.* (1997) showed that for 'one-off' decolorization of an azo dye, using an unnamed white rot, the presence of glucose had little effect. However, where fungal mycelia were to be re-used, the presence of an added carbon source was essential if activity was to be retained. High activity in the absence of an added carbon source has been ascribed to carry-over of nutrient from the growth medium or to storage products or both (E. J. Vantoch-Wood, unpublished data). However, studies with a different unidentified fungus showed that provision of a carbon source was an absolute requirement for decolorization of cotton bleaching effluent, even with mycelia used for the first time (Zhang *et al.*, 1999a).

The need to add carbon sources to effluents before treatment will depend on the nature and composition of the effluent. Some effluents will contain high levels of usable substrates (e.g. carbohydrates), while others will have no useful components. Therefore, effluents from dyeing or chemical/dye manufacture are unlikely to have any usable carbon substrates while others from distilling or paper pulping may contain a range of carbohydrates that are degradable by certain white rots. The necessity to add carbon sources and the type and concentration added depends on the material to be treated and the organism. If in doubt, and certainly for initial experiments, it is wise to add a carbon source, glucose at $5–10\,\mathrm{g}\,\mathrm{l}^{-1}$ being a reasonable first choice.

Nitrogen sources   Early research on ligninolysis by *P. chrysosporium* showed that active ligninolysis was much more effective in the conditions of nitrogen limitation often associated with secondary metabolism (Kirk *et al.*, 1978; Jeffries, Choi & Kirk, 1981). Accordingly, most early research on degradation of xenobiotics by white rots used nitrogen-limiting conditions and, in fact, use was made of the medium of Kirk *et al.* (1978). It is now clear that other fungi differ markedly from *P. chrysosporium* in the production of ligninolytic enzymes. For example, *B. adusta* produces more LiP and MnP in nitrogen-sufficient than in nitrogen-limited media (Mester, Peña & Field, 1996). Similarly Ben Hamman, de la Rubia & Martínez (1997) reported a higher yield of these enzymes from *Phanerochaete flavido-alba* in the presence of excess nitrogen. While some decolorizations by fungi only occur, or are fastest, in nitrogen-limited conditions (Cripps, Bumpus & Aust, 1990; Zhang *et al.*, 1999a) this is not always so and nitrogen-sufficient conditions can be as good or better (Vasdev, Kuhad & Saxena, 1995).

The need for nitrogen can depend on whether the mycelium is to be re-used. Knapp *et al.* (1997), using an unidentified isolate, showed that addition of any nitrogen inhibited decolorization of Orange II when fungal mycelium was first used. Nevertheless, activity declined rapidly in the absence of added nitrogen when the mycelium was re-used. Addition of $0.25\,\mathrm{g\,l^{-1}}$ ammonium chloride to the medium allowed the mycelium to retain high, and unchanged, activity on re-use. When designing media for decolorization, the presence in effluents of usable nitrogen sources should be considered. Another consideration is the possible release of nitrogen during degradation of effluent chromophores. Many dyes (e.g. azos and triphenylmethanes) are nitrogenous and release of ammonia is quite likely. Such liberation of free nitrogen may affect the nitrogen status of the growth medium. White rot fungi seem to be able to use a wide range of inorganic and organic nitrogen sources, and the provision of organic nitrogen does not seem to be advantageous. Generally, researchers use inorganic nitrogen, usually as ammoniun salts. Although many studies employ ammonium tartrate, there does not appear to be any particular reason for choosing this salt, and the chloride (Knapp *et al.*, 1997) and sulfate (Heinfling, Bergbauer & Szewzyk, 1997) salts have been used successfully.

Growth factors In many studies, media are used that contain certain vitamins and other materials. Of the vitamins, thiamine is the most commonly used. The rationale for their addition is seldom given and in many cases white rot fungi have been grown without addition of vitamins on simple or complex media. Kirk *et al.* (1978) showed that the presence of thiamine stimulated lignin degradation. However, considering their expense (and the need, in some cases, for special sterilization procedures), it is difficult to recommend their inclusion in decolorization media.

Trace metals All microbes have certain basic requirements for mineral nutrients but sometimes a need for a specific element can be related to a particular process. In white rot fungi there are requirements for iron, copper and manganese. In some studies these are added to media while in others they are not and must be obtained as trace contaminants of media components or as part of the effluent or other material being tested. LiP and MnP are haem-containing peroxidases and require iron; however, there do not appear to be any studies on the effect of Fe^{2+} or Fe^{3+} on decolorization processes. Laccase contains copper and copper has been shown to stimulate its production in *T. versicolor* (Collins & Dobson, 1997) and *P. ostreatus* (Palmieri *et al.*, 2000). Pointing, Bucher & Vrijmoed (2000) have investigated the effects of the presence and concentration of Cu^{2+}, Cd^{2+} and Zn^{2+} on decolorization of the dye Poly R 478 by several white rots. While decolorization by *P. chrysosporium* and an unknown isolate was inhibited by Cu^{2+}, its addition at $0.1\ mmol\,l^{-1}$ led to a greater degree of decolorization (although at a slower rate) by *T. versicolor*. Further investigation would be of interest, since some dyes (e.g. copper phthalocyanines) contain copper in a chelated form.

MnP uses Mn^{2+} as a substrate rather than a cofactor, and it is known to be recycled during oxidations catalysed by MnP (Mn^{3+} is reduced to Mn^{2+} when it oxidizes lignin and related materials). The presence of manganese ions is, therefore, essential for activity of MnP and has been shown to induce production of MnP and also to repress production of LiP in *P. chrysosporium* (Reddy & D'Souza, 1994). Therefore, it has a major role in controlling not only the production and activity of MnP but also the overall balance of peroxidase activity. Its importance in specific decolorization processes has also been reported. Knapp *et al.* (1997) and Zhang *et al.* (1999a), using two different white rot fungi, demonstrated that added Mn^{2+}, at as little as $2\ mg\,l^{-1}$, was able to promote decolorization, respectively, of an azo dye and a cotton bleaching effluent (the effect being strongest with fungal mycelium that had been re-used a few times). On the negative side, Knapp & Newby (1999) reported that, in the decolorization of a chromophore in an industrial effluent, the presence and concentration of Mn^{2+} was important in controlling the presence of a red-coloured by-product. With cultures of *P. chrysosporium* and *P. ostreatus*, the intensity of the red colour was unaffected by the concentration of Mn^{2+} (0 to $40\ mg\,l^{-1}$) but with *C. versicolor* the presence of Mn^{2+} increased the intensity of this undesirable product. The reason for this is not clear but it is likely that the presence of manganese either promotes production of an enzyme(s) that produces the colour or decreases production of an enzyme that reduces, or prevents, its production. Heinfling *et al.* (1998) recently showed that several dyes were decolourized by MnP from *B. adusta* in a manganese-independent manner and, further, that increasing concentrations of Mn^{2+} inhibited the decolorization. The conclusion has to be that the effects resulting from the presence of manganese (and its concentration) have to be decided on a case by case basis and will depend both on the organism and the substrate involved.

Other additives   A variety of other materials have been shown to promote decolorization. VA is a known mediator of LiP activity and has been shown strongly to promote decolorization of various chromophores by isolated LiP (Young & Yu, 1997; Heinfling *et al.*, 1998). However, the amount of VA required to promote decolorization can vary between dyes (Young & Yu, 1997). Zhang (1997) showed that for Orange II decolorization by strain F29 at low concentrations (5–20 mmol l^{-1}), VA had only a small positive effect while at 50 mmol l^{-1} it was inhibitory. It was also shown that it inhibited decolorization of cotton bleaching effluent by another strain when present at as little as 1 mmol l^{-1}.

Tryptophan (which can stabilise LiP activity) has also been shown to promote dye decolorization by isolated LiP (Collins *et al.*, 1997; Heinfling *et al.*, 1998). Its value in promoting decolorization by whole mycelial cultures does not appear to have been tested. 2,5-Xylidene and 1-hydroxybenzotriazole have been shown to induce laccase synthesis in *T. versicolor* (Collins & Dobson, 1997). The latter has also been shown to be a mediator of indirect oxidation by laccases and to promote decolorization of a range of dyes by isolated laccase from *P. cinnabarinus* (McCarthy *et al.*, 1999). It is possible that 1-hydroxybenzotriazole could have a double effect in decolorizations by live fungal cultures, at least those for which laccase is important, promoting both synthesis and activity of this enzyme. However, this does not appear to have been studied. A range of other aromatics has been shown to act as mediators and to stimulate the ability of laccase (from *T. versicolor*) to oxidize PAHs (Johannes & Majcherczyk, 2000). Even simple, natural aromatics like phenol, aniline and 4-hydroxybenzoic acid (at concentrations as low as 0.1 mmol l^{-1}) were effective. The possibility of using laccase mediators in whole mycelial cultures rather than with isolated enzymes needs to be investigated, especially in the context of decolorization processes.

Agitation and aeration

Lignin degradation requires oxygen, either for the generation (by mycelia) of H_2O_2 for peroxidases or for the direct action of oxidases (e.g. laccase). Ligninolytic fungi are obligate aerobes and so oxygen will obviously be needed for growth and maintenance of culture viability. Decolorization by some of the isolated enzymes from white rot fungi does not necessarily require oxygen. The peroxidases need H_2O_2, not oxygen, as the oxidant, although laccase does use molecular oxygen. Oxygen is also thought to act directly on lignin fragments or other aromatics after their destabilization by redox mediators. The exact requirement for oxygen will depend on the fungus used and also on the enzymes it uses for particular chromophores.

Provision of oxygen can be problematical. Early work on ligninolysis by *P. chrysosporium* suggested that agitation of cultures leads to decreased ligninase activity. Therefore in early research, fungi were grown as static cultures with mycelial mats growing on the surface of liquid medium. To improve availability of oxygen, flasks were flushed occasionally with pure oxygen. This method was adopted in much of the research on decolorization. The problem with static cultures is that oxygen transfer is poor and that there are likely to be problems caused by a lack of homogeneity and poor nutrient distribution (gradient limitations) in culture vessels. One of the likely reasons for low LiP activity in shaken cultures is inactivation caused by shear forces or at the air–liquid interface. Jäger, Croan & Kirk (1985) reported that addition of detergents like Tween 80 or 20 can allow high yields of LiP in agitated cultures. Few studies have directly addressed the question of agitation in decolorization. Kim, Ryu & Shin (1996) and Bakshi *et al.* (1999) report that static cultures of *P. chrysosporium* and *P. ostreatus* gave better decolorization than agitated cultures. On the contrary, Knapp *et al.* (1997), Prasad & Gupta (1997), Sani, Azmi & Banerjee (1998) and Swamy & Ramsay (1999a) all demonstrated, with a range of fungi and chromophores, that agitation of cultures resulted in better decolorization. It is likely, therefore, that decisions on agitation of cultures will be specific to particular systems, although, clearly, agitation of cultures can give as good, or better, results as static cultivation. Detailed studies on the importance of oxygen concentration in decolorization processes do not appear to have been carried out and are required if fungal decolorization is to be a commercially useful process.

Temperature

Most white rot fungi are mesophiles having temperature optima at 27–30°C. Although many studies do not establish temperature optima for decolorizations, those that do usually report similar optimal values. The notable exception is *P. chrysosporium*, which has an unusual optimum at 37–40°C. Although higher temperatures are used for this organism, there is no evidence to suggest that decolorization rates are intrinsically faster than those of other white rots operating at their temperature optima. The main potential advantage of this high temperature optimum does not lie in more rapid catalysis but in the fact that in large-scale reactors removal of excess metabolic heat will present less of a problem at a higher temperature – in other words cooling costs will be less. Where isolated enzymes are used, temperature optima may be considerably higher and this can bring problems of instability. For example, the optimum temperature for decolorization by laccase from *P. cinnabarinus* is 65°C but it is unstable at this temperature, losing approximately 70% of its activity in 6 hours (McCarthy *et al.*, 1999). At 45°C it is less active but maintains 90% activity over a 24 hours period. The compromise between activity and stability has to be assessed individually for each enzyme–substrate combination.

Bioreactors

Much of the research on decolorization by white rots has used batch cultivation. Early work used static culture but recently shaken cultures have been employed. A wide variety of reactor configurations have been assessed but few direct comparative studies have been made. Early designs were somewhat constrained by the dogma that agitation would result in low activity of LiP and, therefore, restricted decolorization. This influenced designs as researchers attempted to find ways of growing fungal biomass without a high degree of agitation. Use was made of rotating biological contactors (RBCs), with the organism growing on the slowly rotating disc (Pellinen *et al.*, 1988a,b; Yin, Joyce & Chang, 1990); percolating filters (Newby, 1994; Messner *et al.*, 1990); and packed beds (Cammarota & Sant'Anna, 1992; Bajpai, Mehna & Bajpai, 1993; Lonergan *et al.*, 1995; Zhang, 1997). More recently fluidized beds using mycelial pellets have been shown to be effective (Pallerla & Chambers, 1996, 1997; Zhang, Knapp & Tapley, 1998, 1999b). There are however few reports in which organisms are grown and used for decolorization in stirred tank reactors (Bajpai *et al.*, 1993) presumably because of concern about the effects of shear forces on enzymes or mycelial structure. Most authors prefer to use aeration, or air-lift techniques, to provide mixing as well as aeration. As yet, most studies using bioreactors have been conducted on a relatively small scale: a few litres at most and often less than 1 litre. Growth of white rot fungi on such a scale is complicated by the propensity of these organisms to grow in clumps, to block pipe-work and to grow on surfaces, pH electrodes, etc. Reports of decolorization in large laboratory-scale bioreactors like the 200 litre packed bed reactor used by Schliephake & Lonergan

(1996) are unusual, and larger-scale tests on these fungi growing in bio-reactors would be useful (see also Lonergan *et al.*, 1995).

Immobilization of fungal mycelia (usually as pellets in alginate or poly-urethane) has been shown to be useful, giving as good, or better, results as free mycelium (Pallerla & Chambers, 1996, 1997; Zhang *et al.*, 1999b). Much work with bioreactors has used batch, or repeated batch culture (fed-batch), rather than continuous culture. Both batch and continuous methods can be effective and both have advantages and disadvantages. On theoretical grounds, a batch process is more likely than a continuous one to achieve complete treatment and may have kinetic advantages. If the effluent is non-toxic then decolorization may approximate to Michaelis–Menten kinetics (Royer *et al.*, 1985; Knapp *et al.*, 1997), al-though in some studies the rate of decolorization was shown to increase linearly with the concentration of colour in the reactor (Royer *et al.*, 1991; Pallerla & Chambers, 1996, 1997). More rapid reaction at high substrate concentrations would favour batch treatment. Where effluents are toxic but the toxic components are degraded by the fungus then continuous treatment may be advantageous (Zhang *et al.*, 1998).

Several papers have reported the repeated use of white rot mycelia over many cycles of decolorization, covering periods of several weeks to a few months, both in fed-batch and continuous reactors (Eaton, Chang & Kirk, 1980; Martín & Manzanares, 1994; Zhang *et al.*, 1998, 1999b; Palma *et al.*, 1999; Swamy & Ramsay, 1999b). The ability to maintain biomass in an active state over prolonged periods bodes well for commercialization of these processes. Most studies have employed aseptic techniques and axenic cultures of white rots. Some, however, have found that effective treatment can occur in non-aseptic conditions (Messner *et al.*, 1990; Royer *et al.*, 1991). Most of these studies relate to wood pulping effluents, which may represent a special case since it is possible that the toxicity of some chlorinated and phenolic components may inhibit excessive growth of organisms other than the white rots. Messner *et al.* (1990) found that heavy contamination (which could be prevented by sanitation) during the colon-ization of a reactor was deleterious, but at later stages low-level contami-nation did not prevent good decolorization. Eaton *et al.* (1982) considered that the combination of temperature, pH and nutrient limitation would be enough to inhibit growth of most organisms in their system for treating Kraft effluent. If systems are not axenic, then competing microbes may be able to use added carbon sources, so denying them to the relatively slow-growing white rots. In the experience of the authors, contamination often leads to treatment failure; failures for this reason have been reported by others (e.g. Royer *et al.*, 1985). The most regularly observed contaminants seem to be yeasts, which can thrive in carbohydrate-based media at pH 5 to 6. Small pink yeasts that appear to be *Rhodotorula* spp. have been observed on several occasions by the authors (unpublished data) and others (Messner *et al.*, 1990; Newby, 1994). It seems possible that for some systems sterile axenic culture is required. Provision of sterile conditions may not be as costly as might first be thought. Several effluents that are candidates for treatment by white rots are essentially sterile when they arise. Both cotton bleaching and wood pulping effluents are produced at high temperature and highly alkaline pH, which will kill most likely contaminants. Some wood pulping effluents also contain sodium sulfite, which has antimicrobial properties. Aseptic handling and storage of such effluents before treatment may be all that is required.

Mycelia of white rots are rugged enough to withstand repeated use; they can be stored for several months at 4°C and still retain 100% activity with ability to decolorize immediately. Zhang (1997) and Zhang *et al.* (1999a) showed that two strains of white rot fungi could be kept for 4 months without loss of activity. This is, potentially, a useful property in an industrial situation in which fungi may be required for treatment at short notice.

There have been several reports of the use of immobilized enzymes (LiP or MnP) for decolorization (Ferrer, Dezotti & Durán, 1991; Dezotti, Innocentini-Mei & Durán, 1995; Peralta-Zamora *et al.*, 1998). Although immobilization may stabilize enzymes, it is usually associated with some initial loss of activity. It seems unlikely, therefore, that immobilized ligninolytic enzymes will offer any advantages over the use of mycelia for large-scale decolorization.

Decolorization studies on specific effluents and materials

Chemical industry effluents

There is little mention in the literature of the use of white rots for treatment of effluents from chemical manufacture, notable exceptions being reports by Knapp & Newby (1999) and Schliephake *et al.* (1993). Knapp & Newby (1999) investigated the ability of several white rots to decolourize an effluent from the manufacture of nitrated stilbene sulfonic acids that contained an azo-linked chromophore. These chemicals are intermediates in the production of optical brighteners and it is rather ironic that their manufacture results in an extremely dark-coloured effluent (A_{390} of neat effluent can be as much as 1000 units). The reduction of A_{390} by 70 to 80%, at effluent concentrations of 5 to 40% (v/v), is encouraging, especially as the process reported used mycelial mats and was not optimized. Unfortunately, with some fungal strains, despite good reduction in A_{390}, there was an increase in A_{480} (see above), which gave the treated effluent a red colour. The report of Schliephake *et al.* (1993) gives few details but shows at least 90% removal of colour from an effluent at 400–450 nm. It is unclear whether the effluent (from pigment manufacture) was neat or diluted, and the nature of the chromophore is uncertain.

Recent work in our laboratory (E. J. Vantoch-Wood, unpublished data) has demonstrated decolorization of a diluted mixed effluent from a chemical factory. Of the four white rots tested, *C. versicolor* and strain F29 gave the greatest rate and extent of decolorization. For these strains, rates increased linearly up to 50% (v/v) of effluent and decolorization was maximal (ca. 85%) in about 20 hours. Even at 95% (v/v) effluent (which was inhibitory to all strains), these strains were capable of about 80% colour (A_{490}) removal in 3 days (initial A_{490} was about 2.5). Little colour removal could be attributed to adsorption. *P. chrysosporium* and another unidentifed isolate performed less well and were inhibited at lower effluent concentrations.

With industrial effluents, there is always the potential for inhibition of the fungi and some are clearly more susceptible than others. In addition to the more obviously toxic components, like phenolics, it should be noted that many industrial effluents contain high levels of mineral salts (mainly sodium chloride or sulfate), which are essentially non-toxic but cause inhibition through osmotic effects. These salts may be by-products of chemical reactions or may be used in recovery processes (like salting out). There appear to be no reports on the effects of salts on decolorization by white rots and attention should be paid to their presence and possible effects. Since dilution is not desirable in effluent treatment systems, it would be useful to have available salt-tolerant white rot fungi. Such organisms are not frequently reported but a recent report on enzyme production and decolorization by marine fungi (Raghukumar *et al.*, 1996) suggests that further investigations on salt-tolerant wood-rotting fungi might be of value.

Cotton bleaching effluents

There are few references to biological treatment of cotton bleaching effluents (Table 10.4). These are highly alkaline and dark black/brown in colour and while activated sludge treatment may remove some BOD, it has

Table 10.4. *Reports on decolorization of cotton bleaching effluents by cultures of white rot fungi and by their isolated enzymes*

Authors	Organism	Method[a]	Comments
Davis & Burns, 1990	*Trametes (Coriolus) versicolor*	B	Compares decolorization by fungal cultures and enzymes
Zhang *et al.*, 1998	Unidentified white rot fungus, strain 7	C	Continuous decolorization in fluidized-bed bioreactor
Zhang *et al.*, 1999a	Unidentified white rot fungus, strain 7	C	Optimization of treatment in batch culture

[a]Method: C, culture; E, enzyme; B, culture and enzyme.

no effect on colour. Their chemical composition is not well described and the nature of the chromophore is uncertain, although it is possibly related to the melanoidins. Davis & Burns (1990) showed that several white rot fungi could decolorize cotton cleaning effluents: *C. versicolor* giving the best (70–80%) reductions. They also showed that soluble and immobilized laccase from this organism could cause decolorization. Zhang (1997) screened several wood-rotting fungi (including strains of *C. versicolor* and *Flammulina velutipes*) for the ability to decolorize cotton bleaching effluent. An unidentified strain, 7, proved to be most effective, giving ca. 95% decolorization. Zhang *et al.* (1999a) optimized decolorization by strain 7 in batch culture; although the fungus gained some nutrition from the effluent (and removed 50–60% of effluent COD), addition of a co-substrate was essential for decolorization. The amount of colour removed and the percentage removal varied with the 'strength' of the batches of effluent and with the proportion of effluent in the medium. The effluent was inhibitory at high concentration in batch cultures (Zhang, 1997). A fluidized bed bioreactor with free mycelial pellets of strain 7, in a continuous process, gave 70–80% removal when concentrated (95% v/v) effluent was treated; decolorization was better with more dilute effluents (Zhang *et al.*, 1998). The continuous process appeared to be superior to batch process for more concentrated effluents. The enzymic mechanisms of decolorization were not determined but Zhang *et al.* (1999a) suggest that MnP is probably important in this system.

Dyes

Classification of dyes

Dyes vary enormously in structure, their main common property being the ability to absorb light in the visible region of the spectrum. The range of structures is bewildering, with many thousand different dyes in commercial production. Dyes can be classified according to their structure (particularly the nature of the chromophore) or the method of application (Gregory, 1993). The main groups of dye, according to their chromophores, are azo, anthraquinone, heterocyclic, metal phthalocyanines and triphenylmethane. The main 'application' classes are reactive, direct, vat, sulfur, disperse, basic, solvent, mordant and acid. Within each of these classes it is possible to find a variety of different chromophores. Common features of the chemical structure other than the chromophore dictate the way in which the dye can be applied and the uses to which it is put. Although there are many natural dyes, the vast majority of commercially used products are synthetic chemicals. Some synthetics are related in their basic structure to natural dyes; for example, indigo carmine is a sulfonated version of the natural indigo. Azo dyes, the most important group in terms both of number of products and of tonnage used, are all synthetic. Although all have azo-linked aromatic groups, the range of structures and applications is vast.

Studies on dye decolorization

Table 10.5 lists recent publications on the decolorization of dyes by white rots or their enzymes. The trivial names of many dyes tell of their colour but generally give little information concerning their structure. This makes it difficult to compare degradability or decolorization of different dyes. The systematic names are also of little help since they are often so complex that again comparison on the basis of names is difficult and the full structure needs to be considered. From the studies reviewed here, it is clear that white rot fungi (or their enzymes) can decolorize members of all the major chomophore groups of dye. It is also clear that while some dyes (e.g. indigo carmine) are readily decolorized by a wide range of organisms, others are much less easily decolorized and then only by a more limited group of isolates (see Knapp *et al.*, 1995).

Relationships between structure and degradability

The structure of dyes has a profound effect on their degradability; however, understanding of structure/degradability relationships in dyes is only superficial. Comparisons of degradability between dyes in different groups are of limited value and there have been few detailed comparisons within particular dye groups. Several authors have shown that within the triphenylmethanes relatively small structural differences are associated with differences in the extent or rate of decolorization (Bumpus & Brock, 1988; Knapp et al., 1995; Vasdev *et al.*, 1995; Sani *et al.*, 1998). Exact relationships have not been defined but the number, position and type of N-alkyl groups all appear to influence decolorization.

Most research on the effects of structure has employed azo dyes and there is much evidence to show that structure of azo dyes affects degradability. Azo dyes contain two aromatic ring structures linked by an azo group, and most studies have considered series of compounds in which one ring structure remains constant while the other ring is varied by incorporating various substituents in different positions. A study by Spadaro, Gold

Table 10.5. *Reports on decolorization of dyes by cultures of white rot fungi and by their isolated enzymes*

Authors	Class(es) of dye (according to chromophores)	Organism	Method[a]	Comments
Azmi, Sani & Banerjee, 1998	TPMs	Various including white rots	B	Review
Bakshi et al., 1999	Various	*Phanerochaete chrysosporium*	C	
Banat et al., 1996	Various including effluents	Various including white rots		Review
Buckley & Dobson, 1998	Polymeric dyes	*Chrysosporium lignorum*	C	Effects of Mn on decolorization by free and immobilized cultures
Bumpus & Brock, 1988	Crystal violet (TPM)	*P. chrysosporium*	B	
Capalash & Sharma, 1992	Azo	*P. chrysosporium*	C	
Chao & Lee, 1994	Azo and heterocyclic	*P. chrysosporium* and others	C	
Chivukula & Renganathan, 1995	Azo	*Pyricularia oryzae*	E	Mechanism of oxidation by laccase; substrate specificity
Chivukula et al., 1995	Azo	*P. chrysosporium*	E	Mechanism of oxidation by LiP
Collins et al., 1997	Azo	*P. chrysosporium, Trametes (Coriolus) versicolor*	E	Azo dye, decolorization by LiP, enhanced by tryptophan
Conneely, Smyth & McMullen, 1999	Metal phthalocyanine	*P. chrysosporium*	C	
Cripps et al., 1990	Azo and heterocyclic	*P. chrysosporium*	B	
Das, Dey & Bhattacharyya, 1995	Azo and triphenylmethane	*P. chrysosporium*	C	Column bioreactor
Dey, Maiti & Bhattacharyya, 1994	Congo Red, Methylene Blue	*Polyporus ostreiformis*	B	Reports on a LiP-producing brown rot
Glenn & Gold, 1983	Polymeric dyes	*P. chrysosporium*	C	
Gogna, Vohra & Sharma, 1992	Rose Bengal	*P. chrysosporium*	C	
Goszczynski et al., 1994	Azo	*P. chrysosporium*	E	Mechanism of oxidation by peroxidases

Reference	Dye/substrate	Organism		Notes
Heinfling et al., 1997	Azo and phthalocyanine	T. versicolor, Bjerkandera adusta	C	
Heinfling et al., 1998	Azo and phthalocyanine	B. adusta, Pleurotus eryngii	E	Oxidation by MnP in Mn-independent reaction
Kim et al., 1996	Remazol Brilliant Blue R	Pleurotus ostreatus	C	
Kirby, McMullen & Marchant, 1995	Artificial textile effluent	P. chrysosporium		
Knapp et al., 1995	Various including azo, TPM, heterocyclic, anthraquinone, phthalocyanine	Various white rots including P. chrysosporium, T. versicolor, P. ostreatus	C	Screens a range of white rots against several different dyes; several strains perform relatively better than P. chrysosporium
Knapp et al., 1997	Orange II (azo)	Unidentified white rot F29	C	Optimization of decolorization
Knapp & Newby, 1999	Azo-linked chromophore	Various white rots including P. chrysosporium, T. versicolor, P. ostreatus	C	Treatment of coloured chemical industry effluent
Lonergan et al., 1995	Remazol Brilliant Blue R	Pycnoporus cinnabarinus	C	Studied physiology and decolorization in 200l packed bed reactor
McCarthy et al., 1999	Various including azo and TPM	P. cinnabarinus	E	Optimization of treatment by laccase
Nagarajan & Annadurai, 1999	Reactive azo (?)	P. chrysosporium	C	
Ollikka et al., 1993	Various including azo, TPM, heterocyclic and polymeric	P. chrysosporium	E	LiP isoenzymes
Ollikka et al., 1998	Crocein Orange G (azo)	P. chrysosporium	E	LiP isoenzymes
Palma et al., 1999	Polymeric anthraquinone	P. chrysosporium	C	Packed bed bioreactor
Pasti-Grigsby et al., 1992	Azo	P. chrysosporium	B	Effects of substitution pattern on degradability
Paszczynski & Crawford, 1991	Azo	P. chrysosporium	B	Degradation of azo dyes with LiP: role of veratryl alcohol
Paszczynski et al., 1991	Azo	P. chrysosporium	B	Improved degradation of recalcitrant azo dyes

Table 10.5. (cont.)

Authors	Class(es) of dye (according to chromophores)	Organism	Method[a]	Comments
Paszczynski et al., 1992	Azo	P. chrysosporium	B	Mineralization of sulfonated azo dyes and sulfanilic acid
Platt, Hadar & Chet, 1985	Polymeric anthraquinone	P. ostreatus	C	Effects of structure on degradability by LiP
Podgornik et al., 1999	Various including azo, anthraquinone, polymeric	P. chrysosporium	E	Considers effects of metal ions (including Cu^{2+}) on dye decolorization by white rots
Pointing et al., 2000	Various including azo, TPM, heterocyclic and polymeric	Various white rots including P. chrysosporium, T. versicolor	C	Correlates dye decolorization rates with laccase activity
Rodriguez et al., 1999	Various	Various white rots including B. adusta, P. chrysosporium, T. versicolor, Trametes hispida, P. ostreatus	E	
Sani et al., 1998	Various	P. chrysosporium	C	Considers effects of agitation
Schliephake et al., 1993	Pigment plant effluent	P. cinnabarinus	C	Packed bed bioreactor for treatment of industrial effluent
Schliephake & Lonergan, 1996	Remazol Brilliant Blue R	P. cinnabarinus	C	Laccase variation in a packed bed bioreactor
Shin, Oh & Kim, 1997	Remazol Brilliant Blue R	P. ostreatus	E	Specificity of peroxidase decolourizing Remazol Brilliant Blue R
Shin & Kim, 1998	Various including azo, TPM, heterocyclic and polymeric	P. ostreatus	E	
Spadaro et al., 1992	Azo	P. chrysosporium	C	Effects of structure on degradability
Spadaro & Renganathan, 1994	Azo	P. chrysosporium	E	Mechanism of oxidation by peroxidases

Reference	Dyes	Organism	Method[a]	Comments
Swamy & Ramsay, 1999a	Various including azo, anthraquinone, phthalocyanine	Various white rots including B. adusta, P. chrysosporium, T. versicolor, T. hispida, P. ostreatus	C	Evaluation of several fungal species and initial optimization
Swamy & Ramsay, 1999b	Various	T. versicolor	C	Decolorization of repeated additions of dyes – effects of nutrients
Tatarko & Bumpus, 1998	Congo Red (azo)	P. chrysosporium	B	
Vasdev & Kuhad, 1994	Polymeric	Cyathus bulleri	C	
Vasdev et al., 1995	TPMs	C. bulleri and other Cyathus spp.	B	
Vyas & Molitoris, 1995	Remazol Brilliant Blue R	P. ostreatus	E	Discusses adsorption and degradation
Wang & Yu, 1998	Azo, anthraquinone, Indigo	T. versicolor	C	Fixed film bioreactor
Yang & Yu, 1996	Red 533 (azo?)	P. chrysosporium	C	LiP-catalysed dye decolorization
Young & Yu, 1997	Various including azo, anthraquinone, phthalocyanine, Indigo	P. chrysosporium, T. versicolor	B	
Zhang et al., 1999b	Orange II (azo)	Unidentified white rot F29	C	Development of packed bed and fluidized-bed reactors operated in continuous and fed-batch modes

LiP, lignin peroxidase; MnP, manganese peroxidase; TPM, triphenylmethane.
[a] Method: C, culture; E, enzyme; B, culture and enzyme.

Table 10.6. *Effect of the structure of azo dyes on their degradation by fungal cultures and enzymes*

Structure of dye
R =

Structure of dye	Decolorization by cultures (%)[a]	Oxidation rate with manganese peroxidase (μmol min^{-1} ml^{-1})[b]	Oxidation rate with ligninase (μmol min^{-1} ml^{-1})[b]	Oxidation rate with laccase (nmol min^{-1} ml^{-1})[c]
	99	16.37	0.42	447
	97	14.33	0.22	703
	96	8.19	0.72	200
	93	3.54	0.00	20
	90	0.02	0.45	0

81	0.32	0.49	N.D.
70	0.00	11.61	N.D.
N.D.	N.D.	N.D.	0
N.D.	N.D.	N.D.	27

N.D. not determined.

[a] Decolorization of dyes at 150 ppm by cultures of *Phanerochaete chrysosporium* (Pasti-Grigsby *et al.*, 1992).
[b] Oxidation rate with enzymes from *P. chrysosporium* (Pasti-Grigsby *et al.*, 1992).
[c] Oxidation rate with laccase from *Pyricularia oryzae* (Chivukula & Renganathan, 1995).

& Renganathan (1992) showed differences in the breakdown of groups of structurally similar dyes by *P. chrysosporium.* It also showed that the two aromatic rings can be degraded to different extents. Pasti-Grigsby *et al.* (1992) produced a series of dyes based on azobenzene 4-sulfonic acid as the 'constant structure' (Table 10.6). While degradation by cultures of *P. chrysosporium* was essentially complete for some dyes, for others it was only partial and complicated by their toxicity. Experiments with enzymes showed large difference between dyes in their susceptibility to decolorization. It is notable that some compounds that were rapidly decolorized by MnP were resistant to LiP and *vice versa.* Degradability was related to the presence and position of hydroxyl groups (*para* to the azo linkage being best for degradability) and to the presence, position and number of electron-donating groups (e.g. methyl and methoxyl). Chivukula & Renganathan (1995) used a similar set of dyes – all were 4-(4′-sulfophenylazo)-phenols substituted on the phenol moiety – to study susceptibility to oxidation by a laccase from *Pyricularia oryzae.* Degradability was promoted by methyl and methoxy groups in the 2 and 6 positions, but not by the same substituents in the 3 and 5 positions. Electron-withdrawing groups (e.g. chloro or nitro) in the 2 and 6 positions prevented degradation. Selected data from these studies are presented in Table 10.6 and it can be seen that the structure of dyes strongly influences their susceptibility to degradation both by pure cultures and isolated enzymes (LiP, MnP and laccase). Unfortunately, only limited data are available from systematic studies and more remains to be discovered. Most of the dyes studied are relatively simple in terms of the number and type of aromatic rings (mostly phenyl) and the type and number of substituents. Many commercial dyes are much more complex and there are no sound data available on structure/degradability relationships in dyes with, for example, multiply substituted naphthyl groups, multiple azo linkages, reactive groups (e.g. chlorotriazine) for covalent linking to cotton or stilbene groups. Chemical structure may affect degradability via steric hindrance, electron distribution and charge. It has also been suggested that ionization potential may have an important role in determining susceptibility to LiP (Podgornik, Grgic & Perdih, 1999), and this is supported by studies on other types of compound (Kay-Shoemake & Watwood, 1996; ten Have *et al.*, 1998). Where degradation proceeds via a redox mediator (e.g. VA radicals or Mn^{3+}) ionization potential is likely to be important.

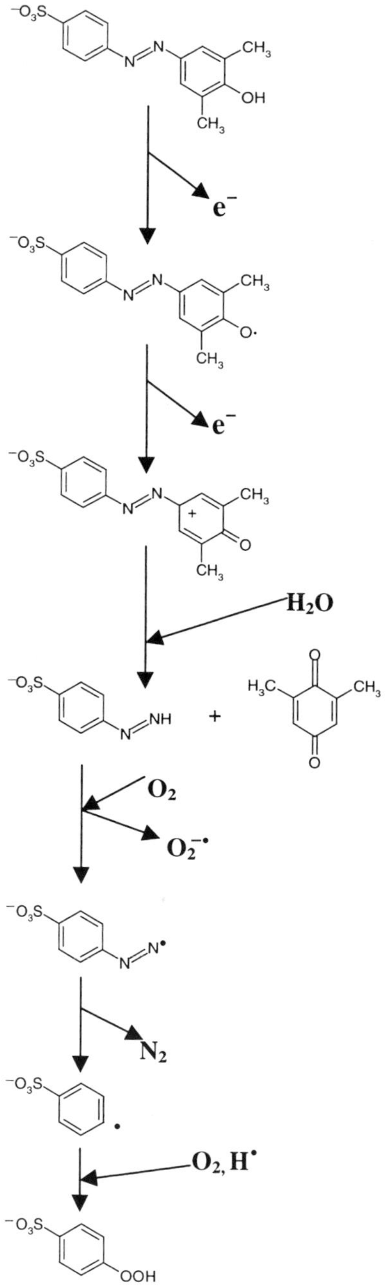

Fig. 10.1. Proposed pathway for the degradation and decolorization of the azo dye (4-(4′-sulfophenylazo)-2,6-dimethylphenol by lignin peroxidase of *Phanerochaete chrysosporium*. (Based on Chivukula *et al.*, 1995.)

Mechanisms for dye decolorization

The mechanisms of decolorization and degradation of dyes by white rots have received scant attention. Bumpus & Brock (1988) have shown that the degradation of several related triphenylmethane dyes appears to start with sequential dealkylation. It is not clear, however, what processes complete their degradation. Evidence from a wide range of sources shows that LiP, MnP and laccase (alone or with the involvement of mediators) are able to degrade some azo dyes. Most fungi, however, will produce at least two of these three enzymes, and possibly others, so when cultures are used the end result will depend on the types of enzyme produced, their relative activities and their specificities. Studies with pure enzymes are of great interest but we cannot extrapolate from them to predict the final products of dye decolorization by whole cultures. Extensive degradation has been reported for some azo dyes (Spadaro *et al.*, 1992; Paszczynski *et al.*, 1992) but all the examples studied are relatively simple in structure and the ultimate fate of more complex and more highly sulfonated dyes following decolorization by white rot fungi is not known.

Pathways for degradation of azo dyes by peroxidases from *P. chrysosporium* have been suggested (Goszczynski *et al.*, 1994; Spadaro & Renganathan, 1994; Chivukula, Spadaro & Renganathan, 1995). The mechanisms propose sequential abstraction of two electrons by peroxidase action followed by attack by water, which results in cleavage of the diazo linkage. All three reports suggest cleavage of the azo dye can occur asymmetrically to give a quinone and a phenyl diazine (Fig. 10.1). The latter decomposes as a result of attack by oxygen to give nitrogen and a phenyl compound (Spadaro & Renganathan, 1994) or a sulfophenyl hydroperoxide (Chivukula *et al.*, 1995) depending on the type of dye involved. A similar pathway has been proposed by Chivukula & Renganathan (1995) for degradation of phenolic dyes by laccase. Goszczynski *et al.* (1994) also suggested that the azo bond can be split symmetrically during peroxidase attack, one nitrogen remaining attached to each aromatic ring. This would give rise to compounds such as amino or nitroso-substituted aromatics or quinone imines.

When LiP is utilized, VA has been shown to stimulate decolorization (Paszczynski & Crawford, 1991; Ollikka *et al.*, 1993; Young & Yu, 1997). The presence, and concentration, of Mn^{2+} is also important and can stimulate decolorization (e.g. Knapp *et al.*, 1997). However, Mn^{2+} can also inhibit decolorization, the effect being dependent on both the dye and the organism. Buckley & Dobson (1998) showed that with *Chrysosporium lignorum*, Mn^{2+} promoted degradation of one polymeric dye (Poly R-478) but inhibited degradation of another (Poly S-119). Inhibition was attributed to inhibition of LiP production by Mn^{2+}. Mn^{2+} can also inhibit dye decolorization by MnP (Heinfling *et al.*, 1998). While many authors consider the peroxidases to be of prime importance in decolorization of dyes, laccases have also been shown to catalyse certain decolorizations (Chivukula & Renganathan, 1995; McCarthy *et al.*, 1999; Rodríguez, Pickard & Vázquez-Duhalt, 1999).

Molasses wastewaters

Molasses, a waste from sugar refining, is widely used as a fermentation feedstock for production of yeast, alcohol and other products. It is highly coloured because of the presence of melanoidin pigments, which are produced by the reaction between sugars and amino acids at elevated temperatures. Very large volumes of highly coloured wastewaters remain after use of molasses in fermentation processes (Fitzgibbon *et al.*, 1995). Melanoidins are recalcitrant and survive normal biological treatment processes (activated sludge and anaerobic digestion) unscathed. Several fungi have been investigated for their ability to decolorize melanoidins. With some non-wood-rotting organisms (e.g. *Rhizoctonia* sp.), it is clear that decolorization is solely adsorptive (Sirianuntapiboon *et al.*, 1995). Where this is the case, there is a limit to re-usability of the mycelium (because of saturation) and adsorbed melanoidin remains to be dealt with. Several investigations have shown that white rots have the ability to degrade melanoidins (Table 10.7). Both *P. chrysosporium* and *T. (C.) versicolor* can achieve a high degree of decolorization, generally 70–85% (Kumar *et al.*, 1998). White rots can also reduce COD (70–90%) of molasses-based effluents (Benito, Miranda & Rodriguez de los Santos, 1997; Kumar *et al.*, 1998). Generally, the provision of a carbon source, usually glucose, greatly improves decolorization (Kumar *et al.*, 1998). Rates and extents of decolorization depend on the organism, the effluent concentration and the intensity of the colour. Decolorizations may take as little as 1 or as much as 10 days, but comparison between studies is difficult. Some effluents can be inhibitory.

The mechanism of decolorization is not completely clear. Watanabe *et al.* (1982) suggested that the enzyme responsible in *Coriolus* sp. 20 is intracellular and reported the involvement of an enzyme with glucose, or sorbose, oxidase activity. They proposed that this enzyme generates 'active oxygen' (O_2^- or H_2O_2), which is responsible for decolorization of

Table 10.7. *Reports on decolorization of molasses-based effluents (containing melanoidins) by cultures of white rot fungi and by their isolated enzymes*

Authors	Material decolorized	Organism	Method[a]	Comments
Aoshima et al., 1985	Molasses pigment[b]	Various including *Trametes (Coriolus) versicolor*	C	Optimization and enzymology
Benito et al., 1997	Waste from alcohol factory	*T. versicolor*	C	Optimization
Dehorter & Blondeau, 1993	Synthesized melanoidins	*T. versicolor*	B	Studies manganese-dependent enzyme
Fahy et al., 1997	Molasses spent wash from distillation	*Phanerochaete chrysosporium*	C	Studies effects of immobilization
Fitzgibbon et al., 1995	Molasses spent wash from distillation	Various including *C. versicolor, P. chrysosporium*	C	Screening organisms for chemical oxygen demand and colour removal
Kasturi-Bai & Ganga, 1996	Distillery effluent	*P. chrysosporium*	C	Compares decolorization by chemical methods with that of white rot fungi
Kumar et al., 1998	Anaerobically digested molasses spent wash from distillery	*P. chrysosporium, T. versicolor*	C	Optimization
Ohmomo et al., 1985a	Molasses waste[b] and synthetic melanoidin	*C. versicolor*	B	Purification and properties of melanoidin-decolorizing enzymes
Ohmomo et al., 1985b	Molasses waste[b]	*C. versicolor*	C	Continuous decolorization and fed batch treatment
Sermkiattipong et al., 1999	Molasses wastewater from a distillery	*C. versicolor*	C	Investigates improved decolorization by mutant strains

| Sirianuntapiboon et al., 1988 | Molasses wastewater | *Mycelia sterilia* D90 | C | Mycelia re-used three times with > 80% colour removal |
| Watanabe et al., 1982 | Synthesized melanoidins | *Coriolus* sp. | C | Decolorization linked to an intracellular enzyme with glucose/sorbose oxidase activity, which produced H_2O_2 |

[a]Method: C, culture; E, enzyme; B, culture and enzyme.
[b]Waste from baker's yeast factory treated with activated sludge.

Table 10.8. *Reports on decolorization of olive milling wastewaters by cultures of white rot fungi and by their isolated enzymes*

Authors	Organism	Method[a]	Comments
Ben Hamman et al., 1999	Phanerochaete flavido-alba	C	Immobilization and re-use of mycelium
D'Annibale et al., 1998	Lentinula edodes	C	Re-use of mycelium
Martirani et al., 1996	Pleurotus ostreatus	B	Importance of manganese peroxidase and laccase
Pérez et al., 1998	Phanerochaete flavido-alba	E	
Sayadi & Ellouz, 1992	Phanerochaete chrysosporium	C	
Sayadi & Ellouz, 1993	Various including P. chrysosporium, Phlebia radiata, Dichomitus squalens, Polyporus frodosus, Trametes (Coriolus) versicolor	C	Comparison of a range of white rot fungi
Sayadi & Ellouz, 1995	P. chrysosporium	B	Assesses relative importance of lignin peroxidase and manganese peroxidase
Sayadi et al., 1996	P. chrysosporium	C	Uses immobilized mycelium
Vinciguerra et al., 1995	L. edodes	C	
Yesilada, Fiskin & Yesilada, 1995	Funalia trogii	C	Re-use of mycelium
Yesilada et al., 1998	F. trogii, C. versicolor	C	Effects of agitation, immobilization and initial chemical oxygen demand

[a]Method: C, culture; E, enzyme; B, culture and enzyme.

melanoidin. Using a strain of *C. versicolor*, Ohmomo *et al.* (1985a) resolved seven different melanoidin-decolourizing enzymes and purified two of them. Enzyme P-III required glucose and oxygen for its melanoidin-decolourizing activity while P-IV decolourizes melanoidin in their absence. A multiplicative effect between the two enzymes was noted. While P-III appears to act via its glucose oxidase activity, the mechanism for decolorization by P-IV is uncertain. Dehorter & Blondeau (1993) reported isolation of a melanoidin-degrading activity from *T. versicolor*. This enzyme system required oxygen (not H_2O_2) and was Mn^{2+} dependent, although the mechanism of action was uncertain. Sermkiattipong *et al.* (1999) have isolated mutants of *C. versicolor* with enhanced melanoidin-degrading activity but the nature of their mutations was not determined.

Olive oil milling wastewater

Olive oil milling effluents are a major problem in many Mediterranean countries; annual production of olive mill wastewater (OMW) in this region is $> 3 \times 10^7 \, m^3$ (Sayadi & Ellouz, 1995). OMW is black in colour, has a high COD and is highly toxic. Both toxicity and colour are associated with the presence of phenolic compounds, often polyphenols like tannins, anthocyanins and catechins (Sayadi & Ellouz, 1995). Conventional biological effluent treatment processes are ineffective because of a combination of recalcitrance to biodegradation and toxicity. The possibility of using white rot fungi for OMW treatment has been investigated by several research groups using a range of different species (Table 10.8). About 60 to 75% (or more) colour removal has been reported by most authors, together with up to 90% reduction in phenolics (Martirani *et al.*, 1996; Ben Hamman, de la Rubia & Martínez, 1999). At its best, good decolorization can occur in 3 days but in other cases 9 or more days may be required. Comparison between studies is difficult because of differences in composition of the OMW studied and the different proportions of OMW in media. Yesilada, Sik & Sam (1998) have studied the effect of OMW concentration on removal of COD, phenols and colour by *Funalia trogii* and *C. versicolor*. Neither organism grew in undiluted effluents. The percentage of all components (especially colour) removed decreased with increasing OMW concentration. For COD and phenols although the percentage removal decreased the amount removed increased with concentration of OMW. However, for colour, both the percentage removed and the amount removed decreased in more concentrated effluents, suggesting inhibition. Some OMW have high sugar contents and it appears that decolorization can occur without an additional carbon source (Martirani *et al.*, 1996; Yesilada *et al.*, 1998). In others, saccharose (D'Annibale *et al.*, 1998), glucose or glycerol (Sayadi & Ellouz, 1992, 1993, 1995) are added.

The ligninolytic enzyme system is the main agent of decolorization and destruction of phenolics. However, organisms differ in the relative importance of different enzymes. Sayadi & Ellouz (1993, 1995) considered that LiP was the most important in cultures of *P. chrysosporium*. However, Ben Hamman *et al.* (1999) and Pérez *et al.* (1998) proposed that MnP and laccase were of prime importance in *P. flavido-alba*. Pérez *et al.* (1998) suggested that laccase was induced by growth on OMW. Martirani *et al.* (1996) showed that purified phenol oxidase from *P. ostreatus* could reduce the phenol content of OMW but this alone did not reduce toxicity. It seems likely that in mycelial cultures (which do decrease toxicity) other enzymes must work in concert with the phenol oxidases. Several studies have demonstrated a change in molecular mass distribution of aromatic compounds in OMW during treatment with white rot fungi (Sayadi & Ellouz, 1993, 1995; Vinciguerra *et al.*, 1995). Low-molecular-mass compounds are largely removed and there is a decrease in the amount of high-molecular-mass materials, which are replaced by compounds of intermediate molecular mass, showing that depolymerization has occurred. No details of the chemical mechanisms for decolorization of OMW are available but it seems safe to assume, given the similarity between the polyphenolics involved and lignin, that the biochemical mechanisms are similar to those involved in lignin degradation. It has been shown that mycelia of *P. ostreatus*, *P. chrysosporium* and *L. edodes* can be re-used several times for OMW treatment and that treatment by immobilized mycelium is also effective (Martirani *et al.*, 1996; Sayadi, Zorgani & Ellouz, 1996; D'Annibale *et al.*, 1998). Clearly, white rot fungi have the potential to decolorize and detoxify OMWs but there is scope for more research to optimize treatment processes. Given the variability of these effluents, attempts to standardize experimental approaches and descriptions would allow better comparison of results.

Paper making and pulping effluents

The potential for white rots to treat paper making and pulping effluents has been recognized for many years (Marton, Syern & Marton, 1969). A variety of effluents are considered here; all are highly coloured (dark brown/black). The colour and chemical composition of the effluents varies considerably according to the pulping process used and the exact stage in the process at which they arise. The type of material being pulped (e.g. wood or an agricultural residue) is also important, as is the type of wood (hard or soft) and the species of tree. In pulping processes, lignocellulosic materials are treated to remove the lignin and the resultant pulp is bleached. The major part of the colour in pulping effluents results from dissolved derivatives or fragments of lignin. Pulping effluents may contain lignin sulfonates or chlorolignins and also lower-molecular-mass phenolics and chlorinated phenolics, together with their partly oxidized derivatives such as quinones. Many of these materials are toxic both to aquatic life and to microorganisms. There are a variety of pulping processes, the best known being the Kraft and the sulfite processes. Most publications on the use of white rots in treating pulping wastes deal with Kraft effluents, some dealing with effluent from particular stages rather than combined plant effluents. The most studied effluent stream is the Kraft E_1 (alkaline extraction process) effluent.

Decolorization studies

Table 10.9 summarizes some reported uses of white rots or their enzymes in treating pulping effluents. Research on decolorization is dominated by two species of white rots: *P. chrysosporium* and *C. versicolor*. Other species, like *B. adusta* or *P. ostreatus*, which are effective in degrading other materials, seem to have received little attention in this context. Most authors routinely report 60–90% colour removal and reports of > 90% can be found (Archibald, Paice & Jurasek, 1990a). Typically, decolorizations are complete in 3 to 6 days but occasionally shorter or longer periods are required.

Nutritional factors

As with lignin, the chromophores in pulping effluents cannot be degraded without the provision of a carbonaceous co-substrate. Levels of such co-substrates are usually low in pulping effluents and so they need to be added, the amount added varying with the organism and the effluent (e.g. Archibald *et al.*, 1990a; Bajpai *et al.*, 1993; Belsare and Prasad, 1988; Srinivasan and Murthy, 1999). Glucose or sucrose are the most usual co-substrates added, (often at $1–10\,\mathrm{g\,l^{-1}}$) but other sugars, polysaccharides, glycerol and ethanol have been reported to promote decolorization (Archibald *et al.*, 1990a). The requirement for nitrogen varies between organisms and also depends on the concentration of usable fixed nitrogen in the effluent. Sometimes addition of nitrogen represses decolorization, particularly with *P. chrysosporium* in which the ligninolytic system is

Table 10.9. *Reports on decolorization of wood pulping/bleaching effluents by cultures of white rot fungi and by their isolated enzymes*

Authors	Source of effluent or material used	Organism	Method[a]	Comments
Archibald et al., 1990a	Kraft effluent	*Trametes (Coriolus) versicolor*	B	Optimization
Archibald et al., 1990b	Kraft effluent	*C. versicolor*	C	
Bajpai & Bajpai, 1994	Pulp and paper mill effluents	Various	B	Review
Bajpai et al., 1993	Kraft effluent	*T. versicolor*	C	Optimization of batch and continuous processes
Belsare & Prasad, 1988	Bagasse-based pulping effluent	*Schizophyllum commune*	C	Optimization of decolorization
Bergbauer et al., 1990	Lignosulfonates and chlorolignins	*S. commune, T. versicolor*	C	Polymerization and depolymerization of wastewater lignins
Bergbauer & Eggert, 1992	Various bleachery effluents	*T. versicolor, Phanerochaete chrysosporium*	C	Compares ability of fungi to degrade effluent from different processes
Cammarota & Sant'Anna, 1992	Kraft effluent	*P. chrysosporium*	C	Uses continuous packed bed reactor with immobilized mycelium
Chambers & Cheng, 1991	C_DEDED effluent	*P. chrysosporium*	C	Continuous hollow fibre membrane reactor
Dezotti et al., 1995	Kraft effluent	*Chrysonilia sitophila*	B	Silica immobilized LiP and fungal culture
Durán et al., 1994	Kraft E_1 effluent	*Lentinus edodes*	C	Combined photochemical and biological process
Eaton et al., 1980	Kraft E_1 effluent	*P. chrysosporium*	C	
Eaton et al., 1982	Kraft E_1 effluent	*P. chrysosporium*	C	

Eposito, Canhos & Durán, 1991	Kraft effluent	Various white rots	C	Screens 51 lignolytic fungi for decolorization in the absence of added carbon source
Ferrer et al., 1991	Kraft effluent	C. sitophila	B	Immobilized enzymes and fungal mycelium
Fukui et al., 1992	Kraft effluent	P. chrysosporium	C	Assesses dechlorination and detoxification as well as decolorization
Galeno & Agosin, 1990	Kraft effluent	Various white rots	C	Assesses several strains; Ramaria strain 158 is as effective as P. chrysosporium
Garg, Chandra & Modi, 1999	Bagasse and gunny bag-based paper mill effluent	T. versicolor	C	
Garg & Modi, 1999	Pulp-paper mill effluents	Various		Review
Jaspers et al., 1994	Kraft effluent	P. chrysosporium	B	Evidence for the role of MnP; effect of culture conditions on MnP production
Jaspers & Penninckx, 1996	Kraft effluent	P. chrysosporium	C	Discusses adsorption of colour to mycelial pellets
Lackner et al., 1991	Chlorolignin from a sulfite pulping plant	P. chrysosporium	E	Role of MnP in decolorization
Lankinen et al., 1991	Kraft E$_1$ effluent	P. chrysosporium, Phlebia radiata, Merulius tremellosus	B	Production of lignin-modifying enzymes during culture on bleach plant effluents
Livernoche et al., 1981	Kraft effluent	C. versicolor	C	Decolorization by mycelium immobilized in alginate
Livernoche et al., 1983	Kraft E$_1$ effluent	Various including C. versicolor	C	Decolorization by mycelium immobilized in alginate
Manzanares et al. 1995	Effluent from pulping of cereal straw	T. versicolor	C	Assesses importance of various enzymes
Martín & Manzanares, 1994	Straw-soda pulping effluent	T. versicolor	C	Optimizes treatment

Table 10.9. (cont.)

Authors	Source of effluent or material used	Organism	Method[a]	Comments
Marwaha et al., 1998	Anaerobically digested black liquor from pulp mill	P. chrysosporium	C	Continuous process using mycelia immobilized on jute rope
Marton et al., 1969	Kraft effluent	Polyporus versicolor	C	Optimizes treatment
Mehna, Bajpai & Bajpai, 1995	Effluent from pulping of agricultural wastes (straws)	T. versicolor	C	Optimizes treatment
Messner et al., 1990	Sulfite plant effluent	P. chrysosporium	C	Sequential batch treatment in the MYCOPOR mycelial trickling filter process
Michel et al., 1991	Kraft effluent	P. chrysosporium	C	Suggests that MnPs are more important than LiPs
Mittar et al., 1992	Pulp and paper mill effluent	Various including P. chrysosporium	C	Optimization of treatment
Modi, Chandra & Garg, 1998	Bagasse-based paper mill effluent	T. versicolor	C	
Pallerla & Chambers, 1996	Kraft effluents	T. versicolor	C	Fluidized bed reactor with urethane foam immobilized mycelium in continuous and sequential batch mode
Pallerla & Chambers, 1997	Paper mill effluents	T. versicolor	C	Continuous bioreactor with calcium alginate-immobilized mycelium
Pellinen et al., 1988a	Kraft effluent	P. chrysosporium	C	Dechlorination of high-molecular-mass chlorolignins in a rotating biological contactor

Reference	Effluent	Organism	Method[a]	Comments
Pellinen et al., 1988b	Kraft effluent	P. chrysosporium	C	MyCoR process, growth in rotating biological contactor: optimization
Peralta-Zamora et al., 1998	Paper industry effluents	P. chrysosporium	E	LiP and MnP immobilized on resin
Pérez et al., 1997	Biologically treated paper mill effluent	Various including Phanerochaete flavido-alba	C	MnP and LiP implicated in decolorization; suggests LiP may be more important
Prasad & Gupta, 1997 Presnell et al., 1992	Pulp and paper mill effluent Kraft effluent	T. versicolor, P. chrysosporium P. chrysosporium	C	Reports influence of effluent fractions on enzyme production
Raghukumar et al., 1996	Paper mill effluent	Marine fungi	C	Reports decolorization and enzyme production by marine fungi
Roy-Arcand, Archibald & Briere, 1991	Kraft effluent	T. versicolor	C	Compares fungal and ozone treatment
Royer et al., 1983	Kraft effluent	C. versicolor	C	Continuous decolorization by immobilized fungi in air lift reactor
Royer et al., 1985	Kraft effluent	C. versicolor	C	Batch and continuous decolorization by fungus; reports colour adsorbed first and then degraded
Royer et al., 1991	Kraft effluent	C. versicolor	C	Maintenance of decolourizing ability in continuous and repeated fed batch reactors
Srinivasan & Murthy, 1999	Bagasse-based pulp mill effluent	T. versicolor	C	Optimization of decolorization
Wang et al., 1992	Kraft effluent	Ganoderma lacidum, C. versicolor, Hericium erinaceum	C	Decolorization and degradation of organic halides
Yin et al., 1990	Kraft effluent	P. chrysosporium	C	Continuous treatment in rotating biological contactor

LiP, lignin peroxidase; MnP, manganese peroxidase.
[a] Method: C, culture; E, enzyme; B, culture and enzyme.

induced in nitrogen-deficient conditions (see above). Nevertheless, the situation can be more complicated, Eaton *et al.* (1980) showed that high nitrogen concentrations repressed decolorization by *P. chrysosporium* when glucose was the co-substrate but promoted it when cellulose was co-substrate. For *C. versicolor*, however, nitrogen-sufficient conditions do not repress decolorization of pulping effluents (e.g. Archibald *et al.*, 1990a; Bajpai *et al.*, 1993; Srinivasan & Murthy, 1999).

Mechanisms of decolorization

The ability to dechlorinate the chloroorganic components in pulping effluents has been demonstrated by several authors (Fukui *et al.*, 1992; Pallerla & Chambers, 1996). Pellinen *et al.* (1988a) have correlated dechlorination and decolorization by *P. chrysosporium* and suggested that these two processes are metabolically linked. Several authors have investigated changes in the molecular mass distribution of lignin fractions during treatment by white rot fungi or their enzymes (Pellinen *et al.*, 1988a; Galeno & Agosin, 1990; Ferrer *et al.*, 1991; Lackner, Sebrotnik & Messner, 1991; Bergbauer & Eggert, 1992; Fukui *et al.*, 1992; Durán *et al.*, 1994; Dezotti *et al.*, 1995). Some distributions showed a marked decrease in all fractions while in others depolymerization was shown by a decrease in high-molecular-mass fractions together with an increase in low-molecular-mass fractions (Galeno & Agosin, 1990). Evidence has also been provided for polymerization (decrease in low- and increase in high-molecular-mass fractions) (Bergbauer, Eggert & Kraipelin, 1990). This can be transitory, since on prolonged treatment the newly formed high-molecular-mass fractions may be degraded (Pellinen *et al.*, 1988a). The distribution and average molecular-mass of the lignin-related fractions can change considerably during the course of a decolorization (Wang, Ferguson & McCarthy, 1992). Bergbauer *et al.* (1990) showed that the occurrence of polymerization or depolymerization depends on the particular strain of fungus used. For example, one strain of *Schizophyllum commune* caused polymerization of lignosulfonates, another strain caused a distinct depolymerization and a third had no effect on the molecular mass distribution. The ability of white rots to alter the distribution of lignin fragments depends on the material being treated; for example a strain of *C. versicolor* caused polymerization of lignosulfonates but depolymerization of chlorolignins (Bergbauer *et al.*, 1990). Changes in molecular mass distribution can also be brought about by isolated enzymes (Ferrer *et al.*, 1991; Lackner *et al.*, 1991; Dezotti *et al.*, 1995) and chemical oxidants such as Mn^{3+} (Lackner *et al.*, 1991). However, there does not yet appear to be any certain correlation between specific changes in molecular mass distribution and the presence and activity of particular enzymes. Interestingly, Presnell *et al.* (1992) showed that the presence of pulping effluents in growth medium changed the profile of extracellular proteins produced by *P. chrysosporium*, changes being dependent on the fraction of effluent used. In *T. versicolor* the presence of pulping liquor in the growth medium could cause a dose-dependent increase in laccase activity (Manzanares, Fajardo & Martín, 1995).

Pérez *et al.* (1997) suggested that both Lip and MnP have a role in decolorization of pulping effluents by *P. flavido-alba*, with LiP seeming to be more important. Michel *et al.* (1991) used mutants of *P. chrysosporium* deficient in both peroxidases or in LiP alone and also investigated the effects of culture conditions on peroxidase and decolorizing activities. They concluded that MnP is the more important of the two enzymes. This is supported by Jaspers, Jimenez & Penninckx (1994), who correlated decolorization by *P. chrysosporium* with the presence of MnP but could not detect LiP in decolorizing cultures. Furthermore, they showed that while purified MnP was capable of decolorizing Kraft effluent, LiP was not. It was suggested that the decolorizing activity of whole cultures depended on the production of enzymes other than MnP by the fungus. Lackner *et al.* (1991) also showed that purified *P. chrysosporium* MnP could decolorize and depolymerize chlorolignins from Kraft effluent. They also found that cell-free peroxidases could not be found in culture fluids of old cultures of *P. chrysosporium* (which could still decolorize). However, it was possible to measure cell-associated MnP (though not LiP) activity. Manzanares *et al.* (1995) reported decolorization of pulping effluents by *T. versicolor* in the absence of detectable LiP activity and concluded that MnP activity was most important in decolorization. Archibald (1992) studied the role of peroxidases in bleaching of Kraft pulp by *T. versicolor* (a process analogous to treatment of Kraft effluent) and demonstrated that LiP was not important. In a related paper, Addleman *et al.* (1995) showed that *T. versicolor* mutants lacking MnP activity could not bleach wood pulp and also that one of these MnP-deficient mutants was unable to decolorize Kraft effluents. These results together suggest that for *T. versicolor* MnP is of prime importance and LiP is unimportant. The finding by Archibald *et al.* (1990b) that decolorization by cultures of *T. versicolor* required oxygen but was not stimulated by added H_2O_2 or inhibited by the presence of catalase (or scavengers of 'active oxygen' species) is somewhat confusing, since the importance of peroxidases (notably MnP) seems certain.

Lankinen *et al.* (1991) suggested that an initial increase in colour (prior to decolorization) of Kraft effluents treated by *Phlebia radiata* and *Merulius tremellosus* could be associated with the high laccase activity of these organisms. However, they were unable to confirm which enzymes were important in decolorization. It should be noted that *T. versicolor* produces high levels of laccase but does not normally produce an increase in the colour of effluents. Bergbauer *et al.* (1990) reported that the strain of *S. commune* that caused depolymerization produced laccase whereas the strain that caused polymerization did not show detectable laccase activity. Furthermore, while their strain of *C. versicolor* produced laccase it could polymerize some lignin derivatives and depolymerize others. Davis & Burns (1990) showed that purified laccase from *C. versicolor* decreased the colour of pulping effluent but that this was associated with the polymerization of phenols and precipitation. While some authors refer to polymerization of lignin-related compounds (Pellinen *et al.*, 1988a; Bergbauer *et al.*, 1990), precipitation as a mechanism for decolorization does not appear to be the norm. Moreover, subsequent depolymerization is also reported. Most strains of *P. chrysosporium* do not produce detectable laccase activity whereas most strains of *C. versicolor* display high laccase activity, and yet both are similarly effective in decolorizing pulping effluents. The exact role of laccase in decolorization of bleaching effluents is, therefore, still uncertain.

Most studies on decolorization of pulping effluents use mycelia rather than isolated enzymes. Some peroxidases have been shown to cause decolorization but usually to a fairly limited degree. It is probable that decolorization is achieved by the concerted activity of a range of enzymes. In studies assessing the activity of enzymes in decolorization it is generally the practice only to measure cell-free enzyme activity. There have been few attempts to measure cell-associated activities, despite the fact that it is well known that the activity of ligninolytic enzymes can be cell associated (Evans *et al.*, 1991, 1994). Mycelium-associated enzymes are probably very important in decolorization, but their relative importance compared with cell-free enzymes has not been established. In decolorization of pulping effluents by white rot mycelia, the total activity may include synergistic and concerted sequential action of enzymes.

Treatment processes

Decolorization studies with pulping effluents have employed batch, semicontinuous (fed batch) and continuous culture. Given that pulping effluents can be toxic, continuous treatment approaches are particularly attractive since materials that are inhibitory but degradable may, in continuous culture, never reach inhibitory levels. Both *P. chrysosporium* and *T. versicolor* have been shown to work effectively as immobilized pellets in, for example, calcium alginate (Livernoche *et al.*, 1981; Royer *et al.*, 1983; Pallerla & Chambers, 1997) and polyurethane foam (Cammarota & Sant'Anna, 1992; Pallerla & Chambers, 1996). Immobilization on discs in a RBC (Pellinen *et al.*, 1988a,b; Yin *et al.*, 1990) and in a semipermeable hollow fibre membrane bioreactor (Chambers & Cheng, 1991) has also been reported. Decolorization by *P. chrysosporium* immobilized by growth on pieces of rope has also proved successful (Marwaha *et al.*, 1998). Use has also been made of naturally grown, mycelial pellets (Bajpai *et al.*, 1993).

Reactor configurations vary considerably but include packed beds (Cammarota & Sant'Anna, 1992; Bajpai *et al.*, 1993), fluidized beds (Royer *et al.*, 1991; Pallerla & Chambers 1996, 1997), trickling filters (Messner *et al.*, 1990) and RBCs (Pellinen *et al.*, 1988a,b; Yin *et al.*, 1990). Many of the reactors used are designed to retain a high biomass in the reactor, none appears to be based on the chemostat principle. Retention of active biomass is used in most major biological effluent treatment processes and ensures a much higher active biomass than would be possible in a chemostat, thus giving more rapid and complete degradation. For white rot fungi, it should also mean that there is a slow growth rate and, consequently, conditions will be nearer to those pertaining in secondary metabolism. Table 10.10 summarizes some of the reports on continuous treatment of pulping effluents. The effectiveness of the continuous reactors reported varies depending, in particular, on the concentration of the effluent and the hydraulic retention time; best results are generally found with long retention times. Some particularly strong effluents can be inhibitory (Royer *et al.*, 1985) but very dilute effluents are also only poorly treated (Royer *et al.*, 1991). In interpreting decolorization data, it must be remembered that a low percentage decolorization of a concentrated effluent may represent more absolute colour removal than a very high percentage decolorization of a dilute effluent. Therefore, estimates of total amounts of colour removed or the rate of colour removal per unit weight of mycelium are useful indicators. In the final analysis, effluents with very little colour are required, and, however much colour has been removed, we still have to consider what is left if a process is to be of practical value.

Table 10.10. *Examples of the performance of white rot fungi in decolourising pulping effluents in continuous bioreactors*

Author	Organism	Decolorization (%)	Hydraulic retention time (hours)	Duration of reactor run (days)	Type of reactor
Bajpai et al., 1993	Trametes (Coriolus) versicolor	93	38	30+	Packed bed with mycelial pellets
Cammarota & Sant'Anna, 1992	Phanerochaete chrysosporium	70	140	66	Packed bed; immobilized on polyurethane particles
Chambers & Cheng, 1991	P. chrysosporium	60^a	24		Semipermeable hollow fibre bioreactor
		38^a	12		
		22^b	8		
		43^a	24		
Marwaha et al., 1998	P. chrysosporium	86	?	~37 21 (deteriorated after 10)	Packed bed; immobilized on pieces of jute rope
Pallerla & Chambers, 1996	T. versicolor	69	24	32	Fluidized bed; polyurethane immobilized
Pallerla & Chambers, 1997	T. versicolor	61	10	?	Fluidized bed; calcium alginate immobilized
		71	16		
		72	24		

[a]Used 25% (v/v) effluent.
[b]Used 50% (v/v) effluent.

Conclusions

White rot fungi as a group have the capacity to decolorize a wide range of dyes and coloured effluents with different chromophores, with some individual species and strains displaying a very broad spectrum of decolorization activity. Although much studied, *P. chrysosporium* is not the only useful white rot fungus, and other species may have similar or superior abilities and may be less prone to inhibition. On balance, strains of *T. versicolor* are probably more likely to be useful in decolorization processes. Note also that certain species or strains can be particularly well suited to specific problems but may be ineffective or unsuitable for others. Similarly, no one particular set of conditions is optimal for all decolorization processes and conditions may need to be tailored to the organism and to the effluent or chromophore being treated. Carbon co-substrate addition, high nitrogen media and agitation of cultures may be useful in some instances. In most cases, the end products of decolorization of dyes are unknown. If decolorization by white rots is to be of practical use, more information is required as to the fate of decolorized materials. There is also a lack of detailed information on the mechanisms of degradation of dyes by fungi and by purified enzymes.

White rot fungi may be repeatedly reused in continuous or fed-batch processes over a prolonged period (up to 3 months, possibly more) and through many cycles of operation. They are able to grow and effect decolorization in a wide range of different reactor types as both freely suspended and immobilized mycelia. However, there are few comparative data to indicate which reactor type is most effective. While there have been several descriptions of kinetic aspects of decolorization by isolated enzymes (Heinfling *et al.*, 1998; Ollikka *et al.*, 1998), there are few kinetic analyses of decolorization processes involving whole fungal mycelia. If progress is to be made with the industrialisation of these processes, then more detailed kinetic analysis and mathematical modelling is required, together with detailed and careful costings for comparison with alternative processes for treatment of coloured effluents. The ability of white rots to treat effluents from wood pulping, cotton bleaching, olive milling and fermentation industries is now well established. White rot fungi also have the potential to treat other industrial effluents, but more research is clearly needed in this area.

References

Addleman, K., Dumonceaux, T., Paice, M. G., Bourbonnais, R. & Archibald, F. S. (1995). Production and characterization of *Trametes versicolor* mutants unable to bleach hardwood kraft pulp. *Applied and Environmental Microbiology,* **61**, 3687–3694.

Alexander, M. (1965). Biodegradation: problems of molecular recalcitrance and microbial fallibility. *Advances in Applied Microbiology,* **7**, 35–80.

Alexander, M. (1994). *Biodegradation and Bioremediation.* San Diego, CA: Academic Press.

Aoshima, I., Tozawa, Y., Ohmomo, S. & Ueda, K. (1985). Production of decolorizing activity for molasses pigment by *Coriolus versicolor* Ps4a. *Agricultural Biology and Chemistry,* **49**, 2041–2045.

Archibald, F. S. (1992). Lignin peroxidase activity is not important in biological bleaching and delignification of unbleached kraft pulp by *Trametes versicolor. Applied and Environmental Microbiology,* **58**, 3101–3109.

Archibald, F. & Roy, B. (1992). Production of manganic chelates by laccase from the lignin-degrading fungus *Trametes (Coriolus). versicolor. Applied and Environmental Microbiology,* **58**, 1496–1499.

Archibald, F., Paice, M. G. & Jurasek, L. (1990a). Decolorization of kraft bleachery effluent chromophores by *Coriolus (Trametes) versicolor. Enzyme and Microbial Technology,* **12**, 846–853.

Archibald, F., Paice, M. G. & Jurasek, L. (1990b). Decolorization of kraft bleached effluent lignins by *Coriolus versicolor,* In *Biotechnology in Pulp and Paper Manufacture,* ed. T. K. Kirk, & H.-M. Chang, pp. 253–261. Boston, MA: Butterworth-Heinemann.

Arjmand, M. & Sandermann, H. (1985). Mineralization of chloroaniline/lignin conjugates and of free chloroanilines by the white rot fungus *Phanerochaete chrysosporium. Journal of Agricultural and Food Chemistry,* **33**, 1055–1060.

Azmi, W., Sani, R. K. & Banerjee, U. C. (1998). Biodegradation of triphenylmethane dyes. *Enzyme and Microbial Technology,* **22**, 185–191.

Bajpai, P. & Bajpai, P. K. (1994). Biological colour removal of pulp and paper mill wastewaters. *Journal of Biotechnology,* **33**, 211–220.

Bajpai, P., Mehna, A. & Bajpai, P. K. (1993). Decolorization of kraft bleach plant effluent with the white rot fungus *Trametes versicolor. Process Biochemistry,* **28**, 377–384.

Bakshi, D. K., Gupta, K. G. & Sharma, P. (1999). Enhanced biodecolourisation of synthetic textile dye effluent by *Phanerochaete chrysosporium* under improved culture conditions. *World Journal of Microbiology and Biotechnology,* **15**, 507–509.

Banat, I. M., Nigam, P., Singh, D. & Marchant, R. (1996). Microbial decolorization of textile-dye-containing effluents. *Bioresource Technology,* **58**, 217–227.

Barr, D. P. & Aust, S. D. (1994). Mechanisms white rot fungi use to degrade pollutants. *Environmental Science and Technology,* **28**, 78–87.

Belsare, D. K. & Prasad, D. Y. (1988). Decolorization of effluent from the bagasse-based pulp mills by white-rot fungus, *Schizophyllum commune. Applied Microbiology and Biotechnology,* **28**, 301–304.

Ben Hamman, O., de la Rubia, T. & Martínez, J. (1997). Effect of carbon and nitrogen limitation on lignin peroxidase and manganese peroxidase production by *Phanerochaete flavido-alba. Journal of Applied Microbiology,*

83, 751–757.

Ben Hamman, O., de la Rubia, T. & Martínez, J. (1999). Decolorization of olive oil mill wastewaters by *Phanerochaete flavido-alba*. *Environmental Toxicology and Chemistry*, **18**, 2410–2415.

Benito, G. G., Miranda, M. P. & Rodriguez de los Santos, D. (1997). Decolorization of wastewater from an alcoholic fermentation process with *Trametes versicolor*. *Bioresource Technology*, **61**, 33–37.

Bergbauer, M. & Eggert, C. (1992). Differences in the persistence of various bleachery effluent lignins against attack by white-rot fungi. *Biotechnology Letters*, **14**, 869–874.

Bergbauer, M., Eggert, C. & Kraepelin, G. (1990). Polymerization and depolymerization of waste water lignins by white-rot fungi. In *Biotechnology in Pulp and Paper Manufacture*, eds. T. K. Kirk & H.-M. Chang, pp. 263–269. Boston, MA: Butterworth-Heinemann.

Bousher, A., Shen, X. & Edyvean, R. G. J. (1997). Removal of coloured organic matter by adsorption onto low-cost waste materials. *Water Research*, **31**, 2084–2092.

Bromley-Challenor, K. C. A., Knapp, J. S., Zhang, Z., Gray, N. C. C., Hetheridge, M. J. & Evans, M. R. (2000). Decolorization of an azo dye by unacclimated activated sludge under anaerobic conditions. *Water Research*, **34**, 4410–4418.

Brown, M. A. & De Vito, S. C. (1993). Predicting azo dye toxicity. *Critical Reviews in Environmental Science and Technology*, **23**, 249–324.

Buckley, K. F. & Dobson, A. D. W. (1998). Extracellular ligninolytic enzyme production and polymeric dye decolourization in immobilized cultures of *Chrysosporium lignorum* CL1. *Biotechnology Letters*, **20**, 301–306.

Bumpus, J. A. & Brock, B. J. (1988). Biodegradation of crystal violet by the white rot fungus *Phanerochaete chrysosporium*. *Applied and Environmental Microbiology*, **54,** 1143–1150.

Cammarota, M. C. & Sant'Anna, G. L. Jr (1992). Decolorization of kraft bleach plant E$_1$ stage effluent in a fungal bioreactor. *Environmental Technology*, **13**, 65–71.

Capalash, N. & Sharma, P. (1992). Biodegradation of textile azo-dyes by *Phanerochaete chrysosporium*. *World Journal of Microbiology and Biotechnology*, **8**, 309–312.

Carliell, C. M., Barclay, S. J., Naidoo, N., Buckley, C. A., Mulholland, D. A. & Senior, E. (1995). Microbial decolourisation of a reactive azo dye under anaerobic conditions. *Water SA*, **21**, 61–69.

Chambers, R. P. & Cheng, M. W. (1991). Biological processing approaches to chlorophenol and color reduction in bleach plant effluents. *Tappi Proceedings, Environmental Conference*, **2**, 627–637.

Chao, W. L. & Lee, S. L. (1994). Decoloration of azo dyes by three white-rot fungi. *World Journal of Microbiology and Biotechnology*, **10**, 556–559.

Chivukula, M. & Renganathan, V. (1995). Phenolic azo dye oxidation by laccase from *Pyricularia oryzae*. *Applied and Environmental Microbiology*, **61**, 4374–4377.

Chivukula, M., Spadaro, J. T. & Renganathan, V. (1995). Lignin peroxidase-catalyzed oxidation of sulfonated azo dyes generates novel sulfophenyl hydroperoxides. *Biochemistry*, **34**, 7765–7772.

Chung, K.-T., Stevens, S. E. & Cerniglia, C. E. (1992). The reduction of azo dyes by the intestinal microflora. *Critical Reviews in Microbiology*, **18**, 175–190.

Collins, P. J. & Dobson, A. D. W. (1997). Regulation of laccase gene transcription in *Trametes versicolor*. *Applied and Environmental Microbiology*, **63**, 3444–3450.

Collins, P. J., Field, J. A., Teunissen, P. & Dobson, A. D. W. (1997). Stabilization of lignin peroxidases in white rot fungi by tryptophan. *Applied and Environmental Microbiology*, **63**, 2543–2548.

Conneely, A., Smyth, W. F. & McMullen, G. (1999). Metabolism of the phthalocyanine textile dye remazol turquoise blue by *Phanerochaete chrysosporium*. *FEMS Microbiology Letters*, **179**, 333–337.

Cooper, P. (ed.) (1995) *Colour in Dyehouse Effluent*. Oxford: Society of Dyers and Colourists.

Cripps, C., Bumpus, J. A. & Aust, S. D. (1990). Biodegradation of azo and heterocyclic dyes by *Phanerochaete chrysosporium*. *Applied and Environmental Microbiology*, **56**, 1114–1118.

D'Annibale, A., Crestini, C., Vinciguerra, V. & Sermanni, G. G. (1998). The biodegradation of recalcitrant effluents from an olive mill by a white-rot fungus. *Journal of Biotechnology*, **61**, 209–218.

Das, S. S., Dey, S. & Bhattacharyya, B. C. (1995). Dye decolorisation in a column bioreactor using wood-degrading fungus *Phanerochaete chrysosporium*. *Indian Chemical Engineering*, Section A, **37**, 176–180.

Davis, S. & Burns, R. G. (1990). Decolorization of phenolic effluents by soluble and immobilized phenol oxidases. *Applied Microbiology and Biotechnology*, **32**, 721–726.

de Jong, E., Field, J. A. & de Bont, J. A. M. (1994). Aryl alcohols in the physiology of ligninolytic fungi. *FEMS Microbiology Reviews*, **13**, 153–188.

Dehorter, B. & Blondeau, R. (1993). Isolation of an extracellular Mn-dependent enzyme mineralizing melanoidins from the white rot fungus *Trametes versicolor*. *FEMS Microbiology Letters*, **109**, 117–122.

Dey, S., Maiti, T. K. & Bhattacharyya, B. C. (1994). Production of some extracellular enzymes by a lignin peroxidase-producing brown rot fungus, *Polyporus ostreiformis*, and its comparative abilities for lignin degradation and dye decolorization. *Applied and Environmental Microbiology*, **60**, 4216–4218.

Dezotti, M., Innocentini-Mei, L. H. & Durán, N. (1995). Silica immobilized enzyme catalyzed removal of chlorolignins from eucalyptus kraft effluent. *Journal of Biotechnology*, **43**, 161–167.

Durán, N., Eposito, E., Innocentini-Mei, L. H. & Canhos, V. P. (1994). A new alternative process for kraft E1 effluent treatment. *Biodegradation*, **5**, 13–19.

Eaton, D., Chang, H.-M. & Kirk, T. K. (1980). Fungal decolorization of kraft bleach plant effluents. *Tappi Journal*, **63**, 103–107.

Eaton, D. C., Chang, H.-M., Joyce, T. W., Jeffries, T. W. & Kirk, T. K. (1982). Method obtains fungal reduction of the color of extraction-stage kraft bleach effluents. *Tappi Journal*, **65**, 89–92.

Eposito, E., Canhos, V. P. & Durán, N. (1991). Screening of lignin-degrading fungi for the removal of color from kraft mill wastewater with no additional extra carbon-source. *Biotechnology Letters*, **13**, 571–576.

Evans, C. S., Gallagher, I. M., Atkey, P. T. & Wood, D. A. (1991). Localisation of degradative enzymes in white-rot decay of lignocellulose. *Biodegradation*, **2**, 93–106.

Evans, C. S., Dutton, M. V. Guillén, F. & Veness, R. G. (1994). Enzymes and small molecular mass agents involved with lignocellulose degradation. *FEMS Microbiology Reviews*, **13**, 235–240.

Fahy, V., FitzGibbon, F. J., McMullan, G., Singh, D. & Marchant, R. (1997). Decolourisation of molasses spent wash by *Phanerochaete chrysosporium, Biotechnology Letters,* **19**, 97–99.

Ferrer, I., Dezotti, M. & Durán, N. (1991). Decolorization of kraft effluent by free and immobilized lignin peroxidases and horseradish peroxidase. *Biotechnology Letters,* **13**, 577–582.

Field, J. A., de Jong, E., Feijoo-Costa, G. & de Bont, J. A. M. (1993). Screening for ligninolytic fungi applicable to the biodegradation of xenobiotics. *Trends in Biotechnology,* **11**, 44–47.

Fitzgibbon, F. J., Nigam, P., Singh, D. & Marchant, R. (1995). Biological treatment of distillery waste for pollution remediation. *Journal of Basic Microbiology,* **5**, 293–301.

Fukui, H., Presnell, T. L., Joyce, T. W. & Chang, H.-M. (1992). Dechlorination and detoxification of bleach plant effluent by *Phanerochaete chrysosporium. Journal of Biotechnology,* **24**, 267–275.

Galeno, G. & Agosin, E. (1990). Screening of white-rot fungi for efficient decolorization of bleach pulp effluents. *Biotechnology Letters,* **12**, 869–872.

Garg, S. K, & Modi, D. R. (1999). Decolorization of pulp-paper mill effluents by white-rot fungi. *Critical Reviews in Biotechnology,* **19**, 85–112.

Garg, S. K., Chandra, H. & Modi, D. R. (1999). Effect of glucose and urea supplements on decolourization of bagasse and gunny bag-based pulp-paper mill effluent by *Trametes versicolor. Indian Journal of Experimental Biology,* **37**, 302–304.

Glenn, J. K. & Gold, M. H. (1983). Decolorization of several polymeric dyes by the lignin-degrading basidiomycete *Phanerochaete chrysosporium. Applied and Environmental Microbiology,* **45**, 1741–1747.

Glenn, J. K., Morgan, M. A., Mayfield, M. B., Kuwahara, M. & Gold, M. H. (1983). An extracellular H_2O_2-requiring enzyme preparation involved in lignin biodegradation by the white rot basidiomycete *Phanerochaete chrysosporium. Biochemical and Biophysical Research Communications,* **114**, 1077–1083.

Gogna, E., Vohra, R. & Sharma, P. (1992). Biodegradation of rose bengal by *Phanerochaete chrysosporium. Letters in Applied Microbiology,* **14**, 58–60.

Goszczynski, S., Paszczynski, A., Pasti-Grigsby, M. B., Crawford, R. L. & Crawford, D. L. (1994). New pathway for degradation of sulfonated azo dyes by microbial peroxidases of *Phanerochaete chrysosporium* and *Streptomyces chromofuscus. Journal of Bacteriology,* **176**, 1339–1347.

Gregory, P. (1993). Dyes and dye intermediates. In *Kirk-Othmer Encyclopedia of Chemical Technology,* Vol. 4, 4th edn, pp. 542–602. New York: Wiley & Sons.

Hatakka, A. (1994). Lignin-modifying enzymes from selected white-rot fungi: production and role in lignin degradation. *FEMS Microbiology Reviews,* **13**, 125–135.

Haug, W., Schmidt, A., Nörtemann, B., Hempel, D. C., Stolz, A. & Knackmuss, H.-J. (1991). Mineralisation of the sulfonated azo dye mordant yellow 3 by a 6-aminonaphthalene-2-sulfonate-degrading bacterial consortium. *Applied and Environmental Microbiology,* **57**, 3144–3149.

Heinfling, A., Bergbauer, M. & Szewzyk, U. (1997). Biodegradation of azo and phthalocyanine dyes by *Trametes versicolor* and *Bjerkandera adusta. Applied Microbiology and Biotechnology.* **48**, 261–266.

Heinfling, A., Martinez, M. J., Martinez, A. T., Bergbauer, M. & Szewzyk, U. (1998). Transformation of industrial dyes by manganese peroxidases from

Bjerkandera adusta and *Pleurotus eryngii* in a manganese-independent reaction. *Applied and Environmental Microbiology,* **64**, 2788–2793.

Higuchi, T. (1980). Lignin structure and morphological distribution in plant cell walls. In *Lignin Biodegradation*, Vol. 1, Ch. 1, eds. T. K. Kirk, T. Higuchi & H.-M. Chang, pp. 1–19. Boca Raton, FL: CRC Press.

Jäger, A., Croan, S. & Kirk, T. K. (1985). Production of ligninases and degradation of lignin in agitated submerged cultures of *Phanerochaete chrysosporium. Applied and Environmental Microbiology*, **50**, 1274–1278.

Jaspers, C. J. & Penninckx, M. J. (1996). Adsorption effects in the decolorization of a kraft bleach plant effluent by *Phanerochaete chrysosporium. Biotechnology Letters,* **18**, 1257–1260.

Jaspers, C. J., Jimenez, G. & Penninckx, M. J. (1994). Evidence of a role of manganese peroxidase in the decolorization of kraft pulp bleach plant effluent by *Phanerochaete chrysosporium*: effects of initial culture conditions on enzyme production. *Journal of Biotechnology,* **37**, 229–234.

Jeffries, T. W., Choi, S. & Kirk, T. K. (1981). Nutritional regulation of lignin degradation by *Phanerochaete chrysosporium. Applied and Environmental Microbiology,* **42**, 290–296.

Johannes, C. & Majcherczyk, A. (2000). Natural mediator in the oxidation of polycyclic aromatic hydrocarbons by laccase mediator systems. *Applied and Environmental Microbiology*, **66**, 524–528.

Kasturi-Bai, R. & Ganga, N. (1996). Treatment strategies for the decolorisation of distillery effluent. *Journal of Industrial Pollution Control,* **12**, 1–8.

Kay-Shoemake, J. L. & Watwood, M. E. (1996). Limitations of the lignin peroxidase system of the white-rot fungus, *Phanerochaete chrysosporium. Appied Microbiology and Biotechnology,* **46**, 438–442.

Khindaria, A., Barr, D. P. & Aust, S. D. (1995). Lignin peroxidases can also oxidise manganese. *Biochemistry,* **34**, 7773–7779.

Kim, B.-S., Ryu, S.-J. & Shin, K.-S. (1996). Effect of culture parameters on the decolorization of Remazol brilliant blue R by *Pleurotus ostreatus. Journal of Microbiology,* **34**, 101–104.

Kirby, N., McMullen, G. & Marchant, R. (1995). Decolourisation of an artificial textile effluent by *Phanerochaete chrysosporium. Biotechnology Letters,* **17**, 761–764.

Kirk, T. K. & Farrell, R. L. (1987). Enzymatic 'combustion': the microbial degradation of lignin. *Annual Review of Microbiology,* **41**, 465–505.

Kirk, T. K., Connors, W. J. & Zeikus, J. G. (1976). Requirement for a growth substrate during lignin decomposition by two wood-rotting fungi. *Applied and Environmental Microbiology,* **32**, 192–194.

Kirk, T. K., Schultz, E., Connors, W. J., Lorenz, L. F. & Zeikus, J. G. (1978). Influence of culture parameters on lignin metabolism by *Phanerochaete chrysosporium. Archives of Microbiology,* **117**, 277–285.

Knapp, J. S. & Newby, P. S. (1995). The microbiological decolourisation of an industrial effluent containing a diazo-linked chromophore, *Water Research,* **49**, 1807–1809.

Knapp, J. S. & Newby, P. S. (1999). The decolourisation of a chemical industry effluent by white rot fungi. *Water Research,* **33**, 575–577.

Knapp, J. S., Newby, P. S. & Reece, L. P. (1995). Decolorization of dyes by wood-rotting basidiomycete fungi. *Enzyme and Microbial Technology,* **17**, 664–668.

Knapp, J. S., Zhang, F. & Tapley, K. N. (1997). Decolourisation of orange II by a wood-rotting fungus. *Journal of Chemical Technology and Biotechnology,*

69, 289–296.

Kudlich, M., Hetheridge, M. J., Knackmuss, H.-J. & Stolz, A. (1999). Autoxidation reactions of different aromatic *ortho*-aminohydroxynaphthalenes which are formed during the anaerobic reduction of sulphonated azo dyes. *Environmental Science and Technology,* **33**, 896–901.

Kumar, V., Wati, L., Nigam, P., Banet, I. M., Yadav, B. S., Singh, D. & Marchant, R. (1998). Decolorization and biodegradation of anaerobically digested sugarcane molasses spent wash effluent from biomethanation plants by white-rot fungi. *Process Biochemistry,* **33**, 83–88.

Lackner, R., Sebrotnik, E. & Messner, K. (1991). Oxidative degradation of high molecular weight chlorolignin by manganese peroxidase of *Phanerochaete chrysosporium. Biochemical and Biophysical Research Communications,* **178**, 1092–1098.

Laing, I. G. (1991). The impact of effluent regulations on the dyeing industry. *Review of Progression in Coloration,* **21**, 56–71.

Lankinen, V. P., Inkeröinen, M. M., Pellinen, J. & Hatakka, A. I. (1991). The onset of lignin-modifying enzymes, decrease of aox and color removal by white-rot fungi grown on bleach plant effluents. *Water Science and Technology,* **24**, 189–198.

Leontievsky, A. A., Vares, T., Lankinen, P., Shergill, J. K., Pozdnyakova, N. N., Myasoedova, N. M., Kalkkinen, N., Golovleva, L. A., Cammack, R., Thurston, C. F. & Hatakka, A. (1997). Blue and yellow laccases of ligninolytic fungi. *FEMS Microbiology Letters,* **156**, 9–14.

Livernoche, D., Jurasek, L., Desroches, M. & Veliky, I. A. (1981). Decolorization of a kraft mill effluent with fungal mycelium immobilized in calcium alginate gel. *Biotechnology Letters,* **3**, 701–706.

Livernoche, D., Jurasek, L., Desrochers, M. & Dorica, J. (1983). Removal of color from kraft mill wastewaters with cultures of white-rot fungi and with immobilized mycelium of *Coriolus versicolor. Biotechnology and Bioengineering,* **25**, 2055–2065.

Lonergan, G. T., Panow, A., Jones, C. L., Schliephake, K., Ni, C. J. & Mainwaring, D. E. (1995). Physiological and biodegradative behaviour of the white rot fungus *Pycnoporus cinnabarinus* in a 200 litre packed-bed bioreactor. *Australian Biotechnology,* **5**, 107–111.

Manzanares, P., Fajardo, S. & Martín, C. (1995). Production of ligninolytic activities when treating paper pulp effluents by *Trametes versicolor. Journal of Biotechnology,* **43**, 125–132.

Martín, C. & Manzanares, P. (1994). A study of the decolourization of straw soda-pulping effluents by *Trametes versicolor. Bioresource Technology,* **47**, 209–214.

Martirani, L., Giardina, P., Marzullo, L. & Sanna, G. (1996). Reduction of phenol content and toxicity in olive oil mill waste waters with the ligninolytic fungus *Pleurotus ostreatus. Water Research,* **30**, 1914–1918.

Marton, J., Syern, A. M. & Marton, T. (1969). Decolorization of kraft black liquor with *Polyporus versicolor,* a white rot fungus. *Tappi Journal,* **52**, 1975–1981.

Marwaha, S. S., Grover, R., Prakash, C. & Kennedy, J. F. (1998). Continuous biobleaching of black liquor from pulp and paper industry using an immobilized cell system. *Journal of Chemical Technology and Biotechnology,* **73**, 292–296.

McCarthy, J. T., Levy, V. C., Lonergan, G. T. & Fecondo, J. V. (1999).

Development of optimal conditions for decolourisation of a range of industrial dyes using *Pycnoporus cinnabarinus* laccase. *Hazardous and Industrial Wastes*, **1**, 489–498.

Mehna, A., Bajpai, P. & Bajpai, P. K. (1995). Studies on decolorization of effluent from a small pulp mill utilizing agriresidues with *Trametes versicolor. Enzyme and Microbial Technology,* **17**, 18–22.

Messner, K., Ertler, G., Jaklin-Farcher, S., Boskovsky, P., Regensberger, U. & Blaha, A. (1990). Treatment of bleach plant effluents by the mycopor system. In *Biotechnology in Pulp and Paper Manufacture*, eds. T. K. Kirk & H.-M. Chang, pp. 245–251. Boston, MA: Butterworth-Heinemann.

Mester, T., Peña, M. & Field, J. A. (1996). Nutrient regulation of extracellular peroxidases in the white rot fungus, *Bjerkandera sp.* strain BOS55. *Applied Microbiology and Biotechnology,* **44**, 778–784.

Michel, F. C. Jr, Dass, B., Grulke, E. A. & Reddy, A. (1991). Role of manganese peroxidases and lignin peroxidases of *Phanerochaete chrysosporium* in the decolorization of kraft bleach plant effluent. *Applied and Environmental Microbiology,* **57**, 2368–2375.

Mittar, D., Khanna, P. K., Marwaha, S. S. & Kennedy, J. F. (1992). Biobleaching of pulp and paper mill effluents by *Phanerochaete chrysosporium. Journal of Chemical Technology and Biotechnology,* **53**, 81–92.

Modi, D. R., Chandra, H. & Garg, S. K. (1998). Decolourization of bagasse-based paper mill effluent by the white-rot fungus *Trametes versicolor. Bioresource Technology,* **66**, 79–81.

Monties, B. (1994). Chemical assesment of lignin biodegradation some qualitative and quantitative aspects. *FEMS Microbiology Reviews,* **13**, 277–284.

Nagarajan, G. & Annadurai, G. (1999). Biodegradation of rective dye (verox red) by the white-rot fungus *Phanerochaete chrysosporium* using box-behnken experimental design. *Bioprocess Engineering,* **20**, 435–440.

Newby, P. S. (1994). Biodegradation of recalcitrant organic chemicals in a coloured effluent. Ph.D. Thesis, University of Leeds, Leeds, UK.

Ohmomo, S., Aoshima, I., Tozawa, Y., Sakurada, N. & Ueda, K. (1985a). Purification and some properties of melanoidin decolorizing enzymes, P-III and P-IV, from mycelia of *Coriolus versicolor* Ps4a. *Agricultural Biology and Chemistry*, **49**, 2047–2053.

Ohmomo, S., Itoh, N., Watanabe, Y., Kaneko, Y., Tozawa, Y. & Ueda, K. (1985b). Continuous decolorization of molasses waste water with mycelia of *Coriolus versicolor* Ps4a. *Agricultural Biology and Chemistry*, **49**, 2551–2555.

Ollikka, P., Alhonmaki, K., Leppanen, V-M., Glumoff, T., Raijola, T. Suominen, I. (1993). Decolorization of azo, triphenyl methane, heterocyclic and polymeric dyes by lignin peroxidase isoenzymes from *Phanerochaete chrysosporium. Applied and Environmental Microbiology*, **59**, 4010–4016.

Ollikka, P., Harjunpää, T., Palmu, K., Mäntsälä, P. & Suominen, I. (1998). Oxidation of crocein orange G by lignin peroxidase isoenzymes. Kinetics and effects of H_2O_2. *Applied Biochemistry and Biotechnology,* **75**, 307–321.

Pallerla, S. & Chambers, R. (1996). New urethane prepolymer immobilized fungal bioreactor for decolorization and dechlorination of kraft bleach effluents. *Tappi Journal,* **79**, 155–161.

Pallerla, S. & Chambers, R. P. (1997). Characterization of a Ca-alginate-immobilized *Trametes versicolor* bioreactor for decolorization and AOX reduction of paper mill effluents. *Bioresource Technology,* **60**, 1–8.

Palma, C., Moreira, M. T., Mielgo, I., Feijoo, G. & Lema, J. M. (1999). Use of a fungal bioreactor as a pretreatment or post-treatment step for continuous decolorisation of dyes. *Water Science and Technology*, **40**, 131–136.

Palmieri, G., Giardina, P., Bianco, C., Fnotanella, B. & Sannia, G. (2000). Copper induction of laccase isoenzymes in the lignolytic fungus *Pleurotus ostreatus*. *Applied and Environmental Microbiology*, **66**, 920–924.

Pasti-Grigsby, M. B., Paszczynski, A., Goszczynski, S., Crawford, D. L. & Crawford, R. L. (1992). Influence of aromatic substitution patterns on azo dye degradability by *Streptomyces* spp. and *Phanerochaete chrysosporium*. *Applied and Environmental Microbiology*, **58**, 3605–3613.

Paszczynski, A. & Crawford, R. L. (1991). Degradation of azo compounds by ligninase from *Phanerochaete chrysosporium*: involvement of veratryl alcohol. *Biochemical and Biophysical Research Communications*, **178**, 1056–1063.

Paszczynski, A., Pasti, M. B., Goszczynski, S., Crawford, D. L. & Crawford, R. L. (1991). New approach to improve degradation of recalcitrant azo dyes by *Streptomyces spp.* and *Phanerochaete chrysosporium*. *Enzyme and Microbial Technology*, **13**, 378–384.

Paszczynski, A., Pasti-Grigsby, M. B., Goszczynski, S., Crawford, R. L. & Crawford, D. L. (1992). Mineralisation of sulfonated azo dyes and sulfanilic acid by *Phanerochaete chrysosporium* and *Streptomyces chromofuscus*. *Applied and Environmental Microbiology*, **58**, 3598–3604.

Pellinen, J., Joyce, T. W. & Chang, H.-M. (1988a). Dechlorination of high-molecular-weight chlorolignin by the white-rot fungus *P. chrysosporium*. *Tappi Journal*, **71**, 191–194.

Pellinen, J., Yin, C., Joyce, T. W. & Chang, H.-M. (1988b). Treatment of chlorine bleaching effluent using a white-rot fungus. *Journal of Biotechnology*, **8**, 67–76.

Peralta-Zamora, P., de Moraes, S. G., Esposito, E., Antunes, R., Reyes, J. & Durán, N. (1998). Decolorization of pulp mill effluents with immobilized lignin and manganese peroxidase from *Phanerochaete chrysosporium*. *Environmental Technology*, **19**, 521–528.

Pérez, J., Saez, T., de la Rubia, T. & Martínez, J. (1997). *Phanerochaete flavido-alba* ligninolytic activities and decolorization of partially bio-depurated paper mill wastes. *Water Research*, **31**, 495–502.

Pérez, J., de la Rubia, T., Hamman, O. B. & Martínez, J. (1998). *Phanerochaete flavido-alba* laccase induction and modification of manganese peroxidase isoenzyme pattern in decolorized olive oil mill waste waters. *Applied and Environmental Microbiology*, **64**, 2726–2729.

Pitter, P. & Chudoba, J. (1990). *Biodegradability of Organic Substances in the Aquatic Environment*. Boca Raton, FL: CRC press.

Platt, M. W., Hadar, Y. & Chet, I. (1985). The decolorization of the polymeric dye poly-blue (polyvinalamine sulfonate-anthraquinone) by lignin degrading fungi. *Applied Microbiology and Biotechnology*, **21**, 394–396.

Podgornik, H., Grgic, I. & Perdih, A. (1999). Decolorization rate of dyes using lignin peroxidases of *Phanerochaete chrysosporium*. *Chemosphere*, **38**, 1353–1359.

Pointing, S. B., Bucher, V. V. C. & Vrijmoed, L. L. P. (2000). Dye decolorization by sub-tropical basidiomycetous fungi and the effect of metals on decolorizing ability. *World Journal of Microbiology and Biotechnology*, **16**, 199–205.

Prasad, G. K. & Gupta, R. K. (1997). Decolourization of pulp and paper mill effluent by two white-rot fungi. *Indian Journal of Environmental Health,* **39**, 89–96.

Presnell, T. L., Fukui, H., Joyce, T. W. & Chang, H-M. (1992). Bleach plant effluent influences enzyme production by *Phanerochaete chrysosporium. Enzyme and Microbial Technology,* **14**, 184–189.

Raghukumar, C., Chandramohan, D., Michel, F. C. Jr. & Reddy, C. A. (1996). Degradation of lignin and decolorization of paper mill bleach plant effluent (BPE) by marine fungi. *Biotechnology Letters,* **18**, 105–106.

Reddy, C. A. (1995). The potential for white-rot fungi in the treatment of pollutants. *Current Opinion in Biotechnology,* **6**, 320–328.

Reddy, C. A. & D'Souza, T. M. (1994). Physiology and molecular biology of the lignin peroxidases of *Phanerochaete chrysosporium. FEMS Microbiology Reviews,* **13**, 137–152.

Rodríguez, E., Pickard, M. A. & Vázquez-Duhalt, R. (1999). Industrial dye decolorization by laccases from ligninolytic fungi. *Current Microbiology,* **38**, 27–32.

Roy-Arcand, L., Archibald, F. S. & Briere, F. (1991). Comparison and combination of ozone and fungal treatments of a kraft bleachery effluent. *Tappi Journal,* **74**, 155–161.

Royer, G., Livernoche, D., Desrochers, M., Jurasek, L., Rouleau, D. & Mayer, R. C. (1983). Decolorization of kraft mill effluent: kinetics of a continuous process using immobilised *Coriolus versicolor. Biotechnology Letters,* **5**, 321–326.

Royer, G., Desrochers, M., Jurasek, L., Rouleau, D. & Mayer, R. C. (1985). Batch and continuous decolorisation of bleached kraft effluents by a white-rot fungus. *Journal of Chemical Technology and Biotechnology,* **35B**, 14–22.

Royer, G., Yerushalmi, L., Rouleau, D. & Desrochers, M. (1991). Continuous decolorization of bleached kraft effluents by *Coriolus versicolor* in the form of pellets. *Journal of Industrial Microbiology,* **7**, 269–278.

Ryan, T. P. & Bumpus, J. A. (1989). Biodegradation of 2,4,5-trichlorophenoxyacetic acid in liquid culture and in soil by the white rot fungus *Phanerochaete chrysosporium. Applied Microbiology and Biotechnology,* **31**, 302–307.

Sani, R. K., Azmi,W. & Banerjee, U. C. (1998). Comparison of static and shake culture in the decolourisation of textile dyes and dye effluents by *Phanerochaete chrysosporium. Folia Microbiologica,* **43**, 85–88.

Sayadi, S. & Ellouz, R. (1992). Decolourization of olive mill waste-waters by the white-rot fungus *Phanerochaete chrysosporium*: involvement of the lignin-degrading system. *Applied Microbial Technology,* **37**, 813–817.

Sayadi, S. & Ellouz, R. (1993). Screening of white rot fungi for the treatment of olive mill waste-waters. *Journal of Chemical Technology and Biotechnology.,* **57**, 141–146.

Sayadi, S. & Ellouz, R. (1995). Roles of lignin peroxidase and manganese peroxidase from *Phanerochaete chrysosporium* in the decolorization olive mill waste waters. *Applied and Environmental Microbiology,* **61**, 1098–1103.

Sayadi, S., Zorgani, F. & Ellouz, R. (1996). Decolorization of olive mill waste-waters by free and immobilized *Phanerochaete chrysosporium* cultures. Effect of the high-molecular-weight polyphenols. *Applied Biochemistry and Biotechnology,* **56**, 265–276.

Schliephake, K. & Lonergan, G. T. (1996). Laccase variation during dye decolourisation in a 200 L packed-bed bioreactor. *Biotechnology Letters,* **18,** 881–886.

Schliephake, K., Lonergan, G. T. Jones, C. L. & Mainwaring, D. E. (1993). Decolourisation of a pigment plant effluent by *Pycnoporus cinnabarinus* in a packed-bed bioreactor. *Biotechnology Letters,* **15,** 1185–1188.

Sermkiattipong, N., Takigami, M., Pongpat, S., Sukusude, O., Charoen, S. & Ito, H. (1999). Decolorization of dark brown pigments in molasses wastewater by mutant strains of *Aspergillus usamii* and *Coriolus versicolor. Biocontrol Science,* **4,** 109–113.

Shaul, G. M., Dempsey, C. R., Dostal, K. A. & Lieberman, R. J. (1986). Fate of azo dyes in the activated sludge process. *Proceedings of Purdue University Industrial Waste Conference,* **41,** 603–611.

Shin, K-S. & Kim, C-J. (1998). Decolorisation of artificial dyes by peroxidase from the white-rot fungus, *Pleurotus ostreatus. Biotechnology Letters,* **20,** 569–572.

Shin, K.-S., Oh, I.-K. & Kim, C.-J. (1997). Production and purification of remazol brilliant blue R decolorizing peroxidase from the culture filtrate of *Pleurotus ostreatus. Applied and Environmental Microbiology,* **63,** 1744–1748.

Sirianuntapiboon, S., Somachai, P., Sihanonth, P., Atthasampunna, P. & Ohmomo, S. (1988). Microbial decolorization of molasses waste water by *Mycelia sterilia D90. Agricultural Biology and Chemistry,* **52,** 393–398.

Sirianuntapiboon, S., Sihanonth, P., Somachai, P., Atthasampunna, P. & Hayashida, S. (1995). An absorption mechanism for the decolorization of melanoidin by *Rhizoctonia* sp. D-90. *Bioscience, Biotechnology and Biochemistry,* **59,** 1185–1189.

Spadaro, J. T. & Renganathan, V. (1994). Peroxidase-catalyzed oxidation of azo dyes: mechanism of disperse yellow 3 degradation. *Archives of Biochemistry and Biophysics,* **312,** 301–307.

Spadaro, J. T., Gold, M. H. & Renganathan, V. (1992). Degradation of azo dyes by the lignin-degrading fungus *Phanerochaete chrysosporium. Applied and Environmental Microbiology,* **58,** 2387–2401.

Srinivasan, S. V. & Murthy, D. V. S. (1999). Colour removal from bagasse-based pulp mill effluent using a white rot fungus. *Bioprocess Engineering,* **21,** 561–564.

Swamy, J. & Ramsay, J. A. (1999a). The evaluation of white rot fungi in the decoloration of textile dyes. *Enzyme and Microbial Technology,* **24,** 130–137.

Swamy, J. & Ramsay, J. A. (1999b). Effects of glucose and NH_4^+ concentrations on sequential dye decoloration by *Trametes versicolor. Enzyme and Microbial Technology,* **25,** 278–284.

Tatarko, M. & Bumpus, J. A. (1998). Biodegradation of congo red by *Phanerochaete chrysosporium. Water Research,* **32,** 1713–1717.

ten Have, R., Rietjens, I. M. C. M., Hartmans, S., Swarts, H. J. & Field, J. A. (1998). Calculated ionisation potentials determine the oxidation of vanillin precursors by lignin peroxidase. *FEBS Letters,* **430,** 390–392.

Thurston, C. F. (1994). The structure and function of fungal laccases. *Microbiology,* **140,** 19–26.

Tien, M. & Kirk, T. K. (1983). Lignin-degrading enzyme from the hymenomycete *Phanerochaete chrysosporium* Burds. *Science,* **221,** 661–663.

Tuor, U., Winterhalter, K. & Fiechter, A. (1995). Enzymes of white-rot fungi involved in lignin degradation and ecological determinants for wood decay.

Journal of Biotechnology, **41**, 1–17.

Vasdev, K. & Kuhad, R. C. (1994). Decolorization of polyR-478 (polyvinylamine sulfonate anthrapyridone) by *Cyathus bulleri. Folia Microbiologica*, **39**, 61–64.

Vasdev, K., Kuhad, R. C. & Saxena, R. K. (1995). Decolorization of triphenylmethane dyes by the birds's nest fungus *Cyathus bulleri. Current Microbiology*, **30**, 269–272.

Vinciguerra, V., D'Annibale, A., Delle Monache, G. & Sermanni, G. G. (1995). Correlated effects during the bioconversion of waste olive waters by *Lentinus edodes. Bioresource Technology*, **51**, 221–226.

Vyas, B. R. M. & Molitoris, H. P. (1995). Involvement of an extracellular H_2O_2-dependent ligninolytic activity of the white rot fungus *Pleurotus ostreatus* in the decolorization of remazol brilliant blue R. *Applied and Environmental Microbiology*, **61**, 3919–3927.

Wang, S.-H., Ferguson, J. F. & McCarthy, J. L. (1992). The decolorization and dechlorination of kraft bleach plant effluent solutes by use of three fungi: *Ganoderma lacidum, Coriolus versicolor* and *Hericium erinaceum. Holzforschung*, **46**, 219–223.

Wang, Y. & Yu, J. (1998). Adsorption and degradation of synthetic dyes on the mycelium of *Trametes versicolor. Water Science and Technology*, **38**, 233–238.

Watanabe, Y., Sugi, R., Tanaka, Y. & Hayashida, S. (1982). Enzymatic decolorization of melanoidin by *Coriolus* sp. No. 20. *Agricultural Biology and Chemistry*, **46**, 1623–1630.

Yang, F.-C. & Yu, J.-T. (1996). Development of a bioreactor system using an immobilized white rot fungus for decolorization. *Bioprocess Engineering*, **16**, 9–11.

Yesilada, Ö., Fiskin, K. & Yesilada, E. (1995). The use of white rot fungus *Funalia trogii* (Malatya) for the decolourization and phenol removal from olive mill wastewater. *Environmental Technology*, **16**, 95–100.

Yesilada, Ö., Sik, S. & Sam, M. (1998). Biodegradation of olive oil mill wastewater by *Coriolus versicolor* and *Funalia trogii*: effects of agitation, initial COD concentration, inoculum size and immobilization. *World Journal of Microbiology and Biotechnology*, **14**, 37–42.

Yin, C.-F., Joyce, T. W. & Chang, H.-M. (1990). Dechlorination of conventional softwood bleaching effluent by sequential biological treatment. In *Biotechnology in Pulp and Paper Manufacture*, eds. T. K. Kirk, & H.-M. Chang, pp. 231–244. Boston, MA: Butterworth-Heinemann.

Youn, H.-D., Hah, Y. C. & Kang, S.-O. (1995). Role of laccase in lignin degradation by white-rot fungi. *FEMS Microbiology Letters*, **132**, 183–188.

Young, L. & Yu, J. (1997). Ligninase-catalysed decolorization of synthetic dyes. *Water Research*, **31**,1187–1193.

Zhang, F. (1997). The treatment of coloured effluents with wood rotting fungi. Ph.D. Thesis, University of Leeds, Leeds, UK.

Zhang, F., Knapp, J. S. & Tapley, K. N. (1998). Decolourisation of cotton bleaching effluent in a continuous fluidized-bed bioreactor using wood rotting fungus. *Biotechnology Letters*, **20**, 717–723.

Zhang, F., Knapp, J. S. & Tapley, K. N. (1999a). Decolourisation of cotton bleaching effluent with wood rotting fungus. *Water Research*, **33**, 919–928.

Zhang, F., Knapp, J. S. & Tapley, K. N. (1999b). Development of bioreactor systems for decolorization of orange II using white rot fungus. *Enzyme and Microbial Technology*, **24**, 48–53.

11

The roles of fungi in agricultural waste conversion

RONI COHEN AND YITZHAK HADAR

Introduction

In recent years, recognition of the importance of biological materials as renewable resources for the production of energy and feed, and as important sources of chemical feedstock for the production of different chemicals, has revived interest in the ancient technology of fungal solid-state fermentation (SSF). In the Orient, commercial SSF is still widely practised for miso, saké, soy sauce and tempe production. Agro-industrial by-products such as lignocellulosic wastes, corn cobs, sugar cane bagasse and coffee pulp, which frequently create serious environmental problems, could potentially be used as low-cost carbohydrate sources for fungal fermentations; these, in turn, would produce biochemical compounds suitable for the food, chemical and pharmaceutical industries.

Fungi are structurally unique organisms that abound in various ecosystems. They are capable of colonizing a wide range of living and dead tissues, including plants, wood and paper products, agricultural plant residues, and live or dead animal tissues. In composting, fungi colonize a mixture of heterogeneous substrates such as municipal solid waste and cattle manure with straw. Many fungi can grow on solid substrates and secrete extracellular enzymes that break down various polymers to molecules that are then reabsorbed by the fungal colony. Consequently any discussion of fungal biodegradation must cover an extraordinary amount of catalytic capability. Exoenzymes derived from filamentous fungi have diverse roles in nature, being involved in the degradation of many types of agricultural matter. Enzymic activities such as oxidation and hydrolysis are involved in the bioconversion of these wastes. The SSF process of filamentous fungi on agricultural wastes can be divided into five main types.

1. Degradation of cellulose and starch for the production of protein-rich animal feed
2. Degradation of lignin in lignocellulose to give cellulase access to

the cellulose for ruminant feed, production of saccharides, feedstock for ethanol and chemicals production

3. Degradation of organic matter for the production of pure enzymes
4. Production of specific biochemicals, such as organic acids and saccharides
5. Conversion of mixed organic waste into a stable organic product through composting.

In this chapter, we discuss the following aspects of fungal bioconversion of agricultural wastes and agro-industrial by-products: fungal growth on solid substrates, the range and nature of available substrates, the mechanism of activity of the extracellular enzymes that are key for utilizing polymeric substrate, and the role of fungi in composting.

Fungal growth on agricultural by-products during solid-state fermentation

Fermentation is the result of substrate modification by microbial activity; it is used to obtain marketable products ranging from foods, beverages and animal feeds to enzymes and antibiotics (Nout, Rinzema & Smits, 1997). Microbial growth in SSF is different from that in submerged cultures because of the different surface phenomena, moisture content, physical structure and chemical composition of the substrate, and the different nutritional environment (Datta, Bettermann & Kirk, 1991). SSF conditions resemble the natural growth habitat for filamentous fungi. It is a process in which the microorganisms grow on a moist, solid substrate in the absence of free water and by which cheap raw materials and residues from the agricultural food industries can be utilized as fermentation products (Pandey, 1992). The necessary moisture in SSF exists in absorbed or complexed forms within the solid matrix; the advantage of this lies in the potential efficiency of the oxygen-transfer process.

Filamentous fungi, grown under SSF conditions, are able to grow and excrete large quantities of a wide variety of enzymes. Compared with submerged culture, fungal growth under SSF conditions is advantageous in that it requires less humidity and oxygen transfer is much more efficient. Product concentrations after extraction are usually larger than those obtained by submerged fermentation and the quantity of liquid waste generated is lower. Moreover, these processes are of special economic interest for countries with an abundance of biomass and agro-industrial residues as these can be used as inexpensive raw materials. Compared with liquid fermentation, the major positive aspects of SSF are (i) better product quality, (ii) higher product yield and concentration, (iii) cheaper processing of many feeds that are solid in nature, and (iv) potential for less wastewater and off-gas production (Nout *et al.*, 1997). Taking into consideration the simplicity of the cultivation equipment and the lower operating costs, wider application of this traditional method is expected with advances in biochemical engineering. SSF can be used advantageously for many biotechnological applications, such as enzyme production (amylase, pectinase, cellulase), mushroom production, bioprocessing crop residues, fibre processing, biopulping, composting, bioremediation and waste recycling. Certain secondary metabolites are particularly produced under SSF conditions, including antibiotics, alkaloids, carotenoids, mycotoxins, quinolines, enzymes and even complex plant growth factors. Table 11.1 gives some examples of the fungi, substrates and products of SSF of agricultural by-products.

Extracellular enzymes secreted during fungal solid-state fermentation and biodegradation

Fungi grown under SSF conditions may be more capable of producing certain enzymes that are not produced or are produced only at low yields in submerged culture. Conversion of the growth substrate into fungal biomass is determined by the capacity of the fungus to synthesize the hydrolytic or oxidative enzymes required to convert the polymeric components of the growth substrate into low-molecular-weight nutrients. Fungal conversion of agricultural wastes involves the secretion of different exoenzymes such as cellulases, xylanases, amylases, glucoamylases, lipases, pectinases, proteases and ligninases. The principles and benefits of the various enzymes used by fungi to facilitate agro-industrial fermentations are summarized below.

Amylase

Amylase hydrolyses starch to glucose. Glucoamylase releases glucose residues from the non-reducing end of starch by hydrolysing α-1,4-linkages. The enzyme can also act as a debranching enzyme in the enzymic saccharification of amylopectin. Alpha-amylase hydrolyses penultimate α-1,4-glycosidic linkages at the non-reducing end of starch to release maltose. Filamentous fungi, such as *Aspergillus oryzae* and *A. niger*, secrete

Table 11.1. *Fungal degradation of agricultural by-products under solid-state fermentation conditions*

Fungus	Substrate	Product	References
Aspergillus niger	Soy and wheat bran, coffee pulp, wheat straw	Pectinase, cellulase, endoglucanase, animal feed	Peñaloza et al., 1985; Roussos et al., 1995; Maldonado & Strasser de Saad, 1998; Castilho Medronho & Alves, 2000; Jecu, 2000
Aspergillus oryzae	Starch processing wastewater, okara	Biomass protein, α-amylase, glucoamylase, nutritional supplementation	Barbesgaard Heldt Hansen & Diderichsen, 1984; Jin et al., 1998; O'Toole, 1999
Aspergillus awamori	Wheat bran	Cellulase, amylase	Silman, 1980
Aspergillus terreus	Okara	Citric acid, nutritional supplementation	O'Toole, 1999
Aspergillus tamarii	Sugar cane bagasse, wheat bran, corn cobs	Xylanase	Kadowaki et al., 1997; Ferreira et al., 1999
Aspergillus foetidusi	Rice starch, wastewater	α-Amylase	Michelena & Castillo, 1984
Rhizopus spp.	Cassava bagasse, soybean, okara	Flavouring compounds, tempe, lactic acid	Soccol, Stonoga & Raimbault, 1994; Christen et al., 2000
Penicillium spp.	Coffee pulp	Animal feed (free of caffeine), xylanase, xylosidase	Roussos et al., 1995
Trichoderma viride, Trichoderma reesei	Sugar beet pulp, apple and tomato pomace, corn cobs, canola meal	Edible protein, glucose, cellulose, xylanase, animal feed	Ward & Perry, 1982; Coughlan et al., 1986; Gattinger Duunjak & Khan, 1990; Avelino et al., 1997
Rhodotorula glutinis	Grape must, glucose syrup, beet molasses, soybean flour extract, maize flour extract	Cellulase, pectinase, chitinase, carotenoids	Buzzini & Martini, 1999
Edible mushrooms: *Agaricus bisporus, Lentinus edodes, Pleurotus* sp., *Volvariella* sp.	Lignocellulose, animal manure, composted straw, sawdust, logs, coffee pulp	Mushroom, fungal biomass	Hayes & Neir, 1975; Martinez-Carrera, 1987; Chang, Buswell & Miles, 1993; Archer & Wood, 1995

Organism	Substrate	Products/Applications	References
White-rot fungi: *Phanerochaete, Phlebia, Trametes, Pleurotus, Ceriporiopsis* spp.	Lignocelluloses, hemicelluloses, pulp mill, kraft effluent, cotton stalks, wheat and corn straw	Animal feed products, oxidative enzyme production (lignin peroxidase, manganese peroxidase laccase), bioremediation	Kirk & Farrell, 1987; Martinez *et al.*, 1994; Kerem & Hadar, 1998
Thermomyces lanuginosus	Corn cobs	Cellulase-free xylanase	Gomes *et al.*, 1993
Thannidium elegans	Apple pomace	γ-Linoleic acid	Stredansky *et al.*, 2000
Saccharomyces cerevisiae	Potatoes	Protein	Hong *et al.*, 1989
Geotrichum candidum	Tomato pomace	Saccharides, enriched feedstuffs	Avelino *et al.*, 1997
Candida spp.	Sugar cane bagasse, xylose	Xylitol, single-cell protein	Meyer, du Preez & Kilian, 1992; Silva *et al.*, 1997

considerable quantities of enzymes that are used extensively in the fermentation industry. *A. oryzae* has been accepted as a host for heterologous
protein expression and enzyme production, and as a commercial source of
α-amylase and glucoamylase (Barbesgaard, Heldt Hansen & Diderichsen,
1984), enzymes that are used for the production of high-glucose syrups
(Lowe, 1992). *Aspergillus foetidus* was selected from nine starch-utilizing
microorganisms for its high amylolytic activity (Michelena & Castillo,
1984). This fungus produced a high level of extracellular α-amylase in
rice-starch medium and degraded the available starch efficiently. This
α-amylase may be used as a saccharifying enzyme for rice starch. *A. foetidus*
also has potential for the treatment of starch-containing wastewaters
(Michelena & Castillo, 1984). Jin *et al.* (1998) described the production of
microbial biomass protein and fungal α-amylase from starch-processing
wastewater by a selected strain of *A. oryzae*. The starch-processing wastewater was treated by removing 95% of the chemical oxidation demand
(COD), 93% of the biological oxidation demand (BOD) and 98% of the
suspended solids, rendering it suitable for farm irrigation (Jin *et al.*, 1998).

Cellulases

Cellulases are a complex mixture of enzyme proteins with different specificities for hydrolysing glycosidic bonds. All cellulases consist of three
major regions: a catalytic core; a heavily glycosylated 'hinge' region, rich in
proline, threonine and serine residues; and a cellulose-binding domain.
Saccharification of agro-industrial materials using fungal cellulases is a
complex process. The three major cellulase enzyme activities are endocellulase, exocellulase and glucosidase. The primary attack on cellulose is by
endoglucanases acting on the amorphous, less-structured regions of cellulose, followed by exo-cleavage by cellobiohydrolase and complete hydrolysis to cellobiose by glucosidases. Cellulolytic fungi occur in all major
fungal taxa. To date, biochemical and molecular studies have focused on
those fungal species capable of effective synthesis and secretion of lignocellulose-degrading enzymes, or those possessing cellulases that are functional under the desired conditions and have a high cellulolytic rate
(Nevalainen & Penttilä, 1995). *Sclerotium rolfsii*, *Phanerochaete chrysosporium*, and *Trichoderma*, *Aspergillus*, *Schizophyllum* and *Penicillium* spp.
are true cellulolytic fungi. These organisms use cellulose as a primary
carbon source and are of industrial interest for their potential to convert
waste consisting of woody cellulosic materials to biofuels. The most detailed enzymological and genetic studies have been performed with soft-rot

fungi such as the *Trichoderma* spp., paticularly *T. reesei* and *T. viride*. Schimenti *et al.* (1983) addressed the preparation of hypercellulolytic mutants (e.g. the selection of *T. reesei* RUT-C30), based on resistance to nystatin.

Various agricultural substrates, such as wheat and rice straw and wheat bran, have been used successfully in SSF for cellulase production. The genus *Aspergillus* is a good producer of cellulases (Oxenboll, 1994). Cellulases and hemicellulases have been utilized for the de-inking of different types of newsprint and office waste paper. Cellulases have also been utilized for the saccharification of various lignocellulosic wastes, such as tomato pomace (Avelino *et al.*, 1997). Cellulases have recently found major application in the fabric/laundry industries. When used in laundry detergents, cellulase acts to remove fine roughened fibrils from the main body of the cotton fabrics, thereby polishing the fibres and releasing associated trapped dirt particles.

Xylanases

Xylan acts as a hardening component in cellulosic structure. Since the structure of xylans is variable, involving linear β-1,4-linked chains of xylose and also branched heteropolysaccharides, a more complex assembly of enzymes is required than for cellulose hydrolysis (Eriksson, Blanchette & Ander, 1990). Complete degradation of xylan requires the concerted action of several different hydrolytic enzymes: endo-1,4-β-xylanase, β-xylosidase, β-glucuronidase and acetylesterase. Endoxylanases launch an endwise attack on the xylan backbone to produce both substituted and non-substituted shorter oligomers, xylobiose and xylose. The potential applications of xylanases, with or without cellulases, include the bioconversion of lignocelluloses to sugar, ethanol and other useful substances; the clarification of juices and wines; the extraction of plant oils, coffee and starch; and the improvement of the nutritional value of silage and green feed. Xylanases are already used in pulp mills to facilitate subsequent chemical bleaching. In recent years, interest has grown in xylanases free of cellulases to remove xylan from lignocellulose selectively without affecting the length of the cellulose fibre. An advantage of this enzyme treatment process is that it minimizes the use of chlorine, thus reducing the concentrations of toxic chloroorganic wastes produced during chemical bleaching (Hoq & Ernst, 1996). By 1995, 10% of Canada's bleached kraft pulp was treated with xylanase (Tolan, Olson & Dines, 1995). SSF processes for xylanase production include the use of agricultural wastes. *Aspergillus tamarii* has been described as a good producer of xylanase under SSF conditions on sugar cane bagasse, wheat bran and corn cob (Ferreira, Boer & Peralta, 1999).

Pectinases

Pectinases are a group of enzymes that degrade pectic substances, the structural polysaccharides present in vegetable cells and responsible for maintaining the integrity of plant tissues. Pectic substances are characterized by long chains of galacturonic acid residues. Pectic enzymes act by breaking glycosidic bonds of the long carbon chains (polygalacturonase, pectin lyase and pectate lyase) and by splitting off methoxyl groups (pectin esterase). The synergetic action of pectic enzymes is used industrially for the extraction, clarification and concentration of fruit juices, for the clarification of wines, for the extraction of oils, flavours and pigments from plant materials and for the preparation of cellulose fibres for linen manufacture.

Acuna et al. (1995) described the production and properties of three pectinolytic activities produced by *A. niger* CH4 in submerged culture and SSF. The highest productivity of endo- and exopectinase and pectin lyase was obtained with SSF. Maldonado & Strasser de Saad (1998) studied the production of pectinesterase and polygalacturonase by *A. niger* in submerged culture and SSF systems. With pectin as a sole carbon source, pectinesterase and polygalacturonase production were four and six times higher, respectively, in a SSF system than in a submerged fermentation system and required a shorter time for enzyme production. Differences in the regulation of enzyme synthesis by *A. niger* depended on the fermentation system, favouring SSF over submerged fermentation for pectinase production. Castilho, Medronho & Alves (2000) described the optimization of *A. niger* SSF on soy and wheat bran for the production of pectinases. The results show that optimizing the extraction conditions is a simple way of obtaining the desired enzyme in a more concentrated form.

Lignin-degrading enzymes

Lignin biodegradation by white rot fungi has obvious ecological significance as well as promising biotechnological applications in industrial processes, such as the bleaching of paper pulp and the remediation of xenobiotics in effluents. The biotechnological approach of using ligninolytic enzymes as a pulping reagent results in easier pulping, with less consumption of chemicals and energy. Lignin biodegradation is an oxidative process, involving enzymes such as lignin peroxidases (LiPs), manganese peroxidases (MnPs) and laccases (Kirk & Farrell, 1987; Eriksson, *et al.*, 1990; Gold & Alic, 1993; Hatakka, 1994). In the 1970s, the main focus was to define laboratory conditions under which white rot fungi, particularly *P. chrysosporium*, would maximally break down lignin. In the 1980s, the biochemistry of lignin-modifying enzymes was predominantly studied because of the discovery of ligninolytic peroxidases. In 1983 and 1984, two extracellular enzymes, LiP and MnP, were discovered in *P. chrysosporium* (Glenn *et al.*, 1983; Tien & Kirk, 1983; Kuwahara *et al.*, 1984). These enzymes were shown to be major components of the lignin degradation system in this organism. The discovery of the enzymes involved in lignin degradation by *P. chrysosporium* led to their biochemical, biophysical and physiological characterization. This work has served as an indispensable background for studies on lignin degradation by other fungi. Knowledge concerning the molecular genetics of white rot fungi, particularly *P. chrysosporium*, has advanced considerably during the 1990s. Standard procedures have been established for auxotroph production, recombination analysis, rapid DNA and RNA purification and genetic transformation by auxotroph complementation and by drug-resistant markers (see Cullen, 1997). In addition to detailed studies on the molecular biology of lignin-modifying peroxidases, major lines of research have involved applying enzymes to the bioconversion of agro-industrial wastes, biopulping and pulp bleaching, and searching for the enzymes responsible for lignin degradation in more preferential lignin degraders: fungi that degrade larger amounts of lignin relative to carbohydrates. This has led to a reassessment of the biotechnological potential of white rot fungi other than *P. chrysosporium*, such as *Ceriporiopsis subvermispora* and *Pleurotus* spp., and an investigation into their ligninolytic enzyme systems (Hatakka, 1994). Many white rot fungi, with the notable exception of *P. chrysosporium*, produce an extracellular laccase (Kirk & Farrell, 1987; Eriksson *et al.*, 1990; Higuchi, 1990; Kerem, Friesem & Hadar, 1992) and extracellular H_2O_2-producing enzymes, such as aryl alcohol oxidase (Gutierrez *et al.*, 1994) and glyoxal oxidase (Kersten & Kirk, 1987). An array of ligninolytic enzymes have been isolated and are currently under extensive study. White rot fungi may be divided on the basis of the typical production patterns of the more common extracellular enzymes into two main groups, those that produce LiP, MnP and laccase and those without LiP (Hatakka, 1994). LiP and MnP have been largely characterized in *P. chrysosporium*, and laccase production has been described in this fungus using a medium containing cellulose as the carbon source (Srinivasan *et al.*, 1995). Most white rot fungi, among them *Pleurotus ostreatus* and *Pleurotus eryngii,* are included in the second group. Despite their lack of LiP, these species are able to degrade lignin preferentially in wheat straw (Martinez *et al.*, 1994) and cotton stalks (Kerem *et al.*, 1992). These fungi also produce aryl alcohol oxidase (Martinez *et al.*, 1994), an enzyme participating in H_2O_2 production and necessary for MnP action. A third type of ligninolytic peroxidase has been described in several species of *Pleurotus* and *Bjerkandera*, characterized by sharing catalytic properties of MnP and LiP (Moreira *et al.*, 1997; Mester & Field, 1998; Ruiz Dueñas, Martinez & Martinez, 1999).

Manganese peroxidase

MnP catalyses the H_2O_2-dependent oxidation of lignin and its derivatives (Pérez & Jeffries, 1992). The oxidation of lignin and other phenols by MnP is dependent on the free manganese ion (Leisola *et al.*, 1987; Paszczynski, Huynh & Crawford, 1986). The primary reducing substrate in the MnP catalytic cycle is Mn^{2+}, which efficiently reduces the enzyme to generate Mn^{3+}; the latter subsequently oxidizes the organic substrate. In many fungi, MnP is thought to play a crucial role in the primary attack on lignin because it generates Mn^{3+}, a strong oxidant. Organic acids such as oxalate and malonate, which are secreted by white rot fungi, stimulate the MnP reaction by stabilizing Mn^{3+} so that it can diffuse from the surface of the enzyme and oxidize the insoluble terminal substrate lignin. The manganese ion participates in the reaction as a diffusible redox coupler rather than as an enzyme activator (Wariishi *et al.*, 1989; Cui & Dolphin, 1990). Although MnP does not oxidize non-phenolic lignin structures during normal turnover with H_2O_2 and Mn^{2+}, these structures are slowly co-oxidized when MnP peroxidizes unsaturated fatty acids, and lipid peroxidation has been suggested to play a role in fungal ligninolysis (Jensen *et al.*, 1996). Recent studies have revealed that *P. eryngii* secretes five peroxidases (two in peptone-containing liquid media and three during lignin degradation under SSF conditions) that efficiently oxidize Mn^{2+} to Mn^{3+} but are different from *P. chrysosporium* MnP because of their manganese-independent activity on aromatic substrates (Martinez *et al.*, 1996).

Laccase

Laccase is a multicopper blue oxidase that catalyses the one-electron oxidation of diphenols and aromatic amines by removing an electron and a proton from a hydroxyl group to form a free radical. It can catalyse the alkyl–phenyl and C_α-C_β cleavage of phenolic lignin dimers (Higuchi, 1990;

Eriksson *et al.*, 1990). The oxidation activity is accompanied by reduction of molecular oxygen to water. An artificial laccase substrate, ABTS (2,2'-azino-bis(3-ethylbenzthiazoline 6-sulfonate)) has been shown to act as a mediator enabling the oxidation of non-phenolic lignin model compounds that are not laccase substrates on their own (Bourbonnais & Paice, 1990).

The most studied laccases appear to be from *Agaricus bisporus, Podospora anserina, Rhizoctonia practicola, Coriolus (Trametes) versicolor, Coriolis hirsutus, P. ostreatus* and *Neurospora crassa* (Leonowicz, Trojanowski & Orlicz, 1978; Leonowicz, Szklarz & Wojtas-Wasilewska, 1985; Yaropolov *et al.*, 1994; Thurston, 1994). The combined use of laccase with a low-molecular-weight mediator for bleaching kraft pulp has gained attention since the initial discovery that non-phenolic lignin compounds can be oxidized, and pulp delignified by laccase in the presence of ABTS (Bourbonnais & Paice, 1990). Since then, a growing number of nitrogen-containing compounds, including 1-hydroxybenzotriazole and violuric acid have also been shown to assist laccase in pulp delignification (Bourbonnais *et al.*, 1997).

Lignin peroxidase

LiP catalyses the H_2O_2-dependent oxidation of a wide variety of non-phenolic lignin model compounds, synthetic lignins and aromatic pollutants (Buswell & Odier, 1987; Higuchi, 1990; Hammel & Moen, 1991). LiP from *P. chrysosporium* has been purified and its mode of action studied in several laboratories (Kirk & Farrell, 1987). The one-electron oxidation of aromatic nuclei generates cation radicals that lead in subsequent spontaneous reactions to extensive degradation of lignin model compounds (Buswell & Odier, 1987).

LiP isozymes of *P. chrysosporium* are encoded by families of structurally similar genes that exhibit complex patterns of regulation, all of which cross-hybridize to various extents (Stewart *et al.*, 1992; Brooks, Sims & Broda, 1993; Gaskell *et al.*, 1994). The *lip* genes of *P. chrysosporium* are differentially expressed, based on the conditions encountered during growth. Despite much detailed work on the reaction mechanism of LiPs, there is little evidence that they can actually cleave polymeric lignin. Experiments with low-molecular-weight lignin model compounds provide strong support for a ligninolytic role but cannot prove it. Investigations with lignin itself as a LiP substrate have, except for some preliminary data, failed to show ligninolysis; however, the oxidation of [^{14}C]- and [^{13}C]-labelled lignins by a purified isozyme of *P. chrysosporium* has been described (Hammel *et al.*, 1993). It has been suggested that LiP catalyses the initial steps of ligninolysis by *P. chrysosporium in vivo*, but that LiP alone is not sufficient for lignin degradation (Pérez & Jeffries, 1992; Hammel *et al.*, 1993).

Agro-industrial wastes used as fermentation substrates

The processing of agro-industrial raw materials produces large amounts of waste, the accumulation of which leads to severe environmental pollution. A large variety of substrates, such as corn cobs, sugar cane bagasse, coffee pulp, pomace and rice hulls, can be used for fungal fermentation, with the aim of producing foods, enzymes, mushrooms and compost. The number of potential wastes and processes varies between regions of the world and crops. Here we discuss several examples of agro-industrial by-products that are currently, or could potentially be, used as fungal fermentation substrates.

Lignocellulosic wastes

Lignocelluloses are defined as plant or wood cell walls, where cellulose is intimately associated with lignin. In addition to these compounds, lignocellulose contains other polysaccharides commonly called hemicelluloses. The three major polymeric components are the polysaccharides cellulose and hemicellulose, and the aromatic polymer of lignin (Eriksson *et al.*, 1990). Cellulose is a linear polymer of glucose linked through β-1,4-linkages and is usually arranged into microcrystalline fibres. Hemicellulose is a matrix polysaccharide of the plant cell wall. It includes all of the plant cell polysaccharides except for pectin and cellulose. Hemicelluloses are usually heteroglycans containing two to four, and occasionally more, different monosaccharides, including D-glucose, D-galactose, D-mannose, and D-xylose. Lignin is an amorphous aromatic polymer found in the cell walls and middle lamellae of vascular plants and surrounding cellulose microfibrils. It gives cell walls their rigidity and protects plants against pathogen attack and mechanical stresses (Eriksson *et al.*, 1990).

Lignocellulosic waste comprises mainly crop residues and wood and forestry wastes. The direct use of lignocellulosic residues as ruminant animal feed, or as a component of such feeds, represents one of its oldest and most widespread applications and, as such, it plays an important role in the ruminant diet (Hadar *et al.*, 1992). However, only a small proportion of the cellulose, hemicellulose and lignin produced as agricultural or forestry by-products is utilized, while most of it is considered waste material. White rot fungi can degrade all of the major components of wood and are generally considered to be the main agents of lignin degradation in nature (Buswell & Odier, 1987). The best-studied organism of this group is *P. chrysosporium* (Kirk & Farrell, 1987; Schoemaker & Leisola, 1990). Most knowledge of the physiological, biochemical and genetic factors associated with lignin biodegradation, as well as of the practical uses of the processes, mainly in the pulp and paper industry, has been obtained from this model organism.

Another group of interesting white rot fungi that can utilize lignocellulose is the edible mushrooms. These saprophytic basidiomycetes have been successfully cultivated at a commercial level worldwide (Wood & Smith, 1987). The button mushroom, *A. bisporus*, is the most popular edible mushroom in Europe, the USA, Asia and Australia (Zadrazil & Kamra, 1997). *Pleurotus* spp. are wood-degrading saprophytic fungi that are widely distributed and cultivated in many countries throughout the world. The cultivation of edible fungi on animal manure and plant by-products has many remarkable ecological advantages in food production. Horse and chicken manure, cereal straw, bagasse, sawdust, sulfite liquor and other residues from the paper industry can be directly converted into fungal substrate. The spent mushroom substrate can be used either as animal feed or as compost. The substrate is usually partially shredded, mixed with water and placed in containers. However, unlike *A. bisporus*, no composting or casing layer is required for *Pleurotus* spp.; since these can decompose lignocellulose efficiently without chemical or biological pretreatment, a large variety of lignocellulosic wastes can be utilized and recycled.

Cassia spp. are tropical shrubs with medicinal (e.g. laxative) properties. The residue of extracted cassia plants was tested for cultivation of *Pleurotus* by Muller (1987). The utilization of *Pleurotus* in the recycling of medicinal plants could be of additional importance when toxic components are present. The reduction of such components by the fungus would simplify the problem of waste disposal (Muller, 1987). Cotton generates a large amount of agricultural waste. Its worldwide importance is illustrated by the 10 million tons of cotton stalks reported for India yearly (Balashubramanya *et al.*, 1989). Those remaining in the field, 5 tons ha^{-1}, are ploughed under the soil surface. In addition to wasting a potential agricultural resource, this treatment could lead to an increase in cotton diseases and pests, as well as to difficulties in cultivation because of slow decomposition in the soil. The major obstacle to using cotton stalks as a mushroom substrate is that they are difficult to preserve. This is mainly because of their high water content and high level of soluble carbohydrates (Silanikove & Levanon, 1986). The substrate is rapidly overgrown by unwanted fungi, resulting in spoilage and aerobic degradation. The anaerobic preservation of cotton stalks and the production of silage were studied by Silanikove & Levanon (1986), Levanon, Danai & Masaphy (1988) and Danai, Levanon & Silanikov (1989). After 1 month of storage, the pH of the preserved cotton stalks stabilized at 5.5 and the material was successfully utilized for commercial *Pleurotus* cultivation, up to 9 months after harvest (Levanon *et al.*, 1988; Danai *et al.*, 1989).

The lignocellulose complex in straw and other plant residues is degraded very slowly by ruminants because of the physical barrier imposed by lignin polymers, preventing free access of hydrolytic enzymes such as cellulases and hemicellulases to their substrates. Normally, the rates of decay of plant debris are proportional to their lignin content. Biological delignification of straw seems to be the most promising way of improving its digestibility (Streeter *et al.*, 1982; Kamra & Zadrazil, 1986; Zadrazil & Reinger, 1988). Several authors have explored this possibility, using mainly wheat straw and *Pleurotus* spp. under different conditions and substrate pretreatments (Zadrazil, 1980; Streeter *et al.*, 1982; Kamra & Zadrazil, 1986; Kerem & Hadar, 1998).

Forest, pulp and paper industry by-products

The forestry, pulp and paper sectors are major contributors of air and water pollutants. The amount of solid waste from the forestry industry is significant. In British Columbia alone, 5 million dry tonnes of excess wood waste are produced annually from lumber milling operations (Murray & Richardson, 1993). Cellulosic wastes obtained from the forest product industry could be bioconverted to ethanol fuel via saccharification–fermentation processes (see Duff & Murray, 1996). For this purpose, lower-quality woods are quite acceptable. Unlike acid hydrolysis, enzymic hydrolysis is highly specific and can produce high yields of relatively pure glucose syrups without generating glucose-degradation products. Enzymic hydrolysis is the one step in the whole wood-to-ethanol process. The nutritional and physiological requirements of cellulolytic fungi and the regulatory mechanisms that control the synthesis of complete cellulase complexes have been defined. Genetically improved strains of *T. reesei* are being used for more rapid synthesis of larger amounts of cellulase complex per unit substrate.

Pulp production by means of mechanical and chemical processes and bleaching treatments are energy-demanding processes and cause environmental pollution. The need for sustainable technologies has brought biotechnology into the realm of pulp and paper making. Suitable biological pretreatment, defined biopulping and biobleaching, in conjunction with less intensive conventional treatments, could help to solve many of the problems associated with currently used processes. Biopulping is defined as the treatment of wood chips with lignin-degrading fungi prior to pulping. White rot fungi alter the wood cell walls, which softens the chips and substantially reduces the electrical energy needed for pulping. The treatment also improves paper strength, reduces the pitch content and reduces the environmental impact of pulping.

Corn cobs

Corn cob is the corn (maize) stalk residue after removal of the grains by the food industry. This agricultural surplus is generally used as animal feed. Corn cobs, a lignocellulosic residue, have been utilized as a growth substrate in cultures to produce xylanolytic enzymes. *A. tamarii* has been found to grow well and to produce high cellulase-free xylanase activity (which is important for the pulp and paper industries) when grown on corn cob powder as the principal substrate (Kadowaki *et al.*, 1997). Ward & Perry (1982) studied the enzymic conversion of corn cobs to glucose with *T. viride*: the treated corn cobs were used successfully for animal feed. Gomes *et al.* (1993) reported on the production of a high level of cellulase-free xylanase by the thermophilic fungus *Thermomyces lanuginosus* grown on corn cobs. This enzyme has been used successfully to enhance the bleaching of kraft pulp.

In our laboratory, *P. ostreatus* was grown on corn cobs in order to improve its quality as a dietary fibre for human consumption. Rats were fed a semisynthetic fibre-free diet or a high-fibre diet (15%) derived from corn cob treated or not treated with the fungus. The fungus-treated corn cobs significantly prevented the development of colon cancer in rats: a high positive correlation was found between tumour grade and p53 protein (tumour suppressor) in the serum or in the cell cytoplasm. Incubation of corn cob with the fungus *P. ostreatus* increased the dietary fibre content up to 78%. Therefore colon cancer development was lower in animals on a diet containing corn cob and such a fibre source may be considered of potential use to the public (Zusman *et al.*, 1997).

Sugar cane bagasse

Bagasse is the residue left after sucrose has been extracted from sugar cane. It has been identified as a renewable resource as a potential carbon substrate for the production of commercial single-cell protein, xylitol and other products. This waste has been used as raw material for hydroxymethylfurfural production, paper pulp, acoustic board, pressed wood and agricultural mulch. Sugar cane bagasse contains 30–35% hemicelluloses, which can be extracted with dilute acids, producing a mixture of monosaccharides, mainly pentoses, with D-xylose as the main compound. The cost of the carbon source is one of the most important determinants for the economic viability of the process. Sugar cane bagasse is an inexpensive and abundant energy source that can be used in several biotechnological processes to obtain products of high economic value.

Xylitol is a five-carbon sugar alcohol with beneficial characteristics, such as a sweetening power comparable to sucrose and with anticancer properties (Makinen & Isokangas, 1988). Furthermore, direct human metabolism of xylitol is independent of insulin. Xylitol is, therefore, well suited as a sucrose substitute for diabetics (Pfeifer *et al.*, 1996). A biotechnological approach for xylitol production appears to be efficient, and a high xylitol concentration can be obtained under controlled physiological conditions. The combination of dilute-acid hydrolysis, which can be carried out efficiently at relatively low cost, and a biotechnological hydrogenation process presents significant advantages since it does not require pure xylose solutions. *Candida guilliermondii* can convert xylose to xylitol with high efficiency: the maximum theoretical yield of xylitol from D-xylose being $0.917\,\mathrm{g\,g^{-1}}$ (Barbosa *et al.*, 1988). A low yield of xylitol could be the result of toxic substances that interfere with metabolism. Silva *et al.* (1997) studied the fermentation of sugar cane bagasse hemicellulosic hydrolysate treated with *C. guilliermondii* to remove the fermentation inhibitors. Efficient fermentation was obtained by using a suitable control for oxygen input. Meyer, du Preez & Kilian (1992) studied the production of single-cell protein from a hemicellulose hydrolysate of sugar cane bagasse. *Candida blankii* was isolated and found to have considerable potential for the production of single-cell protein because of its ability to utilize all of the major carbon substrates in the hydrolysate at low pH and relatively high temperature with a high protein yield.

Coffee pulp

Coffee pulp is the agro-industrial residue produced during the pulping of coffee berries to obtain coffee beans. In coffee-growing regions, coffee pulp is considered to be one of the most abundant agricultural wastes, as well as one of the hardest to handle. According to available data, world green coffee and coffee pulp production during 1989–1990 reached a maximum of 5.52×10^6 tons of green coffee and of 2.76×10^6 tons of coffee pulp (Roussos *et al.*, 1995). To date, most of the by-product remains unused, causing contamination problems. Coffee pulp is essentially rich in carbohydrates, proteins and minerals, and it also contains appreciable amounts of tannins, polyphenols and caffeine (Edwards, 1979). An economical way of dealing with the excessive amount of pulp and of solving the environmental problem would be to use it for animal feed. However, chemical components such as lignin, phenolics and caffeine hamper its utilization by animals. Caffeine and phenolics are known to exert detrimental effects to both the rumen microflora and the host animal. Biological pretreatment could, however, improve the value of this waste product. The presence of proteins, sugars and minerals in coffee pulp and its high humidity favour the rapid growth of microorganisms. There have been many reports describing the composition, conservation, upgrading and utilization of coffee pulp (Christenen, 1981; Orue & Bahar, 1985; Martinez-Carrera, 1987). Peñaloza *et al.* (1985) studied the nutritive improvement of coffee pulp by using *A. niger* under SSF conditions. The fermented product had a higher total amino acid content and a lower cell wall constituent value than the original pulp. A growing chicken's ration containing 10% of the fermented product had a feed efficiency similar to that of the standard ration and significantly higher than that of the same diet containing 10% of the original pulp. Roussos *et al.* (1995) described the isolation of caffeine-degrading fungi from leaves of coffee plants and coffee berries; most of the isolated microorganisms belonged to the genera *Aspergillus*, *Penicillium*, *Trichoderma*, *Fusarium* and *Humicola*. Five *Aspergillus* strains and two *Penicillium* strains could degrade almost 100% of the caffeine in a liquid medium (Roussos *et al.*, 1995).

Sugar beet pulp

Sugar beet pulp, a by-product of sugar extraction, is an important renewable resource, the bioconversion of which appears to be of great biotechnological importance. In the European Union, about 10^8 tonnes of roots are produced annually. The processing of 1 tonne of beets produces about 70 kg exhausted dried pulp, or about 250 kg exhausted pressed pulp, containing approximately 10 and 75% water, respectively. The amount of beet pulp produced and stocked in the sugar industry is, therefore, very high. The lignocellulosic fraction of the dried pulp comprises cellulose, hemicellulose, pectic substances and lignin (Coughlan et al., 1986). The total hydrolysis of beet pulp releases monosaccharides such as L-arabinose and D-galacturonic acid as components of the main chains of cellulose, hemicellulose and pectic substances, respectively (Coughlan et al., 1986). Pulp utilization is confined almost exclusively to use as feed for ruminants, but the low lignin content enhances the suitability of sugar beet pulp for biotransformation processes, suggesting alternative uses. Sugar beet pulp is a suitable substrate for the production of protein using thermophilic microorganisms (Grajek, 1988). Its nutritional upgrading can be improved by white rot fungi (Di Lena, Patroni & Quaglia, 1997). The hexose components may be used in traditional industrial processes, while there is growing interest in the industrial exploitation of pentose sugars and galacturonic acid.

Apple and tomato pomace

Apple pomace is a solid residue from the agro-food industry obtained after crushing and pressing apples during the manufacture of apple juice. It is a cumbersome waste product often considered of little value as animal feedstuff or for pectin production. SSF on apple pomace has been performed for the production of ethanol (Ngadi & Correia, 1992) and citric acid (Hang & Woodams, 1998). Stredansky et al. (2000) described the production of γ-linoleic acid via fermentation of the fungus *Thamnidium elegans* on apple pomace: high-value fungal oil was produced.

Tomato pomace is the residue from tomato processing and is made up of the fruit's dried and crushed skin and seeds. It is a lignocellulose-based substrate, available in large quantities in Mediterranean countries and often creates disposal problems. The bioconversion of tomato pomace has been described (Carvalheiro, Roseiro & Collaco, 1994; Avelino et al., 1997); the process was directed to an assessment of the use of SSF using *Geotrichum candidum* to increase the protein content and improve the digestibility of tomato pomace in order to turn it into raw materials for enriching feedstuffs. The saccharification of tomato pomace yielded glucose, fructose, xylose and cellobiose.

Okara

Okara is the residue left from ground soybeans after extraction of the water-extractable fraction used to produce soy milk and tofu. Huge quantities of okara are produced; for example, in Japan, about 700 000 tons of okara were produced by the tofu production industry in 1986 (Ohno, Ano & Shoda, 1996). Although okara is sometimes used in animal feed, most of it is burned as waste. Okara was fermented using the tempe fungi *Rhizopus oligosporus* and *A. oryzae* to improve its nutritional qualities as a high-fibre, low-energy foodstuff suitable for human consumption. Okara alone has some antinutritional qualities, although fermented okara may provide definite dietary benefits: it can act as a replacement for digestible food in prepared food to reduce caloric intake; it can reduce cholesterol levels in the bloodstream; and, as a food that contains antioxidant activity similar to vitamin E, it can reduce the level of free radicals in the body. Fermentation of okara by *A. niger* and *Aspergillus terreus* is used for the production of citric acid (O'Toole, 1999).

Canola meal

Canola meal is the by-product of oil extraction from canola. Current use of canola meal is limited to animal feed, but it has been shown to cause liver damage in poultry and to taint eggs (Butler, Pearson & Fenwick, 1982). Gattinger, Duvnjak & Khan (1990) reported the production of xylanase, acetyl-xylan esterase and xylosidase by *T. reesei* in a medium containing canola meal as a carbon source; xylanase yields were similar or better than with the refined substrates xylan and glucose.

The role of fungi in composting

When waste materials cannot serve as substrates for any of the pure culture processes described in this chapter, composting is the treatment of choice. Composting is the microbial breakdown of organic waste material in a thermophilic, aerobic environment. The final product is compost or humus, which is stabilized organic matter (Inbar, Chen & Hoitink, 1993). Composting can also be defined as SSF of organic wastes. It differs from the processes described earlier in this chapter in that it treats a heterogeneous substrate and utilizes a mixed population of microorganisms operating in succession. Composting has attracted interest in Western countries because of the large amounts of agricultural, industrial and municipal wastes. These wastes create a severe disposal problem involving a number of environmental aspects related to air, soil, groundwater and surface-water pollution. The organic substrates, bulking agents and amendments used in composting are derived mostly from plant material. A number of organic wastes, such as bark and yard waste, leaf mould, municipal solid waste, sewage sludge, sawdust and farm animal manures have been used for composting.

The composting process can be divided into three phases: (i) mesophilic, during which the temperature rises; (ii) thermophilic; and (iii) cooling. Different types of microorganism are active at different times in the composting pile. Bacteria have the most significant effect on the decomposition process and are the first to take hold in the composting pile. Fungi, which compete with bacteria, play an important role later in the process as the pile dries, since fungi can tolerate low-moisture environments better than bacteria. Raw compost contains about 10^6 colony-forming units (cfu) of mesophilic fungi per gram of raw material and $10^3–10^6$ cfu g^{-1} of thermophilic fungi (Thambirajah, Zulkali & Hashim, 1995). The predominant mesophilic fungi in the raw material are *Geotrichum* spp. and the thermotolerant fungus *Aspergillus fumigatus* (Tuomela *et al.*, 2000).

Mesophilic phase

Most fungi are mesophiles and grow at 5–37°C, optimally at 25–30°C (Dix & Webster, 1995). Mesophiles are dominant throughout the composting mass in the initial phases of the process when temperatures are relatively low. These organisms use available oxygen to transform carbon and obtain energy; during this process, the temperature of the pile rises.

Thermophilic phase

Thermophilic microorganisms prefer temperatures between 45 and 70°C. Thermophiles generate greater quantities of heat than mesophiles, and the temperatures reached during this stage are hot enough to kill most pathogens and weed seeds. The thermophiles continue decomposing the substrate as long as nutrient and energy sources are available. As these sources become depleted, thermophiles die and the temperature of the pile drops. In mushroom compost, thermophilic fungi are responsible for the degradation of lignocellulose (Sharma, 1989). Thermophilic and thermotolerant fungi that have cellulolytic and ligninolytic activity, or which have been found growing in compost, have been studied, for example *Talaromyces emersonii, Thermoascus auranticus, Thermomyces lanuginosus* and *Coprinus* sp. (Tuomela *et al.*, 2000). Chefetz, Chen & Hadar, (1998a) purified laccase from municipal solid waste compost during the thermophilic stage of composting. *Chaetomium thermophilium*, a cellulolytic fungus, was isolated from this stage and exhibited laccase activity when grown at 45°C. The purified laccase exhibited high catalytic activity towards a wide range of phenolic substrates.

Cooling phase

When the temperature has decreased below 60°C, mesophilic and thermotolerant fungi reappear in the compost (Thambirajah *et al.*, 1995). The dominant fungi after peak heating are *Aspergillus* spp. or *T. lanuginosus*, which also dominates at 50°C. *Panaeolus* spp., *Corticium coronilla* and possibly *Mycena* spp. are Basidiomycota occurring in compost. They have all been isolated from compost during the cooling and maturation phase, or from mature compost (Tuomela *et al.*, 2000).

Compost maturity

The process of humification occurs during composting. This process probably includes the polymerization of aromatic compounds as well as lignin modifications. Humic compounds affect soil ecology, structure, fertility and metal complexation, as well as plant growth. According to Stevenson, Fitch & Brar (1993), the formation of polyaromatic humic acid structures is associated with phenoloxidase activity in the soil; laccase is one of these phenoloxidases. Phenol oxidation may be a major driving force in the relatively rapid humification that occurs during composting. Chefetz *et al.* (1998b) showed that fungal laccase can couple the aromatic fraction extracted from compost with guaiacol and it was suggested that this enzyme could be involved in the humification process during composting.

   Another phenomenon related to compost maturity is the build-up of antagonistic microorganism populations capable of suppressing soilborne plant pathogens (Hoitink & Boehm, 1999). In recent years, compost amendments in agriculture have been investigated as part of an integrated system of biological control because of their ability to suppress plant pathogens. These composts have been reported to suppress plant pathogens mainly in container media (Hoitink, Stone & Han, 1997). Nelson & Hoitink (1983) identified specific strains of four *Trichoderma* spp. and isolated *Gliocladium virens* as the most effective fungal hyperparasite of

Rhizoctonia solani present in bark compost. A few of the other 230 fungal species also showed activity, but most were ineffective. Composted grape marc was effective in suppressing disease caused by *Sclerotium rolfsii* in bean and chickpea (Gorodecki & Hadar, 1990). Hadar & Gorodecki (1991) placed sclerotia of *S. rolfsii* on composted grape marc to isolate hyperparasites of these pathogens. Sclerotial viability decreased from 100% to less than 10% within 40 hours. *Penicillium* and *Fusarium* spp. were observed by scanning electron microscopy to colonize the sclerotia. *Trichoderma* populations in grape marc compost were at very low levels (10^2 cfu g^{-1} dry weight).

   Trichoderma hamatum has been identified as an effective biocontrol agent in compost-amended substrates (Kwok *et al.*, 1987). It consistently induced suppression of diseases caused by a broad spectrum of soilborne pathogens when inoculated into compost after peak heating but before substantial recolonization by other mesophilic microorganisms (Hoitink *et al.*, 1997). A verification of fungal densities in fortified potting mixes at various stages after mix formulation is critical to predicting biological control activity. Abbasi *et al.* (1999) described the precise detection and tracing of *T. hamatum* in compost-amended potting mixes using molecular markers. Compost must be of consistent quality to be used successfully in the biological control of diseases of horticultural crops, particularly if used in container media (Inbar *et al.*, 1993). Many compost quality factors must be controlled to obtain consistent effects with these organic amendments. The composition of the organic matter from which the compost is pre-pared, the composting process itself, the stability or maturity of the compost, the quantity of available plant nutrients provided by the com-post, loading rates, time of application and other factors must all be controlled (de Ceuster & Hoitink, 1999).

Conclusions

Fungi can be involved in a large variety of processes utilizing many different substrates derived from agricultural crop residues or agro-indus-trial by-products. When waste materials cannot serve as substrates for pure culture processes, composting is the treatment of choice. Recycling through composting is often considered as a preferred strategy for waste treatment. Fungi are specifically suited to this task because (i) many species are considered safe for either direct consumption or the production of food components; (ii) they grow as hyphae able to invade solid substrates; (iii) they secrete enzymes that degrade polymeric materials to products used subsequently as nutrients; (iv) they can grow at relatively low moisture contents; and (v) they produce many useful products. Only a few processes are currently performed commercially. Making use of the potential of fungi is an important and challenging task that could lead to many useful products from renewable raw materials, thus saving energy and preventing pollution.

References

Abbasi, P. A., Miller, S. A., Meulia, T., Hoitink, H. A. J. & Kim, J. M. (1999). Precise detection and tracing of *Trichoderma hamatum* 382 in compost-amended potting mixes by using molecular markers. *Applied and Environmental Microbiology*, **65**, 5421–5426.

Acuna, A. M. E., Gutierrez, R. M., Viniegra, G. G. & Favela, T. E. (1995). Production and properties of three pectinolytic activities produced by *Aspergillus niger* in submerged and solid-state fermentation. *Applied Microbiology and Biotechnology*, **43**, 808–814.

Archer, D. B. & Wood, D. A. (1995). Fungal exoenzymes. In *The Growing Fungus*, eds. N. A. R. Gow & G. M. Gadd, pp. 137–162. London: Chapman & Hall.

Avelino, A., Teixeira, A. H., Roseiro, J. C. & Amaral, C. M. T. (1997). Saccharification of tomato pomace for the production of biomass. *BioresourceTechnology*, **61**, 159–162.

Balashubramanya, R. H., Pai, Y. D., Shaikh, A. J. & Khandeparkar, V. G. (1989). Biological softening of spent cotton-plant stalks for the preparation of pulp. *Biological Wastes*, **30**, 317–320.

Barbesgaard, P., Heldt Hansen, H. P. & Diderichsen, B. (1984). On the safety of *Aspergillus oryzae*: a review. *Applied Microbiology and Biotechnology*, **36**, 569–572.

Barbosa, M. F. S., de Medeiros, M. B., de Mancilha, I. M., Schneider, H. & Lee, H. (1988). Screening of yeasts for production of xylitol from D-xylose and some factors which affect xylitol yield in *Candida guilliermondii*. *Journal of Industrial Microbiology*, **3**, 241–251.

Bourbonnais, R. & Paice, M. G. (1990). Oxidation of non-phenolic substances, an expanded role for laccase in lignin biodegradation. *FEBS Letters*, **267**, 99–102.

Bourbonnais, R., Paice, M. G., Freiermuth, B., Bodie, E. & Borneman, S. (1997). Reactivities of various mediators and laccases with kraft pulp and lignin model compounds. *Applied and Environmental Microbiology*, **63**, 4627–4632.

Brooks, P., Sims, P. & Broda, P. (1993). Isozyme-specific polymerase chain reaction analysis of differential gene expression: a general method applied to lignin peroxidase genes of *Phanerochaete chrysosporium*. *Biotechnology*, **11**, 830–834.

Buswell, J. A. & Odier, O. (1987). Lignin biodegradation. *Critical Reviews in Biotechnology*, **6**, 1–60.

Butler, E. J., Pearson, A. W. & Fenwick, G. R. (1982). Problems which limit the use of rapeseed meal as a protein source in poultry diets. *Journal of Science of Food Agriculture*, **33**, 866–875.

Buzzini, P. & Martini, A. (1999). Production of carotenoids by strains of

Rhodotorula glutinis cultured in raw materials of agro-industrial origin. *Bioresource Technology*, **71**, 41–44.

Carvalheiro, F., Roseiro, J. C. & Collaco, M. T. A. (1994). Biological conversion of tomato pomace by pure and mixed fungal cultures. *Process Biochemistry*, **29**, 601–605.

Castilho, L. R., Medronho, R. A. & Alves, T. L. M. (2000). Production and extraction of pectinases obtained by solid state fermentation of agroindustrial residues with *Aspergillus niger*. *Bioresource Technology*, **71**, 45–50.

Chang, S., Buswell, J. A. & Miles, P. G. (1993). *Genetics and Breeding of Edible Mushrooms*. New York: Gordon and Breach.

Chefetz, B., Chen, Y. & Hadar, Y. (1998a). Purification and characterization of laccase from *Chaetomium thermophilium* and its role in humification. *Applied and Environmental Microbiology*, **64**, 3175–3179.

Chefetz, B., Adani, F., Genevini, P., Tambone, F., Hadar, Y. & Chen, Y. (1998b). Humic-acid transformation during composting of municipal solid waste. *Journal of Environmental Quality*, **27**, 794–800.

Christen, P., Bramorski, A., Revah, S. & Soccol, C. R. (2000). Characterization of volatile compounds produced by *Rhizopus* strains grown on agro-industrial solid wastes. *Bioresource Technology*, **71**, 211–215.

Christenen, M. S. (1981). Preliminary tests on the suitability of coffee pulp in the diet of common carp (*Cyprinus carpio* L.) and catfish (*Clarias mossambicus* Peters). *Aquaculture*, **25**, 235–242.

Coughlan, M. P., Considine, P. J., O'Rorke, A., Moloney, A. P., Mehra, R. K., Hackett, T. J. & Puls, J. (1986). Energy from biomass: saccharification of beet pulp using fungal enzymes. *Annals of the New York Academy of Sciences*, **469**, 294–303.

Cui, F. & Dolphin, D. (1990). The role of manganese in model systems related to lignin biodegradation. *Holzforschung*, **44**, 279–283.

Cullen, D. (1997). Recent advances on the molecular genetics of ligninolytic fungi. *Journal of Biotechnology*, **53**, 273–289.

Danai, O., Levanon, D. & Silanikov, N. (1989). Cotton straw silage as a substrate for *Pleurotus* sp. cultivation. *Mushroom Science*, **12**, 81–90.

Datta, A., Bettermann, A. & Kirk, T. K. (1991). Identification of a specific manganese peroxidase among ligninolytic enzymes secreted by *Phanerochaete chrysosporium* during wood decay. *Applied and Environmental Microbiology*, **57**, 1453–1460.

De Ceuster, T. J. J. & Hoitink, H. A. J. (1999). Prospects for composts and biocontrol agents as substitutes for methyl bromide in biological control of plant diseases. *Compost Science and Utilization*, **7**, 6–15.

Di Lena, G., Patroni, E. & Quaglia, G. B. (1997). Improving the nutritional value of wheat bran by a white-rot fungus. *International Journal of Food Science Technology*, **32**, 513–519.

Dix, N. J. & Webster, J. (1995). *Fungal Ecology*. London: Chapman & Hall.

Duff, S. J. B. & Murray, W. D. (1996). Bioconversion of forest products industry waste cellulosics to fuel ethanol: a review. *Bioresource Technology*, **55**, 1–33.

Edwards, S. S. (1979). Central America: fungal fermentation of coffee waste. In *Appropriate Technology for Development: A Discussion and Case Histories*, eds. D. Evans & L. Adler, pp. 329–342. Boulder, CO: Westview Press.

Eriksson, K. E. L., Blanchette, R. A. & Ander, P. (1990). *Microbial and Enzymatic Degradation of Wood and Wood Components*. New York:

Springer-Verlag.

Ferreira, G., Boer, C. G. & Peralta, R. M. (1999). Production of xylanolytic enzymes by *Aspergillus tamarii* in solid state fermentation. *FEMS Microbiology Letters*, **173**, 335–339.

Gaskell, J., Stewart, P., Kersten, P. J., Covert, S. F., Reiser, J. & Cullen, D. (1994). Establishment of genetic linkage by allele-specific polymerase chain reaction: application to the lignin peroxidase gene family of *Phanerochaete chrysosporium. Biotechnology*, **12**, 1372–1375.

Gattinger, L. D., Duvnjak, Z. & Khan, A. W. (1990). The use of canola meal as a substrate for xylanase production by *Trichoderma reesei. Applied Microbiology and Biotechnology*, **33**, 21–25.

Glenn, J. K., Morgan, M. A., Mayfield, M. B., Kuwahara, M. & Gold, M. H. (1983). An extracellular H_2O_2-requiring enzyme preparation involved in lignin biodegradation by the white-rot basidiomycete *Phanerochaete chrysosporium. Biochemistry and Biophysical Research Communications*, **114**, 1077–1083.

Gold, M. H. & Alic, M. (1993). Molecular biology of the lignin-degrading basidiomycete *Phanerochaete chrysosporium. Microbiological Reviews*, **57**, 605–622.

Gomes, J., Purkarthofer, H., Hayn, M., Kapplmuller, J., Sinner, M. & Steiner, W. (1993). Production of a high level of cellulase-free xylanase by the thermophilic fungus *Thermomyces lanuginosus* in laboratory and pilot scales using lignocellulosic materials. *Applied Microbiology and Biotechnology*, **39**, 700–707.

Gorodecki, B. & Hadar, Y. (1990). Suppression of *Rhizoctonia solani* and *Sclerotium rolfsii* diseases in container media containing composted separated cattle manure and composted grape marc. *Crop Protection*, **9**, 271–274.

Grajek, W. (1988). Production of protein by thermophilic fungi from sugar-beet pulp in solid-state fermentation. *Biotechnology and Bioengineering*, **32**, 255–269.

Gutierrez, A., Caramelo, L., Prieto, A., Martinez, M. J. & Martinez, A. T. (1994). Anisaldehyde production and aryl-alcohol oxidase and dehydrogenase activities in ligninolytic fungi of the genus *Pleurotus. Applied and Environmental Microbiology*, **60**, 1783–1788.

Hadar, Y. & Gorodecki, B. (1991). Suppression of germination of sclerotia of *Sclerotium rolfsii* in compost. *Soil Biology and Biochemistry*, **23**, 303–306.

Hadar, Y., Kerem, Z., Gorodecki, B. & Ardon, O. (1992). Utilization of lignocellulosic waste by the edible mushroom, *Pleurotus. Biodegradation*, **3**, 189–205.

Hammel, K. E. & Moen, M. A. (1991). Depolymerization of a synthetic lignin *in vitro* by lignin peroxidase. *Enzyme and Microbial Technology*, **13**, 15–18.

Hammel, K. E., Jensen, K. A., Mozuch, M. D., Landucci, L. L., Tien, M. & Pease, E. A. (1993). Ligninolysis by a purified lignin peroxidase. *Journal of Biological Chemistry*, **268**, 12274–12281.

Hang, Y. D. & Woodams, E. E. (1998). Production of citric acid from corncobs by *Aspergillus niger. Bioresource Technology*, **65**, 251–253.

Hatakka, A. I. (1994). Lignin-modifying enzymes from selected white-rot fungi: production and role in lignin degradation. *FEMS Microbiology Reviews*, **13**, 125–135.

Hayes, W. A. & Neir, N. G. (1975). The cultivation of *Agaricus bisporus* and

other edible mushrooms,. In *The Filamentous Fungi*, eds. J. E. Smith and
D. R. Berry, pp. 212–248. London: Edward Arnold.

Higuchi, T. (1990). Lignin biochemistry: biosynthesis and biodegradation. *Wood Science Technology*, **24**, 23–63.

Hoitink, H. A. J. & Boehm, M. J. (1999). Biocontrol within the context of soil microbial communities: a substrate-dependent phenomenon. *Annual Review of Phytopathology*, **37**, 427–446.

Hoitink, H. A. J., Stone, A. G. & Han, D. Y. (1997). Suppression of plant diseases by composts. *HortScience*, **32**, 184–187.

Hong, K., Tanner, R. D., Malaney, G. W. & Danzo, B. J. (1989). Protein entrainment during baker's yeast fermentation on a semi-solid substrate in an air-fluidized bed fermenter. *Bioprocess Engineering*, **4**, 209–216.

Hoq, M. F. & Ernst, W. R. (1996). Influence of organic impurities on chlorine concentration in chlorine dioxide solutions. *Tappi Journal*, **79**, 128–136.

Inbar, Y., Chen, Y. & Hoitink, H. A. J. (1993). Properties for establishing standards for utilization of composts in container media. In *Science and Engineering of Composting*, eds. H. A. J. Hoitink & H. M. Keener, pp. 668–694. Worthington, OH: Renaissance.

Jecu, L. (2000). Solid state fermentation of agricultural wastes for endoglucanase production. *Industrial and Crop Production*, **11**, 1–5.

Jensen, K. A., Bao, W., Kawai, S., Srebotnic, E. & Hammel, K. H. (1996). Manganese-dependent cleavage of nonphenolic lignin structures by *Ceriporiopsis subvermispora* in the absence of lignin peroxidase. *Applied and Environmental Microbiology*, **62**, 3679–3686.

Jin, B., van Leeuwen, H. J., Patel, B. & Yu, Q. (1998). Utilization of starch processing wastewater for production of microbial biomass protein and fungal alpha-amylase by *Aspergillus oryzae*. *Bioresource Technology*, **66**, 201–206.

Kadowaki, M. K., Souza, C. G. M., Simao, R. C. G. & Peralta, R. M. (1997). Xylanase production by *Aspergillus tamarii*. *Applied Biochemistry and Biotechnology*, **66**, 97–106.

Kamra, D. N. & Zadrazil, F. (1986). Influence of gaseous phase, light and substrate pre-treatment on fruit-body formation, lignin degradation and in vitro digestibility of wheat straw fermented with *Pleurotus* spp. *Agricultural Wastes*, **18**, 1–17.

Kerem, Z. & Hadar, Y. (1998). Lignin-degrading fungi: mechanisms and utilization. In *Agricultural Biotechnology*, ed. A. Altman, pp. 351–365. New York: Marcel Dekker.

Kerem, Z., Friesem, D. & Hadar, Y. (1992). Lignocellulose degradation during solid-state fermentation: *Pleurotus ostreatus* versus *Phanerochaete chrysosporium*. *Applied and Environmental Microbiology*, **58**, 1121–1127.

Kersten, P. J. & Kirk, T. K. (1987). Involvement of a new enzyme, glyoxal oxidase, in extracellular H_2O_2 production by *Phanerochaete chrysosporium*. *Journal of Bacteriology*, **169**, 2195–2202.

Kirk, T. K. & Farrell, R. L. (1987). Enzymatic 'combustion': the microbial degradation of lignin. *Annual Review of Microbiology*, **41**, 465–505.

Kuwahara, M., Glenn, J. K., Morgan, M. A. & Gold, M. H. (1984). Separation and characterization of two extracellular H_2O_2-dependent oxidases from ligninolytic cultures of *Phanerochaete chrysosporium*. *FEBS Letters*, **169**, 247–250.

Kwok, O. C. H., Fahy, P. C., Hoitink, H. A. J. & Kuter, G. A. (1987).

Interactions between bacteria and *Trichoderma hamatum* in suppression of *Rhizoctonia* damping-off in bark compost media. *Phytopathology*, 77, 1206–1212.

Leisola, M. S. A., Kozulic, B., Meussdoerffer, F. & Fiechter, A. (1987). Homology among multiple extracellular peroxidases from *Phanerochaete chrysosporium. Journal of Biological Chemistry*, 262, 419–424.

Leonowicz, A., Trojanowski, J. & Orlicz, B. (1978). Induction of laccase in *Basidiomycetes:* apparent activity of the inducible and constitutive forms of the enzyme with phenolic substrates. *Acta Biochimica Polonica*, 25, 369–378.

Leonowicz, A., Szklarz, G. & Wojtas-Wasilewska, M. (1985). The effect of fungal laccase on fractionated lignosulphonates (Peritan Na). *Phytochemistry*, 24, 393–396.

Levanon, D., Danai, O. & Masaphy, S. (1988). Chemical and physical parameters in recycling organic wastes for mushroom production. *Biological Wastes*, 26, 341–348.

Lowe, D. A. (1992). Fungal enzymes. In *Handbook of Applied Mycology*, eds. D. K. Arora, R. P. Elander & K. G. Mukerji, pp. 681–706. New York: Marcel Dekker.

Makinen, K. K. & Isokangas, P. (1988). Relationship between carbohydrate sweeteners and oral diseases. *Progress Food Nutritional Science*, 12, 73–109.

Maldonado, M. C. & Strasser de Saad, A. M. (1998). Production of pectinesterase and polygalacturonase by *Aspergillus niger* in submerged and solid state systems. *Journal of Industrial Microbiology and Biotechnology*, 20, 34–38.

Martinez, A. T., Camarero, S., Guillen, F., Gutierrez, A., Munoz, C., Varela, E., Martinez, M. J., Barrasa, J. M., Ruel, K. & Pelayo, J. M. (1994). Progress in biopulping of non-woody materials – chemical, enzymatic and ultrastructural aspects of wheat straw delignification with ligninolytic fungi from the genus *Pleurotus. FEMS Microbiology Reviews*, 13, 265–274.

Martinez, M. J., Ruiz, D. F. J., Guillen, F. & Martinez, A. T. (1996). Purification and catalytic properties of two manganese peroxidase isoenzymes from *Pleurotus eryngii. European Journal of Biochemistry*, 237, 424–432.

Martinez-Carrera, D. (1987). Design of a mushroom farm for growing *Pleurotus* on coffee pulp. *Mushroom Journal for the Tropics*, 7, 13–23.

Mester, T. & Field, J. A. (1998). Characterization of a novel manganese peroxidase–lignin peroxidase hybrid isozyme produced by *Bjerkandera* species strain BOS55 in the absence of manganese. *Journal of Biological Chemistry*, 273, 15412–15417.

Meyer, P. S., du Preez, J. C. & Kilian, S. G. (1992). Chemostat cultivation of *Candida blankii* on sugar cane bagasse hemicellulose hydrolysate. *Biotechnology and Bioengineering*, 40, 353–358.

Michelena, V. V. & Castillo, F. J. (1984). Production of amylase by *Aspergillus foetidus* on rice flour medium and characterization of the enzyme. *Journal of Applied Bacteriology*, 56, 395–407.

Moreira, M. T., Feijoo, G., Sierraalvarez, R., Lema, J. & Field, J. R. A. (1997). Manganese is not required for biobleaching of oxygen delignified kraft pulp by the white rot fungus *Bjerkandera* sp. strain bos55. *Applied and Environmental Microbiology*, 63, 1749–1755.

Muller, J. (1987). Cultivation of the oyster mushroom, *Pleurotus ostreatus* (Jack. ex Fr.) Kummer, on cassia-substrate. *Mushroom Journal for the Tropics*, 7, 89–95.

Murray, W. D. & Richardson, M. (1993). Development of biological and process technologies for the reduction and degradation of pulp mill wastes that pose a threat to human health. *Critical Reviews in Environmental Science and Technology*, **23**, 157–194.

Nelson, E. B. & Hoitink, H. A. J. (1983). The role of microorganisms in the suppression of *Rhizoctonia solani* in container media amended with composted hardwood bark. *Phytopathology*, **73**, 274–278.

Nevalainen, H. & Penttilä, M. (1995). Molecular biology of cellulolytic fungi. In *The Mycota – Genetics and Biotechnology*, eds. K. Esser & P. A. Lemke, pp. 303–319. Berlin: Springer-Verlag.

Ngadi, M. O. & Correia, L. R. (1992). Kinetics of solid-state ethanol fermentation from apple pomace. *Journal of Food Engineering*, **17**, 97–116.

Nout, M. J. R., Rinzema, A. & Smits, J. P. (1997). Biomass and productivity estimates in solid substrate fermentations. In *The Mycota – Environmental and Microbial Relationships*, eds. K. Esser & P. A. Lemke, pp. 323–345. Berlin: Springer-Verlag.

Ohno, A., Ano, T. & Shoda, M. (1996). Use of soybean curd residue, okara, for the solid state substrate in the production of a lipopeptide antibiotic, iturin A, by *Bacillus subtilis* NB22. *Process Biochemistry*, **31**, 801–806.

Orue, C. & Bahar, S. (1985). Utilization of solid coffee waste as a substrate for microbial protein production. *Journal of Food Science Technology*, **22**, 10–16.

O'Toole, D. K. (1999). Characteristics and use of okara, the soybean residue from soy milk production – a review. *Journal of Agricultural Food Chemistry*, **47**, 363–371.

Oxenboll, K. (1994). *Aspergillus* enzymes and industrial uses. In *The Genus Aspergillus*, ed. K. A. Powell, pp. 147–158. New York: Plenum Press.

Pandey, A. (1992). Recent process development in solid state fermentations. *Process Biochemistry*, **27**, 109–117.

Paszczynski, A., Huynh, V. & Crawford, R. (1986). Comparison of ligninase-I and peroxidase-M2 from the white-rot fungus *Phanerochaete chrysosporium*. *Archives of Biochemistry and Biophysics*, **244**, 750–765.

Peñaloza, W., Molina, M. R., Brenes, R. G. & Bressani, R. (1985). Solid-state fermentation: an alternative to improve the nutritive value of coffee pulp. *Applied and Environmental Microbiology*, **49**, 388–393.

Pérez, J. & Jeffries, T. W. (1992). Roles of manganese and organic acid chelators in regulating lignin degradation and biosynthesis of peroxidases by *Phanerochaete chrysosporium*. *Applied and Environmental Microbiology*, **58**, 2402–2409.

Pfeifer, M. J., Silva, S. S., Felipe, M. G. A., Roberto, I. C. & Mancilha, I. M. (1996). Effect of culture conditions on xylitol production by *Candida guilliermondii* FTI 20037. *Applied Biochemistry and Biotechnology*, **57/58**, 423–430.

Roussos, S., Aquiahuatl, M. D. L. A., Trejo Hernandez, M. D. R., Gaime Perraud, I., Favela, E., Ramakrishna, M., Raimbault, M. & Viniegra Gonzalez, G. (1995). Biotechnological management of coffee pulp: isolation, screening, characterization, selection of caffeine-degrading fungi and natural microflora present in coffee pulp and husk. *Applied Microbiology and Biotechnology*, **42**, 756–762.

Ruiz Dueñas, F. J., Martinez, M. J. & Martinez, A. T. (1999). Molecular characterization of a novel peroxidase isolated from the ligninolytic fungus

Pleurotus eryngii. Molecular Microbiology, **31**, 223–235.

Schimenti, J., Garrett, T., Montenecourt, B. S. & Eveleigh, D. E. (1983). Selection of hypercellulolytic mutants of *Trichoderma reesei* based on resistance to nystatin. *Mycologia*, **75**, 876–880.

Schoemaker, H. E. & Leisola, M. S. A. (1990). Degradation of lignin by *Phanerochaete chrysosporium. Journal of Biotechnology*, **13**, 101–109.

Sharma, H. S. S. (1989). Economic importance of thermophilous fungi. *Applied Microbiology and Biotechnology*, **31**, 1–10.

Silanikove, N. & Levanon, D. (1986). Cotton straw: composition, variability and effect of anaerobic preservation. *Biomass*, **9**, 101–112.

Silman, R. W. (1980). Enzyme formation during solid substrate fermentation in rotating vessels. *Biotechnology and Bioengineering*, **22**, 411–420.

Silva, S. S., Ribeiro, J. D., Felipe, M. G. A. & Vitolo, M. (1997). Maximizing the xylitol production from sugar cane bagasse hydrolysate by controlling the aeration rate. *Applied Biochemistry and Biotechnology*, **63–65**, 557–564.

Soccol, C. R., Stonoga, V. I. & Raimbault, M. (1994). Production of L-lactic acid by *Rhizopus* species. *World Journal of Microbiology and Biotechnology*, **10**, 433–435.

Srinivasan, C., Dsouza, T. M., Boominathan, K. & Reddy, C. A. (1995). Demonstration of laccase in the white rot basidiomycete *Phanerochaete chrysosporium* BKM-f1767. *Applied and Environmental Microbiology*, **61**, 4274–4277.

Stevenson, F. J., Fitch, A. & Brar, M. S. (1993). Stability constants of Cu(II)-humate complexes: comparison of select models. *Soil Science*, **155**, 77–91.

Stewart, P., Kersten, P., Vanden Wymelenberg, A., Gaskell, J. & Cullen, D. (1992). Lignin peroxidase gene family of *Phanerochaete chrysosporium*: complex regulation by carbon and nitrogen limitation and identification of a second dimorphic chromosome. *Journal of Bacteriology*, **174**, 5036–5042.

Stredansky, M., Conti, E., Stredanska, S. & Zanetti, F. (2000). Gamma-linolenic acid production with *Thamnidium elegans* by solid-state fermentation on apple pomace. *Bioresource Technology*, **73**, 41–45.

Streeter, C. L., Conway, K. E., Horn, G. W. & Mader, T. L. (1982). Nutritional evaluation of wheat straw incubated with the edible mushroom *Pleurotus ostreatus. Journal of Animal Science*, **54**, 183–188.

Thambirajah, J. J., Zulkali, M. D. & Hashim, M. A. (1995). Microbiological and biochemical changes during the composting of oil palm empty-fruit-bunches. Effect of nitrogen supplementation on the substrates. *Bioresource Technology*, **52**, 133–144.

Thurston, C. F. (1994). The structure and function of fungal laccases. *Microbiology*, **140**, 19–26.

Tien, M. & Kirk, T. K. (1983). Lignin degrading enzyme from the hymenomycete *Phanerochaete chrysosporium* Burds. *Science*, **221**, 661–662.

Tolan, J. S., Olson, D. & Dines, R. E. (1995). Survey of xylanase enzyme usage in bleaching in Canada. *Pulp and Paper Canada*, **96**, 107–110.

Tuomela, M., Vikman, M., Hatakka, A. & Itavaara, M. (2000). Biodegradation of lignin in a compost environment: a review. *Bioresource Technology*, **72**, 169–183.

Ward, J. W. & Perry, T. W. (1982). Enzymatic conversion of corn cobs to glucose with *Trichoderma viride* fungus and the effect on nutritional value of the corn cobs. *Journal of Animal Science*, **54**, 609–617.

Wariishi, H., Valli, K., Rengathan, V. & Gold, M. H. (1989). Thiol-mediated oxidation of nonphenolic lignin model compounds by manganese peroxidase of *Phanerochaete chrysosporium*. *Journal of Biological Chemistry*, **264**, 14185–14191.

Wood, D. A. & Smith, J. F. (1987). The cultivation of mushrooms. In *Essays in Agricultural and Food Microbiology*, eds. J. R. Norris & G. L. Pettipher, pp. 309–343. Chichester, UK: Wiley & Sons.

Yaropolov, A. I., Skorobogatko, O. V., Vartanov, S. S. & Varfolomeyev, S. D. (1994). Laccase – properties, catalytic mechanism, and applicability. *Applied Biochemistry and Biotechnology*, **49**, 257–280.

Zadrazil, F. (1980). Conversion of different plant wastes into feed by basidiomycetes. *European Journal of Applied Microbiology and Biotechnology*, **9**, 243–248.

Zadrazil, F. & Kamra, D. N. (1997). Edible mushrooms. In *Fungal Biotechnology*, ed. T. Anke, pp. 14–25. London: Chapman & Hall.

Zadrazil, F. & Reinger, P. (1988). *Treatment of Lignocellulosics with White-rot Fungi*. Amsterdam: Elsevier Applied Science.

Zusman, I., Reifen, R., Livni, O., Smirnoff, P., Gurevich, P., Sandler, B., Nyska, A., Gal, R., Tendler, Y. & Madar, Z. (1997). Role of apoptosis, proliferating cell nuclear antigen and p53 protein in chemically induced colon cancer in rats fed corncob fiber treated with the fungus *Pleurotus ostreatus*. *Anticancer Research*, **17**, 2105–2113.

12

Cyanide biodegradation by fungi

MICHELLE BARCLAY
AND CHRISTOPHER J. KNOWLES

Introduction

The aim of this chapter is to review the biodegradation of cyanide and its metal complexes by fungi. However, since the degradation of cyanides by bacteria is in many ways similar to that of fungi, bacterial cyanide metabolism will also be considered. There are also many examples of the degradation and utilization of organic cyanides (nitriles) by both bacteria and fungi, although these are outside the scope of this article and will not be examined. For completion, the ability of fungi to produce cyanide (cyanogenesis) will be briefly discussed, as cyanogenic species have the ability to biotransform or biodegrade cyanide. Reviews that cover more specific aspects of microbial cyanide metabolism include Knowles (1976, 1988), Knowles & Bunch (1986), Raybuck (1992), and Dubey & Holmes (1995).

Cyanide chemistry and toxicity

The identification of cyanide as a poison in bitter almonds and cherry laurel leaves dates back to the early Egyptians (Sykes, 1981). Indeed, hydrogen cyanide (HCN) may account for more human deaths throughout history than any other toxin because of its use in executions and large-scale genocide during World War II (Way, 1981).

Hydrogen cyanide is one of the most rapidly acting metabolic inhibitors known, because of its universal inhibition of respiration. By binding to Fe^{3+} in cytochrome c oxidase, the terminal oxidase of the mitochondrial or bacterial respiratory chain, cyanide inhibits electron transfer to oxygen, and therefore respiration (Stryer, 1988). Other modes of cyanide toxicity include the inhibition of other metalloenzymes as well as some non-metalloenzymes, increased glucose catabolism and decreased thyroid activity (Solomonson, 1981). In humans, HCN can be absorbed through the skin as well as by inhalation. Inhalation of concentrations greater than $0.3\,ppm$ is toxic, whilst $1.0\,mg\,kg^{-1}$ body weight is toxic if injected or ingested (Kirk & Othermer, 1979). Brebion, Gabridine & Huroet (1986) reported that the lethal dose to animals is 9 mg kg^{-1} body weight. Concentrations of HCN in air of 20 ppm can cause slight symptoms (such as irritation of the mucus membranes around the eyes, nose and throat); 50 ppm causes disturbances within 30 to 60 minutes, whilst concentrations of 300 ppm are fatal unless treated promptly (Padiyar *et al.*, 1995). Toxicity, however, is dependent on cyanide speciation. It is free cyanide, (HCN/CN$^-$) and simple cyanides such as KCN that are the most toxic to living systems.

Cyanide in its free form is the weak volatile acid HCN, also known as hydrocyanic acid or prussic acid. It is a colourless gas or liquid with a boiling point of 25.7°C, and has a pK_a value of 9.36 (Chatwin, Zhang & Gridley, 1988). Simple cyanides such as KCN or NaCN easily dissociate to their ionized form, and at pH values below the pK_a of cyanide the majority of simple cyanides are in the form of HCN (Padiyar *et al.*, 1995). Cyanide can also complex to metals, in the form of $[M^{n+}(CN)_x]^{(x-n)-}$. Twenty-eight elements can react with cyanide to form 72 different metallocyanides covering a range of solubilities and stabilities (Fordsmith, 1964). The strongly complexed cyanides generally require heating, digestion and/or severe methods of decomposition to release free cyanide (Chapman, 1992). Weak complexes include those with cadmium, lead, nickel and zinc, which are stable under alkaline conditions but under neutral and acidic conditions the complex is only weakly ionized, resulting in the formation of free cyanide (Pohlandt, Jones & Lee, 1983). Hexacyanoferrates and hexacyanocobaltates are the strongest and, therefore, the most stable cyanide complexes (Chapman, 1992). The stability of these compounds means that they are less likely to dissociate to their free form and are less toxic in relation to other cyanide complexes. However, the relationship between simple/free and complex cyanides in water is dependent on factors such as pH, redox potential and metal concentration (Meeussen, Keizer & de Haan, 1992a).

The iron cyanides include some of the most stable of the metal cyanide complexes. Without the aid of vigorous agents they only release free cyanide as a result of photochemical decomposition. Meeussen *et al.* (1992a) found that although normally considered to be relatively non-toxic because of their stability, iron cyanide complexes (Fe(CN)$_6$$^{4-}$; Fe(CN)$_6$$^{3-}$) are only thermodynamically stable under conditions that would normally be considered extreme in soils and in the environment: high pH, low redox potential and high total cyanide concentration. As a result iron cyanide complexes in groundwater would normally dissociate to release free cyanide. However, further work (Meeussen *et al.* (1992b) showed that below pH 6 the concentration of iron cyanide was controlled by equilibrium with the mineral Prussian Blue (ferriferrocyanide; $Fe_4(Fe(CN)_6)_3$), the solubility of which is strongly dependent on pH and redox potential. Unlike the other iron cyanides, Prussian Blue is more stable at the low pH values that are typical of spent oxides found on former gaswork sites. In other words, iron cyanides are stable at high pH values (like other cyanide complexes) but dissociate at low pH, resulting in the formation of Prussian Blue, which is probably the most stable and least toxic of all the cyanide compounds.

The final cyanide species is the nitrile or organic cyanide RCN, where R is an aliphatic or aromatic carbon compound and is covalently bonded to the cyano (CN^-) group. Nitriles are used widely in industry as 'synthons': precursors in chemical reactions. These processes are based on the chemical properties of the nitriles, which can undergo acid or base hydrolysis to amides or organic acids and hydrogenation to amines under conditions of high pressure temperature. The resulting amides or acids are used extensively in industry including the production of polymers and plastics, animal and chicken feed, and medicaments and antiseptics.

Cyanogenesis

Despite its toxicity, a range of plants, fungi and bacteria are able to generate cyanide. Knowles & Bunch (1986) have discussed the early literature on this topic in detail. Various *Pseudomonas* spp. and *Chromobacterium violaceum* produce cyanide as a secondary metabolite from glycine in a process stimulated by methionine. Hydrogen cyanide is formed from the methylene group of glycine, which is decarboxylated to produce carbon dioxide by the action of the HCN synthase enzyme system:

$$NH_2CH_2COOH \rightarrow HCN + CO_2 + 4[H] \tag{12.1}$$

This enzyme system is membrane bound and requires an electron acceptor; it is presumably linked to respiratory activity. Once produced, cyanide can be bioconverted to β-cyanoalanine by cyanogenic bacteria (Macadam & Knowles, 1984):

$$HCN + \text{cysteine (or } o\text{-acetylserine)} \rightarrow \beta\text{-cyanoalanine}$$
$$+ H_2S \text{ (or acetate)} \tag{12.2}$$

Cyanogenesis by the fungus *Marasmius oreades* was first reported in 1871 (von Lösecke, 1871). It is a widespread phenomenon within the

Basidiomycetes, and also among the Ascomycetes and Zygomycetes. *M. oreades* produces cyanide from glycine in a manner analogous to the cyanogenic bacteria. However, the cyanide is not converted to β-cyanoalanine but to carbon dioxide, possibly via formamide and formate (Bunch & Knowles, 1980).

Several plant diseases involving fungi advance with the liberation of cyanide in the host plant species. For example, winter crown rot or snow mould disease is caused by a cyanogenic psychrophilic basidiomycete that attacks a range of forage plants including *Medicago sativa*. The copperspot disease of *Lutus corniculatus* is caused by *Stemphylium loti*, probably through the fungal β-glucosidase acting on cyanogenic glycosides of the host plant to release cyanide. *S. loti* is not cyanogenic but is able to degrade cyanide to formamide by a cyanide hydratase enzyme, presumably as a defence mechanism (Fry & Millar, 1972); this enzymic activity will be discussed in a later section of this chapter. The fungus *Thielaviopsis basicola* causes black root rot of tobacco. The disease is suppressed by a cyanogenic strain of *Pseudomonas fluorescens* CHAO (Voisard *et al.*, 1989; Laville *et al.*, 1998; Blumer *et al.*, 1999). The authors found that a cyanide-negative mutant, *P. fluorescens* CHA5, constructed by a gene replacement technique, was less able to protect the plant, while complementation of strain CHA5 by the cloned wild-type *hcn*[+] of CHAO restored the ability to suppress black root rot disease. Insertion of *hcn*[+] into a non-cyanogenic *P. fluorescens* strain, P3, enabled it to become cyanogenic and to protect against black root rot. It was concluded that bacterial cyanide production is an important but not the sole factor in suppression of black root rot.

Industrial sources of cyanide in the environment

A number of industries, both past and present, either use cyanide for specific process reasons or generate it as an undesirable consequence of the procedure. In all cases, cyanide waste is a toxic by-product that must be disposed of in a safe and environmentally sound manner. The carbonization of coal and oil, to produce town gas, was developed principally for heating and lighting at the beginning of the nineteenth century. The discovery of natural gas in the 1960s, which can be carried great distances via pipelines, resulted in the decline of town-gas production (Thomas & Lester, 1994). There are as many as 5000 former gaswork sites in the UK alone (Environment Resources Ltd, 1987), the size of which vary considerably from small plants attached to industrial premises to large plants with multiple gas holders. Town gas contained a variety of organic (phenols, polycyclic aromatic hydrocarbons and coal tars) and inorganic impurities (sulfides and cyanide) that had to be removed before the gas could be delivered. Many of these impurities were disposed of on-site, resulting in contamination of the soil and, ultimately, the groundwater.

The majority of the cyanide found on former gaswork sites is in the form of 'spent oxide', which was generated as part of the clean-up process for manufactured gas. Sulfides and cyanides were removed from the gas as it was passed through iron oxide/bog ore by complexation with the iron. The ore could be regenerated two or three times after exposure to the gas but would then become 'spent', hence the name spent oxide. These oxides contain high concentrations of cyanide and sulfide compounds, the latter being a rich source for sulfur-oxidizing bacteria, causing a significant decrease in the pH to around 2 or lower through the production of sulfate. Typical total cyanide content of the soil at gasworks containing tar, clinker or spent oxide is 4.8, 2.3 and $46.5 \, \mathrm{g \, kg^{-1}}$, respectively, the majority of which is complexed to metals. Moreover, a survey of 12 sites in New York state estimated that each site held $70\,000 \, \mathrm{m^3}$ spent oxide, containing up to $50.0 \, \mathrm{g \, kg^{-1}}$ cyanide (Radian Corp., 1987). Attempts by Young & Theis (1991) to determine free cyanide concentration in waste at gasworks were unsuccessful, probably because of the very low levels of free cyanide. There are a number of reasons for the apparent low concentration of free cyanide when the total cyanide concentration is high. Chatwin *et al.* (1988) investigated the natural mechanisms in soil that could result in decreased free cyanide. These authors identified a number of processes, including chelation, adsorption, precipitation, oxidation and volatilization, that occur in unsaturated soil. The low pH of soils containing spent oxide (owing to the action of sulfur-oxidizing bacteria), together with the high vapour pressure of the HCN that can form under these pH conditions, encourages volatilization. Meeussen *et al.* (Meeussen, 1992; Meeussen *et al.*, 1992a–c) have published a number of papers that show complexation between iron and cyanide, in particular the formation of Prussian Blue, as the dominant process in maintaining low concentrations of free cyanide. The bulk of cyanide found in spent oxide is complexed with iron in the form of ferri- or ferrocyanide, or more specifically Prussian Blue, giving spent oxide its characteristic blue colour (Thomas & Lester, 1993).

Both metal recovery and mining industries use cyanide-containing solutions to extract metals such as gold and silver from impure materials (Haddad & Rochester, 1988; Wild, Rudd & Neller, 1994). The ore is mixed with simple cyanides such as NaCN or KCN, which form soluble metal–cyanide complexes, thereby extracting the metal, which can then be recovered by precipitation. Cyanide is also used in the production of chemicals such as acrylonitrile, adiponitrile and methylmethacrylate (Wild *et al.*, 1994), used on a vast scale in the production of polymers. Xenobiotic halogenated aromatic nitriles such as bromoxynitrile, ioxynil and dichlorobenil are also amongst the most widely used pesticides and herbicides (Knowles & Wyatt, 1992). The electroplating industry produces some of the highest cyanide concentrations in wastewaters. The plating baths, containing cyanide to hold ions such as zinc and cadmium in solution, result in the production of wastewaters with total cyanide concentrations of up to $100 \, g l^{-1}$ (Kenfield *et al.*, 1988). Water from the gas scrubbers of blast furnaces in the steel and iron industry contains cyanide, as do the cooling waters. Cyanide is also formed as a by-product of coking in the iron, steel and petroleum industries (Wild *et al.*, 1994).

Cyanide biodegradation

Metabolism of cyanide by fungi

There are over 800 different plant species that produce cyanide when attacked by microorganisms (Mansfield, 1983). In response to the defence mechanism of cyanogenic plants, a number of phytopathogenic fungi have evolved the ability to convert cyanide enzymically to the less toxic formamide using cyanide hydratase (formamide hydrolyase EC 4.2.1.66). Cyanide hydratase was first characterized from *S. loti*, a pathogenic fungus of the cyanogenic plant birdsfoot trefoil (*L. corniculatus* L.) (Fry & Millar, 1972). It has since been identified in a number of other fungi including strains of *Fusarium solani* (Dumestre *et al.*, 1997a; Barclay, Tett & Knowles, 1998a), *Fusarium oxysporum* CCMI 876 (Pereira, Arrabaca & Amaral-Collaco, 1996), *Fusarium lateritium* (Cluness *et al.*, 1993), *Gloeocercospora sorghi* (Fry & Munch, 1975; Wang & Van Etten, 1992; Wang, Matthews & Van Etten, 1992) and most recently *Leptosphaeria maculans* (Sexton & Howlett, 2000). Fry & Millar (1972) noted that the enzyme responsible for the detoxification of cyanide was induced when spores of *S. loti* were incubated with cyanide and that the only product of enzyme activity was formamide. They named the enzyme formamide hydrolyase:

$$HCN + H_2O \rightarrow HCONH_2 \qquad (12.3)$$

In addition to the conversion of cyanide to formamide as a dead-end product by certain fungi, possibly as a detoxification route, some fungi can utilize cyanide as a nitrogen source for growth (Pereira *et al.*, 1996;

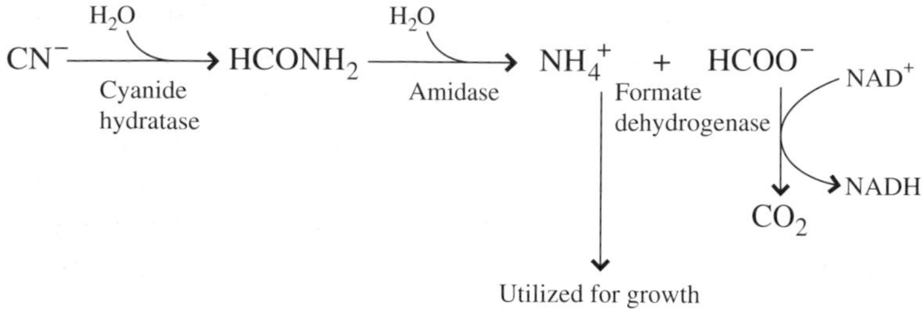

Fig. 12.1. Summary of the pathway for the utilization of cyanide by *Fusarium solani* as determined by Dumestre *et al.* (1997a) and Barclay *et al.* (1998a).

Dumestre *et al.*, 1997a; Barclay *et al.*, 1998b). Pereira *et al.* (1996) isolated three fungi at pH 8 that grew on cyanide as the sole source of nitrogen: *F. oxysporum*, *Trichoderma koningii* and *Gliocladium virens*. Further studies on *F. oxysporum* indicated that it converted cyanide to formamide, presumably by cyanide hydratase. Although not reported, formamide would have to be degraded to enable growth, which could occur via conversion to formate and ammonia, by formamidase.

Dumestre *et al.* (1997a) isolated three bacteria and one fungus, *F. solani* IHEM 8026, from contaminated alkaline wastes and soils. Although all the bacterial isolates could grow in the presence of cyanide at pH values up to 11, tolerance to cyanide was not attributed to the ability to degrade cyanide. It was more likely that the ability of these isolates to survive resulted from the presence of an alternative oxidase, allowing bacterial respiration to continue. However, *F. solani* IHEM 8026 had the capacity to grow under these conditions because of its ability to biodegrade cyanide in alkaline media (pH 9.2–10.7). Cyanide degradation by this fungus was associated with a large increase in biomass. Radiolabelling studies showed that cyanide was converted enzymically to formamide, which was subsequently converted to formic acid and ammonia. Ammonia was presumably utilized as a nitrogen source for growth while formic acid was converted to carbon dioxide, possibly by formate dehydrogenase (Fig. 12.1).

The *F. solani* strain isolated by Barclay *et al.* (1998b) has the ability to degrade and utilize as the sole source of nitrogen for growth both free and metal-complexed cyanides. The pathway for cyanide utilization by this strain of *F. solani*, under neutral and acidic conditions, was found to be identical to that of *F. solani* IHEM (Fig. 12.1) (Barclay *et al.*, 1998a). This isolate was obtained under acidic conditions (pH 4) from soil contaminated with Prussian Blue and other metal cyanide complexes and was part of a consortium comprising two fungi, *F. solani* and *Trichoderma polysporum*. A second *Fusarium* isolate was isolated under similar conditions as part of a consortium comprising three fungi: *F. oxysporum, Scytalidium thermophilium* and *Penicillium miczynski*. These two species of *Fusarium* utilized nickel and iron cyanide as the sole source of nitrogen under neutral and acidic conditions, respectively. Growth on metal–cyanide complexes by these two isolates appeared to relate to the stability of the complex. On the highly stable iron cyanide complex $K_4Fe(CN)_6$, complete removal of cyanide from the growth medium at pH 4 took up to 28 days, while no growth and no cyanide degradation was observed at pH 7, presumably because of the higher stability of this compound under the latter conditions. In comparison, growth on nickel cyanide ($K_2Ni(CN)_4$) and removal of cyanide from the growth medium was complete after 3–5 days at neutral pH (Barclay *et al.*, 1998b).

Cyanide hydratase has been purified from several fungi (Table 12.1). In each it is a large oligomeric protein with a native molecular mass of at least 300 kDa, made up of subunits of 43 to 45 kDa. The K_m of the enzyme is high (4–43 mmol l^{-1}), indicating a low affinity for cyanide, which may limit its commercial potential for treatment of cyanide-containing effluents. The affinity for metal–cyanide complexes is unknown. Comparison of nucleic and amino acid sequences of cyanide hydratase purified for *F. lateritium* and *G. sorghi* revealed that the enzymes are 65% homologous at the level of the gene, and the proteins are 75% identical (Cluness *et al.*, 1993). More recently, the gene for cyanide hydratase has been determined for *L. maculans*. The predicted amino acid sequence encoded by this gene shows that the enzyme has 77% and 82% homology *G. sorghi* and *F. lateritium*, respectively (Sexton & Howlett, 2000). Phylogenetic analysis carried out by Novo *et al.* (1995) grouped cyanide hydratase with the nitrilase enzymes and not the nitrile hydratase group.

The actual mechanism for metal cyanide degradation is unclear. It is not known whether cyanide hydratase acts upon the metal–cyanide complex as a whole, or the free cyanide (CN^-/HCN) that is released through dissociation of the complex. Similarly, it is not known whether the intact complex is taken up into the cell and then dissociates, or if dissociation occurs outside the cell and only free cyanide is taken up. Figure 12.2 shows some of the possible routes for the take up of cyanide/metal–cyanide complexes. There is some evidence that suggests it is the free cyanide and not the metal complex that the enzyme acts upon. First, the amino-terminal sequence of the enzyme purified from a strain of *F. solani* grown on $K_2Ni(CN)_4$ has strong homology to the amino-terminus of cyanide hydratase from both

Table 12.1. *Cyanide hydratase properties from differing fungi and a bacterium*

Species	pH optimum	Native molecular mass (kDa)	Subunit molecular mass (kDa)	K_m (mmol l^{-1})	Reference
Fusarium solani	7.5	> 300	45	4.7	Barclay et al., 1998a
Fusarium lateritium	8.5	> 300	43	43	Cluness et al., 1993
Gloeocercospora sorghi	7–8	> 300	45	12	Wang & Van Etten, 1992
Fusarium solani IHEM8026	7–8	–	–	–	Dumestre et al., 1997b
Fusarium oxysporum	–	–	–	–	Pereira et al., 1996
Stemphylium loti	7–9	> 600	–	–	Fry & Millar, 1972
Pseudomonas fluorescens NCIMB 11764	–	–	–	12	Kunz et al., 1994
Leptosphaeria maculans	–	–	–	–	Sexton & Howlett, 2000

–, not determined.

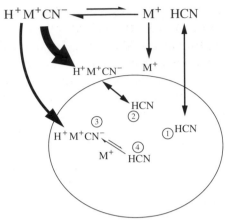

Fig. 12.2. Possible routes of cyanide hydratase activity in *Fusarium solani*. 1. The metal–cyanide complex ($H^+M^+CN^-$) undergoes partial dissociation outside the cell, free cyanide (HCN) then diffuses freely into the cell and cyanide hydratase is induced. The metal (M^+)is taken up and bound to the fungal cell wall. 2. The metal–cyanide complex is biosorbed onto the fungal cell wall; interaction between the cell wall and metal allows release of some cyanide, which diffuses into the cell inducing cyanide hydratase activity. 3. The metal–cyanide complex is taken into the cell, cyanide hydratase activity is induced and acts upon the complex as a whole. 4. Abiotic and/or biotic interactions result in dissociation of the complex allowing HCN to induce cyanide hydratase activity.

F. lateritium and *G. sorghi* (Barclay *et al.*, 1998a). Second, the rate of growth for both *F. solani* and *F. oxysporum* on metal cyanides is related to the stability of the metal–cyanide complex. Although the relationship between growth rate and stability of the metal–cyanide complex could be a consequence of the inability of the enzyme to act upon these stable compounds, it is possible that growth rate is limited by the availability of free cyanide produced from the dissociation of the complex (Barclay *et al.*, 1998b). More recent studies with reverse transcriptase polymerase chain reaction (RT-PCR) also show that the same gene is induced by both nickel and iron cyanide, as well as by KCN (M. Barclay, A. Tabouret, I. P. Thompson, C. J. Knowles & M. J. Bailey, unpublished observations). The fact that the different forms of cyanide induce the same gene suggests that the true substrate for the enzyme may be free cyanide.

Dumestre *et al.* (1997a) have also suggested that the pH optimum for cyanide hydratase activity indicates that the true substrate for the enzyme is HCN and not free cyanide ions, as indicated by pH activity studies

(Table 12.1). This is because of the pK_a of cyanide (9.36), which dictates that at higher pH values (above pH 9.36) the majority of cyanide is in the form of CN^- and not as hydrocyanic acid (HCN). The optimal pH for cyanide hydratase activity in a number of cyanide-degrading organisms was pH 7–9, as shown in Table 12.1. Dumestre *et al.* (1997a) suggested that decreased enzyme activity at pH 10.1 was caused by a lack of availability of substrate, as the enzyme was unable to act upon the free cyanide ions and not by decreased activity of the enzyme under alkaline conditions. Calculation of the initial reaction rates for HCN and CN^- also showed a direct relationship to the specific constants of K_{HCN} and K_{CN-}, again suggesting that decreased activity results from reduced availability of substrate under alkaline conditions (Dumestre, Bousserrhine & Berthelin, 1997b). By analogy with bacterial cyanide-degrading enzymes, it is possible that fungal cyanide hydratase can act on the cyanohydrins of simple keto compounds (Kunz, Chen & Pan, 1998). The utilization of cyanohydrins by bacteria is discussed below.

Cyanide-insensitive respiration

The nature of cyanide toxicity (inhibition of cytochrome oxidase in the mitochondrial or bacterial respiratory chain) dictates that cyanide-tolerant and cyanide-degrading organisms require an alternative mitochondrial respiratory pathway. Indeed, many organisms including plants (Solomos, 1977), some fungi (Edwards & Kwiecinski, 1973; Minagawa & Yoshimoto, 1986) as well as some protists (Chaudhuri, Ajoyi & Hill, 1995) possess cyanide-resistant alternative oxidases. The role of the alternative terminal oxidase in higher plants has been identified as part of the plants' ability to regulate their energy/carbon balance in response to changing environments (McIntosh, 1994). This protects the mitochondrion by preventing over-reduction of the respiratory chain and production of reactive oxygen species, which may cause organelle and cell damage (Day *et al.*, 1995).

In a number of fungi, the cyanide-resistant alternative oxidase is induced by cytochrome oxidase inhibitors, for example cyanide or antimycin A, as reported for *Hansenula anomala* (Minagawa & Yoshimoto, 1986) and *Neurospora crassa* (Edwards & Kwiecinski, 1973). Two explanations for the presence of cyanide-resistant respiration in fungi have been put forward. Lambers (1982) suggested that it acts as an energy overflow, as the activity of the alternative oxidase pathway appears to be largely regulated by the activity of the cytochrome pathway, that is the alternative oxidase only becomes active when the ubiquinone pool is reduced to 50–60% (Day *et al.*, 1995). An alternative, but related, theory is that it replaces the phosphorylating cytochrome pathway in the presence of cytochrome inhibitors (Vanlerberghe & McIntosh, 1997). Cytochrome oxidase inhibitors reduce the flow of electrons through the cytochrome pathway, thereby causing reduction of the ubiquinone pool and resulting in induction of the alternative oxidase. The alternative pathway has also been identified in cyanide-degrading fungi such as *F. oxysporum* (Pereira *et al.*, 1996) and *S. loti* (Rissler & Millar, 1977) and almost certainly provides energy for formamide hydrolyase/cyanide hydratase activity to remove cyanide and allow the 'normal' cytochrome oxidase pathway to operate. Although formation of alternative oxidases has not been reported in other cyanide-degrading fungi, they presumably exist to allow respiration and, therefore, energy production for the induction of cyanide-degrading enzymes and other cellular processes to occur in the presence of cytochrome inhibitors such as cyanide. The genes encoding alternative oxidase enzymes have been sequenced for a number of organisms and are highly conserved (Vanlerberghe & McIntosh, 1997).

Bacterial metabolism

The metabolic pathways for cyanide degradation by *P. fluorescens* NCIMB 11764 have been characterized. The bacterium was isolated by Harris & Knowles (1983a) through its ability to utilize cyanide as a sole source of nitrogen for growth. The authors established the presence of a cyanide oxygenase enzyme in cell-free extracts and found it to be dependent on NADH or NADPH (Harris & Knowles, 1983b; Dorr & Knowles, 1989):

$$NAD(P)H + 2H^+ + HCN + O_2 \rightarrow CO_2 + NH_4^+ + NAD(P)^+$$

$$(12.4)$$

Further studies by Kunz *et al.* (1992) showed concentration-dependent pathways for the degradation of cyanide by this bacterium. Three pathways were identified that involved the conversion of cyanide to ammonia and formate, ammonia and carbon dioxide, or formamide, respectively (Fig. 12.3). During aerobic growth on cyanide at concentrations of $0.5–10 \, \text{mmol} \, \text{l}^{-1}$, the cyanide oxygenase pathway was dominant. However, at higher cyanide concentrations (up to $100 \, \text{mmol} \, \text{l}^{-1}$ KCN), the principal metabolites were formamide, ammonia and formate. These products were also formed during incubation under anaerobic conditions, indicating that metabolism of cyanide to these metabolites was indepen-

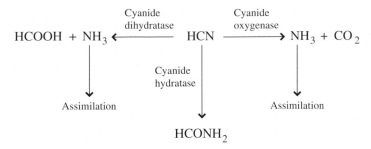

Fig. 12.3. Concentration-dependent pathways for cyanide degradation by *Pseudomonas fluorescens* NCIMB 11764 (Kunz *et al.*, 1992).

dent of oxygen. Because formamide and formate always appeared simultaneously in incubation mixtures and the bacterium was unable to metabolize the formamide that accumulated in the medium, separate pathways for the formation of formamide and for ammonia plus formate were proposed. Kunz, Wang & Chen (1994) concluded that the ability of *P. fluorescens* NCIMB 11764 to grow on cyanide is a result of the direct conversion of cyanide to ammonia, either via cyanide oxygenase or cyanide dihydratase, and not conversion to formamide. Further studies showed that the cyanide oxygenase enzyme had a lower K_m for KCN than the cyanide hydratase/cyanide dihydratase enzymes: 1.2 and $12 \, \text{mmol} \, l^{-1}$, respectively (Kunz *et al.*, 1994).

Cyanide oxygenase is the dominant enzyme at low cyanide concentrations, and the low K_m (relative to the other enzymes) indicates that this enzyme has a higher affinity to cyanide. Further evidence of the importance of this enzyme was obtained using mutant strains of *P. fluorescens* NCIMB 11764 that were unable to grow on cyanide. These mutants lost cyanide oxygenase activity but not the activity of cyanide hydratase/cyanide dihydratase. Radiolabelling studies using $^{18}O_2$ and $H_2^{18}O$ showed that one oxygen atom from O_2 and one from H_2O were incorporated in $C^{18}O_2$ produced during the metabolism of cyanide, confirming the action of a cyanide oxygenase enzyme in *P. fluorescens* NCIMB 11764 (Wang, Kunz & Venables, 1996). The presence of formate dehydrogenase also allows recycling of NADH, necessary for the conversion of cyanide to ammonia and carbon dioxide via cyanide oxygenase (Kunz *et al*, 1994).

Chen & Kunz (1997) have reported a non-enzymic mechanism for cyanide biotransformation in *P. fluorescens* NCIMB 11764, involving an iron-chelating molecule identified as a putative siderophore. Partially purified siderophore preparations removed cyanide at initial rates of up to

$7.6\,\mathrm{mmol\,min^{-1}\,mg^{-1}}$ protein. The authors reported that cyanide was oxidized to ammonia and carbon dioxide by the siderophore and they concluded that both cyanide oxygenase and a putative siderophore component are involved in cyanide utilization. Kunz *et al.* (1998) reported that cyanide utilization was linked to the secretion of keto acids (pyruvate and α-ketoglutarate) into the supernatant under nitrogen-limiting conditions. They suggested that these keto acids reacted with cyanide to give the corresponding cyanohydrin, which is then metabolized to give ammonia and carbon dioxide. The latter reaction was found to be dependent on oxygen and NADH, indicating that cyanide oxygenase is involved and that cyanohydrin and not cyanide is presumably the substrate for this enzyme. Again, mutant strains of *P. fluorescens* depleted of cyanide oxygenase activity that were unable to grow on cyanide were also unable to utilize cyanohydrins. However, the authors did not explain how cell-free extracts can catalyse the conversion of free cyanide to carbon dioxide and ammonia in the absence of keto acids, as reported by Harris & Knowles (1983b) and by Wang *et al.* (1996). Evidence for cyanide oxygenase activity has also been found in a strain of *Escherichia coli*. However, a reaction between glucose and cyanide, referred to as the Kiliani reaction, was necessary for cyanide degradation to take place (Figueira *et al.*, 1996). Hope and Knowles (1991) previously observed this phenomenon in the biodegradation of cyanide under anaerobic conditions by *Klebsiella planticola*.

In addition to KCN (i.e. $\mathrm{HCN/CN^-}$), *P. fluorescens* NCIMB 11764 can utilize cyanide complexes of nickel and copper as sources of nitrogen for growth (Rollinson *et al.*, 1987). Silva-Avalos *et al.* (1990) isolated 10 bacteria able to grow on $\mathrm{Ni(CN)_4}^{2-}$ as the sole source of nitrogen. There are few data on the mechanism of metal cyanide utilization by bacteria. Interestingly, Mudder & Whitlock (1984) have developed a full-scale commercial process to degrade metal-rich cyanide-containing wastes from tailings at the Homestake gold mine, USA, which contains *Pseudomonas paucimobilis* as the predominant bacterium. An *Acinetobacter* strain was isolated from gold mine effluent containing dicyanoaurate (Finnegan *et al.*, 1991). This strain utilized a wide range of metal–cyanide complexes, including those of silver, gold, cadmium, zinc, copper, cobalt and iron. The enzyme system involved is novel as it is expressed in an extracellular protein–lipid complex. Recently, the degradation kinetics of ferrous cyanide complexes by immobilized *P. fluorescens* have been reported (Dursun & Aksu, 2000).

The predominant enzyme reported in fungi is cyanide hydratase; there is one report of this enzyme in bacteria (Kunz *et al.*, 1992). The most common cyanide-degrading enzyme reported in bacteria that have the ability to grow on cyanide as the source of nitrogen is cyanide dihydratase (cyanidase), which catalyses:

$$HCN + 2H_2O \rightarrow NH_3 + HCOOH \qquad (12.5)$$

Meyers *et al.* (1993) purified cyanide dihydratase from *Bacillus pumilus* C1. They found that it consisted of three separate subunits (45.6, 44.6 and 41.2 kDa) with an overall molecular mass of 417 kDa and a K_m for cyanide of 2.56 mmol l^{-1}. Watanabe *et al.* (1998) purified a cyanide dihydratase from a strain of *Pseudomonas stutzeri* that was an aggregated protein made up of 38 kDa subunits. The K_m of this enzyme for cyanide was 1.7 mM. Adjei & Ohta (1999) described a strain of *Burkholderia cepacia* that also probably utilizes cyanide via the cyanide dihydratase pathway. Chapatwala *et al.* (1998) have isolated a *Pseudomonas putida* strain that they claimed could achieve high levels of growth on cyanide as the source of both nitrogen and carbon; growth was associated with release of ammonia and 10% of the carbon into the biomass. The pathway of carbon assimilation was not known but could be via formate as a methylotrophic energy source.

Molecular biology of cyanide degradation

The development of molecular biology techniques has significantly advanced our understanding of the mechanisms of cyanide biodegradation. The sequence homology that has been found between cyanide hydratase of *F. lateritium*, *G. sorghi* and *L. maculans* has been mentioned earlier. Site-directed mutagenesis and cloning of the gene for cyanide hydratase, *chy*, from *F. lateritium* into *E. coli* using the expression vector pGEX-2T revealed that there is an essential cysteine residue at position 163 (Brown, Turner & O'Reilly, 1995). Subsequent mutation of the protein at this position resulted in inactivation of the enzyme. Similar findings have been found for cyanide dihydratase in *P. stutzeri* AK61, where site-directed mutagenesis revealed that Cys-163 has an important role in cyanide dihydratase activity (Watanabe *et al.*, 1998). These findings indicate that the Cys-163 residue is part of the active site and is fundamental to the activity of the enzyme. Novo *et al.* (1995) noted that cyanide hydratase was related to the nitrilase enzymes, which convert nitriles to the corresponding acid and ammonia and which also have a cysteine residue acting as an active site nucleophile.

Wang, Sandrock & Van Etten (1999) have disrupted *chy* in *G. sorghi*, a pathogen of the cyanogenic plant *Sorghum*. These authors showed that gene disruption resulted in increased sensitivity of the fungus to cyanide (in the presence of 1 mmol l^{-1} cyanide the growth of fungal mycelia was only 10% that of the wild type); however, the inability to detoxify cyanide did not affect pathogenicity towards the plant. The authors suggested that if *G. sorghi* does have an alternative oxidase then it is highly sensitive to cyanide as their *chy* mutant had very little tolerance to HCN. It is also possible that although the alternative oxidase allows normal respiration to continue, cyanide hydratase activity is required to detoxify cyanide to prevent inhibition of other enzymes, such as metalloenzymes, so that normal cellular functions can continue. The fact that disruption of *chy* did not affect pathogenicity suggests that this is a function of this fungus that is not affected by cyanide.

Sexton and Howlett (2000) have demonstrated that the genome of the plant pathogenic fungus *L. maculans* contains one copy of the gene for cyanide hydratase. Analysis of the DNA sequence upstream from the gene has demonstrated that the cyanide hydratase promoter contains four putative sites for eukaryotic 'GATA proteins' thought to be involved in the regulation of nitrogen metabolism. Northern blotting and RT-PCR showed that the cyanide hydratase gene in this fungus was induced by potassium cyanide and propionitrile, but not by formamide or acetonitrile (Sexton & Howlett, 2000). Similarily, M. Barclay, I. P. Thompson, C. J. Knowles and M. J. Bailey (submitted for publication) have also recently shown genetic induction of cyanide hydratase in *F. solani*. Amplification of mRNA by RT-PCR using primers specifically designed for *chy* indicated that there was transcription expression 30–60 minutes after the fungus was exposed to cyanide. Prior to and up to 30 minutes after exposure to cyanide there was no evidence of *chy* transcription. Likewise, *chy* mRNA was not detected when cyanide was absent from the biotransformation medium. Metal cyanides such as $K_2Ni(CN)_4$ also induced *chy*. With both free and metal complexed cyanides, RT-PCR indicated that transcription of the gene stopped between 12 and 24 hours after exposure to cyanide. Although free cyanide was totally depleted by this time, metal–cyanide complexes were not so readily degraded and were present in the medium for up to 5 days. This evidence suggests that it is not substrate (cyanide) removal that results in termination of transcription.

Applications of fungi for bioremediation of cyanide-containing wastes

Fungi are particularly amenable for degradative processes because of the ability of individual strains to tolerate a wider range of environmental conditions than bacteria. They are tolerant to both high concentrations of metals and changes in pH. For example, Barclay *et al.* (1998b) reported that a strain of *F. solani* could degrade free and metal-complexed cyanides under neutral and acidic conditions. The ability of *F. solani* IHEM 8026 to degrade cyanide under alkaline conditions, albeit at a reduced rate, could also be advantageous to a bioremediation process by preventing the release of toxic cyanhydric acid through volatilization (Dumestre *et al.*, 1997a).

Patil & Paknikar (1999) have investigated the possibility of using fungal biomass for the biosorption and biodegradation of metal cyanides and found that *Cladosporium cladosporioides* was a highly efficient biosorbent of copper and nickel cyanides. They also tested a range of fungi (including *Fusarium* and *Aspergillus* strains) and found that optimal biosorption occurred at pH 4, while biosorption of metal cyanides did not occur above pH 6. Interestingly, they found that $K_2Ni(CN)_4$ was not biosorbed by either *F. oxysporum* or *Fusarium moniliform* at pH 4. In comparison, Barclay *et al.* (1998b) showed that iron–cyanide complexes were taken up by *F. solani* and *F. oxysporum* biomass at pH 4 during growth; nickel cyanides were taken up by *F. solani* biomass at pH 7 (unpublished results). Patil & Paknikar (1999) suggested that a combined approach using fungal biosorption and bacterial degradation could be used for the treatment of metal cyanide-containing wastewaters. Biosorption techniques maybe particularly useful for processes where recovery of the metal/metal cyanide complexes is desirable.

Work carried out in our laboratory also indicates that fungi, in particular strains of *Fusarium*, could be used for the remediation of cyanide- and metal cyanide-containing wastewaters and soil. Work to date has been carried out on aqueous solutions under neutral and acidic conditions and suggest that it would be feasible to treat cyanide-containing soils as a slurry in an *ex situ* process. Nazly *et al.* (1983) have also shown that *S. loti* mycelia could be immobilized and packed into columns for the treatment of cyanide-containing wastes. However, there is nothing to suggest that *in situ* remediation of cyanide/metal cyanide soils could not take place. Indeed, ICI Biological Products produced dried mycelia from *Fusarium lateritium* that can be sprayed onto cyanide-containing wastes (Richardson & Clarke, 1987).

There are several examples of biological treatment processes involving bacteria. For example, Mudder & Whitlock (1984) developed a process to treat wastes containing cyanides, thiocyanates and metal–cyanide complexes using the bacterium *P. paucimobilis* ATCC 39204 in an attached-growth aerobic biological treatment process developed at the lead operations plant of Homestake Mining Co. This process was further adapted to treat heavy metals under alkaline conditions (pH 9.5) (Whitlock, 1992).

In future, biological cyanide treatment processes may be improved by the development of molecular probes to monitor *in situ* bioremediation. Probes designed for specific genes, for example the functional genes involved in degradative pathways, could be used to assess the genetic potential of an environment (by examining DNA) and as a measure of the degradative activity (by extracting and quantifying mRNA). M. Barclay, I. P. Thompson, C. J. Knowles and M. J. Bailey have shown genetic induction of *chy* in fungi exposed to cyanide and metal cyanide complexes (unpublished results). The use of PCR would allow the presence of *chy* to be detected in the fungus/environment, while the application of RT-PCR to amplify mRNA would show whether the cyanide biodegradation system is active. Quantitative RT-PCR and the manipulation of soil conditions (pH, carbon source, etc.) could then be applied to optimize *in situ* bioremediation.

Conclusions

Despite the toxicity of cyanide, a number of organisms including plants, fungi and bacteria can tolerate and even degrade cyanide and its related compounds. This is largely a consequence of the production of cyanide as a defence mechanism not only by plants against attack and predation from plant pathogenic fungi but also by a number of bacteria and fungi. The advent of the industrial revolution led to the production of vast amounts of wastes containing numerous unwanted and toxic compounds, including cyanides, opening a niche for the exploitation of these cyanide degraders. The simplicity of cyanide and its natural occurrence in the environment suggests that bioremediation is ideal for the treatment of cyanide-containing wastes. Indeed, both bacteria and fungi can detoxify cyanide to ammonia and carbon dioxide via one or two simple enzymic steps (Kunz *et al.*, 1992; Barclay *et al.*, 1998a). However, fungi have the advantage over bacteria in that they can tolerate a wider range of pH conditions. Such fungi can degrade, and in some cases utilize as a source of nitrogen, both free and metal-complexed cyanide in both acidic and alkaline environments (Wang & Van Etten, 1992; Cluness *et al.*, 1993; Pereira *et al.*, 1996; Dumestre *et al.*, 1997a; Barclay *et al.*, 1998b). Modern techniques will help to further both our understanding of the structure and function of enzymes such as cyanide hydratase and the development of future technologies, for example, the development of molecular probes. Although, the mechanism involved in the degradation of free and metal-complexed cyanides is yet to be resolved, preliminary studies strongly suggest that the manipulation of fungi for the treatment of cyanide-containing waste could be a far simpler and environmentally friendly way of handling such toxins.

References

Adjei, M. D. & Ohta, Y. (1999). Isolation and characterization of a cyanide-utilizing *Burkholderia cepacia* strain. *World Journal of Microbiology and Biotechnology*, **15**, 699–704.

Barclay, M., Tett, V. A. & Knowles, C. J. (1998a). Metabolism and enzymology of cyanide/metallocyanide biodegradation by *Fusarium solani* under neutral and acidic conditions. *Enzyme and Microbial Technology*, **23**, 321–330.

Barclay, M., Hart, A., Knowles, C. J., Meeussen, J. C. L. & Tett, V. A. (1998b). Biodegradation of metal cyanides by mixed and pure cultures of fungi. *Enzyme and Microbial Technology*, **22**, 223–231.

Blumer, C., Heeb, S., Pessi, G. & Haas, D. (1999). Global GacA-steered control of cyanide and exoprotease production in *Pseudomonas fluorescens* involves specific ribosome binding sites. *Proceedings of the National Academy of Sciences USA*, **96**, 14073–14078.

Brebion, G., Gabridine, R. & Huroet, B. (1986). Biology of cyanides: possibilities of treatment of wastewater containing cyanides. *Eau*, **53**, 463–470.

Brown, D. T., Turner, P. D. & O'Reilly, C. (1995). Expression of the cyanide hydratase enzyme from *Fusarium lateritium* in *Escherichia coli* and identification of an essential cysteine residue. *FEMS Microbiology Letters*, **134**, 143–146.

Bunch, A. W. & Knowles, C. J. (1980). Cyanide production and degradation during growth of the snow mould fungus. *Journal of General Microbiology*, **116**, 9–16.

Chapatwala, K. D., Babu, G. R. V., Vijaya, O. K., Kumar, K. P. & Wolfram, J. H. (1998). Biodegradation of cyanides, cyanates and thiocyanates to ammonia and carbon dioxide by immobilized cells of *Pseudomonas putida*. *Journal of Industrial Microbiology and Biotechnology*, **20**, 28–33.

Chapman, D. (1992). *Water Quality Assessments: A Guide to the Use of Biota, Sediments and Water in Environmental Monitoring*. London: Chapman & Hall.

Chatwin, T. D., Zhang, J. & Gridley, G. M. (1988). Natural mechanisms in soil to mitigate cyanide releases. *Leachate Material Handling*, **24**, 467–473.

Chaudhuri M., Ajoyi, W. & Hill G. C. (1995). Identification and partial purification of a stage-specific 33 kDa mitochondrial protein as the alternative oxidase of the *Trypanosoma brucei* blood-stream trypomastigotes. *Journal of Eukaryotic Microbiology*, **42**, 467–472.

Chen, J. L. & Kunz, D. A. (1997). Cyanide utilization in *Pseudomonas fluorescens*

NCIMB 11764 involves a putative siderophore. *FEMS Microbiology Letters*, **156**, 61–67.

Cluness, M. J., Turner, P. D., Clements, E., Brown, D. T. & O'Reilly, C. (1993). Purification and properties of cyanide hydratase from *Fusarium lateritium* and analysis of the corresponding *chy1* gene. *Journal of General Microbiology*, **139**, 1807–1815.

Day, D. A., Whelan, J., Millar, A. H., Siedow, J. N. & Wiskich, J. T. (1995). Regulation of the alternative oxidase in plants and fungi. *Australian Journal of Plant Physiology*, **22**, 497–509.

Dorr, P. K. & Knowles, C. J. (1989). Cyanide oxygenase and cyanase activities of *Pseudomonas fluorescens* NCIMB 11764. *FEMS Microbiology Letters*, **60**, 289–294.

Dubey, S. K. & Holmes, D. S. (1995). Biological cyanide destruction mediated by microorganisms. *World Journal of Microbiology and Biotechnology*, **11**, 257–265.

Dumestre, A., Chone, T., Portal, J. M., Gerard, M. & Berthelin, J. (1997a). Cyanide degradation under alkaline conditions by a strain of *Fusarium solani* isolated from contaminated soils. *Applied and Environmental Microbiology*, **63**, 2729–2734.

Dumestre, A., Bousserrhine, N. & Berthelin, J. (1997b). Biodegradation des cyanures libres par le champignon *Fusarium solani*: relation avec le pH et la distribution des especes cyanurees en solution. *Academie des sciences, Geosciences de Surface*, **325**, 133–138.

Dursun, A. Y. & Aksu, Z. (2000). Biodegradation kinetics of ferrous (II) cyanide complex ions by immobilized *Pseudomonas fluorescens* in a packed bed column reactor. *Process Biochemistry*, **35**, 615–622.

Edwards, L. E. & Kwiecinski, F. (1973). Altered mitochondrial respiration in a chromosomal mutant of *Neurospora crassa*. *Journal of Bacteriology*, **116**, 610–618.

Environment Resources Ltd (1987). *Problems Arising from the Redevelopment of Gasworks and Similar Sites*, 2nd edn. London: HMSO.

Figueira, M. M., Ciminelli, V. S. T., de Andrade, M. C. & Linardi, V. R. (1996). Cyanide degradation by an *Escherichia coli* strain. *Canadian Journal of Microbiology*, **42**, 519–523.

Finnegan, I., Toerien, S., Abbot, F. & Raubenheimer, H. G. (1991). Identification and characterization of an *Acinetobacter* sp. capable of assimilation of a range of cyano-metal complexes, free cyanide ions and simple organic nitriles. *Applied Microbiology and Biotechnology*, **36**, 142–144.

Fordsmith, M. H. (1964). *The Chemistry of Complexed Cyanides: A Literature Survey*. London: HMSO.

Fry, W. E. & Millar, R. L. (1972). Cyanide degradation by an enzyme from *Stemphylium loti*. *Archives of Biochemistry and Biophysics*, **161**, 468–474.

Fry, W. E. & Munch, D. C. (1975). Hydrogen cyanide detoxification by *Gloeocercospora sorghi*. *Physiology and Plant Pathology*, **7**, 23–33.

Haddad, P. R. & Rochester, N. E. (1988). Ion-interaction reversed-phased chromatographic method for the determination of gold (I) cyanide in mine process liquors using automated sample preconcentration. *Journal of Chromatography*, **439**, 23–26.

Harris, R. E. & Knowles, C. J. (1983a). Isolation and growth of a *Pseudomonas* species that utilizes cyanide as a source of nitrogen. *Journal of General Microbiology*, **129**, 1005–1011.

Harris, R. E. & Knowles, C. J. (1983b). The conversion of cyanide to ammonia by extracts of a strain of *Pseudomonas fluorescens* that utilizes cyanide as a source of nitrogen for growth. *FEMS Microbiology Letters*, **20**, 337–341.

Hope, K. M. & Knowles, C. J. (1991). The anaerobic utilisation of cyanide in the presence of sugars by microbial cultures can involve an abiotic process. *FEMS Microbiology Letters*, **80**, 217–220.

Kenfield, C. F., Gin, R., Semmens, M. J. & Cussler, E. L. (1988). Cyanide recovery across hollow fibre gas membranes. *Environmental Science and Technology*, **22**, 1151–1155.

Kirk, B. S. & Othermer, D. F. (1979). *Encyclopedia of Chemical Technology*, Vol. 7, 3rd edn, ed. J. I. Kraschwitz. p. 133. New York: Wiley and Sons.

Knowles, C. J. (1976). Microorganisms and cyanide. *Bacteriological Reviews*, **40**, 652–680.

Knowles, C. J. (1988). Cyanide utilization and degradation by microorganisms. In *Ciba Foundation Symposium: Cyanide Compounds in Biology*, eds. D. Evered & S. Harnett. pp. 3–15. Chichester, UK: Wiley & Sons.

Knowles, C. J. & Bunch, A. W. (1986). Microbial cyanide metabolism. *Advances in Microbial Physiology*, **27**, 73–111.

Knowles, C. J. & Wyatt, J. W. (1992). The degradation of cyanide and nitriles. In *Society of General Microbiology Symposium*, No. 48, *Microbial Control of Pollution*, eds. J. C. Fry, G. M. Gadd, R. A. Herbert, C. J. Jones & I. A. Watson-Craik. pp. 113–128. Cambridge, UK: Cambridge University Press.

Kunz, D. A., Nagappan, O., Silva-Avalos, J. & DeLong, G. T. (1992). Utilization of cyanide as a nitrogenous substrate by *Pseudomonas fluorescens* NCIMB 11764: evidence for multiple pathways of metabolic conversion. *Applied and Envronmental Microbiology*, **58**, 2022–2029.

Kunz, D. A., Wang, C. & Chen, J. (1994). Alternative routes of enzymatic cyanide metabolism in *Pseudomonas fluorescens* NCIMB 11764. *Microbiology*, **140**, 1705–1712.

Kunz, D. A., Chen, J. L. & Pan, G. (1998). Accumulation of I-keto acids as essential components in cyanide assimilation by *Pseudomonas fluorescens* NCIMB 11764. *Applied and Environmental Microbiology*, **64**, 4452–4459.

Lambers, H. (1982). Cyanide-resistant respiration: a nonphosphorylating electron transport pathway acting as an energy overflow. *Physiologia Plantarum*, **55**, 478–485.

Laville, J., Blumer, C., von Schroetter, C., Gaia, V., Defago, G., Keel, C. & Haas, D. (1998). Characterisation of the *hcnABC* gene cluster encoding hydrogen cyanide synthase and anaerobic regulation by ANR in the strictly aerobic biocontrol agent *Pseudomonas fluorescens* CHAO. *Journal of Bacteriology*, **180**, 3187–3196.

Macadam, A. M. & Knowles, C. J. (1984). Purification and properties of cyano-L-alanine synthase from the cyanide-producing bacterium, *Chromobacterium violaceum*. *Biochimica et Biophysica Acta*, **786**, 123–132.

Mansfield, J. W. (1983). Antimicrobial compounds. In *Biochemical Plant Pathology*, ed. J. A. Callow, pp. 337–365. London: Wiley & Sons.

McIntosh, L. (1994). Molecular biology of the alternative oxidase. *Plant Physiology*, **105**, 781–786.

Meeussen, J. C. L. (1992). Chemical speciation and behaviour of cyanide in contaminated soils. PhD Thesis, Wageningen Agricultural University, Wageningen, the Netherlands.

Meeussen, J. C. L., Keizer, M. G. & de Haan, F. A. M. (1992a). The chemical stability and decomposition rate of iron cyanide complexes in soil solutions. *Environmental Science & Technology*, **26**, 511–516.

Meeussen, J. C. L., Keizer, M. G., van Riemsdijk, W. H. & de Haan, F. A. M. (1992b). Dissolution behaviour of iron cyanide (Prussian Blue) in contaminated soils. *Environmental Science and Technology*, **26**, 1832–1838.

Meeussen, J. C. L., Keizer, M. G. & Lukassen, W. D. (1992c). Determination of total and free cyanide in water after distillation. *Analyst*, **117**, 1009–1012.

Meyers, P. R., Rawlings, D. E., Woods, D. R. & Lindsey, G. G. (1993). Isolation and characterization of a cyanide dihydratase from *Bacillus pumilus* C1. *Journal of Bacteriology*, **175**, 6105–6112.

Minagawa, N. & Yoshimoto, A. (1986). Induction of cyanide-resistant respiration *in Hansenula anomala*. *Journal of Biochemistry*, **101**, 1141–1146.

Mudder, T. I. & Whitlock, J. L. (1984). Biological treatment of cyanidation waste waters. *Minerals and Metallurgical Processing*, 161–165.

Nazly, N., Knowles, C. J., Beardsmore, A., Naylor., W. T. & Corcoran, E. G. (1983). Detoxification of cyanide by immobilised fungi. *Journal of Chemical Technology and Biotechnology*, **B33**, 119–126.

Novo, C., Tata, R., Clementa, A. & Brown, P. R. (1995). *Pseudomonas aeruginosa* aliphatic amidase is related to the nitrilase/cyanide hydratase enzyme family and Cys[166] is predicted to be the active site nucleophile of the catalytic mechanism. *FEBS Letters*, **367**, 275–279.

Padiyar, V., Mishra, V. S., Mahajani, V. V. & Joshi, J. B. (1995). Cyanide detoxification – a review. *Transactions of the Institution of Chemical Engineers*, **73**, 33–51.

Patil, Y. B. & Paknikar, K. M. (1999). Removal and recovery of metal cyanides using a combination of biosorption and biodegradation processes. *Biotechnology Letters*, **21**, 913–919.

Pereira, P. T., Arrabaca, J. D. & Amaral-Collaco, M. T. (1996). Isolation, selection and characterization of a cyanide-degrading fungus from an industrial effluent. *International Biodeterioration and Biodegradation*, **45**, 45–52.

Pohlandt, C., Jones, E. A. & Lee, A. F. (1983). A critical evaluation of methods applicable to the determination of cyanides. *Journal of the South African Institute of Mining and Metallurgy*, **83**, 11–19.

Radian Corp. (1987). *Survey of Town Gas and By-product Production in the US*. [1800–1950 *US EPA* 600/7-85-004]. Washington, DC: US Environmental Protection Agency.

Raybuck, S. A. (1992). Microbes and microbial enzymes for cyanide degradation. *Biodegradation,* **3**, 3–18.

Richardson, K. R. & Clarke, P. M. (1987). Production of cyanide hydratase. European Patent 0 234 760.

Rissler, J. F. & Millar, R. L. (1977). Contribution of a cyanide-insensitive alternative respiratory system to increase in formamide hydrolyase activity and to growth in *Stemphylium loti in vitro*. *Plant Physiology*, **60**, 857–861.

Rollinson, G., Jones, R., Meadows, M. P., Harris, R. E. & Knowles, C. J. (1987). The growth of a cyanide-utilising strain of *Pseudomonas fluorescens* in liquid culture on nickel cyanide as a source of nitrogen. *FEMS Microbiology Letters,* **40**, 199–205.

Sexton, A. C. & Howlett, B. J. (2000). Characterisation of a cyanide hydratase gene in the phytopathogenic fungus *Leptosphaeria maculans*. *Molecular and General Genetics,* **263**, 463–470.

Silva-Avalos, J., Richmond, M. G., Nagappan, O. & Kunz, D. A. (1990).

Degradation of the metal-cyano complex tetranickelate (II) by cyanide utilizing bacterial isolates. *Applied and Environmental Microbiology*, **56**, 3664–3670.

Solomonson, L. P. (1981). Cyanide as a metabolic inhibitor. In *Cyanide in Biology*, eds. B. Vennesland, E. E. Conn, C. J. Knowles, J. Westley & F. Wissing, pp. 11–29. London: Academic Press.

Solomos, T. (1977). Cyanide-resistant respiration in higher plants. *Annual Review of Plant Physiology*, **28**, 279–297.

Stryer, L. (1988). Oxidative phosphorylation. In *Biochemistry*, 3rd edn, pp. 397–429. New York: Freeman.

Sykes, A. H. (1981). Early studies on the toxicology of cyanide. In *Cyanide in Biology*, eds. B. Vennesland, E. E. Conn, C. J. Knowles, J. Westley & F. Wissing. London: Academic Press.

Thomas, A. O. & Lester, J. N. (1993). Gaswork sites as sources of pollution and land contamination: an assessment of past and present public perceptions of their physical impact on the surrounding environment. *Environmental Technology*, **14**, 801–814.

Thomas, A. O. & Lester, J. N. (1994). The reclamation of disused gasworks sites: new solutions to an old problem. *Science of the Total Environment*, **152**, 239–260.

Vanlerberghe, G. C. & McIntosh, L. (1997). Alternative oxidase: from gene to function. *Annual Review of Plant Physiology and Plant Molecular Biology*, **48**, 703–734.

Voisard, C., Keel, C., Hass, D. & Defago, G. (1989). Cyanide production by *Pseudomonas fluorescens* helps suppress black root rot of tobacco under gnotobiotic conditions. *EMBO Journal*, **8**, 351–358.

von Lösecke, A. (1871). Archives de Pharmazie, **147**, 2–36.

Wang, C. S., Kunz, D. A. & Venables, B. J. (1996). Incorporation of molecular oxygen and water during enzymatic oxidation of cyanide by *Pseudomonas fluorescens* NCIMB 11764. *Applied and Environmental Microbiology*, **62**, 2195–2197.

Wang, P. & Van Etten, H. D. (1992). Cloning and properties of a cyanide hydratase gene from the phytopathogenic fungus *Gloeocercospora sorghi*. *Biochemical and Biophysical Research Communications*, **187**, 1048–1054.

Wang, P., Matthews, D. E. & Van Etten, H. D. (1992). Purification and characterization of cyanide hydratase from the phytopathogenic fungus *Gloeocercospora sorghi*. *Archives of Biochemistry and Biophysics*, **298**, 569–575.

Wang, P., Sandrock, R. W. & Van Etten, H. D. (1999). Disruption of the cyanide hydratase gene in *Gloeocercospora sorghi* increases its sensitivity to the phytoanticipin cyanide but does not affect its pathogenicity on the cyanogenic plant sorghum. *Fungal Genetics and Biology*, **28**, 126–134.

Watanabe, A., Yano, K., Ikebukuro, K. & Karube, I. (1998). Cyanide hydrolysis in a cyanide-degrading bacterium, *Pseudomonas stutzeri* AK61, by cyanidase. *Microbiology*, **144**, 1677–1682.

Way, J. L. (1981). Pharmacological aspects of cyanide and its antagonism. In *Cyanide in Biology*, eds. B. Vennesland, E. E. Conn, C. J. Knowles, J. Westley & F. Wissing, pp. 29–50. London: Academic Press.

Whitlock, J. L. (1992). Method for the biological removal of cyanides, thiocyanate and toxic heavy metals from highly alkaline environments. *US Patent* 5,169–532.

Wild, S. R., Rudd, R. & Neller, A. (1994). Fate and effects of cyanide during wastewater treatment processes. *Science of the Total Environment*, **156**, 93–107.

Young, T. C. & Theis, T. L. (1991). Determination of cyanide in manufactured gas plant purifier wastes. *Environmental Technology*, **12**, 1063–1069.

13

Metal transformations

GEOFFREY M. GADD

Introduction

Fungi are of fundamental importance as decomposer organisms and plant symbionts (mycorrhizas) and can comprise the largest pool of biomass (including other microorganisms and invertebrates) in the soil (Wainwright, 1988; Metting, 1992). They can be dominant in acidic conditions, where the mobility of toxic metals may be increased (Morley et al., 1996), and this, combined with their explorative filamentous growth habit and high surface area to mass ratio, ensures that fungi are integral bioactive components of major environmental cycling processes for metals and other elements including carbon, nitrogen, sulfur and phosphorus (Gadd & Sayer, 2000). There are examples where fungal isolates from soils with high metal contents exhibit higher metal tolerance than isolates from agricultural soils (Amir & Pineau, 1998), while adaptive and constitutive mechanisms of metal resistance are well known in free-living (Gadd, 1993a; Gadd & Sayer, 2000) and mycorrhizal fungi (Meharg & Cairney, 2000). Metals and their compounds, derivatives and radionuclides, interact with fungi in a variety of ways depending on the metal species, organism and environmental conditions, while fungal metabolism can dramatically influence speciation and, therefore, mobility and toxicity (Gadd, 1993a; Gadd & Sayer, 2000). Antagonistic effects between different metal species may also be a significant phenomenon in free-living (Amir & Pineau, 1998) and symbiotic fungi (Hartley et al., 1997). Solubilization mechanisms, for example complexation with organic acids, other metabolites and siderophores, can mobilize metals into forms available for cellular uptake and leaching from the system (Francis, 1994). Immobilization can result from sorption onto cell components or exopolymers, transport and intra- and extracellular sequestration or precipitation (Morley & Gadd, 1995; Gadd, 1996, 1997; Sayer & Gadd, 1997; Gadd & Sayer, 2000). These processes of metal solubilization and immobilization are key components of biogeochemical cycles for toxic metals, whether indigenous or introduced, and therefore fundamental determinants of fungal survival (Morley *et al.*, 1996; White, Sayer & Gadd, 1997; Ramsay, Sayer & Gadd, 1999): metals are directly and indirectly involved in all aspects of fungal growth, physiology and morphogenesis (Ramsay *et al.*, 1998). Further-more, several processes are of relevance to environmental bioremediation (White *et al.*, 1997; Gadd, 1997, 2000a; Gadd & Sayer, 2000). Almost all metal–microbe interactions have been examined in the context of environ-mental biotechnology as a means for removal, recovery or detoxification of inorganic and organic metal or radionuclide pollutants, and fungi have been important components of this research particularly in the areas of biosorption, heterotrophic leaching and methylation (Gadd, 1990, 2000a; Lovley & Coates, 1997; Losi & Frankenberger, 1997; Francis, 1998; Eccles, 1999; Gadd & Sayer, 2000). Note also that recent advances in the molecular biology and genetics of metal ion accumulation and intracellu-lar fate using yeast models are directly relevant to such processes in plants, and therefore relevant to phytoremediation (Eide, 1997; Guerinot & Eide, 1999). As in many other areas of bioremediation, much research on fungal metal bioremediation is laboratory based, with fewer developments to pilot/demonstration scale, and little commercial operation. However, there are some examples of large-scale bioremediation where microorgan-isms, including fungi, have been used to treat metalloid-contaminated soil (Thompson-Eagle & Frankenberger, 1992; Losi & Frankenberger, 1997) and it should be noted that microbial metal removal/transformation pro-cesses are intrinsic though less-appreciated components of traditional means of water/sewage treatment as well as reed bed, lagoon, wetlands and emerging phytoremediation technologies (Gadd, 2000a). This chapter seeks to outline some of the mechanisms by which fungi can interact with and transform toxic metal species between soluble and insoluble forms, the environmental significance of these processes, and their potential for bio-remediation. Fungal biosorption and heterotrophic leaching are detailed elsewhere in this volume and will only be mentioned briefly here.

Solubilization

Mechanisms of metal solubilization

Fungal solubilization of insoluble metal compounds, including certain oxides, phosphates, sulfides and mineral ores, occurs by several mechan-isms. Solubilization can occur by protonation of the anion of the metal compound, decreasing its availability to the cation, with the proton-translocating ATPase of the plasma membrane and the production of organic acids being sources of protons (Gadd, 1993a; Sayer, Raggett & Gadd, 1995; Karamushka, Sayer & Gadd, 1996; Morley *et al.*, 1996; Gadd & Sayer, 2000). In addition, organic acid anions are frequently capable of soluble complex formation with metal cations, thereby increasing mobility (White *et al.*, 1997). The incidence of metal-solubilizing ability among natural soil fungal communities appears to be high; in one study approximately one-third of the isolates tested were able to solubilize at least one of $Co_3(PO_4)_2$, ZnO or $Zn_3(PO_4)_2$, and approximately one-tenth were able to solubilize all three (Sayer *et al.*, 1995). A further mechanism of metal solubilization is the production of low-molecular-weight iron-chelating siderophores which solubilize iron(III). Siderophores are the most common means of acquisition of iron by bacteria and fungi, the most common fungal siderophore being ferrichrome (Crichton, 1991).

Environmental significance of metal solubilization by fungi

Solubilization of insoluble metal compounds is an important aspect of fungal physiology for the release of anions, such as phosphate, and essential metal cations into forms available for intracellular uptake and into biogeochemical cycles. Most phosphate fertilizers are applied in a solid form (e.g. calcium phosphate) and this has to be solubilized before it is available to plants and other organisms (Cunningham & Kuiack, 1992). In fact, increased phosphate uptake by mycorrhizal plants is believed to be a result of the high phosphate-solubilizing abilities of the mycorrhizas (Lapeyrie, Ranger & Vairelles, 1991). As well as phosphate, soil fungi, including mycorrhizas, may increase inorganic nutrient availability to plants and other microorganisms by increasing the mobility of essential metal cations, and other anions, such as sulfate (Gharieb, Sayer & Gadd, 1998; Gharieb & Gadd, 1999). Individual organic acids in the soil solution can exceed millimolar concentrations, with high concentrations being recorded in the vicinity of plant roots and fungal hyphae (Cromack *et al.*, 1979). It is known that weatherable minerals under many European coniferous forests contain a network of pores that are believed to be formed by fungal organic acids (Jongmans *et al.*, 1997). The pores (approximately 3–10 µm in diameter) sometimes contain hyphae and are widespread in feldspars and horneblendes from a variety of locations. Organic acid concentrations in the soil ranged from micromolar to millimolar levels and included citrate and oxalate (Jongmans *et al.*, 1997).

Solubilization of insoluble toxic metal compounds in the environment may have adverse effects if potentially toxic metal ions are released into soil and/or water systems from metal-contaminated locations. Metal–citrate complexes are highly mobile and not readily degraded, degradation depending on the type of complex formed rather than the toxicity of the metal cation involved. Hence, the presence of citric acid in the terrestrial environment will leach potentially toxic metals from soil (Francis, Dodge & Gillow, 1992). Some rock phosphate fertilizers contain cadmium, and solubilization of these to release the phosphate could also release cadmium ions, increasing its availability to the soil biota (Leyval, Surtiningish & Berthelin, 1993).

Significance of fungal–metal solubilization for bioremediation

As described above, many species of fungi are able to remove metals from industrial wastes and by-products, low-grade ores and metal-bearing minerals (Burgstaller & Schinner, 1993; Tzeferis, Agatzini & Nerantzis, 1994; Drever & Stillings, 1997; Sayer, Kierans & Gadd, 1997; Sayer & Gadd, 1997), and the relevance of this process to metal recovery and recycling and/or bioremediation is dealt with elsewhere (Chapter 14) (Burgstaller & Schinner, 1993; White et al., 1997; Brandl, Bosshard & Wegmann, 1999; Gadd, 1999). Although heterotrophic leaching by fungi can occur as a result of several processes, including the production of siderophores (in the case of iron), in most fungal strains, leaching appears to occur mainly by the production of organic acids (Burgstaller & Schinner, 1993; Sayer & Gadd, 1997; Dixon-Hardy et al., 1998; Gadd, 1999; Gadd & Sayer, 2000). If necessary, the resulting metal–citrate (or other metal–organic acid complexes) could eventually be degraded for ultimate metal recovery (Francis, 1994), while interaction of leaching technologies with biosorption is also a possibility (Chapter 14) (Gadd, 1993a; Tobin, White & Gadd, 1994).

Heterotrophic solubilization can also have consequences for other remedial treatments for contaminated soils. Pyromorphite ($Pb_5(PO_4)_3Cl$) is a stable lead mineral and can form in urban and industrially contaminated soils. Since such insolubility reduces lead bioavailability, the formation of pyromorphite has been suggested as a remediation technique for lead-contaminated land, if necessary by means of phosphate addition (Cotter-Howells, 1996). However, pyromorphite can be solubilized by phosphate-solubilizing fungi, for example Aspergillus niger, and plants grown with pyromorphite as a sole phosphorus source accumulated both phosphorus and lead (Sayer et al., 1999). Further, during the fungal transformation of pyromorphite, the biogenic production of lead oxalate dihydrate was observed for the first time (Sayer *et al.*, 1999). These phenomena emphasize the importance of considering microbial processes in developing remediation techniques for metal-contaminated soils. Such mechanisms of lead solubilization or immobilization as lead oxalate may have obvious consequences for metal mobility between environmental compartments and organisms.

Related to heterotrophic solubilization is fungal translocation, for example of caesium, zinc and cadmium, which can lead to metal or radionuclide concentration in the mycelium and/or in fruiting bodies. Whether the concentration factors observed *in vitro* can be reproduced in the field and whether such amounts can contribute to soil bioremediation remains uncertain (Gray, 1998). However, one study concluded that the fungal component of soil can immobilize the total Chernobyl radiocaesium fall-out received in upland grasslands (Dighton, Clint & Poskitt, 1991), though grazing of fruit bodies by animals may lead to radiocaesium transfer along the food chain (Bakken & Olsen, 1990). Another possibility relates to the use of copper-tolerant wood-decay fungi to degrade copper-treated wood products, although results obtained so far have shown only slight effects on the copper content of wood before or after decay (DeGroot & Woodward , 1998).

Immobilization

Metal immobilization by fungi may be metabolism independent, occurring whether the biomass is dead or alive, or metabolism dependent, comprising processes that sequester, precipitate, internalize or transform the metal species as well as the production of extracellular metabolites, both organic and inorganic (Gadd, 1993a; Morley *et al.*, 1996). Metal immobilization by fungi, including mycorrhizas (Chapter 15), is also of significance to plant productivity and, therefore, phytoremediation. Although some mycorrhizas are inhibited by toxic metals, some ecotypes (e.g. *Glomus claroideum*) appear to show adaptation to increased metal concentrations (Del Val, Barea & Azcon-Aguilar, 1999; see Meharg & Cairney, 2000). Despite the wide variation in responses (Hartley *et al.*, 1997; Bardi, Perotto & Bonfante, 1999), amelioration of metal phytotoxicity by mycorrhizas has been frequently described (Colpaert & van Tichelen, 1996; Kaldorf *et al.*, 1999; see Meharg & Cairney, 2000). It should be noted that some developing phytoremediation practices, such as choice of plant species and soil amendments, may have an impact on mycorrhizal diversity, reproduction and function (Pawlowska *et al.*, 2000). The sequestering ability of the fungal partner may reduce metal bioavailability and reduce or prevent translocation to the plant (Bradley, Burt & Read, 1981; Galli, Schuepp & Brunold, 1994; Martino *et al.*, 2000). This relationship, though, might not always be beneficial as it is possible that fungal accumulation of toxic metals might increase the apparent root metal concentration if soil physicochemical conditions alter or when the fungi die, degrade or otherwise release the accumulated metals (Morley *et al.*, 1996).

Physico-chemical mechanisms of metal immobilization

Fungal cell walls are complex macromolecular structures predominantly consisting of chitin, chitosan and glucans, but also containing other polysaccharides, proteins, lipids and pigments such as melanin (Peberdy, 1990; Gadd, 1993a). This variety of structural components ensures many different functional groups are able to bind metal ions to varying degrees depending on their chemical proclivities (Gadd, 1990; Bardi *et al.*, 1999). Such uptake of metals by fungal biomass is commonly encompassed within the term 'biosorption' (see Chapter 15). Fungi and their by-products have received considerable attention as possible biosorbent materials for metal-contaminated aqueous solutions, because of the ease with which they are grown and the availability of fungal biomass as an industrial waste product, for example *A. niger* (citric acid production) and *Saccharomyces cerevisiae* (brewing) (Gadd, 1990; Kapoor, Viraraghavan & Cullimore, 1999). Many studies have shown their efficacy in sorbing metal contaminants either as living or dead biomass, in pelleted whole-cell or dissembled forms, and as freely suspended or immobilized sorbents in batch and continuous processes (Tobin *et al.*, 1994; Mogollon *et al.*, 1998; Yetis *et al.*, 1998; Karamushka & Gadd, 1999; Lo *et al.*, 1999; Yin *et al.*, 1999; Zhou, 1999). They are also amenable to removal of the loaded metals by acids, alkalis or chelating agents, retaining a high percentage of their original sorption capacity after repeated regenerative stages (Gadd & White, 1992; Gadd, 1993a). However, although there have been several attempts to commercialize biosorption using microbial biomass, success has been limited and there appears to be no adoption of biosorption as a commercially viable treatment method to date. This lack of commercial development is somewhat perplexing, although the lack of specificity and lower robustness of biomass-based systems compared with ion-exchange resins is often cited as a reason (Eccles, 1999).

Physiological mechanisms of metal immobilization

Transport and intracellular fate

Many metals are essential for fungal growth and metabolism, and for metals such as sodium, magnesium, potassium, calcium, manganese, iron, cobalt, nickel, copper and zinc, mechanisms exist for their acquisition from the external environment by transport systems of varying specificity (Gadd, 1993a; Gadd & Sayer, 2000). Inessential toxic metals, although generally of low abundance in the environment unless redistributed by anthropogenic activities, can often compete with physiologically essential ions for such transport systems. Caesium, for example, competes for potassium transport systems and can substitute for it in potassium-regulated enzymes, with deleterious results (Perkins & Gadd, 1993, 1996; Avery, 1995). Most work on metal ion transport in fungi has concerned potassium and calcium ions in view of their importance in fungal growth, metabolism and differentiation, although the transport of other essential metal species has now received considerable attention, albeit primarily in the *S. cerevisiae* model system (Kosman, 1994; Gadd & Lawrence, 1996; Eide & Guerinot, 1997; Yamaguchilwai *et al.*, 1997; Zhao & Eide, 1997; Gitan *et al.*, 1998; Hassett, Romeo & Kosman, 1998; Jensen & Winge, 1998; Zhao *et al.*, 1998; Guerinot & Eide, 1999; Joshi, Serpe & Kosman, 1999; McDaniels *et al.*, 1999; Serpe, Joshi & Kosman, 1999; Bird *et al.*, 2000; Blackburn *et al.*, 2000). A number of intracellular fates are possible for transported metals, which include sequestration by metal-binding molecules and/or compartmentation in organelles such as the vacuole (Ramsay & Gadd, 1997; Gharieb & Gadd, 1998; Gadd & Sayer, 2000). In addition, intracellular metal concentrations may also be regulated by transport, including efflux mechanisms (Gadd, 1993a; Macreadie, Sewell & Winge, 1994; Gadd & Sayer, 2000), and regulated expression of genes involved in uptake, intracellular sequestration and protection against reactive oxygen intermediates (Winge *et al.*, 1997). Such mechanisms are involved in normal metal homeostasis within cells but also have a role in detoxification of potentially toxic metals. More detailed analysis of metal ion transport, intracellular compartmentation, molecular biology and regulation of gene expression is available in the literature (Gadd, 1993a; Kosman, 1994; Macreadie *et al.*, 1994; Eide & Guerinot, 1997; Eide, 1997, 1998; Winge *et al.*, 1997; Winge, Jensen & Srinivasan, 1998; Winge, 1998; Guerinot & Eide, 1999; Gadd & Sayer, 2000). Note that studies with *S. cerevisiae* can greatly aid understanding of metal metabolism in plants, which has applied implications regarding phytoremediation (Eide, 1997; Guerinot & Eide, 1999).

Intracellular metal-binding molecules

Specific, low-molecular-weight (6000–10 000) metal-binding proteins, termed metallothioneins, can be produced by animals, plants and microorganisms in response to toxic metals (Howe, Evans & Ketteridge, 1997). Other metal-binding proteins, phytochelatins and related peptides, containing glutamic acid and cysteine at the amino terminus, have been identified in plants, algae and several microorganisms (Rauser, 1995; Ha *et al.*, 1999; Cobbett, 2000). The metal-binding abilities of these and similar molecules may have potential for bioremediation (Ow, 1996).

For copper and cadmium, intracellular detoxification in fungi predominantly depends on their sequestration in the cytosol by induced metal-binding molecules (Hayashi & Mutoh, 1994; Macreadie *et al.*, 1994; Ow *et al.*, 1994; Rauser, 1995; Ow, 1996; Presta & Stillman, 1997). These include low-molecular-weight cysteine-rich proteins (metallothioneins (MT)) and peptides derived from glutathione (phytochelatins) (Mehra & Winge, 1991; Macreadie *et al.*, 1994; Ow *et al.*, 1994; Wu, Sung & Juang, 1995; Inouhe *et al.*, 1996; Hunter & Mehra, 1998; Ha *et al.*, 1999; Cobbett, 2000). The latter peptides have the general structure $(\gamma \text{Glu–Cys})_n$–Gly where n may be up to 11 (Ow *et al.*, 1994). As well as being termed phytochelatins, such peptides are also known as cadystins and metal γ-glutamyl peptides, although the chemical structure $(\gamma \text{EC})_n$G is a more precise description. Although $(\gamma \text{EC})_n$G induction has been reported with a wide variety of metals, including silver, gold, mercury, nickel, lead, tin and zinc, metal binding has only been shown for a few, primarily cadmium and copper (Ow *et al.*, 1994). For cadmium, two types of complex exist in *Schizosaccharomyces pombe* and *Candida glabrata*. A low-molecular-weight complex consists of $(\gamma \text{EC})_n$G and cadmium, whereas a higher-molecular-weight complex also contains acid-labile sulfide (Murasugi, Wada & Hayashi, 1983; Ow *et al.*, 1994). The $(\gamma \text{EC})_n$G–Cd–S^{2-} complex has greater stability and higher cadmium-binding capacity than the low-molecular-weight complex and has a structure consisting of a CdS crystallite core and an outer layer of glutathione and $(\gamma \text{EC})_n$G peptides (Dameron *et al.*, 1989; Torres-Martinez *et al.*, 1999). The higher binding capacity of the sulfide-containing complex confers a greater degree of tolerance to cadmium (Ow *et al.*, 1994). In *S. pombe*, evidence has also been presented for subsequent vacuolar localization of $(\gamma \text{EC})_n$G–Cd–S^{2-} complexes, illustrating a link between cytosolic sequestration and vacuolar compartmentation (Ortiz *et al.*, 1992, 1995; Ow, 1993, 1996). Finally, it should be noted that phytochelatin–CdS complexes can behave like semiconductor nano-crystallites, providing a further possibility for industrial applications (Bae & Mehra, 1998). Incorporation of sulfide into zinc–histidine resulted in histidine–ZnS nanocrystals, which possessed photocatalytic properties: efficient degradation of paraquat and *p*-nitrophenol was demonstrated in the presence of ultraviolet irradiation (Kho *et al.*, 2000). Glutathione- and cysteine-capped ZnS nanocrystals were also efficient in degrading test model compounds (Torres-Martinez *et al.*, 1999) and photoreducing the dyes Methyl Viologen, Basic Fuchsin and Naphthol Blue Black, respectively (Bae & Mehra, 1998).

Although the main function of *S. cerevisiae* MT (yeast MT) is cellular copper homeostasis, induction and synthesis of MT as well as amplification of the genes for MT leads to enhanced copper resistance in both *S. cerevisiae* and *C. glabrata* (Macreadie *et al.*, 1994). Production of MT has been detected in both copper- and cadmium-resistant strains of *S. cerevisiae* (Tohoyama *et al.*, 1995; Inouhe *et al.*, 1996; Presta & Stillman, 1997). However, *S. cerevisiae* cannot rely on MT synthesis as one of the copper-resistance mechanisms when grown with Cd^{2+} (Okuyama *et al.*, 1999). Relatively little work has been carried out on MT or $(\gamma EC)_n G$ peptides in filamentous fungi (see Gadd, 1993a; Galli *et al.*, 1994; Howe *et al.*, 1997).

Eukaryotic MTs, including yeast MT, and other metal-binding peptides have been expressed in *Escherichia coli* as fusions to membrane or membrane-associated proteins such as LamB, an outer membrane protein that functions as a coliphage surface receptor and is involved in maltose/maltodextrin transport. Such *in vivo* expression of MTs provides a means of designing biomass with specific and/or increased metal-binding properties (Pazirandeh, Wells & Ryan, 1998; Sousa *et al.*, 1998; Valls *et al.*, 1998; Chen *et al.*, 1999). Expression of yeast and mammalian MTs increased the ability of *E. coli* to bind Cd^{2+} 15–20-fold, this increase correlating well with the number of metal-binding centres contributed by the MT moiety of the protein fusions (Sousa *et al.*, 1998). Metal-binding peptides of sequences Gly-His-His-Pro-His-Gly (HP) and Gly-Cys-Gly-Cys-Pro-Cys-Gly-Cys-Gly (CP) were engineered into LamB protein and expressed in *E. coli*. Cd^{2+} : HP and Cd^{2+} : CP were 1 : 1 and 3 : 1 respectively. Surface display of CP increased the Cd^{2+}-binding ability of *E. coli* fourfold. Some competition of Cu^{2+} with Cd^{2+} for HP resulted from strong Cu^{2+} binding to HP, indicating that the relative metal binding affinities of inserted peptides and the wall to metal ion ratio were important in the design of peptide sequences and their metal specificities (Kotrba *et al.*, 1999). Another gene encoding for a *de novo* peptide sequence containing the metal-binding motif Cys-Gly-Cys-Cys-Gly was chemically synthesized and expressed in *E. coli* as a fusion with the maltose-binding protein. Such cells possessed enhanced binding of Cd^{2+} and Hg^{2+} compared with cells lacking the peptide (Pazirandeh *et al.*, 1998).

Related to the application of metal-binding molecules is the identification of genes encoding phytochelatin synthases, phytochelatins playing major roles in metal detoxification in plants and fungi (Clemens *et al.*, 1999; Ha *et al.*, 1999; Cobbett, 2000). This provides molecular evidence for the role of phytochelatin molecules in metal tolerance since heterologous expression of *PCS* genes dramatically enhanced metal tolerance. Future research with microorganisms and plants may allow testing of the potential of *PCS* genes for bioremediation (Clemens *et al.*, 1999). Additionally, in the area of phytoremediation, plant metal tolerance may be improved by transfer of fungal metal-resistance genes as well as by increasing phytoextractive properties (Ow, 1996; Bae & Mehra, 1997). For example, the yeast gene *MT* has been transferred into cauliflower, which enabled selection of a cadmium-tolerant transgenic cauliflower that accumulated more cadmium in the leaves than non-transformed plants (Hasegawa *et al.*, 1997).

As well as possible use in metal recovery, metal-binding molecules may also have applications for the detection and measurement of toxic metals ions in water. A rapid method for quantification of metal ions has been developed using a chemically synthesized phytochelatin as a mediator: the metal-binding property of the phytochelatin and quantification of the thiol group were used to measure metal ions at low concentrations (Satofuka *et al.*, 1999). Other mechanisms for metal immobilization within cells include precipitation, for example by reduction, sulfide production or association with polyphosphate (Gadd, 1990, 1993a).

Extracellular metal-binding molecules

A diverse range of specific (see above) and non-specific metal-binding compounds are produced by fungi, some of which are associated with exterior surfaces and/or released into the environment. The most well-known extracellular metal-binding compounds are siderophores, which are low-molecular-weight ligands (500–1000) possessing a high affinity for iron(III) (Neilands, 1981). They scavenge for iron(III) and complex and solubilize it, making it available to the organisms. Although primarily produced as a means of obtaining iron, siderophores are also able to bind other metals such as magnesium, manganese, chromium(III), gallium(III) and radionuclides such as plutonium(IV) (Birch & Bachofen, 1990). Non-specific extracellular metal-binding compounds range in size from small molecules such as organic acids and alcohols to macromolecules such as polysaccharides, and all may affect metal bioavailability and toxicity (Gadd & Griffiths, 1978). Extracellular polymeric substances are produced by many fungi, as well as by bacteria and algae, and may bind significant amounts of potentially toxic metals (Geesey & Jang, 1990). Extracellular polysaccharides may bind charged metal species, as well as adsorb or entrap particulate matter such as metal sulfides and oxides.

Oxalate production

Organic acids are released into the soil by both plant roots and fungal hyphae, with citric and oxalic acids being most commonly reported (see previously and Chapter 14) (Fox & Comerford, 1990; Jones & Kochian, 1996; Gadd, 1999). However, while most metal citrates are highly mobile, the production of oxalic acid by fungi provides a means of immobilizing soluble metal ions, or complexes, as insoluble oxalates (Chang, 1993; Sayer & Gadd, 1997; Gadd, 1999; Gadd & Sayer, 2000). *A. niger* can form oxalate crystals when grown on agar amended with a wide range of metal compounds including insoluble phosphates of calcium, cadmium, cobalt, copper, manganese, strontium and zinc (Sayer & Gadd, 1997) and pow-dered metal-bearing minerals (Sayer *et al.*, 1997; Gharieb, Sayer & Gadd, 1998; Gharieb & Gadd, 1999). The formation of oxalates containing potentially toxic metal cations may provide a mechanism whereby oxalate-producing fungi can tolerate environments containing high concentrations of toxic metals. Most metal oxalates are insoluble, some exceptions being those of sodium, potassium, lithium and iron (Strasser, Burgstaller & Schinner, 1994). Copper oxalate (moolooite) has been observed around hyphae growing on wood treated with copper as a preservative (Murphy & Levy, 1983; Sutter, Jones & Walchi, 1984). The copper appeared on the surface of the wood and around hyphae as copper oxalate, which was reported to be non-toxic because of its insolubility. Copper oxalate has also been observed in lichens growing on copper-rich rocks, where it is thought that the precipitation of copper oxalate could be a detoxification mechanism; up to 5% (dry weight) copper was fixed in the lichen thallus as copper oxalate (Purvis, 1984; Purvis & Halls, 1996). In contrast to the formation of toxic metal oxalates, however, there are many reports of the formation of calcium oxalate (whewellite, calcium oxalate monohydrate; weddellite, calcium oxalate dihydrate) (Gharieb & Gadd, 1999). Oxalate, ubiquitous in the terrestrial environment, can reach concentrations of 10^{-3}–10^{-6} mol l^{-1} in the soil (Allison, Daniel & Cornick, 1995). Calcium oxalate crystals are often found around free-living hyphae (Tait *et al.*, 1999) and around mycorrhizal roots, where they are thought to have a major role in calcium detoxification (Lapeyrie, Chilvers & Bhem, 1987; Jones *et al.*, 1992) as high concentrations of free calcium ions can be toxic (Gadd, 1993a, 1995).

Metal and metalloid transformations

Mechanisms of transformation

Several species of fungi, including unicellular and filamentous forms, can transform metals, metalloids and organometallic compounds by reduction, methylation and dealkylation, again processes of environmental and biotechnological importance since transformation of a metal or metalloid may modify its mobility and toxicity (Tamaki & Frankenberger, 1992; Thompson-Eagle & Frankenberger, 1992; Gadd, 1993b; Losi & Frankenberger, 1997). For example, methylated selenium derivatives are volatile and less toxic than inorganic forms while reduction of metalloid oxyanions, such as selenite or tellurite to amorphous elemental selenium or tellurium respectively, results in immobilization and detoxification (Thompson-Eagle & Frankenberger, 1992; Morley et al., 1996).

Reduction

Reduction of Ag(I) to Ag(0) during fungal growth on AgNO$_3$-containing media results in blackened colonies with metallic silver(0) precipitated in and around cell walls (Kierans *et al.*, 1991). Both enzymic and non-enzymic copper(II)-reducing systems have been purified from *Debaryomyces hansenii* cell walls, with the enzyme being involved in the control of copper(II) uptake (Wakatsuki *et al.*, 1991a,b). In fact, it is now known that reductive processes, for example, of copper(II) to copper(I) and iron(III) to iron(II) are integral prerequisites for high-affinity transport of these metals (Kosman, 1994; Lesuisse & Labbe, 1995; Eide & Guerinot, 1997). Reduction of mercury(II) to mercury(0) by fungi has also been demonstrated (Yannai, Berdicevsky & Duek, 1991) but, in contrast to bacteria, there is little detailed characterization of this system. The ability of fungi to reduce metalloids is perhaps more clearly demonstrated. Reduction of selenate (Se^{VI}) and selenite (Se^{IV}) to elemental selenium can be catalysed by numerous fungal species, resulting in a red coloration of colonies (Ramadan *et al.*, 1988). Both extracellular and intracellular deposition of selenium(0) has been demonstrated (Gharieb, Wilkinson & Gadd, 1995). Tellurite (TeO_3^{2-}) reduction to tellurium(0) results in black or dark grey colonies (Smith, 1974). In a *Fusarium* sp. and *Penicillium citrinum*, transmission electron microscopy revealed the deposition of large black granules, apparently in vacuoles, which corresponded with the reduction of tellurite to elemental tellurium (Gharieb, Kierans & Gadd, 1999).

Methylation

The biological methylation (biomethylation) of metalloids has been demonstrated in filamentous fungi and yeasts, and this frequently results in their volatilization (Gadd, 1993b). Arsenic and selenium have received most attention with the biochemical pathway for the fungal production of trimethylarsine from arsenite first being suggested by Challenger (1945). Subsequent studies have shown that several fungi, such as *Gliocladium roseum*, *Candida humicola* and *Penicillium* sp., can convert monomethylarsenic acid to trimethylarsine (Huysmans & Frankenberger, 1991; Tamaki & Frankenberger, 1992). The pathway for arsenic methylation involves the transfer of methyl groups as carbonium ions (CH_3^+) by S-adenosylmethionine (Tamaki & Frankenberger, 1992; Gadd, 1993b). Numerous fungi can convert both selenite (SeO_3^{2-}) and selenate (SeO_4^{2-}) to methyl derivatives such as dimethylselenide ($(CH_3)_2Se$) and dimethyldiselenide ($(CH_3)_2Se_2$) (Thompson-Eagle, Frankenberger & Karlson, 1989; Thompson-Eagle, Frankenberger & Longley, 1991; Brady, Tobin & Gadd, 1996). Inorganic forms of selenium (SeO_3^{2-}, SeO_4^{2-}) appear to be methylated more rapidly than organic forms, such as selenium-containing amino acids, with the mechanism for selenium methylation appearing to be similar to that for arsenic (Gadd, 1993b). The ability of fungi, as with bacteria, to transform metals and metalloids has been utilized in the bioremediation of contaminated land and water. Selenium methylation results in volatilization, a process that has been used to remove selenium from the San Joaquin Valley and Kesterson Reservoir, California, using evaporation pond management and primary pond operation (Thompson-Eagle *et al.*, 1991; Thompson-Eagle & Frankenberger, 1992). Incoming selenium-contaminated drainage water was evaporated to dryness and the process repeated until the sediment selenium concentration approached $100 \, mg \, kg^{-1}$ on a dry weight basis. The volatilization process was then optimized until the selenium concentration in the sediment fell to acceptable limits, using parameters such as carbon source, moisture, temperature and aeration (Thompson-Eagle & Frankenberger, 1992). There are few detailed studies on fungal biomethylation of other metals and metalloids. Mercury biomethylation by fungal species has been reported (Yannai *et al.*, 1991), while there is evidence of dimethyltelluride ($(CH_3)_2Te$) and dimethylditelluride ($(CH_3)_2Te_2$) production from both TeO_3^{2-} and TeO_4^{2-} by a *Penicillium* sp. (Huysmans & Frankenberger, 1991). Although the production of volatile tellurium by a *Fusarium* sp. occurred over the entire growing period, amounts were extremely small, indicating that this process may not be an important detoxification mechanism (Gharieb *et al.*, 1999).

Dealkylation

Fungal organometal transformations may be envisaged for the removal of alkylleads or organotins from water (Macaskie & Dean, 1987; 1990; Gadd, 2000b). Degradation of organometallic compounds can be carried out by fungi, either by direct biotic action (enzymes) or by facilitating abiotic degradation, for instance by alteration of pH and excretion of metabolites. Organotin compounds, such as tributyltin oxide and tributyltin naphthenate, may be degraded to mono- and dibutyltins by fungal action, inorganic tin(II) being the ultimate degradation product (Barug, 1981; Orsler & Holland, 1982). Organomercury compounds may be detoxified by conversion to mercury(II) by fungal organomercury lyase, the mercury(II) being subsequently reduced to mercury(0) by mercuric reductase; this system is broadly analogous to that found in mercury-resistant bacteria (Tezuka & Takasaki, 1988). Trimethyllead degradation has been demonstrated in an alkyllead-tolerant yeast (Macaskie & Dean, 1987) and in the wood-decay fungus *Phaeolus schweintzii* (Macaskie & Dean, 1990).

Conclusions

Mechanisms of fungal solubilization and immobilization of metal(loid)s, radionuclides and related substances are of potential for bioremediation. While biosorption has received little development in an industrial context, work on metal leaching from contaminated matrices, metal(loid) transformation and bioprecipitation shows promise of 'field' development as well as providing fundamental scientific insights into metal–microbe interactions. In addition to the biotechnological significance, it should be emphasized that this work also provides understanding of the roles of fungi in affecting metal mobility and transfer between different biotic and abiotic locations and their importance in the biogeochemistry of metal(loid) cycling in the environment.

Metal transformations 373

Acknowledgements

G. M. G. gratefully acknowledges financial support from the BBSRC (GR/J48214, 94/BRE13640, 94/MAF12243, 94/SPC02812), and NATO (Envir.Lg.950387 Linkage Grant) for some of the work described.

References

Allison, M. J., Daniel, S. L. & Cornick, N. A. (1995). Oxalate degrading bacteria. In Calcium Oxalate in Biological Systems, ed. S. R. Khan, pp. 131–168. Boca Raton, FL: CRC Press.

Amir, H. & Pineau, R. (1998). Effects of metals on the germination and growth of fungal isolates from New Caledonian ultramafic soils. *Soil Biology and Biochemistry*, **30**, 2043–2054.

Avery, S. V. (1995). Caesium accumulation by microorganisms – uptake mechanisms, cation competition, compartmentalization and toxicity. *Journal of Industrial Microbiology*, **14**, 76–84.

Bae, W. & Mehra, R. K. (1997). Metal-binding characteristics of a phytochelatin analog (Glu–Cys)$_2$Gly. *Journal of Inorganic Biochemistry*, **68**, 201–210.

Bae, W. & Mehra, R. K. (1998). Cysteine-capped ZnS nanocrystallites: preparation and characterization. *Journal of Inorganic Biochemistry*, **70**, 125–135.

Bakken, L. R. & Olsen, R. A. (1990). Accumulation of radiocaesium in fungi. *Canadian Journal of Microbiology*, **36**, 704–710.

Bardi, L., Perotto, S. & Bonfante, P. (1999). Isolation and regeneration of protoplasts from two strains of the ericoid mycorrhizal fungus *Oidiodendron maius*: sensitivity to chemicals and heavy metals. *Mycological Research*, **154**, 105–111.

Barug, G. (1981). Microbial degradation of bis(tributyltin) oxide. *Chemosphere*, **10**, 1145–1154.

Birch, L. & Bachofen, R. (1990). Complexing agents from microorganisms. *Experientia*, **46**, 827–834.

Bird, A., Evans Galea, M. V., Blankman, E., Zhao, H., Luo, H., Winge, D. R. & Eide, D. J. (2000). Mapping the DNA binding domain of the Zap1 zinc-responsive transcriptional activator. *Journal of Biological Chemistry*, **275**, 16160–16166.

Blackburn, N. J., Ralle, M., Hassett, R. & Kosman, D. J. (2000). Spectroscopic analysis of the trinuclear cluster in the Fet3 protein from yeast, a multinuclear copper oxidase. *Biochemistry*, **39**, 2316–2324.

Bradley, R., Burt, A. J. & Read, D. J. (1981). Mycorrhizal infection and resistance to heavy metals. *Nature*, **292**, 335–337.

Brady, J. M., Tobin, J. M. & Gadd, G. M. (1996). Volatilization of selenite in aqueous medium by a *Penicillium* species. *Mycological Research*, **100**, 955–961.

Brandl, H., Bosshard, R. & Wegmann, M. (1999). Computer-munching microbes: metal leaching from electronic scrap by bacteria and fungi. *Process Metallurgy*, **9B**, 569–576.

Burgstaller, W. & Schinner, F. (1993). Leaching of metals with fungi. *Journal of Biotechnology*, **27**, 91–116.

Challenger, F. (1945). Biological methylation. *Chemical Reviews*, **36**, 15–61.

Chang, J. C. (1993). Solubility product constants. In *CRC Handbook of Chemistry and Physics*, Sect. 8, ed. D. R. Lide, p. 39. Boca Raton, FL: CRC Press.

Chen, W., Brühlmann, F., Richins, R. D. & Mulchandani, A. (1999). Engineering of improved microbes and enzymes for bioremediation. *Current Opinion in Biotechnology*, **10**, 137–141.

Clemens, S., Kim, E. J., Neumann, D. & Schroeder, J. I. (1999). Tolerance to toxic metals by a gene family of phytochelatin synthases from plants and yeast. *EMBO Journal*, **18**, 3325–3333.

Cobbett, C. S. (2000). Phytochelatin biosynthesis and function in heavy-metal detoxification. *Current Opinion in Plant Biology*, **3**, 211–216.

Colpaert, J. V. & van Tichelen, K. K. (1996). Mycorrhizas and environmental stress. In *Fungi and Environmental Change*, eds. J. C. Frankland, N. Magan and G. M. Gadd, pp. 109–128. Cambridge, UK: Cambridge University Press.

Cotter-Howells, J. (1996). Remediation of contaminated land by formation of heavy metal phosphates. *Applied Geochemistry*, **11**, 335–342.

Crichton, R. R. (1991). Inorganic biochemistry of iron metabolism. Chichester, UK: Ellis Horwood.

Cromack, K. Jr, Sollins, P., Graustein, W. C., Speidel, K., Todd, A. W., Spycher, G., Li, C. Y. & Todd, R. L. (1979). Calcium oxalate accumulation and soil weathering in mats of the hypogeous fungus *Hysterangium crassum*. *Soil Biology and Biochemistry*, **11**, 463–468.

Cunningham, J. E. & Kuiack, C. (1992). Production of citric and oxalic acids and solubilization of calcium phosphates by *Penicillium bilaii*. *Applied and Environmental Microbiology*, **58**, 1451–1458.

Dameron, C. T., Reese, R. N., Mehra, R. K., Kortan, A. R., Carrol, J. P., Steigerwald, M. L., Brus, L. E. & Winge, D. R. (1989) Biosynthesis of cadmium sulfide quantum semiconductor crystallites. *Nature*, **338**, 596–597.

DeGroot, R. C. & Woodward, B. (1998). *Wolfiporia cocos* – a potential agent for composting or bioprocessing Douglas-fir wood treated with copper-based preservatives. *Material und Organismen*, **32**, 195–215.

Del Val, C., Barea, J. M. & Azcon-Aguilar, C. (1999). Assessing the tolerance to heavy metals of arbuscular mycorrhizal fungi isolated from sewage sludge-contaminated soils. *Applied Soil Ecology*, **11**, 261–269.

Dighton, J., Clint, D. M. & Poskitt, J. (1991). Uptake and accumulation of ^{137}Cs by upland grassland soil fungi: a potential pool of Cs immobilization. *Mycological Research*, **95**, 1052–1056.

Dixon-Hardy, J. E., Karamushka, V. I., Gruzina, T. G., Nikovska, G. N. & Sayer, J. A. (1998). Influence of the carbon, nitrogen and phosphorus source on the solubilization of insoluble metal compounds by *Aspergillus niger*. *Mycological Research*, **102**, 1050–1054.

Drever, J. I. & Stillings, L. L. (1997). The role of organic acids in mineral weathering. *Colloids and Surfaces* **120**, 167–181.

Eccles, H. (1999). Treatment of metal-contaminated wastes: why select a biological process? *Trends in Biotechnology*, **17**, 462–465.

Eide, D. J. (1997). Molecular biology of iron and zinc uptake in eukaryotes. *Current Opinion in Cell Biology*, **9**, 573–577.

Eide, D. J. (1998). The molecular biology of metal ion transport in *Saccharomyces cerevisiae*. *Annual Review of Nutrition*, **18**, 441–469.

Eide, D. & Guerinot, M. L. (1997). Metal ion uptake in eukaryotes. *American Society for Microbiology News*, **63**, 199–205.

Fox, T. R. & Comerford, N. B. (1990). Low-molecular weight organic acids in selected forest soils of the southeastern USA. *Soil Science Society of America Journal*, **54**, 1139–1144.

Francis, A. J. (1994). Microbial transformations of radioactive wastes and environmental restoration through bioremediation. *Journal of Alloys and Compounds*, **213–214**, 226–231.

Francis, A. J. (1998). Biotransformation of uranium and other actinides in radioactive wastes. *Journal of Alloys and Compounds*, **271–273**, 78–84.

Francis, A. J., Dodge, C. J. & Gillow, J. B. (1992). Biodegradation of metal citrate complexes and implications for toxic metal mobility. *Nature*, **356**, 140–142.

Gadd, G. M. (1990). Fungi and yeasts for metal binding. In *Microbial Mineral Recovery*, eds. H. L. Ehrlich & C. L. Brierley, pp. 249–275. New York: McGraw-Hill.

Gadd, G. M. (1993a). Interactions of fungi with toxic metals. *New Phytologist*, **124**, 25–60.

Gadd, G. M. (1993b). Microbial formation and transformation of organometallic and organometalloid compounds. *FEMS Microbiology Reviews*, **11**, 297–316.

Gadd, G. M. (1995). Signal transduction in fungi. In *The Growing Fungus*. eds. N. A. R. Gow & G. M. Gadd, pp. 183–210. London: Chapman & Hall.

Gadd, G. M. (1996). Influence of microorganisms on the environmental fate of radionuclides. *Endeavour*, **20**, 150–156.

Gadd, G. M. (1997). Roles of microorganisms in the environmental fate of radionuclides. In *CIBA Foundation Symposia No. 203: Health Impacts of Large Releases of Radionuclides*, eds. J. V. Lake, G. R. Bock & G. Cardew, pp. 94–108. Chichester, UK: Wiley & Sons.

Gadd, G. M. (1999). Fungal production of citric and oxalic acid: importance in metal speciation, physiology and biogeochemical processes. *Advances in Microbial Physiology*, **41**, 47–92.

Gadd, G. M. (2000a). Bioremedial potential of microbial mechanisms of metal mobilization and immobilization. *Current Opinion in Biotechnology*, **11**, 271–279.

Gadd, G. M. (2000b). Heavy metal pollutants: environmental and biotechnological aspects. In: *The Encyclopedia of Microbiology*, 2nd edn, ed. J. Lederberg, pp. 607–617. San Diego, CA: Academic Press.

Gadd, G. M. & Griffiths, A. J. (1978). Microorganisms and heavy metal toxicity. *Microbial Ecology*, **4**, 303–317.

Gadd, G. M. & Lawrence, O. S. (1996). Demonstration of high-affinity Mn^{2+} uptake in *Saccharomyces cerevisiae* – specificity and kinetics. *Microbiology*, **142**, 1159–1167.

Gadd,G. M. & Sayer, G. M. (2000). Fungal transformations of metals and metalloids. In *Environmental Microbe–Metal Interactions*, ed. D. R. Lovley, pp. 237–256. Washington, DC: American Society for Microbiology.

Gadd, G. M. & White, C. (1992). Removal of thorium from simulated acid process streams by fungal biomass: potential for thorium desorption and reuse of biomass and desorbent. *Journal of Chemical Technology and Biotechnology*, **55**, 39–44.

Galli, U., Schuepp, H. & Brunold, C. (1994). Heavy metal binding by mycorrhizal fungi. *Physiologia Plantarum*, **92**, 364–368.

Geesey, G. & Jang, L. (1990). Extracellular polymers for metal binding. In *Microbial Mineral Recovery*, ed. H. L. Ehrlich & C. L. Brierley, pp. 223–275. New York: McGraw-Hill.

Gharieb, M. M. & Gadd, G. M. (1998). Evidence for the involvement of vacuolar activity in metal(loid) tolerance: vacuolar-lacking and -defective mutants of *Saccharomyces cerevisiae* display higher sensitivity to chromate, tellurite and selenite. *Biometals*, **11**, 101–106.

Gharieb, M. M. & Gadd, G. M. (1999). Influence of nitrogen source on the solubilization of natural gypsum ($CaSO_4.2H_2O$) and the formation of calcium oxalate by different oxalic and nitric acid-producing fungi. *Mycological Research*, **103**, 473–481.

Gharieb, M. M., Wilkinson, S. C. & Gadd, G. M. (1995). Reduction of selenium oxyanions by unicellular, polymorphic and filamentous fungi: cellular location of reduced selenium and implications for tolerance. *Journal of Industrial Microbiology*, **14**, 300–311.

Gharieb, M. M., Sayer, J. A. & Gadd, G. M. (1998). Solubilization of natural gypsum ($CaSO_4.2H_2O$) and the formation of calcium oxalate by *Aspergillus niger* and *Serpula himantioides*. *Mycological Research*, **102**, 825–830.

Gharieb, M. M., Kierans, M. & Gadd, G. M. (1999). Transformation and tolerance of tellurite by filamentous fungi: accumulation, reduction, and volatilization. *Mycological Research*, **103**, 299–305.

Gitan, R. S., Luo, H., Rodgers, J., Broderius, M. & Eide, D. (1998). Zinc-induced inactivation of the yeast ZRT1 zinc transporter occurs through endocytosis and vacuolar degradation. *Journal of Biological Chemistry*, **273**, 28617–28624.

Gray S. N. (1998). Fungi as potential bioremediation agents in soil contaminated with heavy or radioactive metals. *Biochemical Society Transactions*, **26**, 666–670.

Guerinot, M. L. & Eide, D. (1999). Zeroing in on zinc uptake in yeast and plants. *Current Opinion in Plant Biology*, **2**, 244–249.

Ha, S. B., Smith, A. P., Howden, R., Dietrich, W. M., Bugg, S., O Connell, M. J., Goldsbrough, P. B. & Cobbett, C. S. (1999). Phytochelatin synthase genes from *Arabidopsis* and the yeast *Schizosaccharomyces pombe*. *Plant Cell*, **11**, 1153–1163.

Hartley, C., Cairney, J. W. G., Sanders, F. E. & Meharg, A. A. (1997). Toxic interactions of metal ions (Cd^{2+}, Pb^{2+}, Zn^{2+} and Sb^{3+}) on *in vitro* biomass production of ectomycorrhizal fungi. *New Phytologist*, **137**, 551–562.

Hasegawa, I., Terada, E., Sunairi, M., Wakita, H., Shinmachi, F., Noguchi, A., Nakajima, M. and Yazaki, J. (1997). Genetic improvement of heavy metal tolerance in plants by transfer of the yeast metallothionein gene (CUP1). *Plant and Soil*, **196**, 277–281.

Hassett, R. F., Romeo, A. M. & Kosman, D. J. (1998). Regulation of high affinity iron uptake in the yeast *Saccharomyces cerevisiae* – role of dioxygen and Fe(II). *Journal of Biological Chemistry*, **273**, 7628–7636.

Hayashi, Y. & Mutoh, N. (1994). Cadystin (phytochelatin) in fungi. In *Metal Ions in Fungi*, eds. G. Winkelmann and D. R. Winge, pp. 311–337. New York: Marcel Dekker.

Howe, R., Evans, R. L. & Ketteridge, S. W. (1997). Copper binding proteins in ectomycorrhizal fungi. *New Phytologist*, **135**, 123–131.

Hunter, T. C. & Mehra, R. K. (1998). A role for HEM2 in cadmium tolerance. *Journal of Inorganic Biochemistry*, **69**, 293–303.

Huysmans, K. D. & Frankenberger, W. T. (1991). Evolution of trimethylarsine by a *Penicillium* sp. isolated from agricultural evaporation pond water. *Science of the Total Environment*, **105**, 13–28.

Inouhe, M., Sumiyoshi, M., Tohoyama, H. & Joho, M. (1996). Resistance to cadmium ions and formation of a cadmium-binding complex in various wild-type yeasts. *Plant Cell Physiology*, **37**, 341–346.

Jensen, L. T. & Winge, D. R. (1998). Identification of a copper-induced intramolecular interaction in the transcription factor Mac1 from *Saccharomyces cerevisiae*. *EMBO Journal*, **17**, 5400–5408.

Jones, D. L. & Kochian, L. V. (1996). Aluminium-organic acid interactions in acid soils. *Plant and Soil*, **182**, 221–228.

Jones, D., McHardy, W. J., Wilson, M. J. & Vaughan, D. (1992). Scanning electron microscopy of calcium oxalate on mantle hyphae of hybrid larch roots from a farm forestry experimental site. *Micron and Microscopica Acta*, **23**, 315–317.

Jongmans, A. G., van Breeman, N., Lundstrom, U., van Hees, R. A. W., Finlay, R. D., Srinivasan, M., Unestam, T., Giesler, R., Melkerud, P.-A. & Olsson, M. (1997). Rock-eating fungi. *Nature* **389**, 682–683.

Joshi, A., Serpe, M. & Kosman, D. J. (1999). Evidence for (Mac1p)₂·DNA ternary complex formation in Mac1p-dependent transactivation at the *CTR1* promoter. *Journal of Biological Chemistry*, **274**, 218–226.

Kaldorf, M., Kuhn, A. J., Schroder, W. H., Hildebrandt, U. & Bothe, H. (1999). Selective element deposits in maize colonized by a heavy metal tolerance conferring arbuscular mycorrhizal fungus. *Journal of Plant Physiology*, **154**, 718–728.

Kapoor, A., Viraraghavan, T. & Cullimore, D. R. (1999). Removal of heavy metals using the fungus *Aspergillus niger*. *Bioresource Technology*, **70**, 95–104.

Karamushka, V. I. & Gadd, G. M. (1999). Interaction of *Saccharomyces cerevisiae* with gold: toxicity and accumulation. *Biometals*, **12**, 289–294.

Karamushka, V. I., Sayer, A. A. & Gadd, G. M. (1996). Inhibition of H^+ efflux from *Saccharomyces cerevisiae* by insoluble metal phosphates and protection by calcium and magnesium: inhibitory effects a result of soluble metal cations? *Mycological Research*, **100**, 707–713.

Kho, R., Nguyen, L., Torres-Martinez, C. L. & Mehra, R. K. (2000). Zinc-histidine as nucleation centers for growth of ZnS nanocrystals. *Biochemical and Biophysical Research Communications*, **272**, 29–35.

Kierans, M., Staines, A . M., Bennett, H. & Gadd, G. M. (1991). Silver tolerance and accumulation in yeasts. *Biology of Metals*, **4**, 100–106.

Kosman, D. J. (1994). Transition metal ion uptake in yeasts and filamentous fungi. In *Metal Ions in Fungi*, eds. G. Winkelmann and D. R. Winge, pp. 1–38. New York: Marcel Dekker.

Kotrba, P., Doleckova, L., de Lorenzo, V. & Ruml, T. (1999). Enhanced bioaccumulation of heavy metal ions by bacterial cells due to surface display of short metal binding peptides. *Applied and Environmental Microbiology*, **65**, 1092–1098.

Lapeyrie, F., Chilvers, G. A. & Bhem, C. A. (1987). Oxalic acid synthesis by the mycorrhizal fungus *Paxillus involutus*. *New Phytologist*, **106**, 139–146.

Lapeyrie, F., Ranger, J. & Vairelles, D. (1991). Phosphate solubilizing activity of ectomycorrhizal fungi *in vitro. Canadian Journal of Botany*, **69**, 342–346.

Lesuisse, E. & Labbe, P. (1995). Effects of cadmium and of YAP1 and CAD1/YAP2 genes on iron metabolism in the yeast *Saccharomyces cerevisiae. Microbiology*, **141**, 2937–2943.

Leyval, C., Surtiningish, T. & Berthelin, J. (1993). Mobilization of P and Cd from rock phosphates by rhizosphere microorganisms (phosphate dissolving bacteria and ectomycorrhizal fungi). *Phosphorus, Sulphur and Silicon*, **77**, 133–136.

Lo, W. H., Chua, H., Lam, K. H. & Bi, S. P. (1999). A comparative investigation on the biosorption of lead by filamentous fungal biomass. *Chemosphere*, **39**, 2723–2736.

Losi, M. E. & Frankenberger, W. T. (1997). Bioremediation of selenium in soil and water. *Soil Science*, **162**, 692–702.

Lovley, D. R. & Coates, J. D. (1997). Bioremediation of metal contamination. *Current Opinion in Biotechnology*, **8**, 285–289.

Macaskie, L. E. & Dean, A. C. R. (1987). Trimethyllead degradation by an alkyllead-tolerant yeast. *Environmental Technological Letters*, **8**, 635–640.

Macaskie, L. E. & Dean, A. C. R. (1990). Trimethyl lead degradation by free and immobilized cells of an *Arthrobacter* sp. and by the wood decay fungus *Phaeolus schweintzii. Applied Microbiology and Biotechnology*, **38**, 81–87.

Macreadie, I. G., Sewell, A. K. & Winge, D. R. (1994). Metal ion resistance and the role of metallothionein in yeast. In *Metal Ions in Fungi*, eds. G. Winkelmann and D. R. Winge, pp. 279–310. New York: Marcel Dekker.

Martino, E., Turnau, K., Girlanda, M., Bonfante, P. & Perotto, S. (2000). Ericoid mycorrhizal fungi from heavy metal polluted soils: their identification and growth in the presence of zinc ions. *Mycological Research*, **104**, 338–344.

McDaniels, C. P. J., Jensen, L. T., Srinivasan, C., Winge, D. R. & Tullius, T. D. (1999). The yeast transcription factor Mac1 binds to DNA in a modular fashion. *Journal of Biological Chemistry*, **274**, 26962–26967.

Meharg, A. A. & Cairney, J. W. G. (2000). Co-evolution of mycorrhizal symbionts and their hosts to metal-contaminated environments. *Advances in Ecological Research*, **30**, 69–112.

Mehra, R. K. & Winge, D. R. (1991). Metal ion resistance in fungi: molecular mechanisms and their related expression. *Journal of Cellular Biochemistry*, **45**, 30–40.

Metting, F. B. (1992). Structure and physiological ecology of soil microbial communities. In *Soil Microbial Ecology, Applications and Environmental Management*, ed. F. B. Metting, pp. 3–25. New York: Marcel Dekker.

Mogollon, L., Rodriguez, R., Larrota, W., Ramirez, N. & Torres, R. (1998). Biosorption of nickel using filamentous fungi. *Applied Biochemistry and Biotechnology*, **70**, 593–601.

Morley, G. F. & Gadd, G. M. (1995). Sorption of toxic metals by fungi and clay minerals. *Mycological Research*, **99**, 1429–1438.

Morley, G. F., Sayer, J. A., Wilkinson, S. C., Gharieb, M. M. & Gadd, G. M. (1996). Fungal sequestration, solubilization and transformation of toxic metals. In *Fungi and Environmental Change*, eds. J. C. Frankland, N. Magan & G. M. Gadd, pp. 235–256. Cambridge, UK: Cambridge University Press.

Murasugi, A., Wada, C. & Hayashi, Y. (1983). Occurrence of acid labile sulfide in cadmium binding peptide 1 from fission yeast. *Journal of Biochemistry*, **93**, 661–664.

Murphy, R. J. & Levy, J. F. (1983). Production of copper oxalate by some copper tolerant fungi. *Transactions of the British Mycological Society*, **81**, 165–168.

Neilands, J. B. (1981). Microbial iron compounds. *Annual Review of Biochemistry*, **50**, 715–731.

Okuyama, M., Kobayashi, Y., Inouhe, M., Tohoyama, H. and Joho, M. (1999). Effects of some heavy metal ions on copper-induced metallothionein synthesis in the yeast *Saccharomyces cerevisiae*. *Biometals*, **12**, 307–314.

Orsler, R. J. & Holland, G. E. (1982). Degradation of tributyltin oxide by fungal culture filtrates. *International Biodeterioration Bulletin* **18**, 95–98.

Ortiz, D. F., Kreppel, D. F., Speiser, D. M., Scheel, G., McDonald, G. & Ow, D. W. (1992). Heavy-metal tolerance in the fission yeast requires an ATP-binding cassette-type vacuolar membrane transporter. *EMBO Journal*, **11**, 3491–3499.

Ortiz, D. F., Ruscitti, T., McCue, K. F. & Ow, D. W. (1995). Transport of metal-binding peptides by HMT1, a fission yeast ABC-type vacuolar membrane protein. *Journal of Biological Chemistry*, **270**, 4721–4728.

Ow, D. W. (1993). Phytochelatin-mediated cadmium tolerance in *Schizosaccharomyces pombe*. *In Vitro Cellular and Developmental Biology – Plant*, **29P**, 13–219.

Ow, D. W. (1996). Heavy metal tolerance genes: prospective tools for bioremediation. *Resources, Conservation and Recycling*, **18**, 135–149.

Ow, D. W., Ortiz, D. F., Speiser, D. M. & McCue, K. F. (1994). Molecular genetic analysis of cadmium tolerance in *Schizosaccharomyces pombe*. In *Metal Ions in Fungi*, eds. G. Winkelmann & D. R. Winge, pp. 339–359. New York: Marcel Dekker.

Pawlowska, T. E., Chaney, R. L., Chin, L. & Charvat, I. (2000). Effects of metal phytoextraction practices on the indigenous community of arbuscular mycorrhizal fungi at a metal-contaminated landfill. *Applied and Environmental Microbiology*, **66**, 2526–2530.

Pazirandeh, M., Wells, B. M. & Ryan, R. L. (1998). Development of bacterium-based heavy metal biosorbents: enhanced uptake of cadmium and mercury by *Escherichia coli* expressing a metal binding motif. *Applied and Environmental Microbiology*, **64**, 4086–4092.

Peberdy, J. F. (1990). Fungal cell walls – a review. In *Biochemistry of Cell Walls and Membranes in Fungi*, eds. P. J. Kuhn, A. P. J. Trinci, M. J. Jung, M. W. Coosey & L. E. Copping, pp. 5–30. Berlin: Springer-Verlag.

Perkins, J. & Gadd, G. M. (1993). Accumulation and intracellular compartmentation of lithium ions in *Saccharomyces cerevisiae*. *FEMS Microbiology Letters*, **107**, 255–260.

Perkins, J. & Gadd, G. M. (1996). Interactions of Cs^+ and other monovalent cations (Li^+, Na^+, K^+, Rb^+, NH_4^+) with K^+-dependent pyruvate-kinase and malate-dehydrogenase from the yeasts *Rhodotorula rubra* and *Saccharomyces cerevisiae*. *Mycological Research*, **100**, 449–454.

Presta, A. and Stillman, M. J. (1997). Incorporation of copper into the yeast *Saccharomyces cerevisiae*. Identification of Cu(I)-metallothionein in intact yeast cells. *Journal of Inorganic Biochemistry*, **66**, 231–240.

Purvis, O. W. (1984). The occurrence of copper oxalate in lichens growing on copper sulphide-bearing rocks in Scandinavia. *Lichenologist*, **16**, 197–204.

Purvis, O. W. & Halls, C. (1996). A review of lichens in metal-enriched environments. *Lichenologist*, **28**, 571–601.

Ramadan, S. E., Razak, A. A., Youseff, Y. A. & Sedky, N. M. (1988). Selenium metabolism in a strain of *Fusarium*. *Biological Trace Element Research*, **18**, 161–170.

Ramsay, L. M. & Gadd, G. M. (1997). Mutants of *Saccharomyces cerevisiae* defective in vacuolar function confirm a role for the vacuole in toxic metal ion detoxification. *FEMS Microbiology Letters*, **152**, 293–298.

Ramsay, L. M., Sayer, J. A. & Gadd, G. M. (1999). Stress responses of fungal colonies towards toxic metals. In *The Fungal Colony*, eds. N. A. R. Gow, G. D. Robson & G. M. Gadd. pp. 178–200. Cambridge, UK: Cambridge University Press.

Rauser, W. E. (1995). Phytochelatins and related peptides. *Plant Physiology*, **109**, 1141–1149.

Satofuka, H., Amano, S., Atomi, H., Takagi, M., Hirata, K., Miyamoto, K. and Imanaka, T. (1999). Rapid method for detection and detoxification of heavy metal ions in water environments using photochelatin. *Journal of Bioscience and Bioengineering*, **88**, 287–292.

Sayer, J. A. & Gadd, G. M. (1997). Solubilization and transformation of insoluble inorganic metal compounds to insoluble metal oxalates by *Aspergillus niger*. *Mycological Research*, **101**, 653–661.

Sayer, J. A., Raggett, S. L. & Gadd, G. M. (1995). Solubilization of insoluble metal compounds by soil fungi: development of a screening method for solubilizing ability and metal tolerance. *Mycological Research*, **99**, 987–993.

Sayer, J. A., Kierans, M. & Gadd, G. M. (1997). Solubilisation of some naturally occurring metal-bearing minerals, limescale and lead phosphate by *Aspergillus niger*. *FEMS Microbiology Letters*, **154**, 29–35.

Sayer J. A., Cotter-Howells, J. D., Watson, C., Hillier, S. & Gadd, G. M. (1999). Lead mineral transformation by fungi. *Current Biology*, **9**, 691–694.

Serpe, M., Joshi, A. & Kosman, D. J. (1999). Structure-function analysis of the protein-binding domains of Mac1p, a copper-dependent transcriptional activator of copper uptake in *Saccharomyces cerevisiae*. *Journal of Biological Chemistry*, **274**, 29211–29219.

Smith, D. G. (1974). Tellurite reduction in *Schizosaccharomyces pombe*. *Journal of General Microbiology*, **83**, 389–392.

Sousa, C., Kotrba, P., Ruml, T., Cebolla, A. & de Lorenzo, V. (1998) Metalloadsorption by *Escherichia coli* cells displaying yeast and mammalian metallothioneins anchored to the outer membrane protein LamB. *Journal of Bacteriology*, **180**, 2280–2284.

Strasser, H., Burgstaller, W. & Schinner, F. (1994). High yield production of oxalic acid for metal leaching purposes by *Aspergillus niger*. *FEMS Microbiology Letters*, **119**, 365–370.

Sutter, H.-P., Jones, E. B. G. & Walchi, O. (1984). Occurrence of crystalline hyphal sheaths in *Poria placenta* (Fr.) Cke. *Journal of the Institute of Wood Science*, **10**, 19–23.

Tait, K., Sayer, J. A., Gharieb, M. M. & Gadd, G. M. (1999). Fungal production of calcium oxalate in leaf litter microcosms. *Soil Biology and Biochemistry*, **31**, 1189–1192.

Tamaki, S. & Frankenberger, W. T. (1992). Environmental biochemistry of arsenic. *Reviews of Environmental Contamination and Toxicology*, **124**, 79–110.

Tezuka, T. & Takasaki, Y. (1988). Biodegradation of phenylmercuric acetate by organomercury-resistant *Penicillium* sp. MR-2. *Agricultural Biology and Chemistry*, **52**, 3183–3185.

Thompson-Eagle, E. T. & Frankenberger, W. T. (1992). Bioremediation of soils contaminated with selenium. In *Advances in Soil Science*, eds. R. Lal & B. A. Stewart, pp. 261–309. New York: Springer.

Thompson-Eagle, E. T., Frankenberger, W. T. & Karlson, U. (1989). Volatilization of selenium by *Alternaria alternata*. *Applied and Environmental Microbiology*, **55**, 1406–1413.

Thompson-Eagle, E. T., Frankenberger, W. T. & Longley, K. E. (1991). Removal of selenium from agricultural drainage water through soil microbial transformations. In *The Economics and Management of Water and Drainage in Agriculture*, eds. A. Dinar & D. Zilberman, pp. 169–186. New York: Kluwer.

Tobin, J. M., White, C. & Gadd, G. M. (1994). Metal accumulation by fungi – applications in environmental biotechnology. *Journal of Industrial Microbiology*, **13**, 126–130.

Tohoyama, H., Inouhe, M., Joho, M. & Murayama, T. (1995). Production of metallothionein in copper-resistant and cadmium-resistant strains of *Saccharomyces cerevisiae*. *Journal of Industrial Microbiology*, **14**, 126–131.

Torres-Martinez, C. L., Nguyen, L., Kho, R., Bae, W., Bozhilov, K., Klimov, V. & Mehra, R. K. (1999). Biomolecularly capped uniformly sized nanocrystalline materials: glutathione-capped ZnS nanocrystals. *Nanotechnology*, **10**, 340–354.

Tzeferis, P.G., Agatzini, S. & Nerantzis, E. T. (1994). Mineral leaching of non-sulphide nickel ores using heterotrophic micro-organisms. *Letters in Applied Microbiology*, **18**, 209–213.

Valls, M., González-Duarte, R., Atrian, S. & De Lorenzo, V. (1998) Bioaccumulation of heavy metals with protein fusions of metallothionein to bacterial OMPs. *Biochimie*, **80**, 855–861.

Wainwright, M. (1988). Metabolic diversity of fungi in relation to growth and mineral cycling in soil – a review. *Transactions of the British Mycological Society*, **90**, 159–170.

Wakatsuki, T., Hayakawa, S., Hatayama, T., Kitamura, T. & Imahara, H. (1991a). Solubilization and properties of copper reducing enzyme systems from the yeast cell surface in *Debaryomyces hansenii*. *Journal of Fermentation and Bioengineering*, **72**, 79–86.

Wakatsuki, T., Hayakawa, S., Hatayama, T., Kitamura, T. & Imahara, H. (1991b). Purification and some properties of copper reductase from cell surface of *Debaryomyces hansenii*. *Journal of Fermentation and Bioengineering*, **72**, 158–161.

White, C., Sayer, J. A. & Gadd, G. M. (1997). Microbial solubilization and immobilization of toxic metals: key biogeochemical processes for treatment of contamination. *FEMS Microbiology Reviews*, **20**, 503–516.

Winge, D. R. (1998). Copper-regulatory domain involved in gene expression. *Progress in Nucleic Acid Research and Molecular Biology*, **58**, 165–195.

Winge, D. R., Graden, J. A., Posewitz, M. C., Martins, L. J., Jensen, L. T. & Simon, J. R. (1997). Sensors that mediate copper-specific activation and repression. *Journal of Biological Inorganic Chemistry*, **2**, 2–10.

Winge, D. R., Jensen, L. T. & Srinivasan, C. (1998). Metal-ion regulation of gene expression in yeast. *Current Opinion in Chemical Biology*, **2**, 216–221.

Wu, J. S., Sung, H. Y. & Juang, R. J. (1995). Transformation of cadmium-binding complexes during cadmium sequestration in fission yeast. *Biochemical, Molecular and Biological Interactions*, **36**, 1169–1175.

Yamaguchilwai, Y., Serpe, M., Haile, D., Yang, W. M., Kosman, D. J., Klausner, R. D. & Dancis, A. (1997). Homeostatic regulation of copper uptake in yeast via direct binding of MAC1 protein to upstream regulatory sequences of FRE1 and CTR1. *Journal of Biological Chemistry*, **272**, 17711–17718.

Yannai, S., Berdicevsky, I. & Duek, L. (1991). Transformations of inorganic mercury by *Candida albicans* and *Saccharomyces cerevisiae*. *Applied and Environmental Microbiology*, **57**, 245–247.

Yetis, U., Ozcengiz, G., Dilek, F. B., Ergen, N. & Dolek, A. (1998). Heavy metal biosorption by white-rot fungi. *Water Science and Technology*, **38**, 323–330.

Yin, P. H., Yu, Q. M., Jin, B. & Ling, Z. (1999). Biosorption removal of cadmium from aqueous solution by using pretreated fungal biomass cultured from starch wastewater. *Water Research*, **33**, 1960–1963.

Zhao, H. & Eide, D. J. (1997). Zap1p, a metalloregulatory protein involved in zinc-responsive transcriptional regulation in *Saccharomyces cerevisiae*. *Molecular and Cellular Biology*, **17**, 5044–5052.

Zhao, H., Butler, C., Rodgers, J., Spizzo, T., Duesterhoeft, S. & Eide, D. (1998). Regulation of zinc homeostasis in yeast by binding of the ZAP1 transcriptional activator to zinc-responsive promoter elements. *Journal of Biological Chemistry*, **273**, 28713–28720.

Zhou, J. L. (1999). Zn biosorption by *Rhizopus arrhizus* and other fungi. *Applied Microbiology and Biotechnology*, **51**, 686–693.

14

Heterotrophic leaching

HELMUT BRANDL

Introduction

Bacterial leaching of metals (bioleaching, biomining) from mineral resources has a very long historical record (Rossi, 1990; Ehrlich, 1999). Metals have been mobilized from sulfide minerals using processes that involved autotrophic sulfur-oxidizing microorganisms, for example *Thiobacillus* spp., although the involvement of microorganisms in this process was demonstrated only in the 1920s (Rudolfs & Helbronner, 1922; Waksman & Joffe, 1922). In 1947, *Thiobacillus ferrooxidans* was identified in acid mine drainage as part of a microbial community that also included several fungi (e.g. *Spicaria* sp.) (Colmer & Hinkle, 1947). Several industrial processes have been developed based on these findings for the mining of cobalt, copper, nickel, uranium, zinc and gold (Bosecker, 1997; Rawlings, 1997). However, all industrial applications to obtain metals from a series of solid materials depend on the activities of sulfur-oxidizing microorganisms.

Bioleaching is mainly based on three mechanisms. Besides proton-induced metal solubilization and metal reduction or oxidation, metals can also be mobilized from solid materials by ligand-induced metal solubilization. Organic acids from heterotrophic microorganisms represent such ligands. This is particularly important in the biohydrometallurgical treatment of silicate, carbonate and oxide minerals since these materials cannot be directly attacked by sulfur-oxidizing microorganisms. Further developments should enable heterotrophic leaching to be used to extract metals from non-sulfide ores (Ehrlich, 1999). The broad diversity of heterotrophic organisms provides a huge industrial potential that has been hardly investigated.

Fungi are known for their ability to form a broad spectrum of organic acids (Bigelis & Arora, 1992; Zidwick, 1992). This metabolic potential can be used not only for the mobilization of metals from various sources such as primary ores but also to recycle metals from waste materials such as fly ash, galvanic sludge and electronic scrap (Burgstaller & Schinner, 1993) and, possibly for the bioremediation of metal-contaminated soils. Bio-leaching allows metal cycling by a process close to natural biogeochemical cycles and can contribute to sustainable development, reducing the demand for non-renewable resources such as ores or fossil fuel.

Historical background

One of the first reports where heterotrophic leaching by fungi and bacteria might have been involved in the mobilization of metals is given by the Roman writer Gaius Plinius Secundus (23–79 AD). In his work on natural sciences, he describes how copper minerals are obtained using a leaching process (Liber XXXIII, v. 86; König, 1989).

> Chrysocolla umor est in puteis, quos diximus, per venam auri defluens crassescente limo rigoribus hibernis usque in duritiam pumicis. Laudatiorem eandem in aerariis metallis et proxinam in argentariis fieri conpertum est. Invenitur et in plumbariis, vilior etiam auraria. In omnibus autem his metallis fit et cura multum infra naturalem illam inmissis in venam aquis leniter hieme tota usque in Iunium mensem; dein siccatis Iunio et Iulio, ut plane intellegatur nihil aliud chrysocolla quam vena putris.

The translation reads approximately as follows.

> Chrysocolla [hydrated copper silicate] is a liquid in the before mentioned gold mines running from the gold vein. In cold weather during the winter the sludge freezes to the hardness of pumice. It is known from experience that the most wanted is formed in copper mines, the following in silver mines. The liquid is also found in silver mines although it is of minor value. In all these mines chrysocolla is also artificially produced by slowly passing water through the mine during the winter until the month of June; subsequently, the water is evaporated in June and July. It is clearly demonstrated that chrysocolla is nothing but a decomposed vein.

This report leads to the suggestion that the metal is solubilized by micro-organisms other than thiobacilli because chrysocolla is a silicate mineral. Therefore, copper might have been mobilized by heterotrophic organisms including fungi. However, the technical descriptions of the process by Plinius remain somewhat unclear and have other possible interpretations.

Fungi can also attack solid materials such as minerals or building materials (Sterflinger, 2000). An early description of this is found in the Old Testament (Leviticus 14: 36, 14: 37)

> Then the priest shall command that they empty the house, before the priest go into it to see the plague, that all that is in the house be not made

unclean: and afterward the priest shall go in to see the house.

And he shall look on the plague, and, behold, if the plague be in the walls of the house with hollow strakes, greenish or reddish, which in sight are lower than the wall . . .

This describes the growth of red and green organisms (fungi? lichens?) on the walls of houses, creating pits and cavities (see also Krumbein & Schönborn-Krumbein, 1987). Fungi can be enriched and isolated from weathered sandstone buildings and monuments (de la Torre *et al.*, 1991; Palmer, Siebert & Hirsch, 1991; Hirsch, Eckhardt & Palmer, 1995). It has been demonstrated that organic acids are formed *in situ* even in conditions of low nutrient availability (Palmer *et al.*, 1991). Early reports describe the solubilization of fine powdered glass and a variety of minerals by oxalic acid, which is a fungal metabolite (Slater, 1856). Phosphates of iron, silver, zinc and copper, arsenates of iron, silver and copper, chromates of zinc, bismuth, barium, mercury and lead are also decomposed by oxalic acid. Slater (1856) assumed that the influence of lichens containing oxalic acid was a major factor in effecting the disintegration and decomposition of rocks.

Fungal biodiversity

A large number of fungi can mobilize elements from solid substrates (Table 14.1). *Aspergillus niger* and *Penicillium simplicissimum* are the strains most commonly used for biohydrometallurgical treatments. Generally, *Aspergillus* and *Penicillium* spp. are known to be very effective producers of organic acids either from the citric acid (TCA) cycle (e.g. citric and oxalic) or derived from glucose (e.g. gluconic acid) (Bigelis & Arora, 1992). This is commercially exploited in large-scale production of citric acid, which is a billion-dollar business (Demain, 2000).

Three different techniques have been employed to study and to achieve metal solubilization by fungi. In the first, microorganisms are grown on the surface of solidified media that contains powdered minerals (plate technique) (Sayer, Raggett & Gadd, 1995). During fungal growth the minerals are solubilized as acids are formed, which allows an optical measure (clear zone) of metal leaching. In a one-step (batch) process organisms are grown in the presence of solid substrates for a certain time period (a few days up to several months). Mobilized metals are determined in the supernatant. However, growth can be reduced or inhibited by the presence of toxic elements, which will reduce leaching efficiencies and metal yields. In addition, for a technical application, the biomass/mineral phase mixture might

Table 14.1. *Filamentous fungi and yeasts for the mobilization of elements from solid substrates*

Organism	Strain	Carbon source	Leaching agent	Process type[a]	Temperature (°C)[b]	Duration (d)[b]	Solid substrate	Elements leached	Reference
Absidia orchidis		Glucose	Glutarate, tartrate, succinate	Batch	27	5	Albite, hornblende, orthoclase, oligoclase	K, Ca	Müller & Förster, 1964
Actinomucor sp.		Glucose	Succinate	Batch	27	5	Albite, hornblende, orthoclase, oligoclase	K, Ca	Müller & Förster, 1964
Alternaria sp.		Glucose		Batch		1095	Copper–molybdenum -ore	Cu	Kovalenko & Malakhova, 1990
Alternaria sp.		Glucose, sucrose	Citrate, oxalate	Batch	32	30	Kaolinite, illite	Al	Borovec, 1990
Alternaria sp.	1L1.5	Glucose		Plate	25	21	Sandstone		Palmer *et al.*, 1991
Aspergillus sp.		Glucose		Batch		1095	Copper–molybdenum ore	Cu	Kovalenko & Malakhova, 1990
Aspergillus sp.		Malt extract	Oxalate	Two-step		12	Glass, TV picture tubes	Pb	Weissmann, Drewello & Müller, 1999
Aspergillus sp.		Sucrose	Citrate, gluconate, oxalate	Two-step	30	1	Soil	Zn, Pb, Cu	Burri, 1999
Aspergillus sp.	1L3.6	Glucose		Plate	25	21	Sandstone		Palmer *et al.*, 1991
Aspergillus sp.	A1	Sucrose, glucose, molasses	Citrate, oxalate	Batch	30	40–48	Lateritic nickel ore	Ni, Fe	Tzeferis, Agatzini & Nerantzis, 1994
Aspergillus sp.	A1	Glucose, sucrose	Citrate, oxalate	Batch, two-step	30, 95	48	Nickeliferous laterites	Ni, Fe, Co	Tzeferis, 1994a
Aspergillus sp.	A1	Sucrose, glucose	Citrate, oxalate	Batch, two-step	30, 95	48, 0.25	Hematitic lateritic ore	Ni, Co, Fe, Ca, Mg	Tzeferis, 1994b
Aspergillus sp.	A1, CMI 31821	Sucrose	Citrate, oxalate	Batch	30	48	Lateritic nickel ore	Ni, Co, Fe, Mg	Alibhai *et al.*, 1993
Aspergillus sp.	A2	Sucrose, glucose, molasses	Citrate, oxalate	Batch	30	40–48	Lateritic nickel ore	Co, Ni, Fe	Tzeferis *et al.*, 1994
Aspergillus sp.	A2	Glucose, sucrose	Citrate, oxalate	Batch, two-step	30, 95	48	Nickeliferous laterites	Ni, Fe, Co	Tzeferis, 1994a
Aspergillus sp.	A2	Sucrose, glucose	Citrate, oxalate	Batch, two-step	30, 95	48, 0.25	Hematitic lateritic ore	Ni, Co, Fe, Ca, Mg	Tzeferis, 1994b

Organism	Strain	Carbon source	Organic acids	Process	Temperature	Duration	Material	Metals	Reference
Aspergillus sp.	A2, CMI 246753	Sucrose	Citrate, oxalate	Batch	30	48	Lateritic nickel ore	Co, Ni, Fe	Alibhai *et al.*, 1993
Aspergillus sp.	A3	Sucrose, glucose, molasses	Citrate, oxalate	Batch	30	40–48	Lateritic nickel ore	Ni, Fe	Tzeferis *et al.*, 1994
Aspergillus sp.	A3	Sucrose, glucose, molasses	Citrate	Batch, two-step	30, 90	48	Lateritic nickel ore	Ni, Fe, Co, Ca, Mg	Tzeferis *et al.*, 1991
Aspergillus sp.	A3	Glucose, sucrose	Citrate, oxalate	Batch, two-step	30, 95	48	Nickeliferous laterites	Ni, Fe, Co	Tzeferis, 1994a
Aspergillus sp.	A3	Sucrose, glucose	Citrate, oxalate	Batch, two-step	30, 95	48, 0.25	Hematitic lateritic ore	Ni, Co, Fe, Ca, Mg	Tzeferis, 1994b
Aspergillus sp.	A3, CMI 75353	Sucrose	Citrate, oxalate	Batch	30	48	Lateritic nickel ore	Co, Ni, Fe	Alibhai *et al.*, 1993
Aspergillus awamori		Glucose		Batch	28	7	Bauxite	Si	Ogurtsova *et al.*, 1989
Aspergillus brunneo-uniseriatus	ATCC SD 1076		Citrate, oxalate	Batch	21–35	0.75–4	Gold-bearing ore, carbonaceous ore, carbonaceous pyritic ore	Au	Portier, 1991
Aspergillus foetidus		Glucose	Citrate, oxalate, fumarate	Batch	30	7	Ferromanganese sea nodules	Cu	Khan, Gupta & Kumar Saxena, 1997
A. foetidus		Sucrose	Citrate, gluconate, oxalate	Two-step	30	0.3	Soil	Zn, Cu	Frei, 1999
A. foetidus		Sucrose	Citrate, gluconate, oxalate	Two-step	30	1	Soil	Zn, Pb, Cu	Burri, 1999
Aspergillus fumaricus		Glucose, sucrose	Citrate, oxalate	Batch	32	30	Kaolinite, illite	Al	Borovec, 1990
Aspergillus fumigatus		Glucose		Batch	30	13	Silicate nickel ore	Al, Ni, Cr, Fe, Mg	Bosecker, 1989
Aspergillus japonicus		Glucose	Citrate, oxalate, fumarate	Batch	30	7	Ferromanganese sea nodules	Cu	Khan *et al.*, 1997
Aspergillus niger		Sucrose	Citrate, gluconate, oxalate	Two-step	30	4	Quartz sand, hematite	Fe	Strasser *et al.*, 1993a
A. niger		Glucose, sucrose, molasses	Citrate, oxalate	Batch	22	20	Oxidized copper ore, oxidized lead-zinc ore	Cu, Zn	Dave, Natarajan & Bhat, 1981
A. niger		Molasses	Citrate, oxalate	Batch, two-step	30, 90	14, 0.2	Limonite, goethite, hematite, quartz sand, kaolinite, clay	Al, Ca, Fe, Mn	Groudev, 1987

Table 14.1. (cont.)

Organism Strain	Carbon source	Leaching agent	Process type[a]	Temperature (°C)[b]	Duration (d)[b]	Solid substrate	Elements leached	Reference
A. niger	Malt extract		Plate	25	1–2	Cuprite, galena, rhodochrosite, limescale	Ca, Cu, Mn, Pb	Sayer, Kierans & Gadd, 1997
A. niger	Malt extract		Plate	25	10	Phosphates of aluminium, calcium, cobalt, manganese, zinc; calcium carbonate, zinc oxide	Al, Ca, Co, Mn, Zn	Sayer et al., 1995
A. niger	Sucrose	Citrate, gluconate	Batch, two-step	30	16–35, 1	Fly ash	Al, Cd, Cr, Cu, Fe, Mn, Ni, Pb, Zn	Bosshard et al., 1996
A. niger	Sucrose		Batch	26	14	Spodumene	Al, Li, Si	Karavaiko et al., 1980
A. niger	Glucose	Citrate, gluconate, oxalate	Batch		70–101	Biotite, orthoclase, amphibolite	Al, K, Mg, Fe, Ca, Si	Eckhardt, 1979a
A. niger	Glucose	Gluconate	Batch		84	Biotite, orthoclase, amphibolite	Ca, K, Si, Mg, Fe, Al	Eckhardt, 1979b
A. niger	Glucose	Oxalate	Batch, two-step	20	5, 98	Chrysolite	Mg, Fe, Si	Wilson, Jones & McHardy, 1981
A. niger	Glucose, galactose, fructose, mannose, starch, carboxymethyl-cellulose	Oxalate, gluconate, malate, citrate, fumarate, succinate	Batch	27	14–21	Sandstone		de la Torre et al., 1991
A. niger	Molasses, glucose, sucrose	Citrate, oxalate, gluconate, tartrate, histidine, proline, glycine, aspartate, glutamate, arginine, threonine, leucine, serine	Two-step	28, 20	1	Oxides of copper, zinc, aluminium	Zn, Cu, Al	Golab & Orlowska, 1988

Organism	Carbon source	Lixiviant	Process			Substrate	Metals	Reference
A. niger	Glucose, sucrose	Citrate, oxalate	Batch	32	30	Bauxite, kaolinite, halloysite, montmorillonite, illite, vermiculite, serpentine, chrysotile	Al, Si, Fe, Mn	Borovec, 1990
A. niger	Glucose		Plate	25	10	Phosphates of cobalt, zinc; zinc oxide	Co, Zn	Dixon-Hardy et al., 1998
A. niger	Malt extract	Oxalate	Plate	25		Phosphates of manganese, cadmium, copper, zinc, cobalt; zinc oxide	Zn, Cd, Cu, Mn, Co	Sayer & Gadd, 1997
A. niger	Xylose, glucose, galactose, rhamnose, mannose, inositol, sucrose, trehalose, maltose, raffinose	Oxalate, citrate	Plate, batch		14	Phosphates of calcium, magnesium	Ca, Mg, P	Rose, 1957
A. niger	Potato-dextrose		Batch	30	20	Lateritic nickel ore	Ni	Sukla, Panchanadikar & Kar, 1993
A. niger	Potato-dextrose, malt extract, corn steep liquor, cane molasses		Batch	37	20	Lateritic nickel ore	Ni, Fe, Co	Sukla & Panchanadikar, 1993
A. niger	Sucrose	Oxalate, citrate	Batch, two-step	21, 90	7, 30	Spodumene	Li	Ilgar, Guay & Torma, 1993
A. niger	Sucrose, sufite liquor	Citrate	Batch	30	28	Malachite, chrysocolla, lydite	Cu	Kiel, 1977
A. niger	Sucrose	Citrate, gluconate, oxalate	Batch	30	21	Electronic waste	Al, Cu, Ni, Pb, Sn, Zn	Brandl et al., 1999
A. niger	Glucose	Citrate, oxalate, tartrate	Batch	37	5	Albite, hornblende, orthoclase, oligoclase	K, Ca	Müller & Förster, 1964
A. niger	Glucose, sugar cane molasses		Batch	22	28–399	Manganese ore	Mn	Noble et al., 1991

Table 14.1. (*cont.*)

Organism	Strain	Carbon source	Leaching agent	Process type[a]	Temperature (°C)[b]	Duration (d)[b]	Solid substrate	Elements leached	Reference
Aspergillus niger		Sucrose	Citrate, gluconate, oxalate	Two-step	30	7	Plant biomass	Cd, Cu, Zn	Ryser, 2000
A. niger		Sucrose	Citrate, gluconate, oxalate	Two-step	30	0.3	Soil	Zn, Cu	Frei, 1999
A. niger		Sucrose	Citrate, gluconate, oxalate	Two-step	30	1	Soil	Zn, Pb, Cu	Burri, 1999
A. niger		Glucose		Batch	22	36–68	Glass	K	Drewello, Nüssler & Weissmann, 1994
A. niger		Glucose, sucrose, molasses		Batch	22	20	Oxidized copper ore, oxidized lead–zinc ore	Cu, Zn	Dave et al., 1981
A. niger	1	Sucrose, sulfite liquor, malt	Citrate	Batch, plate	30	7	Low-grade copper ore; oxides of yttrium, lanthanum, neodymium, dysprosium	Cu, Y, La, Nd, Dy	Kiel & Schwartz, 1980
A. niger	2	Sucrose, sulfite liquor, malt	Citrate	Batch, plate	30	7	Low-grade copper ore; oxides of yttrium, lanthanum, neodymium, samarium, dysprosium	Cu, Y, La, Nd, Sm, Dy	Kiel & Schwartz, 1980
A. niger	21	Malt	Citrate	Plate	30	7	Yttrium, lanthanum, neodymium, samarium, dysprosium	Y, La, Nd, Sm, Dy	Kiel & Schwartz, 1980
A. niger	21	Glucose	Citrate	Batch	30	19–29	Lateritic nickel ore	Ni	Bosecker, 1986

Organism	No.	Carbon source	Acid	Mode	Temp.	pH	Substrate	Elements	Reference
A. niger	22	Sucrose, sulfite liquor, malt	Citrate	Batch, plate	30	4–12	Low-grade copper ore; yttrium, lanthanum, neodymium, samarium, dysprosium	Cu, Y, La, Nd, Sm, Dy	Kiel & Schwartz, 1980
A. niger	23	Glucose		Batch	28	7	Bauxite	Al, Fe, Si	Ogurtsova et al., 1989
A. niger	3	Sucrose, sulfite liquor, malt	Citrate	Batch, plate	30	9–10	Low-grade copper ore; yttrium, lanthanum, neodymium, samarium, dysprosium	Cu, Y, La, Nd, Sm, Dy	Kiel & Schwartz, 1980
A. niger	36	Malt	Citrate	Plate	30	7	Yttrium, lanthanum, neodymium, samarium, dysprosium	Y, La, Nd, Sm, Dy	Kiel & Schwartz, 1980
A. niger	4	Sucrose, sulfite liquor, malt	Citrate	Batch, plate	30	4	Low-grade copper ore; yttrium, lanthanum, neodymium, samarium, dysprosium	Cu, Y, La, Sm, Dy	Kiel & Schwartz, 1980
A. niger	4	Glucose		Batch	28	7	Bauxite	Al, Fe, Si	Ogurtsova et al., 1989
A. niger	4	Sucrose	Citrate	Batch	30	35	Bauxite	Al, Fe, Si	Karavaiko et al., 1989
A. niger	49	Malt	Citrate	Plate	30	7	Yttrium, lanthanum, samarium, dysprosium	Y, La, Sm, Dy	Kiel & Schwartz, 1980
A. niger	5	Sucrose, sulfite liquor, malt	Citrate	Batch, plate	30	7	Low-grade copper ore; yttrium, lanthanum, neodymium, samarium, dysprosium	Cu, Y, La, Nd, Sm, Dy	Kiel & Schwartz, 1980

Table 14.1. (*cont.*)

Organism	Strain	Carbon source	Leaching agent	Process type[a]	Temperature (°C)[b]	Duration (d)[b]	Solid substrate	Elements leached	Reference
A. niger	6	Sucrose, sulfite liquor, malt	Citrate	Batch, plate	30	4	Low-grade copper ore; yttrium, lanthanum, neodymium, samarium, dysprosium	Cu, Y, La, Nd, Sm, Dy	Kiel & Schwartz, 1980
A. niger	62	Malt	Citrate	Plate	30	7	Yttrium, lanthanum, samarium, dysprosium	Y, La, Sm, Dy	Kiel & Schwartz, 1980
A. niger	65	Malt	Citrate	Plate	30	7	Yttrium, lanthanum	Y, La	Kiel & Schwartz, 1980
A. niger	66	Malt	Citrate	Plate	30	7	Yttrium, dysprosium	Y, Dy	Kiel & Schwartz, 1980
A. niger	7	Sucrose, sulfite liquor, malt	Citrate	Batch, plate	30	10–14	Low-grade copper ore; neodymium, samarium, dysprosium	Cu, Nd, Sm, Dy	Kiel & Schwartz, 1980
A. niger	8	Sucrose, sulfite liquor, malt	Citrate	Batch, plate	30	7–10	Low-grade copper ore; samarium, dysprosium	Cu, Sm, Dy	Kiel & Schwartz, 1980
A. niger	8	Glucose	Citrate	Batch	28	7	Bauxite	Al, Fe, Si	Ogurtsova et al., 1989
A. niger	A3		Citrate, oxalate	Batch	22	27	Lateritic nickel ore	Ni	Tarasova.. Khavski & Dudeney, 1993
A. niger	A-92	Sucrose	Citrate, oxalate	Batch		7–14	Biotite, muscovite, phlogopite	K	Weed, Davey & Cook, 1969
A. niger	ATCC 1015	Sucrose	Citrate	Batch, two-step	30	5–10	Garnierite, calamine	Ni, Zn	Castro et al., 2000
A. niger	ATCC 6275	Sucrose	Citrate, gluconate, succinate, malate	Batch, two-step	30	15	Mining residues	Cu, Fe, Ni, Zn	Mulligan, Galvez-Cloutier, Renaud, 1999

Organism	Strain	Substrate	Acid	Process			Material	Metal	Reference
A. niger	ATCC 6277	Glucose		Batch	28	5	Biotite, muscovite, glauconite, microcline	K	Eno & Reuzer, 1955
A. niger	ATCC 9142	Sucrose, sulfite liquor	Citrate	Plate	30	10–20	Low-grade copper ore	Cu	Kiel & Schwartz, 1980
A. niger	ATCC 9142	Sucrose		Two-step batch, two-step	25, 60	30, 0.25–1	Fly ash	Al, Fe	Singer et al., 1982
A. niger	ATCC 10108	Glucose, potato dextrose	Oxalate		28	0.2	Red mud	Al	Vachon et al., 1994
A. niger	ATCC 201373	Malt extract	Oxalate	Plate	25	8	Gypsum	Ca	Gharieb, Sayer & Gadd, 1998
A. niger	ATCC 201373	Malt extract	Oxalate	Plate	25	8	Gypsum	Ca	Gharieb et al., 1998
A. niger	ATCC 201373	Malt extract	Oxalate	Plate	25	10	Pyromorphite	Pb	Sayer et al., 1999
A. niger	ATCC 201373	Sucrose	Citrate, oxalate, gluconate	Plate	25	8	Gypsum	Ca	Gharieb & Gadd, 1999
A. niger	BS	Glucose	Citrate	Batch	30	19–29	Lateritic nickel ore	Ni	Bosecker, 1986
A. niger	CBS 246–65	Sucrose	Citrate, oxalate	Two-step	60	0.2–0.5	Kaolinite	Fe	Cameselle et al., 1995
A. niger	CMI 31821	Molasses	Citrate	Batch, two-step, fluid bed		10–16	Nepheline	Al, K, Na	King & Dudeney, 1987
A. niger	DSM 821	Glucose, malt extract	Citrate	Batch	30	17–34	Lateritic nickel ore	Ni	Bosecker, 1987
A. niger	DSM 821	Glucose	Citrate	Batch	30	19–29	Lateritic nickel ore	Ni	Bosecker, 1986
A. niger	DSM 823	Glucose	Citrate, oxalate, α-ketoglutarate	Two-step	30	21	Silicate manganese ore	Al, Fe, Mg, Mn	Bosecker, 1993
A. niger	DSM 823	Glucose, malt extract	Citrate	Batch	30	17–34	Lateritic nickel ore	Ni	Bosecker, 1987
A. niger	DSM 823	Glucose	Citrate	Batch	30	19–29	Lateritic nickel ore	Ni	Bosecker, 1986
A. niger	G-6	Citrate	Molasses, glucose	Two-step	30, 90	5, 0.2	kaolinite, halloysite, illite, montmorillonite, vermiculite, serpentine, chrysolite, genthite	Al, Fe	Groudev & Groudeva, 1986

Table 14.1. (cont.)

Organism	Strain	Carbon source	Leaching agent	Process type[a]	Temperature (°C)[b]	Duration (d)[b]	Solid substrate	Elements leached	Reference
A. niger	G-8	Citrate, oxalate	Molasses, glucose	Two-step	30, 90	5, 0.2	Kaolinite, halloysite, illite, montmorillonite, vermiculite, serpentine, chrysotile, genthite	Al, Fe	Groudev & Groudeva, 1986
A. niger	M	Glucose	Citrate, fumarate, oxalate	Plate, batch	21–25	7–14	Apophyllite, biotite, diorite, genthite, granite, harmotome, heulandite, leucite, muscovite, natrolite, nepheline, olivine, phlogopite, saponite, serpentine, stilbite, vermiculite, wollastonite; soil; clay, silt sand	Al, Fe, Mg, Mn, Si	Henderson & Duff, 1963
A. niger	N.82	Glucose	Citrate		21–25	7–14	Muscovite, natrolite	Al	Henderson & Duff, 1963
A. niger	VMK F-1119	Glucose		Batch	28	7	Bauxite	Al, Fe, Si	Ogurtsova et al., 1989
Aspergillus ochraceus	46	Glucose		Batch	28	7	Bauxite	Al, Fe, Si	Ogurtsova et al., 1989
Aspergillus oryzae Aspergillus terreus		Glucose, sucrose Xylose, glucose, galactose, rhamnose, mannose, inositol, sucrose, trehalose, maltose, raffinose	Citrate, oxalate Oxalate, citrate	Batch Plate, batch	32	30 14	Kaolinite, illite Phosphates of calcium and magnesium	Al Ca, Mg, P	Borovec, 1990 Rose, 1957
A. terreus	VMK F-1862	Glucose		Batch	28	7	Bauxite	Al, Fe, Si	Ogurtsova et al., 1989

Organism	Strain	Carbon source	Complexing agent	Process	Temp (°C)	Time (days)	Material	Elements	Reference
Aspergillus versicolor	VMK F-837	Glucose		Batch	28	7	Bauxite	Al, Fe, Si	Ogurtsova et al., 1989
Aspergillus wentii		Glucose, sucrose	Citrate, oxalate	Batch	32	30	Kaolinite, illite	Al	Borovec, 1990
Aureobasidium pullulans		Sucrose	Citrate	Batch	30	35	Bauxite	Al, Fe, Si	Karavaiko et al., 1989
Botrytis sp.		Glucose		Plate	21–25	7–14	Rock, weathered stone		Henderson & Duff, 1963
Botrytis sp.		Glucose	Citrate, oxalate	Batch	25	7	Silicates of calcium, magnesium, zinc	Ca, Mg, Zn, Si	Webley, Henderson & Taylor, 1963
Botrytis cinerea		Glucose		Batch	22	36–68	Glass	K	Drewello et al., 1994
Brettanomyces lambicus		Glucose		Batch	30	14	Filter dust	Zn, Cu, Pb	Wenzl, Burgstaller & Schinner, 1990
Candida sp.		Sucrose	Citrate	Batch	30	35	Bauxite	Al, Fe, Si	Karavaiko et al., 1989
Candida sp.	Y-76	Sucrose		Batch		7–14	Biotite, muscovite, phlogopite	K	Weed et al., 1969
Candida ethanolica		Glucose		Batch	28	7	Bauxite	Si	Ogurtsova et al., 1989
Candida lipolytica		Molasses		Batch	30	14	Limonite, goethite, hematite	Fe	Groudev, 1987
Candida tropicalis		Glucose		Batch	28	7	Bauxite	Si	Ogurtsova et al., 1989
Candida utilis		Glucose		Batch	28	7	Bauxite	Al, Fe, Si	Ogurtsova et al., 1989
Candida valida	VMK Y-2326	Glucose		Batch	28	7	Bauxite	Si	Ogurtsova et al., 1989
C. valida	VMK Y-2328	Glucose		Batch	28	7	Bauxite	Si	Ogurtsova et al., 1989
C. valida	VMK Y-247	Glucose		Batch	28	7	Bauxite	Si	Ogurtsova et al., 1989
Cephalosporium sp.		Glucose	Citrate	Plate	21–25	7–14	Rock, weathered stone		Henderson & Duff, 1963
Cephalosporium sp.		Glucose	Citrate, oxalate	Batch	25	7	Silicates of calcium, magnesium, zinc	Ca, Mg, Zn, Si	Webley et al., 1963
Chaetomium cochlioides	C-3	Sucrose		Batch		7–14	Biotite, muscovite, phlogopite	K	Weed et al., 1969

Table 14.1. (cont.)

Organism	Strain	Carbon source	Leaching agent	Process type[a]	Temperature (°C)[b]	Duration (d)[b]	Solid substrate	Elements leached	Reference
Cladosporium sp.		Glucose		Batch		1095	Copper–molybdenum ore	Cu	Kovalenko & Malakhova, 1990
Cladosporium sp.	1L1.9	Glucose		Plate	25	21	Sandstone		Palmer et al., 1991
Cladosporium cladosporioides		Glucose	Citrate, gluconate, oxalate	Batch		180	Sandstone, limestone, granite	Al, Ca, Fe, K, Mg, Mn, Na, Zn	de la Torre et al., 1993
Cladosporium cucumerinum		Glucose, sucrose	Citrate, oxalate	Batch	32	30	Kaolinite, illite	Al	Borovec, 1990
Cladosporium herbarum		Glucose	Citrate, gluconate, oxalate	Batch		70–101	Biotite, orthoclase, amphibolite	Al, K, Mg, Fe, Ca, Si	Eckhardt, 1979a
C. herbarum		Glucose	Gluconate	Batch		84	Biotite, orthoclase, amphibolite	Ca, K, Si, Mg, Fe, Al	Eckhardt, 1979b
Cladosporium resinae	IMI 296264	Malt extract		Plate	25	10	Zinc oxide, cobalt phosphate	Zn, Co	Sayer et al., 1995
C. resinae	VMK F-1701	Glucose		Batch	28	7	Bauxite	Al, Fe, Si	Ogurtsova et al., 1989
Cladosporium sphaerospermum		Glucose		Batch	22	36–68	Glass	K	Drewello et al., 1994
Coriolus (Trametes) versicolor	C7B 863A	Malt extract		Plate	25	10	Zinc oxide, zinc phosphate	Zn	Sayer et al., 1995
C. versicolor	C7B 863A	Malt extract	Oxalate	Plate	27	10	Pyromorphite	Pb	Sayer et al., 1999
Cytospora sp.		Glucose, galactose, fructose, mannose, starch, carboxymethyl-cellulose	Oxalate, citrate, succinate	Batch		14–21	Sandstone		de la Torre et al., 1991
Exophiala jeanselmei		Glucose, galactose, fructose, mannose, maltose, sucrose, starch, glycogen, cellulose		Plate	28	14	Sandstone		Petersen et al., 1988

Organism	Carbon source	Organic acid(s)	Process	Temperature	Duration (days)	Substrate	Elements	Reference
Fusarium sp.	Malt extract	Oxalate	Batch	27	14–21	Sandstone		Gomez-Alarcon, Munoz & Flores, 1994
Fusarium sp.	Glucose		Plate	21–25	7–14	Rock, weathered stone		Henderson & Duff, 1963
Fusarium sp.	Glucose, fructose, mannose, starch, carboxymethyl-cellulose	Oxalate, gluconate, citrate, fumarate, succinate	Batch	27	14–21	Sandstone		de la Torre et al., 1991
Fusarium sp.	Glucose		Batch	28	28	Silicate nickel ore	Al, Ni, Cr, Fe, Mg	Bosecker, 1989
Fusarium bulbigenum	Glucose		Batch	28	7	Bauxite	Al, Fe, Si	Ogurtsova et al., 1989
Fusarium cerealis	Glucose		Batch	22	36–68	Glass	K	Drewello et al., 1994
Fusarium moniliforme	Glucose, sucrose	Citrate, oxalate	Batch	32	30	Kaolinite, illite	Al	Borovec, 1990
Fusarium reticulatum	Malt extract	Oxalate	Batch	27	14–21	Sandstone		Gomez-Alarcon et al., 1994
F. reticulatum	Glucose, galactose, fructose, mannose, starch, carboxymethyl-cellulose	Oxalate, gluconate, citrate, fumarate, succinate	Batch	27	14–21	Sandstone		de la Torre et al., 1991
Glomus etunicatum			Batch	20–25	49	Soil	Zn	Banks et al., 1994
Hormodendrum sp.	Glucose	Citrate, oxalate	Batch	25	7	Silicates of calcium, magnesium, zinc	Ca, Mg, Zn, Si	Webley et al., 1963
Hormonendron sp.	Glucose		Plate	21–25	7–14	Rock, weathered stone		Henderson & Duff, 1963
Kluyveromyces marxianus	Glucose		Batch	28	7	Bauxite	Si	Ogurtsova et al., 1989
Monilia sp.	Glucose	Citrate, oxalate	Batch	25	7	Silicates of calcium, magnesium, zinc	Ca, Mg, Zn, Si	Webley et al., 1963
Mucor sp.	Glucose		Plate	21–25	7–14	Rock, weathered stone		Henderson & Duff, 1963
Mucor sp.	Glucose	Gluconate	Batch		84	Biotite, orthoclase, amphibolite	Ca, K, Si, Mg, Fe, Al	Eckhardt, 1979b
Mucor sp.	Glucose		Batch		1095	Copper–molybdenum ore	Cu	Kovalenko & Malakhova, 1990

Table 14.1. (cont.)

Organism	Strain	Carbon source	Leaching agent	Process type[a]	Temperature (°C)[b]	Duration (d)[b]	Solid substrate	Elements leached	Reference
Mucor sp.		Glucose	Citrate, oxalate	Batch	25	7	Silicates of calcium, magnesium, zinc	Ca, Mg, Zn, Si	Webley et al., 1963
Mucor hiemalis		Glucose, mannose, starch, carboxymethyl-cellulose	Fumarate, succinate	Batch	27	14-21	Sandstone		de la Torre et al., 1991
M. hiemalis	27	Glucose	Citrate, oxalate	Batch	28	7	Bauxite	Al, Fe, Si	Ogurtsova et al., 1989
Mucor mucedo		Glucose, sucrose	Citrate, oxalate	Batch	32	30	Kaolinite, illite	Al	Borovec, 1990
Mucor piriformis		Molasses		Batch	30	14	Limonite, goethite, hematite	Fe	Groudev, 1987
Mucor racemosus		Glucose	Citrate, succinate	Batch	27	5	Albite, hornblende, orthoclase, oligoclase	K, Ca	Müller & Förster, 1964
Paecilomyces sp.	1L1.10	Glucose	Oxalate, gluconate, citrate, fumarate, succinate	Plate	25	21	Sandstone		Palmer et al., 1991
Paecilomyces farinosus		Glucose, fructose, mannose, starch		Batch	27	14-21	Sandstone		de la Torre et al., 1991
Paecilomyces varioti		Glucose, sucrose, molasses	Citrate, oxalate	Batch	22	20	Oxidized copper ore; oxidized lead–zinc ore	Cu, Zn	Dave et al., 1981
Papularia sp.		Glucose	Citrate, oxalate	Batch	25	7	Silicates of calcium, magnesium, zinc	Ca, Mg, Zn, Si	Webley et al., 1963
Paxillus involutus		Sucrose	Citrate oxalate, gluconate	Plate	25	8	Gypsum	Ca	Gharieb & Gadd, 1999
Penicillium sp.		Molasses		Batch	30	14	Limonite, goethite, hematite	Fe	Groudev, 1987
Penicillium sp.		Glucose	Gluconate	Batch		84	Biotite, orthoclase, amphibolite	Ca, K, Si, Mg, Fe, Al	Eckhardt, 1979b
Penicillium sp.		Glucose, starch, carboxymethyl-cellulose	Oxalate, gluconate, citrate, fumarate, succinate	Batch	27	14-21	Sandstone		de la Torre et al., 1991

Organism	Strain	Carbon source	Acid	Process	Temp (°C)	Duration (days)	Substrate	Metals	Reference
Penicillium sp.		Sucrose	Citrate	Batch	30	13–20	Converter filter residue	Zn	Schinner & Burgstaller, 1989
Penicillium sp.		Glucose		Batch		1095	Copper–molybdenum ore	Cu	Kovalenko & Malakhova, 1990
Penicillium sp.		Glucose	Citrate, oxalate	Batch	25	7	Silicates of calcium, magnesium, zinc	Ca, Mg, Zn, Si	Webley *et al.*, 1963
Penicillium sp.		Glucose		Batch, two-step	25–35	35, 7	Manganese-containing silver ore	Mg, Ag	Gupta & Ehrlich, 1989
Penicillium sp.		Glucose	Citrate	Batch	30	19–29	Lateritic nickel ore	Ni	Bosecker, 1986
Penicillium sp.		Sucrose	Citrate	Batch	30	35	Bauxite	Al, Fe, Si	Karavaiko *et al.*, 1989
Penicillium sp.		Glucose		Batch		11–28	Silicate nickel ore	Al, Ni, Cr, Fe, Mg	Bosecker, 1989
Penicillium sp.		Glucose, galactose, fructose, mannose, maltose, sucrose, starch, glycogen, cellulose		Plate	28	14	Sandstone		Petersen *et al.*, 1988
Penicillium sp.		Glucose		Batch	37	20	Limonite lateritic ore	Co, Mn, Fe, Ni	Sukla *et al.*, 1995
Penicillium sp.	2L1.13	Glucose		Plate	25	21	Sandstone		Palmer *et al.*, 1991
Penicillium sp.	B1	Sucrose	Citrate, oxalate	Batch	30	48	Lateritic nickel ore	Co, Ni, Fe	Alibhai *et al.*, 1993
Penicillium sp.	C.2	Glucose	Citrate	Plate	21–25	7–14	Genthite, muscovite, natrolite, nepheline, olivine, phlogopite	Al, Mg	Henderson & Duff, 1963
Penicillium sp.	C1	Sucrose	Citrate, oxalate	Batch	30	48	Lateritic nickel ore	Co, Ni, Fe	Alibhai *et al.*, 1993
Penicillium sp.	C1.5	Glucose	Citrate	Plate	21–25	7–14	Rock, weathered stone	Co, Ni, Fe	Henderson & Duff, 1963
Penicillium sp.	D1	Sucrose	Citrate, oxalate	Batch	30	48	Lateritic nickel ore	Co, Ni, Fe	Alibhai *et al.*, 1993
Penicillium sp.	DSM 62867	Glucose	Citrate	Batch	30	19–29	Lateritic nickel ore	Ni, Fe	Bosecker, 1986
Penicillium sp.	E1	Sucrose	Citrate, oxalate	Batch	30	48	Lateritic nickel ore	Co, Ni, Fe	Alibhai *et al.*, 1993
Penicillium sp.	F1	Sucrose, glucose, molasses	Citrate, oxalate	Batch	30	40–48	Lateritic nickel ore	Ni, Fe	Tzeferis *et al.*, 1994

Table 14.1. (*cont.*)

Organism	Strain	Carbon source	Leaching agent	Process type[a]	Temperature (°C)[b]	Duration (d)[b]	Solid substrate	Elements leached	Reference
Penicillium sp.	F1	Sucrose	Citrate, oxalate	Batch	30	48	Lateritic nickel ore	Co, Ni, Fe	Alibhai *et al.*, 1993
Penicillium sp.	F1	Glucose, sucrose	Citrate, oxalate	Batch, two-step	30, 95	48	Nickeliferous laterites	Ni, Fe, Co	Tzeferis, 1994a
Penicillium sp.	F1	Sucrose, glucose	Citrate, oxalate	Batch, two-step	30, 95	48, 0.25	Hematitic lateritic ore	Ni, Co, Fe, Ca, Mg	Tzeferis, 1994b
Penicillium sp.	M1	Glucose	Citrate	Plate	21–25	7–14	Rock, weathered stone	Mg	Henderson & Duff, 1963
Penicillium sp.	P14	Sucrose, glucose, molasses	Citrate, oxalate	Batch	30	40–48	Lateritic nickel ore	Ni, Fe	Tzeferis *et al.*, 1994
Penicillium sp.	P14	Glucose, sucrose	Citrate, oxalate	Batch, two-step	30, 95	48	Nickeliferous laterites	Ni, Fe, Co	Tzeferis, 1994a
Penicillium sp.	P14	Sucrose, glucose	Citrate, oxalate	Batch, two-step	30, 95	48, 0.25	Hematitic lateritic ore	Ni, Co, Fe, Ca, Mg	Tzeferis, 1994b
Penicillium sp.	P2	Sucrose, glucose, molasses	Citrate, oxalate	Batch	30	40–48	Lateritic nickel ore	Ni, Fe	Tzeferis *et al.*, 1994
Penicillium sp.	P2	Sucrose, glucose, molasses	Citrate	Batch, two-step	30, 90	48	Lateritic nickel ore	Ni, Fe, Co, Ca, Mg	Tzeferis *et al.*, 1991
Penicillium sp.	P2	Glucose, sucrose	Citrate, oxalate	Batch, two-step	30, 95	48	Nickeliferous laterites	Ni, Fe, Co	Tzeferis, 1994a
Penicillium sp.	P2	Sucrose, glucose	Citrate, oxalate	Batch, two-step	30, 95	48, 0.25	Hematitic lateritic ore	Ni, Co, Fe, Ca, Mg	Tzeferis, 1994b
Penicillium sp.	P24	Sucrose, glucose, molasses	Citrate, oxalate	Batch	30	40–48	Lateritic nickel ore	Ni, Fe	Tzeferis *et al.*, 1994
Penicillium sp.	P24	Glucose, sucrose	Citrate, oxalate	Batch, two-step	30, 95	48	Nickeliferous laterites	Ni, Fe, Co	Tzeferis, 1994a
Penicillium sp.	P24	Sucrose, glucose	Citrate, oxalate	Batch, two-step	30, 95	48, 0.25	Hematitic lateritic ore	Ni, Co, Fe, Ca, Mg	Tzeferis, 1994b
Penicillium sp.	P6	Sucrose, glucose, molasses	Citrate, oxalate	Batch	30	40–48	Lateritic nickel ore	Ni, Fe	Tzeferis *et al.*, 1994
Penicillium sp.	P6	Glucose, malt extract	Citrate	Batch	30	17–34	Lateritic nickel ore	Ni, Fe, Mg	Bosecker, 1987
Penicillium sp.	P6	Glucose, sucrose	Citrate, oxalate	Batch, two-step	30, 95	48	Nickeliferous laterites	Ni, Fe, Co	Tzeferis, 1994a

Organism	Strain	Carbon source	Organic acid	Culture	Temperature (°C)	Value	Substrate	Elements	Reference
Penicillium sp.	P6	Glucose	Citrate	Batch	30	19–29	Lateritic nickel ore	Ni, Fe	Bosecker, 1986
Penicillium sp.	P6	Sucrose, glucose	Citrate, oxalate	Batch, two-step	30, 95	48, 0.25	Hematitic lateritic ore	Ni, Co, Fe, Ca, Mg	Tzeferis, 1994b
Penicillium sp.	WPF-57	Sucrose, starch, glucose, fructose, lactose, galactose	Citrate	Batch	23	5	Cuprous sulfide, chalcocite, native copper	Cu	Wenberg, Erbisch & Volin, 1971
Penicillium sp.	WPF-61	Sucrose, starch, glucose, fructose, lactose, galactose	Citrate	Batch	23	5	Cuprous sulfide, chalcocite, native copper	Cu	Wenberg et al., 1971
Penicillium aurantiogriseum		Glucose		Batch	22	36–68	Glass	K	Drewello et al., 1994
P. bilaii		Malt extract	Oxalate	Plate	25	10	Pyromorphite	Pb	Sayer et al., 1999
		Sucrose	Citrate, oxalate, gluconate	Plate		8	Gypsum	Ca	Gharieb & Gadd, 1999
Penicillium brevicompactum		Glucose	Gluconate	Batch	28	84	Biotite, orthoclase, amphibolite	Ca, K, Si, Mg, Fe, Al	Eckhardt, 1979b
P. brevicompactum		Glucose		Batch		7	Bauxite	Al, Fe, Si	Ogurtsova et al., 1989
P. brevicompactum		Glucose		Batch	22	36–68	Glass	K	Drewello et al., 1994
Penicillium chrysogenum		Glucose, sucrose	Citrate, oxalate	Batch	32	30	Kaolinite, illite	Al	Borovec, 1990
P. chrysogenum		Glucose		Batch	22	36–68	Glass	K	Drewello et al., 1994
P. chrysogenum	15	Glucose		Batch	28	7	Bauxite	Al, Fe, Si	Ogurtsova et al., 1989
P. chrysogenum	16	Glucose		Batch	28	7	Bauxite	Al, Fe, Si	Ogurtsova et al., 1989
P. chrysogenum	18	Glucose		Batch	28	7	Bauxite	Al, Fe, Si	Ogurtsova et al., 1989
P. chrysogenum	19	Glucose		Batch	28	7	Bauxite	Al, Fe, Si	Ogurtsova et al., 1989
P. chrysogenum	28	Glucose		Batch	28	7	Bauxite	Al, Fe, Si	Ogurtsova et al., 1989
P. chrysogenum	29	Glucose		Batch	28	7	Bauxite	Al, Fe, Si	Ogurtsova et al., 1989
P. chrysogenum	IMI 178514	Malt extract		Plate	25	10	Zinc oxide, cobalt phosphate	Zn, Co	Sayer et al., 1995

Table 14.1. (cont.)

Organism	Strain	Carbon source	Leaching agent	Process type[a]	Temperature (°C)[b]	Duration (d)[b]	Solid substrate	Elements leached	Reference
Penicillium citrinum		Glucose, sucrose	Citrate, oxalate	Batch	32	30	Kaolinite, illite	Al	Borovec, 1990
P. citrinum	ATCC SD 1077		Citrate, oxalate	Batch	21–35	0.75–4	Gold-bearing ore, carbonaceous ore, carbonaceous pyritic ore	Au	Portier, 1991
Penicillium corylophilum		Malt extract	Fumarate, oxalate	Batch	27	14–21	Sandstone		Gomez-Alarcon et al., 1994
Penicillium expansum		Glucose	Citrate, gluconate, oxalate	Batch		70–101	Biotite, orthoclase, amphibolite	Al, K, Mg, Fe, Ca, Si	Eckhardt, 1979a
P. expansum		Glucose	Gluconate	Batch		84	Biotite, orthoclase, amphibolite	Ca, K, Si, Mg, Fe, Al	Eckhardt, 1979b
P. expansum		Glucose, sucrose	Citrate, oxalate	Batch	32	30	Kaolinite, illite	Al	Borovec, 1990
Penicillium frequentans		Glucose	Citrate, gluconate, oxalate	Batch		180	Sandstone, limestone, granite	Al, Ca, Fe, K, Mg, Mn, Na, Zn	de la Torre et al., 1993
P. frequentans		Glucose, galactose, fructose, mannose, starch, carboxymethyl-cellulose	Oxalate, gluconate, citrate, fumarate, succinate	Batch	27	14–21	Sandstone		de la Torre et al., 1991
P. frequentans		Glucose	Succinate, tartrate, citrate, oxalate	Batch	27	5–6	Albite, hornblende, orthoclase, oligoclase	K, Ca	Müller & Förster, 1964
Penicillium funiculosum		Glucose		Batch		23	Silicate nickel ore	Al, Ni, Cr, Fe, Mg	Bosecker, 1989
Penicillium granulatum		Glucose	Gluconate	Batch		84	Biotite, orthoclase, amphibolite	Ca, K, Si, Mg, Fe, Al	Eckhardt, 1979b
Penicillium griseofulvum		Malt extract	Oxalate, succinate	Batch	27	14–21	Sandstone		Gomez-Alarcon et al., 1994
Penicillium lanosum		Glucose, starch	Oxalate, gluconate, malate, citrate, fumarate, succinate	Batch	27	14–21	Sandstone		de la Torre et al., 1991

Organism	Strain	Carbon source	Organic acids	Mode	Temp (°C)	Duration	Substrate	Elements	Reference
Penicillium lilacinum		Glucose	Succinate, citrate, tartrate	Batch	27	5–6	Albite, hornblende, orthoclase, oligoclase	K, Ca	Müller & Förster, 1964
Penicillium luteum		Glucose, sucrose	Citrate, oxalate	Batch	32	30	Kaolinite, illite	Al	Borovec, 1990
Penicillium martensii		Glucose	Tartrate, succinate, oxalate, citrate	Batch	27	5–6	Albite, hornblende, orthoclase, oligoclase	K, Ca	Müller & Förster, 1964
Penicillium nigricans		Glucose	Succinate, citrate, tartrate, oxalate	Batch	27	5–6	Albite, hornblende, orthoclase, oligoclase	K, Ca	Müller & Förster, 1964
Penicillium notatum				Batch	25	10–30	Thin foils of erbium, cobalt, copper, zinc, cadmium, aluminium, tin; stainless steel	Er, Co, Cu, Zn, Cd, Al, Sn, Fe	Siegel, Siegel & Clark, 1983
P. notatum		Sucrose		Batch	26	14	Spodumene	Al, Li, Si	Karavaiko *et al.*, 1980
P. notatum		Glucose, sucrose	Citrate, oxalate	Batch	32	30	Kaolinite, illite	Al	Borovec, 1990
P. notatum		Sucrose		Batch	24–26	210	Pegmatite, spodumene	Li, Si, Al, Fe	Avakyan *et al.*, 1981
P. notatum	ATCC 6275	Glucose, potato dextrose		Batch, two-step	28	0.2	Red mud	Al	Vachon *et al.*, 1994
Penicillium rugulosum	IR-94MF1	Sucrose	Citrate, gluconate	Batch	28	7	Phosphates of calcium, iron, aluminium, apatite	Fe, Al	Reyes *et al.*, 1999
Penicillium simplicissimum		Sucrose	Citrate	Batch	30	9–14	Filter dust	Zn	Burgstaller *et al.*, 1992
P. simplicissimum		Sucrose	Citrate	Batch	30	6	Filter dust, zinc oxide	Zn	Franz, Burgstaller & Schinner, 1991
P. simplicissimum		Sucrose	Citrate, amino acids	Batch	30		Filter dust	Zn	Müller *et al.*, 1995
P. simplicissimum		Sucrose	Citrate	Batch	30	8	Electrofilter dust	Ni, V	Strasser, Brunner & Schinner, 1993b
P. simplicissimum		Glucose, sucrose	Citrate, oxalate	Batch	32	30	Kaolinite, illite	Al	Borovec, 1990
P. simplicissimum		Glucose		Batch		20–28	Silicate nickel ore	Al, Ni, Cr, Fe, Mg	Bosecker, 1989
P. simplicissimum		Sucrose	Citrate, oxalate, gluconate	Plate	25	8	Gypsum	Ca	Gharieb & Gadd, 1999

Table 14.1. (cont.)

Organism	Strain	Carbon source	Leaching agent	Process type[a]	Temperature (°C)[b]	Duration (d)[b]	Solid substrate	Elements leached[a]	Reference
P. simplicissimum	1	Citrate	Molasses, glucose	Two-step	30, 90	5, 0.2	Kaolinite, halloysite, illite, montmorillonite, vermiculite, serpentine, chrysotile, genthite	Al, Fe	Groudev & Groudeva, 1986
P. simplicissimum	6	Citrate, oxalate	Molasses, glucose	Two-step	30, 90	5, 0.2	Kaolinite, halloysite, illite, montmorillonite, vermiculite, serpentine, chrysotile, genthite	Al, Fe	Groudev & Groudeva, 1986
P. simplicissimum	As	Malt extract		Plate	25	10	Phosphates of cobalt, zinc; zinc oxide	Zn, Co	Sayer et al., 1995
P. simplicissimum	ATCC 10495	Sucrose		Batch	30	28–42	Labradorite, augite, illmenite, olivine	Al	Mehta, Torma & Murr, 1979
P. simplicissimum	ATCC 48705	Glucose, potato-dextrose	Citrate	Batch, two-step	28	0.2	Red mud	Al	Vachon et al., 1994
P. simplicissimum	CBS 288.53	Sucrose	Citrate, gluconate, oxalate	Batch	30	21	Electronic waste	Al, Cu, Ni, Pb, Sn, Zn	Brandl et al., 1999
P. simplicissimum	DSM 62867	Glucose, malt extract	Citrate	Batch	30	17–34	Lateritic nickel ore	Ni	Bosecker, 1987
P. simplicissimum	Ls	Malt extract		Plate	25	10	Phosphates of cobalt, zinc; calcium carbonate, zinc oxide	Ca, Co, Zn	Sayer et al., 1995
P. simplicissimum	P23	Citrate, oxalate		Batch	22	27	Lateritic nickel ore	Ni	Tarasova et al., 1993
P. simplicissimum	P6	Glucose		Batch	30	29	Silicate nickel ore	Ni, Fe, Mg	Bosecker, 1989
P. simplicissimum	WB-28	Glucose		Batch		7	Basalt, granodiorite, granite, quartzite	Ti	Silverman & Munoz, 1971

Organism	Strain	Carbon source	Organic acids	Mode	Temp.	Time	Substrate	Metals	Reference
P. simplicissimum	WB-28	Glucose		Batch, two-step	30	7	Basalt, granodiorite, granite, quartzite, dunite, peridotite, andesite, rhyolite	Si, Fe, Al, Mg	Silverman & Munoz, 1970
Phialophora sp. (*Margarinomyces*)		Glucose	Citrate, oxalate	Batch	25	7	Silicates of calcium, magnesium, zinc	Ca, Mg, Zn, Si	Webley *et al.*, 1963
Phoma sp.		Glucose, galactose, fructose, mannose, maltose, sucrose, starch, glycogen, cellulose		Plate	28	14	Sandstone		Petersen *et al.*, 1988
Phoma sp.	2L1.3	Glucose	Citrate, oxalate	Plate	25	21	Sandstone		Palmer *et al.*, 1991
Pullularia sp.		Glucose		Batch	25	7	Silicates of calcium, magnesium, zinc	Ca, Mg, Zn, Si	Webley *et al.*, 1963
Rhizoctonia sp.	R-160	Sucrose		Batch		7–14	Biotite, muscovite, phlogopite	K	Weed *et al.*, 1969
Rhizopus arrhizus	IMI 57412	Malt extract	Oxalate	Plate		10	Pyromorphite	Pb	Sayer *et al.*, 1999
Rhizopus japonicus	VKM F-1043	Glucose		Batch	28	7	Bauxite	Al, Fe, Si	Ogurtsova *et al.*, 1989
Rhizopus nigricans		Glucose, sucrose	Citrate, oxalate	Batch	32	30	Kaolinite, illite	Al	Borovec, 1990
R. nigricans		Glucose	Glycolate, tartrate	Batch	27	5	Albite, hornblende, orthoclase, oligoclase	Al, Fe, K, Ca	Müller & Förster, 1964
Rhodotorula rubra		Glucose		Batch	28	7	Bauxite	Al, Fe, Si	Ogurtsova *et al.*, 1989
Saccharomyces cerevisiae		Glucose, sulfite waste liquor	Formate, oxalate	Batch, fed-batch	25–30	5–22	Electronic waste	Cu, Pb, Sn	Hahn, Willscher & Straube, 1993
S. cerevisiae		Glucose		Batch	28	7	Bauxite	Al, Fe	Ogurtsova *et al.*, 1989
S. cerevisiae		Glucose		Batch	28	52–227	Orthoclase, microcline, oligoclase, labradorite, nepheline, leucite, muscovite, olivine, augite, hornblende, turmalin, apatite	Ca, Fe, K, Si, Mg	Basselik, 1913

Table 14.1. (*cont.*)

Organism	Strain	Carbon source	Leaching agent	Process type[a]	Temperature (°C)[b]	Duration (d)[b]	Solid substrate	Elements leached	Reference
Sclerotium rolfsii		xylose, glucose, galactose, rhamnose, mannose, inositol, sucrose, trehalose, maltose, raffinose	Oxalate, citrate	Plate, batch		14	Phosphates of calcium and magnesium	Ca, Mg, P	Rose, 1957
Scopulariopsis brevicaulis		Glucose, sucrose	Citrate, oxalate	Batch	32	30	Kaolinite, illite	Al	Borovec, 1990
Serpula himantioides		Malt extract	Oxalate	Plate	25	8	Gypsum	Ca	Gharieb *et al.*, 1998
S. himantioides		Malt extract	Oxalate	Plate	25	8	Gypsum	Ca	Gharieb *et al.*, 1998
S. himantioides		Malt extract	Oxalate	Plate	25	10	Gypsum	Ca	Tait *et al.*, 1999
Spicaria sp.		Glucose	Oxalate	Plate	21–25	7–14	Genthite, leucite, muscovite, natrolite, nepheline, olivine, phlogopite, saponite, stilbite	Al, Mg	Henderson & Duff, 1963
Sporotrichum sp.		Glucose	Citrate, oxalate	Batch	25	7	Silicates of calcium, magnesium, zinc	Ca, Mg, Zn, Si	Webley *et al.*, 1963
Trichoderma sp.		Glucose		Plate	21–25	7–14	Genthite, natrolite	Al, Mg	Henderson & Duff, 1963
Trichoderma sp.		Glucose	Citrate, oxalate	Batch	25	7	Silicates of calcium, magnesium, zinc	Ca, Mg, Zn, Si	Webley *et al.*, 1963
Trichoderma sp.	T-12	Sucrose		Batch		7–14	Biotite, muscovite, phlogopite	K	Weed *et al.*, 1969
Trichoderma lignorum		Sucrose		Batch	24–26	210	Pegmatite, spodumene	Li, Si, Al, Fe	Avakyan *et al.*, 1981
Trichoderma pseudokoningii		Glucose, galactose, fructose, mannose, starch, carboxymethyl-cellulose	Oxalate, gluconate, citrate, fumarate, succinate	Batch	27	14–21	Sandstone		de la Torre *et al.*, 1991

Organism	Strain	Carbon source	Organic acids	Culture	Temp (°C)	Time (days)	Mineral/substrate	Elements	Reference
Trichoderma viride		Glucose, galactose, fructose, mannose, starch, carboxymethyl-cellulose	Gluconate, citrate, fumarate, succinate	Batch	27	14–21	Sandstone		de la Torre et al., 1991
T. viride		Glucose, sucrose	Citrate, oxalate	Batch	32	30	Kaolinite, illite	Al	Borovec, 1990
T. viride	ATCC 32098	Glucose, potato dextrose		Batch, two-step	28	0.2	Red mud	Al	Vachon et al., 1994
Truncatella sp.		Glucose, starch, carboxymethyl-cellulose	Oxalate, citrate, fumarate, succinate	Batch	27	14–21	Sandstone		de la Torre et al., 1991
Yarrowia lipolytica		Glucose, sulfite waste liquor	Formate, oxalate	Batch, fed-batch	25–30	5–22	Catalyst, electronic waste	Cu, Pb, Sn	Hahn et al., 1993
Zygorhynchus moelleri	Z-41	Sucrose		Batch		7–14	Biotite, muscovite, phlogopite	K	Weed et al., 1969
Zygorhynchus sp.		Glucose	Glutarate	Batch	27	5	Albite, hornblende, orthoclase, oligoclase	K, Ca	Müller & Förster, 1964
Unidentified strain		Glucose	Oxalate	Batch	24	45	Forest soil	Al, Fe, Mg	Berthelin, Kogblevi & Domergues, 1974
Unidentified strain			Oxalate				Yellow quartzite	Ca	Feldmann et al., 1997
Unidentified strain		Malt extract, glucose		Batch	25	15–31	Scheelite	W, Si, Fe, Al, Ca, Mg	Guedes de Carvalho et al., 1990
Unidentified strain 14		Malt extract		Plate	25	10	Cobalt phosphate	Co	Sayer et al., 1995
Unidentified strain 2		Malt extract		Plate	25	10	Phosphates of zinc, cobalt; zinc oxide	Zn, Co	Sayer et al., 1995
Unidentified strain 27		Malt extract		Plate	25	10	Zinc oxide, cobalt phosphate	Zn, Co	Sayer et al., 1995
Unidentified strain 29		Malt extract		Plate	25	10	Phosphates of zinc, cobalt; zinc oxide	Zn, Co	Sayer et al., 1995
Unidentified strain 2L1.12		Glucose		Plate	25	21	Sandstone		Palmer et al., 1991
Unidentified strain 2L1.12P		Glucose		Plate	25	21	Sandstone		Palmer et al., 1991
Unidentified strain 2L1.2		Glucose		Plate	25	21	Sandstone		Palmer et al., 1991
Unidentified strain 3		Malt extract		Plate	25	10	Phosphates of zinc, cobalt; zinc oxide	Zn, Co	Sayer et al., 1995

Table 14.1. (*cont.*)

Organism	Strain	Carbon source	Leaching agent	Process type[a]	Temperature (°C)[b]	Duration (d)[b]	Solid substrate	Elements leached	Reference
Unidentified strain	31	Malt extract		Plate	25	10	Zinc oxide, cobalt phosphate	Zn, Co	Sayer et al., 1995
Unidentified strain	33	Malt extract		Plate	25	10	Zinc oxide, cobalt phosphate	Zn, Co	Sayer et al., 1995
Unidentified strain	34	Malt extract		Plate	25	10	Phosphates of zinc, cobalt; zinc oxide	Zn, Co	Sayer et al., 1995
Unidentified strain	35	Malt extract		Plate	25	10	Zinc oxide, cobalt phosphate	Zn, Co	Sayer et al., 1995
Unidentified strain	36	Malt extract		Plate	25	10	Zinc oxide, cobalt phosphate	Zn, Co	Sayer et al., 1995
Unidentified strain	37	Malt extract		Plate	25	10	Zinc oxide, cobalt phosphate	Zn, Co	Sayer et al., 1995
Unidentified strain	4	Malt extract		Plate	25	10	Zinc oxide, zinc phosphate	Zn	Sayer et al., 1995
Unidentified strain	40	Malt extract		Plate	25	10	Phosphates of zinc, cobalt; zinc oxide	Zn, Co	Sayer et al., 1995
Unidentified strain	51	Malt extract		Plate	25	10	Zinc oxide, cobalt phosphate	Zn, Co	Sayer et al., 1995
Unidentified strain	9	Malt extract		Plate	25	10	Cobalt phosphate	Co	Sayer et al., 1995
Unidentified strain	KAT 4	Glucose, sulfite waste liquor	Formate, oxalate	Batch, fed-batch	25–30	24–35	Catalyst, electronic waste	Cu, Pb, Sn	Hahn et al., 1993
Unidentified strain		Potato-dextrose broth		Batch	33	40	Argentite, proutite, polybasite, pearceite	Ag	Salameh et al., 1999
Unidentified strain (yeast)		Glucose		Batch		1095	Copper–molybdenum ore	Cu	Kovalenko & Malakhova, 1990
Unidentified strain (yeast)	1L3.1	Glucose		Plate	25	21	Sandstone		Palmer et al., 1991
Unidentified strain (yeast)		Potato-dextrose broth		Batch	33	40	Argentite, proutite, polybasite, pearceite	Ag	Salameh et al., 1999

| Unidentified strain(s) | Plant biomass (lucerne) | Batch | 180 | Oxides of cobalt, copper, lead, nickel, zinc, manganese, iron | Co, Cu, Pb, Ni, Zn, Mn, Fe | Bloomfield & Kelso, 1971 |
| Unidentified strains | Glucose | Batch | 2–90 | Glass | K, Ca, Fe, Mn, P | Krumbein, Urzi & Gehrmann, 1991 |

[a]Plate, microorganisms are grown on the surface of solidified media; batch, microorganisms are in direct contact with solid substrates in liquid medium; two-step, in a first step, microorganisms are grown in the absence of solid substrates and the cell-free supernatant is used for leaching in a second step.
[b]Where two processes are given the temperature and leaching duration is given first for the batch and then for the two-step process.

pose a problem and limit downstream processing. Solubilized metals can be also adsorbed by the fungal biomass. The third technique is a two-step process in which fungi are grown in the absence of metal-containing solids to allow optimal production of organic acids in high concentrations. Subsequently, the biomass is separated, for example by filtration, and the supernatant is used for the second leaching step. This method allows the leaching temperatures to be increased and, therefore, the treatment duration is reduced. However, as yet there are no industrial-scale applications.

Most biohydrometallurgical treatments have been performed at temperatures of 25–35 °C. For industrial applications (especially field applications), processes at lower temperatures might be important to reduce costs for cooling and heating and to allow the development of processes such as soil bioremediation and *in situ* metal leaching processes in colder geographic regions. Thermophilic fungi have not been exploited for bioleaching purposes.

Various solid substrates have been microbiologically processed to leach elements of interest (Table 14.1). In many cases natural minerals and ores were treated; for example, metals from a series of silicate minerals were mobilized (Table 14.1). In contrast, only a few reports are available on the treatment of metal-containing industrial waste materials by fungi (e.g. Singer, Navrot & Shapira, 1982; Burgstaller, Strasser & Schinner, 1990; Bosshard, Bachofen & Brandl, 1996; Drewello & Weissmann, 1997; Krebs *et al.*, 1997). Consequently there are a large number of potential applications for fungi biohydrometallurgical waste treatment and the mobilization and recovery of metals.

Metal leaching from fly ash

Electric filters are used to retain fly ash during municipal solid waste incineration, resulting in a solid concentrate containing a wide variety of elements mostly as oxides (Bosshard *et al.*, 1996). High concentrations of heavy metals, for example cadmium, copper, nickel, lead and zinc, pose an environmental hazard requiring post-treatment (immobilization with cement) and disposal in controlled landfills. This is not entirely satisfactory. Some of the elements (aluminium, zinc) are present in concentrations that allow an economical metal recovery. Others (e.g. silver, nickel, zirconium) are present at relatively low concentrations, comparable to low-grade ores, which makes conventional recovery difficult. In these cases, microbial processes are the only possible technique to obtain metal values from these materials. As these residues have low toxicities, a biological recovery

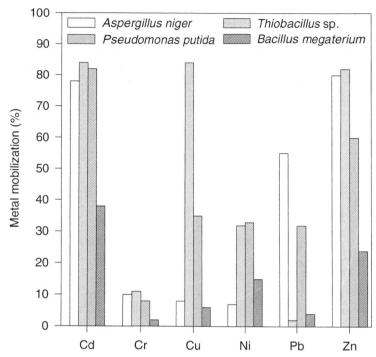

Fig. 14.1. Metal mobilization from fly ash originating from municipal waste incineration in suspension $(20\,g\,l^{-1})$ after 14 to 21 days of incubation at 30°C by *Aspergillus niger* (from Brandl, 1999), *Bacillus megaterium*, *Pseudomonas putida* and *Thiobacillus* sp. The original metal concentration in the fly ash was, 0.5, 0.6, 1, 0.1, 8, and $31\,g\,kg^{-1}$ for cadmium, chromium, copper, nickel, lead, and zinc, respectively.

process is potentially possible. The expected shortage of non-renewable resources is a stimulation for the development of new or improved technologies to supply raw materials. The use of microbiological (fungal as well as bacterial) leaching processes offers a possibility of obtaining metals from mineral resources such as fly ash (Bosecker, 1994; Torma, 1988).

In the presence of fly ash, *A. niger* produces mainly gluconic acid, whereas in its absence citric acid is formed (Bosshard *et al.*, 1996). Citric acid formation is enhanced in low-manganese medium through inhibition of the enzymes of the TCA cycle apart from citrate synthase. In addition, low pH has been reported to favour citric acid formation whereas higher pH values, which might result in the presence of the highly alkaline fly ash, stimulate oxalic and gluconic acid formation (Grewal & Kalra, 1995).

Figure 14.1 compares the effectiveness of metal leaching from a fly ash suspension by *A. niger* with other heterotrophic organisms (*Bacillus megaterium*, *Pseudomonas putida*) as well as autotrophic *Thiobacillus* spp. Good mobilization was achieved for cadmium, lead and zinc. *A. niger* seems to be especially suited for the mobilization of lead compared with the other organisms. Almost 100% of the lead is mobilized when suspensions with higher fly ash concentrations are treated (Bosshard *et al.*, 1996).

Metal leaching from electronic waste material

Relatively short lifetimes of electrical and electronic equipment are leading to an increased amount of waste materials. In Switzerland, approximately 110 000 tonnes of electrical appliances have to be disposed yearly, in Germany it is ten times more (1.5 million tonnes). Specialized companies are responsible for recycling and disposal. The equipment is dismantled, manually sorted and subjected to a mechanical separation process. Dust-like material is generated by shredding and other separation steps during mechanical recycling of electronic wastes: approximately 4% of the 2400 tonnes of scrap treated yearly by a specialized company (IMMARK AG; Kaltenbach, Switzerland) is collected as fine-grained powdered material. Whereas most of the electronic scrap can be recycled and processed in metal manufacturing industries, dust residues have to be disposed in landfills or incinerated. However, these residues contain metals in concentrations that might be of economical value, for example 237, 80, 15, 20, 23 and $26 \, \mathrm{g \, kg^{-1}}$ dust for aluminium, copper, nickel, lead, tin and zinc, respectively; there are also small amounts of silver and gold. If a suitable treatment and recovery process was available, this material might serve as a secondary metal resource.

Only limited data are available on the bioleaching of metals from electronic waste materials (Krebs *et al.*, 1997). Heterotrophic bacteria and fungi, namely *Bacillus* spp., *Saccharomyces cerevisiae*, *Yarrowia lipolytica*, have been used to mobilize lead, copper and tin from printed circuit boards (Hahn, Willscher & Straube, 1993). Aluminium, copper, lead, nickel, tin and zinc have been leached from dust-like residues from the mechanical recycling of used electronic equipment by *A. niger* and *P. simplicissimum* (Brandl, Bosshard & Wegmann, 1999). In general, *A. niger* and *P. simplicissimum* are able to grow on electronic scrap at concentrations of $< 10 \, \mathrm{g \, l^{-1}}$ scrap in the medium (Brandl *et al.*, 1999). After a prolonged adaptation time of 6 weeks and longer, fungi also grew at concentrations up to $100 \, \mathrm{g \, l^{-1}}$. During growth on sucrose, various organic acids (citric, gluconic, oxalic acid) are formed. After an incubation period of 21 days,

A. niger formed $3 \, \text{mmol} \, \text{l}^{-1}$ oxalate and $180 \, \text{mmol} \, \text{l}^{-1}$ citrate, whereas *P. simplicissimum* formed $5 \, \text{mmol} \, \text{l}^{-1}$ oxalate and $20 \, \text{mmol} \, \text{l}^{-1}$ citrate over the same period (Brandl *et al.*, 1999). Differences can be seen, therefore, in the leaching pattern of the two fungal strains. In general, *P. simplicissimum* is able to mobilize more metals than *A. niger* under the same conditions. However, *A. niger* has a certain preference for mobilization of copper: 65% of copper and tin present in electronic scrap was mobilized, and $> 95\%$ of aluminium, nickel, lead and zinc. A two-step process seems appropriate to increase leaching efficiencies for an industrial application. In a first step, organisms would be grown in the absence of electronic scrap; this would be followed by a second step where the formed metabolites are used for metal solubilization. This has already been suggested for metal mobilization from fly ash by both fungi and bacteria. There are several advantages with this approach: (i) biomass is not in direct contact with the metal-containing waste and might be recycled; (ii) waste material is not contaminated by microbial biomass; (iii) acid formation in the absence of waste material can be optimized; and (iv) higher waste concentrations can be applied compared with a one-step process, resulting in increased metal yields. In addition to the two-step process, microorganisms might be selected by adaptation experiments to tolerate electronic scrap at elevated concentrations and to leach metals selectively from such materials.

Metal leaching from soil

Fungi are a fundamental and vital part of the soil microbial flora, accounting for the bulk of microbial biomass in the soil. They are extensively involved in the biogeochemical cycling of elements (Foster, 1949). At the beginning of the twentieth century, the important role of fungi in soil as producers of a series of acids was recognized (Heinze, 1906). Insoluble carbonates and magnesium minerals as well as phosphate-based fertilizers are solubilized and bioavailability for plants is increased. Carbon dioxide formed during respiration can also contribute to the dissolution of minerals. *A. niger* has been used in a growth biotest to assess the bioavailability of elements such as potassium, magnesium or phosphorus in soils (Niklas & Poschenrieder, 1932; Mehlich, Truog & Fred, 1933). The method involved the mobilization of a certain compound that was absent in the growth medium and which the fungus, therefore, needed to solubilize from the soil. Besides different soil samples, the release of potassium from biotite, muscovite, glauconite and microcline was also investigated using this technique.

The solubilization of major soil mineral elements such as aluminium, calcium, iron, manganese, silicon, sodium or titanium from carbonates, oxides, phosphates, silicates or sulfides is the result of interactions between rocks and the biosphere and is termed weathering (Berthelin, 1983; Robert & Berthelin, 1986; Hirsch *et al.*, 1995). Fungi can act as geological agents affecting these biogeochemical processes through metal speciation as well as metal solubilization and immobilization (Gadd, 1999; Sterflinger, 2000). Fungal metal-leaching activities in soil depend on several factors such as the soil nutrient status (mineral availabilities), particle size or mineral type (which influences resistance to weathering through crystal structure) (Berthelin, 1983; Tan, 1986). The effects of bacteria and fungi on soil minerals occur through mechanisms such as acidolysis, complexolysis, and/or redoxolysis. A series of organic acids can be found in soil originating from fungal (as well as bacterial) metabolism, resulting in organic acidolysis and complex and chelate formation (Berthelin, 1983; Tan, 1986). A kinetic model of the coordination chemistry of weathering has been developed that describes the dissolution of oxides by protonation of the mineral surface as well as by the surface concentration of suitable complex-forming ligands such as oxalate, malonate, citrate and succinate (Furrer & Stumm, 1986). Proton-induced and ligand-induced mineral solubilization occurs simultaneously in the presence of ligands under acidic conditions. Redoxolysis is of minor importance in fungi. However, inorganic metal sulfides can be oxidized by *A. niger* or *Trichoderma harzianum*, but only to a small extent (Wainwright & Grayston, 1986). It has been hypothesized that these organisms cannot be effectively applied to leach metals from sulfides because of fungal metal uptake or metal adsorption.

Only recently have leaching activities of fungi been exploited for the bioremediation of metal-contaminated soils (Roane, Pepper & Miller, 1996; Gadd *et al.*, 1998; Gray, 1998). Surprisingly, the use and technical application of this metabolic potential is hardly investigated in spite of broad fungal biodiversity and the knowledge accumulated from studies on the effect of organic acids on soil minerals (Table 14.1).

Conclusions

Research developments and technical applications of the metabolic potential of fungi for metal leaching from metal-containing solids could exploit a number of different areas of fungal metabolisms, and biohydrometallurgical applications could consider leaching by heterotrophic microorganisms, including fungi (Ehrlich, 1999). However, there are several drawbacks in pursuing this goal. Usually, conditions for thiobacilli-mediated leaching are very restrictive (extremely low pH) and consequently can proceed without sterilization of growth media and solid materials. In contrast, leaching activities of heterotrophic organisms take place over a much broader pH range, giving rise to possibilities of contamination with unwanted microbial flora. As a result, sterility might become a problem both for large-scale applications and from an economical point of view (i.e. energy costs). Another economic restriction is the organic carbon sources required for fungal growth and acid formation (see Table 14.1). Generally, these carbon sources represent a major cost factor. The availability of an inexpensive carbon source, for example sugar molasses or by-products from the food industry, is a prerequisite for the development of heterotrophic leaching processes.

The use of thermophilic fungi may overcome some of these limitations. In general, thermophilic fungi have simple nutritional requirements and are often characterized by high growth rate and fast substrate turnover (Satyanarayana, Johri & Klein, 1992). In addition, they are more resistant to stress conditions. Contamination problems are minimized because of their high cultivation temperature. However, thermophilic fungi have not yet been evaluated for applications in metal mobilization.

Other advantages of using fungi for metal mobilization would ensue from their differing metal leaching patterns from those of the sulfur-oxidizing bacteria; depending on the strains used, growth conditions and metal-containing solids, certain metals, such as lead and tin, are solubilized that cannot be leached by thiobacilli. A stepwise process that combines fungal and bacterial versatility might result in very effective metal mobilization and increase recovery yields.

Another broad area that still requires to be evaluated is fungal treatment of metal-containing waste materials from industrial processes. These materials can serve as secondary raw materials and reduce the demand for primary resources. However, a prerequisite is the development of a process that is technically and economically feasible as well as ecologically justifiable. As mentioned above, such processes might also be applied for the bioremediation of soils contaminated with heavy metals. In contrast to fungal remediation of wastewaters and other aqueous solutions using biosorptive capacities to remove metals (Ahmann, 1997; Kambe-Honjoh *et al.*, 1998; Eccles, 1999), metal solubilization has hardly been investigated. Combining both fungal mobilizing and immobilizing abilities will lead to an integrated biologically based process, as already demonstrated for bacteria involved in the sulfur cycle (sulfur oxidizers, sulfate reducers) (White, Sayer & Gadd, 1997; White, Sharman & Gadd, 1998).

Acknowledgements

Parts of the data, concepts and ideas were adopted from Diploma and PhD theses, or result from other research projects of Regula Bosshard, Philipp Bosshard, Christoph Brombacher, Pascal Burri, Lorenz Diethelm, Roland Frei, Walter Krebs, Lukas Lüchinger, Janine Ryser, and Manuel Wegmann. All projects were carried out at the University of Zürich (Institute of Environmental Sciences, Institute of Plant Biology). Financial support was provided by the Swiss National Science Foundation within the Priority Program Environment.

References

Ahmann, D. (1997). Bioremediation of metal-contaminated soil. *SIM News*, **47**, 218–233.

Alibhai, K. A. K., Dudeney, A. W. L., Leak, D. J., Agatzini, S. & Tzeferis, P. (1993). Bioleaching and bioprecipitation of nickel and iron from laterites. *FEMS Microbiology Reviews*, **11**, 87–96.

Avakyan, Z. A., Karavaiko, G. I., Mel'nikova, E. O., Krutsko, V. S. & Ostroushko, Y. I. (1981). Role of microscopic fungi in weathering of rocks and minerals from a pegmatite deposit. *Microbiology*, **50**, 115–120.

Banks, M. K., Schwab, A. P., Fleming, G. R. & Hetrick, B. A. (1994). Effects of plants and soil microflora on leaching of zinc from mine tailings. *Chemosphere*, **8**, 1691–1699.

Basselik, K. (1913). Über Silikatzersetzung durch Bodenbakterien und Hefen. *Zeitschrift für Gährphysiologie*, **3**, 15–42.

Berthelin, J. (1983). Microbial weathering processes. In *Microbial Geochemistry*, ed. W. E. Krumbein, pp. 223–262. Oxford: Blackwell.

Berthelin, J., Kogblevi, A. & Dommergues, Y. (1974). Microbial weathering of brown forest soil: influence of partial sterilization. *Soil Biology and Biochemistry*, **6**, 393–399.

Bigelis, R. & Arora, D. K. (1992). Organic acids of fungi. In *Handbook of Applied Mycology*, Vol. 4, *Fungal Biotechnology*, eds. D. K. Arora, R. P. Elander & K. G. Mukerji, pp. 357–76. New York: Marcel Dekker.

Bloomfield, C. & Kelso, W. I. (1971). The mobilisation of trace elements by aerobically decomposing plant material under simulated soil conditions. *Chemistry and Industry*, **2/71**, 59–61.

Borovec, Z. (1990). Úprava bauxitu a keramických surovin bakteriemi a mikroskopickými houbami. [Treatment of bauxite and ceramic raw materials by means of bacteria and microscopic fungi.] *Ceramics Silikaty*, **34**, 163–168.

Bosecker, K. (1986). Leaching of lateritic nickel ores with heterotrophic microorganisms. In *Fundamental and Applied Biohydrometallurgy*, eds. R. W.

Lawrence, R. M. R. Branion & H. G. Ebner, pp. 367–382. Amsterdam: Elsevier.

Bosecker, K. (1987). Laugung lateritischer Nickelerze mit heterotrophe Mikroorganismen. *Acta Biotechnologica*, **7**, 389–399.

Bosecker, K. (1989). Bioleaching of valuable metals from silicate ores and silicate waste products. In *Biohydrometallurgy*, eds. J. Salley, R. G. L. McGready & P. L. Wichlacz, pp. 15–24. Ottawa: CANMET.

Bosecker, K. (1993). Bioleaching of silicate manganese ores. *Geomicrobiology Journal*, **11**, 195–203.

Bosecker, K. (1994). Mikrobielle Laugung (Leaching). In *Handbuch der Biotechnologie*, eds. P. Präwe, U. Faust, W. Sittig & D. A. Sukatsch, pp. 837–858. Munich: Oldenbourg Verlag.

Bosecker, K. (1997). Bioleaching: metal solubilization by microorganisms. *FEMS Microbiology Reviews*, 20, 591–604.

Bosshard, P. P., Bachofen, R. & Brandl, H. (1996). Metal leaching of fly ash from municipal waste incineration by *Aspergillus niger*. *Environmental Science and Technology*, **30**, 3066–3070.

Brandl H. (1999). Bioleaching: mobilization and recovery of metals from solid waste materials. In *Bacterial-Metal/Radionuclide Interaction: Basic Research and Bioremediation*, eds. Selenska-Pobell & H. Nitsche, pp. 20–22. Wissenschaftlich-Technische Berichte TZR-252. Forschungszentrum Rossendorf.

Brandl, H., Bosshard, R. & Wegmann, M. (1999). Computer-munching microbes: metal leaching from electronic scrap by bacteria and fungi. In *Biohydrometallurgy and the Environment Toward the Mining of the 21st Century*, Vol. 9B, eds. R. Amils & A. Ballester, pp. 569–576. Amsterdam: Elsevier.

Burgstaller, W. & Schinner, F. (1993). Leaching of metals with fungi. *Journal of Biotechnology*, **27**, 91–116.

Burgstaller, W., Strasser, H. & Schinner, F. (1990). Leaching with fungi: solubilization of metals from ores and industrial wastes by means of organic acids. In *5th European Congress on Biotechnology*, eds. C. Christiansen, L. Munck & J. Villadsen, pp. 569–572. Copenhagen: Munsgaard.

Burgstaller, W., Strasser, H., Wöbking, H. & Schinner, F. (1992). Solubilization of zinc oxide from filter dust with *Penicillium simplicissimum*: bioreactor leaching and stoichiometry. *Environmental Science and Technology*, **26**, 340–346.

Burri, P. (1999). Mobilisierung von Schwermetallen aus Böden und Minenabraum. Diploma thesis, University of Zürich.

Cameselle, C., Nuñez, M. J., Lema, J. M. & Pais, J. (1995). Leaching of iron from kaolins by a spent fermentation liquor: influence of temperature, pH, agitation, and citric acid concentration. *Journal of Industrial Microbiology*, **14**, 288–292.

Colmer, A. R. & Hinkle, M. E. (1947). The role of microorganisms in acid mine drainage. *Science*, **106**, 253–256.

Dave, S. R., Natarajan, K. A. & Bhat, J. V. (1981). Leaching of copper and zinc from oxidised ores by fungi. *Hydrometallurgy*, **7**, 235–242.

de la Torre, M. A., Gomez-Alarcon, G., Melgarejo, P. & Saiz-Jimenez, C. (1991). Fungi in weathered sandstone from Salamanca cathedral, Spain. *Science of the Total Environment*, **197**, 159–168.

de la Torre M. A., Gomez-Alarcon, G., Vizcaino, C. & Garcia M. A. (1993). Biochemical mechanisms of stone alteration carried out by filamentous fungi living in monuments. *Biogeochemistry*, **19**, 129–147.

Demain, A. L. (2000). Microbial biotechnology. *Trends in Biotechnology*, **18**, 26–31.

Dixon-Hardy, J. E., Karamushka, V. I., Gruzina, T. G., Nikovska, G. N., Sayer, J. A. & Gadd, G. M. (1998). Influence of the carbon, nitrogen and phosphorus source on the solubilization of insoluble metal compounds by *Aspergillus niger*. *Mycological Research*, 102, 1050–1054.

Drewello, R. & Weissmann, R. (1997). Microbially influenced corrosion of glass. *Applied Microbiology and Biotechnology*, **47**, 337–346.

Drewello, R., Nüssler, M. & Weissmann, R. (1994). Korrosion von Modellgläsern durch mikrobielle Stoffwechselprodukte. *Werkstoffe & Korrosion*, **45**, 122–124.

Eccles, H. (1999). Treatment of metal-contaminated wastes: why select a biological process? *Trends in Biotechnology*, **17**, 462–465.

Eckhardt, F. E. W. (1979a). Über den Einfluss niederer Pilze auf die Mineral- und Gesteinsverwitterung. *Mitteilung der Deutschen Bodenkundlichen Gesellschaft*, **29**, 391–398.

Eckhardt, F. E. W. (1979b). Über die Einwirkung heterotropher Mikroorganismen auf die Zersetzung silikatischer Minerale. *Zeitschrift für Pflanzenernährung und Bodenkunde*, **142**, 434–445.

Ehrlich, H. L. (1999). Past, present and future of biohydrometallurgy. In *Biohydrometallurgy and the Environment Toward the Mining of the 21st Century*, Vol. 9A, eds. R. Amils & A. Ballester, pp. 3–12. Amsterdam: Elsevier.

Eno, C. F. & Reuzer, H. W. (1955). Potassium availability from biotite, muscovite, greensand, and microcline as determined by growth of *Aspergillus niger*. *Soil Science*, **80**, 199–209.

Feldmann, R., Neher, J., Jung, W. & Graf, F. (1997). Fungal quartz weathering and iron crystallite formation in an Alpine environment, Piz Alv, Switzerland. *Eclogae Geologicae Helvetiae* **90**, 541–556.

Foster, J. W. (1949). *Chemical Activities of Fungi*. New York: Academic Press.

Franz, A., Burgstaller, W. & Schinner, F. (1991). Leaching with *Penicillium simplicissimum*: influence of metals and buffers on proton extrusion and citric acid production. *Applied and Environmental Microbiology*, **57**, 769–774.

Frei, R. (1999). Entfernung von Schwermetallen aus Boden mit Hilfe von Pilzen. Diplom thesis, University of Zürich.

Furrer, G. & Stumm, W. (1986). The coordination chemistry of weathering: I. Dissolution kinetics of δ-Al_2O_3 and BeO. *Geochimica et Cosmochimica Acta*, **50**, 1847–1860.

Gadd, G. M. (1999). Fungal production of citric and oxalic acid: importance in metal speciation, physiology and biogeochemical processes. *Advances in Microbial Physiology*, **41**, 47–92.

Gadd, G. M., Gharieb, M. M., Ramsay, L. M., Sayer, J. A., Whatley, A. R. & White, C. (1998). Fungal processes for bioremediation of toxic metal and radionuclide pollution. *Journal of Chemical Technology and Biotechnology*, **71**, 364–366.

Gharieb, M. M. & Gadd, G. M. (1999). Influence of nitrogen source on the solubilization of natural gypsum ($CaSO_4.2H_2O$) and the formation of calcium oxalate by different oxalic and citric acid-forming fungi. *Mycological Research*, **103**, 473–481.

Gharieb, M. M., Sayer, J. A. & Gadd, G. M. (1998). Solubilization of natural gypsum ($CaSO_4 \cdot 2H_2O$) and the formation of calcium oxalate by *Aspergillus niger* and *Serpula himantioides*. *Mycological Research*, **102**, 825–830.

Golab, Z. & Orlowska, B. (1988). The effect of amino and organic acids produced by selected microorganisms on metal leaching. *Acta Microbiologica Polonica*, **37**, 83–93.

Gomez-Alarcon, G., Munoz M. L. & Flores, M. (1994). Excretion of organic acids by fungal strains isolated from decayed sandstone. *International Biodeterioration and Biodegradation*, **34**, 169–180.

Gray, S. N. (1998). Fungi as potential bioremediation agents in soil contaminated with heavy or radioactive metals. *Biochemical Society Transactions*, **26**, 666–670.

Grewal, H. S. & Kalra, K. L. (1995). Fungal production of citric acid. *Biotechnology Advances*, **13**, 209–234.

Groudev, S. N. (1987). Use of heterotrophic microorganisms in mineral biotechnology. *Acta Biotechnologica*, **7**, 299–306.

Groudev, S. N. & Groudeva, V. I. (1986). Biological leaching of aluminum from clays. *Biotechnology and Bioengineering Symposium*, **16**, 91–99.

Guedes de Carvalho, R. A., Celeste Cruz, M., Ceu Goncalves, M., Manuela Moura, M. & Neves, O. (1990). Bioleaching of tungsten ores. *Hydrometallurgy*, **24**, 263–267.

Gupta, A. & Ehrlich, H. L. (1989). Selective and non-selective bioleaching of manganese from a manganese-containing silver ore. *Journal of Biotechnlogy*, **9**, 287–304.

Hahn, M., Willscher, S. & Straube, G. (1993). Copper leaching from industrial wastes by heterotrophic microorganisms. In *Biohydrometallurgical Technologies*, eds. A. E. Torma, J. E. Wey & V. L. Laksmanan, pp. 99–108. Warrendale, UK: Minerals, Metals and Materials Society.

Heinze, B. (1906). Sind Pilze imstande, den elementaren Stickstoff der Luft zu verarbeiten und den Boden an Gesamtstickstoff anzureichern? *Annales Mycologici*, **4**, 41–63.

Henderson, M. E. K. & Duff, R. B. (1963). The release of metallic and silicate ions from minerals, rocks, and soils by fungal activity. *Journal of Soil Science*, **14**, 236–246.

Hirsch, P., Eckhardt, F. E. W. & Palmer, R. J. Jr (1995). Fungi active in weathering of rock and stone monuments. *Canadian Journal of Botany*, **73**(Suppl. 1), S1384–S1390.

Ilgar, E., Guay, R. & Torma, A. E. (1993). Kinetics of lithium extraction from spodumene by organic acids produced be *Aspergillus niger*. In *Biohydrometallurgical Technologies*, eds. A. E. Torma, J. E. Wey & V. L. Lakshmanan, pp. 293–301. Warrendale, UK: Minerals, Metals and Materials Society.

Kambe-Honjoh, H., Hata, S., Ohsumi, K. & Kitamoto, K. (1998). A novel screening method for microorganisms that effectively remove heavy metals from aqueous solutions. *Biotechnology Techniques*, **12**, 91–95.

Karavaiko, G. I., Krutsko, V. S., Mel'nikova, E. O., Avakyan, Z. A. & Ostroushko, Y. I. (1980). Role of microorganisms in spodumene degradation. *Microbiology*, **49**, 402–406.

Karavaiko, G. I., Avakyan, Z. A., Ogurtsova, L. V. & Safonova, O. F. (1989). Microbial processing of bauxite. In *Biohydrometallurgy*, eds. J. Salley, R. G. L. McGready & P. L. Wichlacz, pp. 93–102. Ottawa: CANMET.

Khan, S., Gupta, R. & Kumar Saxena, R. (1997). Bioleaching of copper from ferromanganese sea nodule of Indian ocean. *Current Science*, **73**, 602–605.

Kiel, H. (1977). Laugung von Kupferkarbonat- und Kupfersilikat-Erzen mit heterotrophen Mikroorganismen. In *Conference on Bacterial Leaching 1977*, ed. W. Schwartz, pp. 261–270. Weinheim: Verlag Chemie.

Kiel, H. & Schwartz, W. (1980). Leaching of a silicate and carbonate ore with heterotrophic fungi and bacteria, producing organic acids. *Zeitschrift für Allgemeine Mikrobiologie*, **20**, 627–636.

King, A. B. & Dudeney, A. W. L. (1987). Bioleaching of nepheline. *Hydrometallurgy*, **19**, 69–81.

König, R. (1989). *C. Plinii Secundi – Naturalis Historiae*, Libri XXXVII. Munich: Artemis.

Kovalenko, E. V. & Malakhova, P. T. (1990). Microbial succession in compensated sulfide ores. *Microbiology*, **59**, 227–232.

Krebs, W., Brombacher, C., Bosshard, P. P., Bachofen, R. & Brandl, H. (1997). Microbial recovery of metals from solids. *FEMS Microbiology Reviews*, **20**, 605–617.

Krumbein, W. E. & Schönborn-Krumbein, C. E. (1987). Biogene Bauschäden. *Bautenschutz Bausanierung*, **10**, 14–23.

Krumbein, W. E., Urzi, C. E. & Gehrmann, C. (1991). Biocorrosion and biodeterioration of antique and medieval glass. *Geomicrobiology Journal*, 9, 139–160.

Mehlich, A., Truog, E. & Fred, E. B. (1933). The *Aspergillus niger* method of measuring available potassium in soil. *Soil Science*, **35**, 259–277.

Mehta, A. P., Torma, A. E. & Murr, L. E. (1979). Effect of environmental parameters on the efficiency of biodegradation of basalt rock by fungi. *Biotechnology and Bioengineering*, **21**, 875–885.

Müller, B., Burgstaller, W., Strasser, H., Zanella, A. & Schinner, F. (1995). Leaching of zinc fron an industrial filter dust with *Penicillium*, *Pseudomonas* and *Corynebacterium*: citric acid is the leaching agent rather than amino acids. *Journal of Industrial Microbiology*, **14**, 208–212.

Müller, G. & Förster, I. (1964). Der Einfluss mikroskopischer Bodenpilze auf die Nährstofffreisetzung aus primären Materialien, als Beitrag zur biologischen Verwitterung. II. Mitteilung. *Zentralblatt für Bakteriologie II*, **118**, 594–621.

Mulligan, C. N., Galvez-Cloutier, R. & Renaud, N. (1999). Biological leaching of copper mine residues by *Aspergillus niger*. In *Biohydrometallurgy and the Environment Toward the Mining of the 21st Century*, Vol. 9A, eds. R. Amils & A. Ballester, pp. 453–461. Amsterdam: Elsevier.

Niklas, H. & Poschenrieder, H. (1932). Die Ausführung der *Aspergillus* methode zur Prüfung auf Kali. *Ernährung der Pflanze*, **28**, 86–88.

Noble, E. G., Baglin, E. G., Lamphire, D. L. & Eisele, J. A. (1991). Bioleaching of manganese form ores using heterotrophic microorganisms. In *Mineral Bioprocessing*, eds. R. W. Smith & M. Misra, pp. 233–245. Warrendale, UK: Minerals, Metals and Materials Society.

Ogurtsova, L. V., Karavaiko, G. I., Avakyan, Z. A. & Korenevskii, A. A. (1989). Activity of various microorganisms in extracting elements from bauxite. *Microbiology*, **58**, 774–780.

Palmer, R. J. J., Siebert, J. & Hirsch, P. (1991). Biomass and organic acids in sandstone of a weathering building: production by bacterial and fungal isolates. *Microbial Ecology*, **21**, 253–266.

Petersen, K., Kuroczkin, J., Strzelczyk, A. B. & Krumbein, W. E. (1988). Distribution and effects of fungi on and in sandstones. In *Biodeterioration 7*, eds. D. R. Houghton, R. N. Smith & H. O. W. Eggins, pp. 123–128. London: Elsevier.

Portier, R. J. (1991). Biohydrometallurgical processing of ores, and microorganisms therefore. US patent 5,021,088.

Rawlings, D. E. (1997). *Biomining: Theory, Microbes and Industrial Processes.* Berlin: Springer.

Reyes, I., Bernier, L., Simard, R. R. & Antoun, H. (1999). Effect of nitrogen source on the solubilization of different inorganic phosphates by an isolate of *Penicillium rugulosum* and two UV-induced mutants. *FEMS Microbiology Ecology*, **28**, 281–290.

Roane, T. M., Pepper, I. L. & Miller, R. M. (1996). Microbial remediation of metals. In *Bioremediation: Principles and Applications*, eds. R. L. Crawford & D. L. Crawford, pp. 312–340. Cambridge, UK: University Press.

Robert, M. & Berthelin, J. (1986). Role of biological and biochemical factors in soil mineral weathering. In *Interactions of Soil Minerals with Natural Organics and Microbes*, eds. P. M. Huang & M. Schnitzer, pp. 453–495. Madison, WI: Soil Science Society of America.

Rose, R. E. (1957). Techniques for determining the effect of microorganisms on insoluble inorganic phosphates. *New Zealand Journal of Science and Technology*, **B38**, 773–780.

Rossi, G. (1990). *Biohydrometallurgy*. Hamburg: McGraw-Hill.

Rudolfs, W. & Helbronner, A. (1922). Oxidation of zinc sulfide by microorganisms. *Soil Science* **14**, 459–464.

Ryser, J. (2000). Metallmobilisierung durch Pilze aus pfanzenabfall der Phytoremediation. Diploma thesis, University of Zürich.

Salameh, M., Özcengiz, G., Atalay, Ü., Özbayoglu, G. & Alaeddinoglu, N. G. (1999). Enhanced recovery of silver from Artvin-Kafkasör ore by microbial treatment. In *Biohydrometallurgy and the Environment Toward the Mining of the 21st Century*, Vol. 9A, eds. R. Amils & A. Ballester, pp. 493–499. Amsterdam: Elsevier.

Satyanarayana, T., Johri, B. N. & Klein, J. (1992). Biotechnological potential of thermophilic fungi. In *Handbook of Applied Mycology*, Vol. 4, *Fungal Biotechnology*, eds. D. K. Arora, R. P. Elander & K. G. Mukerji, pp. 729–761. New York: Marcel Dekker.

Sayer, J. A. & Gadd, G. M. (1997). Solubilization and transformation of insoluble inorganic metal compounds to insoluble metal by *Aspergillus niger*. *Mycologcial Research*, 101, 653–661.

Sayer, J. A., Raggett, S. A. & Gadd, G. M. (1995). Solubilization of insoluble metal compounds by soil fungi: development of a screening method for solubilizing ability and metal tolerance. *Mycological Research*, **99**, 987–993.

Sayer, J. A., Kierans, M. & Gadd, G. M. (1997). Solubilisation of some naturally occurring metal-bearing minerals, limescale and lead phosphate by *Aspergillus niger*. *FEMS Microbiology Letters*, **154**, 29–35.

Sayer, J. A., Cotter-Howells, J. D., Watson, C., Hillier, S. & Gadd, G. M. (1999). Lead mineral transformation by fungi. *Current Biology*, **9**, 691–694.

Schinner, F. & Burgstaller, W. (1989). Extraction of zinc from industrial waste by a *Penicillium* sp. *Applied and Environmental Microbiology*, **55**, 1153–1156.

Siegel, S. M., Siegel, B. Z. & Clark, K. E. (1983). Bio-corrosion: solubilization and accumulation of metals by fungi. *Water, Air and Soil Pollution*, **19**, 229–236.

Silverman, M. P. & Munoz, E. F. (1970). Fungal attack on rock: solubilization and altered infrared spectra. *Science*, **169**, 985–987.

Silverman, M. P. & Munoz, E. F. (1971). Fungal leaching of titanium from rock. *Applied Microbiology*, **22**, 923–924.

Singer, A., Navrot, J. & Shapira, R. (1982). Extraction of aluminum from fly-ash by commercial and microbiologically-produced citric acid. *European Journal of Applied Microbiology and Biotechnology*, **16**, 228–230.

Slater, J. W. (1856). On some reactions of oxalic acid. *Chemical Gazette*, **14**, 130–131.

Sterflinger, K. (2000). Fungi as geological agents. *Geomicrobiology Journal*, **17**, 97–124.

Strasser, H., Pümpel, T., Brunner, H. & Schinner, F. (1993a). Veredelung von Quarzsand durch mikrobielle Laugung von Eisenoxid-Verunreinigungen. *Archiv für Lagerstättenforschung der Geologischen Bundesanstalt*, **16**, 103–107.

Strasser, H., Brunner, H. & Schinner, F. (1993b). The feasibility of nickel and vanadium recovery from a filter dust of a combustion furnace. In *Biohydrometallurgical Technologies*, eds. A. E. Torma, J. E. Wey & Lakshmanan, pp. 157–162. Warrendale, UK: The Minerals, Metals and Materials Society.

Sukla, L. B. & Panchanadikar, V. V. (1993). Bioleaching of lateritic nickel ore using an indigenous microflora. In *Biohydrometallurgical Technologies*, eds. A. E. Torma, J. E. Wey & V. L. Lakshmanan, pp. 373–381. Warrendale, UK: Minerals, Metals and Materials Society.

Sukla, L. B., Panchanadikar, V. V. & Kar, R. N. (1993). Microbial leaching of lateritic nickel ore. *World Journal of Microbiology and Biotechnology*, **9**, 255–257.

Sukla, L. B., Kar, R. N., Panchanadikar, V. V., Choudhury, S. & Mishra, R. K. (1995). Bioleaching of lateritic nickel ore using *Penicillium* sp.. *Transactions of the Indian Institute of Metals*, **48**, 103–106.

Tait, K., Sayer, J. A., Gharieb, M. M., Gadd, G. M. (1999). Fungal production of calcium oxalate in leaf litter microcosms. *Soil Biology and Biochemistry*, **31**, 1189–1192.

Tan, K. H. (1986). Degradation of soil minerals by organic acids. In *Interactions of Soil Minerals With Natural Organics and Microbes*, eds. P. M. Huang & M. Schnitzer, pp. 1–27. Madison, WI: Soil Science Society of America.

Tarasova, I. I., Khavski, N. N. & Dudeney, A. W. L. (1993). The effects of ultrasonics on the bioleaching of laterites. In *Biohydrometallurgical Technologies*, ed. A. E. Torma, J. E. Wey & V. L. Lakshmanan, pp. 357–361. Warrendale, UK: The Minerals, Metals and Materials Society.

Torma, A. E. (1988). Leaching of metals. In *Biotechnology*, Vol. 6B, eds. H. J. Rehm & G. Reed, pp. 369–399. Weinheim: VCH.

Tzeferis, P. G. (1994a). Fungal leaching of nickeliferous laterites. *Folia Microbiologica*, **39**, 137–140.

Tzeferis, P. G. (1994b). Leaching of a low grade hematitic laterite ore using fungi and biologically produced acid metabolites. *International Journal of Mineral Processing*, **42**, 267–283.

Tzeferis, P. G., Agatzini, S., Nerantzis, E. T., Dudeney, A., Leak, D. & Alibhai, K. (1991). Bioleaching of Greek non-sulphide nickel ores using microorganism assisted leaching process. *Mededelingen van de Faculteit Landbouwwetenschapen Rijksuniversiteit te Gent*, **56**, 1797–1802.

Tzeferis, P. G., Agatzini, S. & Nerantzis, E. T. (1994). Mineral leaching of non-sulphide nickel ores using heterotrophic micro-organisms. *Letters in Applied Microbiology*, **18**, 209–213.

Vachon, P., Tyagi, R. D., Auclair, J. C. & Wilkinson, K. J. (1994). Chemical and biological leaching of aluminum from red mud. *Environmental Science and Technology*, **28**, 26–30.

Wainwright, M. & Grayston, S. J. (1986). Oxidation of heavy metal sulphides by *Aspergillus* and *Trichoderma harzianum*. *Transactions of the British Mycological Society*, **86**, 269–272.

Waksman, S. A. & Joffe, I. S. (1922). Microörganisms concerned in the oxidation of sulfur in the soil. II. *Thiobacillus thiooxidans*, a new sulfur-oxidizing organism isolated from the soil. *Journal of Bacteriology*, **7**, 239–256.

Webley, D. M., Henderson, M. E. K. & Taylor, I. F. (1963). The microbiology of rocks and weathered stones. *Journal of Soil Science*, **14**, 102–112.

Weed, S. B., Davey, C. B. & Cook, M. G. (1969). Weathering of mica by fungi. *Soil Science Society of America Proceedings*, 33, 702–706.

Weissmann, R., Drewello, R. & Müller, N. (1999). Mikrobielle Entfernung von Schwermetallen aus Gläsern, Glasuren und Emails. In *BayFORREST Berichtsheft 9*, eds. P. A. Wilderer & D. C. Tartler, pp. 323–334. Munich: Bayerischer Forschungsverbund Abfallforschung und Reststoffverwertung.

Wenberg, G. M., Erbisch, F. H. & Volin, M. E. (1971). Leaching of copper by fungi. *Transactions of the Society of Mining Engineers*, **250**, 207–212.

Wenzl, R., Burgstaller, W. & Schinner, F. (1990). Extraction of zinc, copper and lead from a filter dust by yeasts. *Biorecovery*, 2, 1–13.

White, C., Sayer, J. A. & Gadd, G. M. (1997). Microbial solubilization and immobilization of toxic metals: key biogeochemical processes for treatment of contamination. *FEMS Microbiology Reviews*, **20**, 503–516.

White, C., Sharman, A. K. & Gadd, G. M. (1998). An integrated process for the bioremediation of soil contaminated with toxic metals. *Nature Biotechnology*, **16**, 572–575.

Wilson, M. J., Jones, D. & McHardy, W. J. (1981). The weathering of serpentinite by *Lecanora atra*. *Lichenologist*, **13**, 167–176.

Zidwick, M. J. (1992). Organic acids. In *Biotechnology of Filamentous Fungi*, eds. D. B. Finkelstein & C. Ball, pp. 303–334. Boston, MA: Butterworth-Heinemann.

15

Fungal metal biosorption

JOHN M. TOBIN

Introduction

Interactions with microorganisms have long been recognized as playing a key role in determining the cycling and ultimate fate of metals in the environment. On the one hand, bioleaching from naturally occurring ores or synthetic sources may result in the release and dispersion of metals while, on the other hand, microbial sorption or accumulation processes concentrate and tend to remove metal species from the surrounding environment. The bioconcentration occasioned by the latter processes may also represent an entry path into the food chain, with potentially fatal consequences for higher organisms.

Microbial metal sorption or accumulation processes may be classified as either dependent or independent of metabolism (Blackwell, Singleton & Tobin, 1995). The former occurs in most, if not all microbial forms, sorption depending on the physicochemical nature of the microbial cell wall. Metal sorption or uptake (typically from the surrounding solution) results from chemical and/or physical binding of metal ions to cell wall functional groups and is, in the main, unchanged if the cells are living, denatured or dead. Metabolism-dependent processes are generally slower and involve active metal transport into and localization within the cell interior (Blackwell & Tobin, 1999). In many instances non-active binding occurs first and it is the initially bound metal that is subsequently transported to the cell interior.

The term biosorption has variously been applied to both the overall process of metal uptake by biological materials and the non-metabolic sorption process. Strictly, however, biosorption refers to non-active binding. The term bioaccumulation more accurately describes metabolism-dependent microbial metal uptake (Gadd, 1990, Blackwell *et al.*, 1995). Biosorption has been defined as a 'non-directed physiochemical interaction that may occur between metal species and cellular components of biological species' (Meyer & Wallis, 1997). Moreover, for most

424

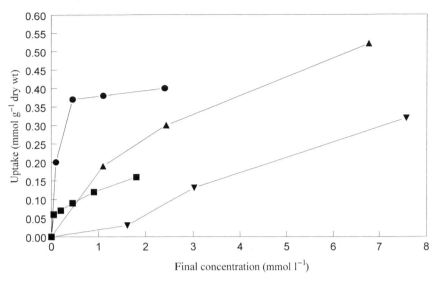

Fig. 15.1. Representative biosorption isotherms. Biosorption of Cu^{2+} by *Rhizopus arrhizus* at pH 4 (●); Cr^{3+} by *Mucor meihi* at pH 4 (▲); Sr^{2+} by *R. arrhizus* at pH 4 (■); and Cr^{3+} by *M. meihi* at pH 2 (▼).

filamentous fungal species, metabolism-dependent metal uptake is small or negligible compared with biosorption values (bioaccumulation has most commonly been investigated in yeasts). Hence, for the purpose of this review, the term biosorption will be generally adopted.

Fungal metal biosorption

Current interest in fungal metal biosorption dates principally from the 1980s although Adams & Holmes (1935) published related work much earlier (Volesky & Holan, 1995). Studies of the potential of *Rhizopus arrhizus* demonstrated that uptake levels exceeding those of commercial ion-exchange resins for uranium and thorium were attainable (Tsezos & Volesky, 1981). Biosorption levels were reported in terms of adsorption isotherms, which show metal binding (in millimoles or milligrams metal per unit mass of cells) as a function of the equilibrium or residual metal concentration (mmolar) following a contacting or sorption time. Typical biosorption isotherms for the uptake of Cr^{3+} by *Mucor meihi* and Cu^{2+} by *R. arrhizus* are shown in Fig. 15.1. Similar work conducted with other *Rhizopus* spp. and fungal genera showed that a broad range of cationic metal ions were biosorbed to similar levels, typically of the order of 0.1–1.0 mmol metal g^{-1} dry weight (Tobin, Cooper & Neufeld,

1984; Treen-Sears, Martin & Volesky, 1984; de Rome & Gadd, 1987).

The possibility of using fungi as a means of treating metal/radionuclide-bearing effluents became well accepted (Gadd & White, 1989; Siegel, Galun & Siegel, 1990). However, a review of the recent literature reveals that less than 20% of the papers published from 1998 to 2000 in the general area of biosorption were devoted to fungal metal biosorption. This reflects, in part, the increasing interest in applying the sorptive properties of biomass systems to removal of organics and textile dyes from industrial effluents (Bustard, McMullan & McHale, 1998; Hong, Hwang & Chang, 2000) and also the use of bacterial and algal biomass for biosorption studies (Schiewer, 1999; Stringfellow & Alvarez-Cohen, 1999; Figueira, Volesky & Ciminelli, 2000). Moreover, many recent published works tend only to confirm or consolidate earlier findings. There exists a plethora of research devoted to characterization of biosorption in different biomass–metal systems under varying experimental conditions using isotherm fitting as the only modelling input. Currently, the state-of-the-art work in fungal metal biosorption remains research based rather than application based. Despite the development of a number of techniques and patented processes (Veglio & Beolchini, 1997), actual industrial application of the technology remains to be proven. This review addresses some of the recent advances in the understanding and development of biosorption processes. It is not an exhaustive survey of recent publications and, where appropriate, reference is made to earlier works and biosorption by non-fungal systems.

Biosorption levels and mechanisms

Recent studies of metal biosorption have largely confirmed the sorption potential of fungal and other biomass types reported in earlier studies (Pillichshammer *et al.*, 1995; Yin *et al.*, 1999; Zhou, 1999). Maximum uptake levels for cationic metals fall generally in the range 0.1–1.0 mmol metal g^{-1} dry weight. Selected biosorption values from recent studies are presented in Table 15.1 (compilations of biosorption data from earlier investigations are available in reviews by Volesky & Holan (1995), Singleton & Tobin (1996) and Veglio & Beolchini (1997) amongst others). As seen in Table 15.1, a range of metals are biosorbed by most fungal biomass types although some trends are evident. Uranium, or more specifically the uranyl ion (UO_2^{2+}), has consistently been found in these and earlier studies to be taken up to a high level by most biomasses examined. Uptake levels in excess of 0.8 mmol g^{-1} dry weight or 200 mg g^{-1} dry weight are

Table 15.1. *Selected fungal biosorbents and maximum uptake values* (q_M)

Fungal biomass	Cation	Maximum uptake (mmol g^{-1})	Reference
Rhizopus arrhizus	Cd^{2+}	0.56	Yin *et al*., 1999
(pre-treated)	Pb^{2+}	0.61	
	Cu^{2+}	0.60	
	Zn^{2+}	0.53	
Aspergillus oryzae	Cd^{2+}	0.38	Yin *et al*., 1999
R. arrhizus	Zn^{2+}	0.21	Zhou, 1999
Mucor racemosus	Zn^{2+}	0.20	Zhou, 1999
	Zn^{2+}	0.18	Zhou, 1999
Mucor hiemalis	Cr^{3+}	0.42	Pillichshammer *et al*., 1995
R. arrhizus	Cr^{3+}	0.21	Pillichshammer *et al*., 1995
R. arrhizus	Cu^{2+}	0.40	Brady *et al*., 1999
Rhizopus oligosporus	Pb^{2+}	1.1	Ariff *et al*., 1999
Aspergillus fumigatus	UO_2^{2+}	0.81	Bhainsa & D'Souza, 1999
Mucor meihi	Cr^{3+}	1.15	Tobin and Roux, 1998

commonly reported. Similarly, lead has repeatedly been found to be biosorbed to high levels. Besides these, several divalent cations exhibit somewhat lower sorption levels, in the range 0.1–0.5 mmol g^{-1} dry weight. There are few recent studies of monovalent cation biosorption, though earlier work reported that while Ag^+ is appreciably biosorbed (to a maximum level of 0.5 mmol g^{-1} dry weight), alkali metals were not taken up (Tobin *et al*., 1984).

In terms of biomass type, members of the order Mucorales consistently exhibit high levels of uptake for a variety of cations. Specifically, biosorption to *Rhizopus* and *Mucor* genera is well documented and uptake values are generally amongst the highest across a range of metals (Tobin & Roux, 1998). Despite early reports of high potential, *Penicillium* and *Aspergillus* spp. were reported not to perform well as biosorbents (Volesky & Holan, 1995). However, more recent studies have provided contrasting evidence, as seen in Table 15.1. Despite their popularity in biosorption investigations, yeasts (notably *Saccharomyces cerevisiae* and *Candida maltosa*) are not amongst the best-performing biosorbents though they are frequently employed as model organisms (Rapoport & Muter, 1995; Donmez & Aksu, 1999; Tobin & Cooney, 1999).

There is agreement in the literature as to the general mechanisms underlying biosorption although the complexities of cellular structure have limited understanding of the detail. Many potential binding sites are present in fungal cell walls, including chitin, amino, carboxyl, phosphate, sulfhydryl and other functional groups (Norris & Kelly, 1977; Tobin, Cooper & Neufeld, 1990), which may act individually or synergistically to bind cations. While chitin and chitosan were identified to be key binding sites in early studies (Tsezos & Volesky, 1982), carboxyl, phosphate and other moieties are now recognized to be of principal importance (Tobin *et al.*, 1990; Volesky & Holan, 1995; Fourest, Serre & Roux, 1996). Recent X-ray studies of lead bound to cell walls of *Penicillium chrysogenum* indicated that phosphoryl groups accounted for up to 95% of binding, with carboxyl groups making up 5%. At low concentrations, the carboxyl groups were preferentially bound, however, because of their greater affinity for lead ions (Sarrat *et al.*, 1999). In contrast, carboxyl groups are reported to account for up to 55 and 70% of zinc binding by *P. chrysogenum* and *Trichoderma reesei*, respectively (Fourest *et al.*, 1996).

The binding to these sites is usually ascribed to interactions that include ion exchange, adsorption, complexation, coordination, crystallization and precipitation (Tobin *et al.*, 1990). However, the complexity and likely combination of these processes means that the specifics of the binding are unresolved. The general view is of an initial binding step involving either ion exchange (as evidenced by ion release) or coordination (as evidenced by proton release) followed, in certain conditions but not always, by a crystallization/precipitation step. Various reasons have been advanced to explain the amount of the initial binding step, including the effects of ionic size (radius) and the chemical nature of the ions and cell wall functional groups involved (the 'hard and soft theory') (Avery & Tobin, 1993).

Environmental factors that may influence the metal-binding capacity of fungi include the pH of the solution, temperature, concentration of biosorbent and concentration of metal co-ions, both cationic and anionic. In most cases, the effects of these parameters are consistent with a straightforward coordination/complexation mechanism of cation biosorption. Optimum pH values for biosorption are usually in the 4–7 range. Below this, increasing competition by H^+ for the binding sites diminishes cation uptake levels (Fourest & Roux, 1992). Above pH 7, hydrolysis effects tend to cause metal precipitation, which may increase the apparent removal from solution but is not a biosorption phenomenon. Temperature has little effect on biosorption levels within normal ranges, 5–30°C (Veglio & Beolchini, 1997) but cation competition usually diminishes the uptake of each of the competing ions, though total biosorption levels may be unchanged. There have been no recent studies of the simultaneous biosorption of metal cations and anions.

Biosorption of metal anions

Although several toxic metals and metalloids, including arsenic, selenium, chromium, molybdenum and vanadium, occur in effluents in anionic form, the majority of biosorption studies have focused on metal cations. Early investigations of fungal anion biosorption demonstrated the marked pH dependence of the process (Tobin *et al.*, 1984). Non-living biomass of *R. arrhizus* biosorbed molybdate and vanadate anions to 0.38 and 0.45 mmol g^{-1} dry weight, respectively, from a solution of pH 4.5 but at pH 5.5 negligible biosorption occurred. More recently, biosorption of the anions of platinum, palladium and molybdenum to the biopolymer chitosan has been reported (Guibal *et al.*, 1999a; Guibal, Milot & Roussy, 2000a).

Interest in the biosorption of chromium species has intensified because of potential chromium contamination as a result of chrome tanning industries, leather, electroplating, mining, paints, pigments and wood preservation, film and photography (Pillichshammer *et al.*, 1995; Niyogi, Abraham & Ramakrishna, 1998). Although both Cr^{III} and Cr^{VI} ions are released in the effluent from these industries, Cr^{VI} is more toxic and carcinogenic. Dead biomass of *R. arrhizus*, *Rhizopus nigricans*, *Aspergillus niger* and *Aspergillus oryzae* has been reported to have good Cr^{VI} uptake capacities, with 100% removal of Cr^{VI} from solutions of 100 and 200 mg l^{-1} (Bai & Abraham, 1998). Immobilization of *R. arrhizus* in sodium alginate (Prakasham *et al.*, 1999) and in polyvinyl alcohol matrices (Niyogi *et al.*, 1998) had a negligible effect on chromate biosorption capacities, although these are significantly lower than those of other biosorbents (see Table 15.2). However, as is the case with cations, comparisons between maximum uptake values are problematic. These values are extremely case specific and vary widely depending on metal concentration, solution pH and the form of the biosorbent.

The pH dependence of chromate and anion biosorption in general is well established. The optimum pH for Cr^{VI} removal typically lies in the range 1–2. Binding decreases with increasing pH, and negligible biosorption is not uncommon at neutral values. At low pH, functional groups on the biosorbent are protonated and anion binding by electrostatic attraction occurs. However, other factors may be involved. At low pH, reduction of Cr^{VI} to Cr^{III} has been reported in the presence of biosorbent material, though the mechanism is poorly understood (Sharma & Forster, 1993). In addition, changes in speciation with varying pH may alter both the ionic size and the charge of the moiety being biosorbed. Variations in chromate speciation predicted by MINEQL software have

Table 15.2. *Fungal and plant-derived biosorbents for Cr[VI] and maximum uptake values*

Biomass	Maximum uptake	Reference
Rhizopus arrhizus (free)	23.07 mg g^{-1}	Prakasham *et al.*, 1999
R. arrhizus (alginate-immobilized)	23.88 mg g^{-1}	Prakasham *et al.*, 1999
Mucor hiemalis	0.16 mmol g^{-1}	Pillichshammer *et al.*, 1995
R. arrhizus (immobilized in polyvinyl alcohol)	43.7 mg g^{-1}	Niyogi *et al.*, 1998
R. arrhizus (free)	46 mg g^{-1}	Niyogi *et al.*, 1998
Sphagnum moss peat	68 mg g^{-1}	Sharma & Forster, 1993
Milled peat biomass	0.58 mmol g^{-1}	Dean & Tobin, 1999

been correlated with peat biomass biosorption levels (Dean & Tobin, 1999).

Uptake of metal complexes and organic molecules

A notable recent development has been the focus on metal complexes and organic molecules in biosorption studies. The range of compounds investigated includes biocides (Benoit, Barrusio & Calvet, 1998; Tobin & Cooney, 1999), phenol derivatives (Aksu & Yener, 1998; Ning, Kennedy & Fernandes, 1998), textile dyes (Bustard *et al.*, 1998) and metal complexes arising from ore processing, bleaching and tanning plants (Aksu & Calik, 1999; Christov, van Driessel & du Plessis, 1999; Gomez, Figueira & Camargos, 1999). The adsorption of organochlorines and other organic moieties by both living and dead fungal cells and activated sludge has been investigated (Tsezos & Bell, 1989; Young & Banks, 1998). Pure *R. arrhizus* cultures exhibit only reversible biosorption, with no evidence of the biotransformation reactions seen with activated sludge. The biosorption data were well defined by Freundlich type isotherms (Tsezos & Wang, 1991). Speciation of metal complexes is key in determining biosorption characteristics. Anionic cyanide complexes would be expected to exhibit greater biosorption levels from low pH environments, although the converse has been reported (Aksu *et al.*, 1999).

Fungal interactions with organometals are well documented and are the basis of their use as fungicides. However, there are few quantitative reports of biosorption levels. Triphenyltin (TPhT) and tributyltin (TBT) compounds exhibited the greatest uptake to cyanobacteria, algae and yeasts amongst a family of organotin compounds (Avery, Codd & Gadd, 1993;

Tobin & Cooney, 1999) with maximum loadings in the range 0.5–$0.6\,\text{mmol}\,\text{g}^{-1}$ dry weight. These values are equivalent to typical maximum cation biosorption values as shown in Table 15.1. The organic character of organotins and other organometals increases their liposolubility and consequently biosorption levels. TPhT and TBT uptake by yeast was at least a factor of two greater than inorganic tin uptake under equivalent conditions, although monomethyltin and trimethyltin uptake was found to be negligible (Tobin & Cooney, 1999).

Biomass immobilization

Problems encountered in industrial-scale processing of microbial cells include both the relative fragility of the cells and their small density difference with water. Consequently, some form of immobilization within a strengthening matrix is desirable to ensure stability and effective separation from processed solutions (Tsezos, Noh & Baird, 1988; Yin *et al.*, 1999). Immobilization techniques used for biosorption studies mirror those of other biotechnology applications and are widely used with algal, bacterial and biopolymer as well as fungal biosorbents. It should be emphasized that the incorporation of non-biological immobilizing material into the biosorbent particle may increase, decrease or have no effect on the metal-sorbing capacity of the biosorbent (Hu & Reeves, 1997; Brady, Tobin & Roux, 1999). In terms of the biosorption kinetics, the inert material represents an additional mass transfer resistance that may significantly retard the biosorption rate, as discussed below.

The growth of fungi (*Aspergillus, Rhizopus* and *Penicillium* spp.) in pellet form through manipulation of the growth conditions has also been investigated (Treen-Sears *et al.*, 1984; de Rome & Gadd, 1991; Pumpel & Schinner, 1993; Huang & Chiu, 1994). Despite promising results, this has not been followed up in recent studies. Instead, the use of immobilizing matrices has been extensively examined (Tobin, White & Gadd, 1994; Singleton & Tobin, 1996; Veglio & Beolchini, 1997). Three principal techniques have been applied, which are described below.

Immobilization as a biofilm on inert matrices

Immobilization by growth to inert materials such as coal, sand and foam particles has been shown for fungal species (Zhou & Kiff, 1991). Problems associated with this approach include poor stability of the resulting biosorbent and limitations in reactor performance. However, enzymically mediated bacterial accumulation of uranium by a *Citrobacter* sp. has demonstrated the potential of biofilm biosorption in continuous flow reactors (Macaskie, 1990; Macaskie *et al.*, 1995).

Entrapment in polymeric matrices

Both natural and synthetic gelling agents have been employed in the production of biosorbent gel beads. Natural polymers such as alginate and carrageenan, commonly used in immobilized living cell systems, produce porous biosorbents with little reduction in biosorption kinetics. Functional groups within these matrices, however, may alter the biomass biosorption capacity. Moreover poor stability and low cell loadings reduce overall efficiency (de Rome & Gadd, 1991; Tobin, l'Homme & Roux, 1993). Greater biosorbent stability is achieved by entrapment in synthetic polymer gels. Materials used include polyacrylamide, polysulfone, polyethylimine (Veglio & Beolchini, 1997) and silica gels; the last were used to develop the patented, algae-based AlgaSORB process (Brierley, 1990). Immobilization in synthetic polymers may cause little or no loss in metal-binding capacity. Various polyurethane-based protocols yielded bacterial biosorbents retaining 70–90% binding efficiency (Hu & Reeves, 1997). Immobilization of *R. arrhizus* in polyvinyl formal caused no loss in copper-uptake capacity (Brady *et al.*, 1999). In both cases the biosorbent was suitable for use in packed bed bioreactor configurations.

Chemical cross-linking

Common chemical cross-linking agents include formaldehyde, formaldehyde–urea mixtures, sulfone and divinylsulfone. Cross-linking of *Penicillium* biomass with formaldehyde and *Rhizopus* biomass with bis(ethenyl)sulfone increased metal uptake capacity by 10–30% compared with untreated cells (Holan & Volesky, 1995). The majority of recent biosorption studies involving cross-linking is devoted to algal or biopolymer systems (Guibal, Milot & Roussy, 1999b; Figueira *et al.*, 2000; Guibal, Vincent & Navarro-Mendoza, 2000b).

Continuous flow reactors

To date, most biosorption studies have been conducted in batch systems with varying agitation rates. These configurations are simple to set up and control and are useful in establishing both equilibrium and kinetic data.

For the treatment of large volumes of dilute metal-bearing solutions, however, continuous flow reactors are preferable. The development of immobilized biosorbents facilitates the use of continuous systems. Packed bed systems are the most commonly applied continuous flow configuration and their use has been described elsewhere (Tobin *et al.*, 1994; Kratochvil & Volesky, 1998). Bioreactor residence time is a key parameter that determines the extent of metal biosorption. Despite the rapidity of the biosorption process in general, insufficient contact times markedly decrease removal levels (Tobin *et al.*, 1994). Reductions in uptake efficiency of 30–70% of batch system values have been reported in systems using fungal mycelial pellets in packed columns (de Rome & Gadd, 1991). Similarly, copper-uptake capacities were reduced by approximately 20% for *R. arrhizus* biosorbent when used in upflow packed columns, although the decrease was attributed to greater hydrogen ion competition (Brady *et al.*, 1999).

Some recent works have examined the use of multistage (stirred batch and continuous) reactors using fungal, algal and bacterial systems (Sag & Kutsal, 1995; Chang & Chen, 1999; Ozer, Ozer & Ekiz, 1999). Hollow-fibre microfiltration processes (Chang & Chen, 1999) or centrifugation (Sag & Kutsal, 1995; Ozer *et al.*, 1999) was used for biomass–liquid separation and the systems were amenable to simple mathematical modelling.

Mathematical modelling

A number of models of varying degrees of complexity have been developed for fungal and other types of biosorption. In the main, these parallel mathematical models of traditional sorption processes and can be grouped as equilibrium or kinetic models.

Equilibrium models

Data from the majority of equilibrium biosorption studies are reported in terms of equilibrium isotherms (Tsezos & Volesky, 1981; de Rome & Gadd, 1987; Brady & Tobin, 1994), as shown in Fig. 15.1. These are readily modelled by Langmuir, Freundlich, or Brunauer, Emmett and Teller (BET) type adsorption models (Langmuir, 1918; Freundlich, 1926; Brunauer, Emmett & Teller, 1938). For example, the Langmuir model relates metal biosorbed (mmoles per gram dry weight cells) to the equilibrium metal concentration in solution through a saturation curve type function:

$$q = q_M \frac{b \ C_e}{1 + b \ C_e} \tag{15.1}$$

where q is the metal biosorbed, q_M is the maximum sorption capacity, C_e is the equilibrium metal concentration in solution and b is the Langmuir constant related to binding strength.

In most cases, agreement between experimental data and models is very good where only biosorption is involved, although in view of the generality of the models this is not surprising. Where more complex interactions occur, such as microbially mediated metal precipitation or metabolic metal uptake, discrepancies are expected. Scatchard analyses have also been applied to fungal biosorption data and result in characteristic curved plots (Huang, Huang & Morehart, 1990; Avery & Tobin, 1993) that are indicative of sorption to multiple and different binding sites in the biomass.

Where more than one metal is present in solution, as is the case in many industrial and other effluents, competitive uptake or biosorption of each metal may be expected. Binary isotherms, of which a number are available, are required to model these systems (McKay & al Duri, 1989). For example, the Langmuir isotherm equation may be extended to:

$$q_1 = \frac{q_{M1} a_1 C_1}{1 + a_1 C_1 + a_2 C_2} \tag{15.2}$$

and

$$q_2 = \frac{q_{M2} a_2 C_2}{1 + a_1 C_1 + a_2 C_2} \tag{15.3}$$

where q_1 and q_2 are the quantities of metals 1 and 2 biosorbed at equilibrium concentrations C_1 and C_2, respectively, q_{M1} and q_{M2} are the maximum metal loadings and a_1 and a_2 are the Langmuir constants related to sorption energies of metals 1 and 2 respectively (Veglio & Beolchini 1997; Ho & McKay, 2000).

This type of approach has been adopted in recent two-metal biosorption studies, where the uptake of each metal has been plotted as a function of both metal concentrations (Chong & Volesky, 1995; Schiewer & Volesky, 1995). By adding a second concentration axis to the standard isotherm plot, the isotherm curve becomes a three-dimensional sorption isotherm surface (Kratochvil & Volesky 1998). While curve-fitting to experimental data is required to evaluate the model constants, the resulting systems can be used to predict the effects of untested co-ion concentrations on metal biosorption levels. More recently, model modifications that allow for metal-binding competition effects (Jain & Snoeyink, 1973) have been adapted to multicomponent biosorption modelling with a reported significant improvement in correlation with empirical data (Ho & McKay, 2000). However, to date, these methods have been applied principally to algal and other non-fungal biosorption systems.

An alternative approach to modelling two-component biosorption equilibria has been the adaptation of speciation software packages such as REDEQL or MINEQL to include binding of metals by biosorption to fungal and other surfaces (Tobin, Cooper & Neufeld, 1987, 1988; Kuyucak & Volesky, 1988). In these studies, Langmuir parameters describing metal–biomass biosorption were individually derived by experiment and were incorporated into the speciation program. Model predictions of two-component biosorption processes generally showed good agreement with experimental data, although the technique did not account for biosorption of non-metals or metals complexed with other solution components.

Kinetic models

Models ranging from simple empirical models to those based on mass transfer have been applied to biosorption kinetics in batch and continuous flow systems. From a chemical engineering viewpoint, four principal mass transfer factors that may affect biosorption rates are recognized: mass transfer from the bulk solution, external film transport through the hydrodynamic layer surrounding the biomass, intraparticle diffusion, and the biosorption step itself, which is considered to be relatively fast. If the solution is well mixed, mass transfer resistance from the bulk solution is negligible and most modelling studies focus on either film resistance or intraparticle diffusion, or a combination of these.

Batch systems

A simple kinetic model based solely on external film mass transfer has been recently used successfully to describe lead and zinc biosorption by waste industrial fungal biomass (Puranik, Modak & Paknikar, 1999). The authors noted that the dried biomass was hard and non-porous, indicating that rate effects of intraparticle diffusion and sorption were negligible. In contrast, in formulating a model of the biosorption of uranium to *R. arrhizus* cells immobilized by a proprietary technique, the hydrodynamic

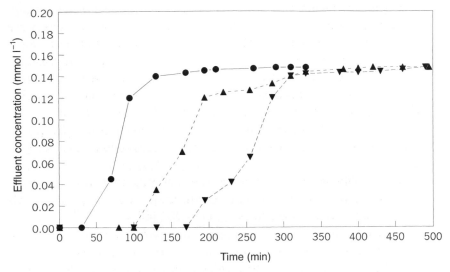

Fig. 15.2. Breakthrough curves for copper effluent from a laboratory scale *Rhizopus arrhizus* packed column. Flow rates are 11.6 (●) 5.9 (▲) and 3.4 (▼) ml min^{-1}.

boundary layer (external film resistance), the non-biomass layer and the intraparticle transport resistance were considered to be the major factors controlling the biosorption rate (Tsezos *et al.*, 1988). The modelling results showed that, while film resistance could in fact be neglected, diffusion through the non-biomass layer is as significant as that in the biomass itself. Similarly, a pore-diffusion model has been used to describe the biosorption of uranium by polyurethane-immobilized bacterial cells (Hu & Reeves, 1997).

Continuous flow reactors

Packed bed or column reactors are the configuration of choice for high-volume sorption processes including biosorption. Performance is reported in terms of breakthough curves showing column effluent concentration as a function of either operating time or volume of solution treated. Typical breakthrough curves are shown in Fig. 15.2 for copper removal by *R. arrhizus* biosorbent immobilized in polyvinyl formal.

Conventional chemical engineering literature contains numerous approaches to modelling adsorption in packed or fixed bed reactors. Most theoretical treatments involve consideration of intraparticle diffusion processes and the mass transfer resistance of the external film, as described above. Intraparticle diffusion is commonly described by either the homogeneous solid diffusion model (HSDM) or a pore diffusion model. The former assumes that adsorption occurs to the outer adsorbent surface and that diffusion of the adsorbed species into the adsorbent pellet follows. The pore diffusion model assumes that diffusion occurs within the fluid phase that occupies pores within the adsorbent pellet (Weber & Chakravorti, 1974). The models can also be modified to treat binary adsorption kinetics (Glover, Young & Bryson, 1990). However, because of their complexity as well as problems associated with the heterogeneity of biological adsorbents, few microbial biosorption studies have applied these models.

One simplifying approach that has been adopted in the investigation of metal sorption to biomaterials has been the use of single resistance models (McKay and Poots, 1980; McKay, Blair & Findon, 1986; Allen *et al.*, 1992). Under appropriate conditions, the relative effects of the external film resistance and the intraparticle diffusion can be separated and evaluated. For uranium sorption to modified chitosan, intraparticle diffusion was found to be the rate-controlling factor (Guibal *et al.*, 1993). A more complex two-resistance model has successfully been applied to the sorption of copper and mercury by chitosan (McKay *et al.*, 1986) and to dye sorption by peat (Allen & McKay, 1987).

Although many recent studies of microbial biosorption contain breakthrough curve results, comparatively few include mathematical analysis. Brady *et al.* (1999) adopted an empirical approach using a simple two-parameter model (Belter, Cussler & Hu, 1988) to relate the column effluent concentration (C_e) to the column residence time (t) through two empirically found parameters: a characteristic time (t_0) and a standard deviation (δ):

$$\frac{C_e}{C_i} = \frac{1}{2}(1 + erf(t - \frac{t_0}{\delta t_0 \sqrt{2}}))$$

(15.4)

where *erf* is the the error function and C_i is the inlet concentration. When the two parameters are evaluated by curve fitting to experimental breakthrough curves, the model provided a basis for scale-up to operating conditions outside the experimental range (Brady *et al.*, 1999).

Classical approaches to mathematical modelling of traditional fixed bed ion exchange/sorption processes have been adapted to biosorption work with varying degrees of success. The Bohart–Adams (1920) model, originally developed for sorption of chlorine onto granular activated carbon, has been applied to algal biosorption of cadmium (Volesky & Prasetyo, 1994), plant biosorption of zinc (Zhao & Duncan, 1998) and dye sorption to chitosan (McKay, Blair, & Gardner, 1984). However, it does not allow for multicomponent sorption or variation in effluent parameters and is of limited use in column design or scale-up (Zhao & Duncan, 1998).

Other models for fixed bed adsorption have been reviewed in recent biosorption literature but without practical application. The equilibrium column model is suitable for multicomponent systems and predicts sorbent usage rates but neglects the mass transfer resistance of diffusion in and out of the sorbent particles (Kratochvil & Volesky, 1998). More complete models have been developed for ion-exchange or activated carbon column modelling. These generally involve two or more mass transfer resistance terms and result in complex differential equations requiring solution by computer (Brausch & Schlunder, 1975; McKay & Bino, 1990; Tan & Spinner, 1994). In general, these have not been applied to fungal metal biosorption.

Conclusions

The potential of fungal biosorption for removal of cations from aqueous effluent has been confirmed by numerous recent investigations. Further, the sorption capacity for metallic anions and metal complexes has been demonstrated to be comparable to that of commercial available sorbents. Recent research has also increased our understanding of the underlying mechanisms and the range of potential applications. Immobilization techniques have been developed for fungal and other biosorbents that cause minimal loss of sorption capacity or retardation of uptake rates. Mathematical models have been successfully applied to both equilibrium and kinetic fungal biosorption data, although in most cases these have been empirical. Nonetheless, although research in the field is advancing, practical application is still sparse. The industrial use of traditional ion exchange or adsorption processes is well established and biosorption processes have yet to be shown to be competitive. Enhanced sorption capacities or specificity, possibly achieved through genetic engineering techniques, will increase the attractiveness of biosorption, though advances in this direction are slow. However, growing awareness of the dangers of metal contamination and increasingly strict regulation of effluents are likely to increase the demand for novel and complementary water treatment systems in the future. Biological and biosorption processes have the potential to meet this demand.

References

Adams, B. A. & Holmes, E. L. (1935). Adsorptive properties of synthetic resins. 1. *Journal of the Chemical Society*, **54**, 1–6.

Aksu, Z. & Calik, A. (1999). Comparative study of the biosorption of iron(III)–cyanide complex anions to *Rhizopus arrhizus* and *Chlorella vulgaris*. *Separation Science Technology*, **34**, 817–832.

Aksu, Z. & Yener, J. (1998). Investigation of the biosorption of phenol and monochlorinated phenols on dried activated sludge. *Process Biochemistry*, **33**, 649–655.

Aksu, Z., Calik, A., Dursun, A. Y. & Demircan, Z. (1999). Biosorption of iron(III)–cyanide complex anions to *Rhizopus arrhizus*: application of adsorption isotherms. *Process Biochemistry*, **34**, 483–491.

Allen, S. J. & McKay, G. (1987). The adsorption of acid dye onto peat from aqueous solution – solid diffusion model. *Journal of Separation Process Technology*, **8**, 18–25.

Allen, S. J., Brown, P., McKay, G. & Flynn, O. (1992). An evaluation of single resistance transfer models in the sorption of metal ions to peat. *Journal of Chemical Technology and Biotechnology*, **54**, 271–276.

Ariff, A. B., Mel, M., Hasan, M. A. & Karim, M. I. A. (1999). The kinetics and mechanism of lead(II) biosorption by powderized *Rhizopus oligosporus*. *World Journal of Microbiology and Biotechnology*, **15**, 255–260.

Avery, S. V. & Tobin, J. M. (1993). Mechanisms of adsorption of hard and soft metal ions to *Saccharomyces cerevisiae* and the influence of hard and soft anions. *Applied and Environmental Microbiology*, **59**, 2851–2856.

Avery, S. V., Codd, G. A. & Gadd, G. M. (1993). Biosorption of tributyltin and other organotin compounds by cyanobacteria and microalgae. *Applied Microbiology and Biotechnology*, **39**, 812–817.

Bai, R. S. & Abraham, T. E. (1998). Studies on biosorption of chromium(VI) by dead fungal biomass. *Journal of Scientific and Industrial Research*, **57**, 821–824.

Belter, P. A., Cussler, E. L. & Hu, W. S. (1988). *Bioseparations, Downstream Processing for Biotechnology*. New York: Wiley & Sons.

Benoit, P., Barrusio, E. & Calvet, R. (1998). Biosorption characterisation of herbicides, 2,4-D atrazine, and two chlorophenols on fungal mycelium. *Chemosphere*, **37**, 1271–1282.

Bhainsa, K. C. & D'Sousa, S. F. (1999). Biosorption of uranium (VI) by *Aspergillus fumigatus*. *Biotechnology Techniques*, **13**, 695–699.

Blackwell, K. J. & Tobin, J. M. (1999). Cadmium accumulation and its effects on intracellular ion pools in a brewing strain of *Saccharomyces cerevisiae*. *Journal of Industrial Microbiology and Biotechnology*, **23**, 204–208.

Blackwell, K. J., Singleton, I. & Tobin, J. M. (1995). Metal cation uptake by yeast: a review. *Applied Microbiology and Biotechnology*, **43**, 579–584.

Bohart, G. S. & Adams, E. Q. (1920). Some aspects of the behaviour of charcoal with respect to chlorine. *Journal of the American Chemical Society*, **42**, 523–544.

Brady, J. M. & Tobin, J. M. (1994). Adsorption of metal ions by *Rhizopus arrhizus* biomass: characterisation studies. *Enzyme and Microbial Technology*, **16**, 671–675.

Brady, J. M., Tobin, J. M. & Roux, J.-C. (1999). Continuous fixed bed biosorption of Cu^{2+} ions: application of a simple mathematical model.

Journal of Chemical Technology and Biotechnology, **74**, 71–77.

Brausch, V. & Schlunder, E. U. (1975). Scale-up of activated carbon columns for water purification based on the results from batch tests-II. *Chemical Engineering Science*, **30**, 539–548.

Brierley, C. L. (1990). Bioremediation of metal-contaminated surfaces and groundwaters. *Geomicrobiology Journal*, **8**, 201–223.

Brunauer, S., Emmett, P. H & Teller, E. (1938). Adsorption of gases in multimolecular layers. *Journal of the American Chemical Society*, **60**, 309–319.

Bustard, M., McMullan, G. & McHale, A. P. (1998). Biosorption of textile dyes by biomass derived from *Kluyveromyces marxianus* IMB3. *Bioprocess Engineering*, **19**, 427–430.

Chang, J. S. & Chen, C. C. (1999). Biosorption of lead, copper and cadmium with continuous hollow-fiber microfiltration processes. *Separation Science and Technology*, **34**, 1607–1627.

Chong, K. H. & Volesky, B. (1995). Description of 2-metal biosorption equilibria by Langmuir-type isotherms. *Biotechnology and Bioengineering*, **47**, 451–460.

Christov, L. P., van Driessel, B. & du Plessis C. A. (1999). Fungal biomass from *Rhizomucor pusillus* as adsorbent of chromophores from a bleach plant effluent. *Process Biochemistry*, **35**, 91–95.

Dean, S. A. & Tobin, J. M. (1999). Uptake of chromium cations and anions by milled peat. *Resources, Conservation and Recycling*, **27**, 151–156.

de Rome, L. & Gadd, G. M. (1987). Copper adsorption by *Rhizopus arrhizus*, *Cladosporium resinae*, and *Penicillium italicum*. *Applied Microbiology and Biotechnology*, **26**, 84–90.

de Rome, L. & Gadd, G. M. (1991). Use of pelleted and immobilised yeast and fungal biomass for heavy metal recovery. *Journal of Industrial Microbiology*, **7**, 97–104.

Donmez, G. & Aksu, Z. (1999). The effect of copper (II) ions on the growth and bioaccumulation properties of some yeasts. *Process Biochemistry*, **35**, 135–142.

Figueira, M. M., Volesky, B. & Ciminelli, V. S. T. (2000). Biosorption of metals in brown seaweed biomass. *Water Research*, **34**, 196–204.

Fourest, E. & Roux, J. C. (1992). Heavy metal biosorption by fungal mycelial by-products: mechanism and influence of pH. *Applied Microbiology and Biotechnology*, **37**, 399–403.

Fourest, E., Serre, A. & Roux, J. C. (1996). Contribution of carboxyl groups to heavy metal binding sites in fungal walls. *Toxicological and Environmental Chemistry*, **54**, 1–10.

Freundlich, H. (1926). *Colloid and Capillary Chemistry*. London: Methuen.

Gadd, G. M. (1990). Heavy metal accumulation by bacteria and other microorganisms. *Experientia*, **46**, 834–840.

Gadd, G. M. & White, C. (1989). Removal of thorium from simulated acid process streams by fungal biomass. *Biotechnology and Bioengineering*, **33**, 592–597.

Glover, M. R. L., Young, B. D. & Bryson, A. W. (1990). Modelling the binary adsorption of gold and zinc cyanides onto a strong-base anion exchange resin. *International Journal of Mineral Processing*, **30**, 217–228.

Gomez, N. C. M., Figueira, M. M. & Camargos, E. R. S. (1999). Cyano-metal complexes uptake by *Aspergillus niger*. *Biotechnology Letters*, **21**, 487–490.

Guibal, E., Saucedo, I., Roussy, J., Roulph, C. & Le Cloirec, P. (1993). Uranium sorption by glutamate glucan: a modified chitosan. Part II: kinetic studies. *Water SA*, **19**, 119–126.

Guibal, E., Larkin, T., Vincent, T. & Tobin, J. M. (1999a). Chitosan sorbents for platinum recovery from dilute solutions. *Industrial and Engineering Chemistry Research*, **38**, 4011–4022.

Guibal, E., Milot, C. & Roussy, J. (1999b). Molybdate sorption by cross-linked chitosan beads: dynamic studies. *Water and Environmental Research*, **71**, 10–17.

Guibal, E., Milot, C. & Roussy, J. (2000a). Influence of hydrolysis mechanisms on molybdate sorption isotherms using chitosan. *Separation Science and Technology*, **35**, 1021–1038.

Guibal, E., Vincent, T. & Navarro-Mendoza, R. (2000b). Synthesis and characterisation of a thiourea-derivative of chitosan for platinum recovery. *Journal of Applied Polymer Science*, **75**, 119–134.

Ho, Y. S. & McKay, G. (2000). Correlative biosorption equilibria model for a binary batch system. *Chemical Engineering Science*, **55**, 817–825.

Holan, Z. R. & Volesky, B. (1995). Accumulation of cadmium, lead and nickel by fungal and wood biosorbents. *Applied Biochemistry and Biotechnology*, **53**, 133–146.

Hong, H. B., Hwang, S. H. & Chang, Y. S. (2000). Biosorption of 1,2,3,4-tetrachlorodibenzo-P-dioxin and polychlorinated dibenzofurans by *Bacillus pumilus*. *Water Research*, **34**, 349–353.

Hu, M. Z.-C. & Reeves, M. (1997). Biosorption of uranium by *Pseudomonas aeruginosa* Strain CSU immobilised in a novel matrix. *Biotechnological Progress* **13**, 60–70.

Huang, C. P. & Chiu, H. H. (1994). Removal of trace Cd(II) from aqueous solutions by fungal adsorbents: an evaluation of self-immobilization of *Rhizopus oryzae*. *Water Science and Technology*, **30**, 245–253.

Huang, C. P., Huang, C. P. & Morehart, A. L. (1990). The removal of Cu(II) from dilute aqueous solutions by *Saccharomyces cerevisiae*. *Water Research*, **24**, 433–439.

Jain, S. J. & Snoeyink, L. V. (1973). Adsorption from biosolute systems on active carbon. *Journal of Water Pollution Control Federation*, **45**, 2463–2470.

Kuyucak, N. & Volesky, B. (1988). The mechanism of cobalt biosorption. *Biotechnology and Bioengineering*, **33**, 823–831.

Kratochvil, D. & Volesky, B. (1998). Advances in the biosorption of heavy metals. *Trends in Biotechnology*, **16**, 291–300.

Langmuir, I. (1918). The adsorption of gases on plane surfaces of glass, mica and platinum. *Journal of the American Chemical Society*, **40**, 1361–1403.

Macaskie, L. E. (1990). An immobilised cell bioprocess for the removal of heavy metals from aqueous flows. *Journal of Chemical Technology and Biotechnology*, **49**, 357–359.

Macaskie, L. E., Empson, R. M., Lin, F. & Tolley, M. R. (1995). Enzymatically-mediated uranium accumulation and uranium recovery using a *Citrobacter* sp. immobilised as a biofilm within a plug-flow reactor. *Journal of Chemical Technology and Biotechnology*, **63**, 1–16.

McKay, G. & al Duri, B. (1989). Prediction of multicomponent adsorption data using empirical correlations. *Chemical Engineering Journal*, **41**, 9–15.

McKay, G. & Bino, M. J. (1990). Simplified optimisation procedure for fixed bed adsorption systems. *Water, Air and Soil Pollution*, **51**, 33–41.

McKay, G. & Poots, V. J. (1980). Kinetics and diffusion processes in colour removal from effluent using wood as an adsorbant. *Journal of Chemical Technology and Biotechnology*, **30**, 279–292.

McKay, G., Blair, H. S. & Gardner, J. (1984). The adsorption of dyes onto chitin fixed bed columns and batch adsorbers. *Journal of Applied Polymer Science*, **27**, 4251–4261.

McKay, G., Blair, H. S. & Findon, A. (1986). Sorption of metal ions by chitosan. In *Immobilisation of Ions by Biosorption*, eds. H. Eccles and S. Hunt, pp. 59–69. Chichester: Ellis Horwood.

Meyer, A. & Wallis, F. M. (1997). The use of *Aspergillus niger* (strain 4) biomass for lead uptake from aqueous systems. *Water SA*, **23**, 187–192.

Ning, Z., Kennedy, K. J. & Fernandes, L. (1998). Effect of physicochemical conditions on anaerobic sorption of chlorophenols. *Environmental Technology*, **19**, 231–236.

Niyogi, S., Abraham, T. E. & Ramakrishna, S. V. (1998). Removal of chromium (VI) ions from industrial effluents by immobilised biomass of *Rhizopus arrhizus*. *Journal of Scientific and Industrial Research*, **57**, 809–816.

Norris, P. R. & Kelly, D. P. (1977). Accumulation of cadmium and copper by *Saccharomyces cerevisiae*. *Journal of General Microbiology*, **99**, 317–324.

Ozer, A., Ozer, D. & Ekiz, H. I. (1999). Application of Freundlich and Langmuir models to multistage purification process to remove heavy metal ions by using *Schizomeris leibeinii*. *Process Biochemistry*, **34**, 919–927.

Pillichshammer, M., Pumpel, T., Poder, R., Eller, K., Klima, J. & Schinner. F. (1995). Biosorption of chromium to fungi. *Biometals*, **8**, 117–121.

Prakasham, R. S., Merrie, J. S., Sheela, R., Saswathi, N. & Ramakrishna, S. V. (1999). Biosorption of chromium(VI) by free and immobilised *Rhizopus arrhizus*. *Environmental Pollution*, **104**, 421–427.

Pumpel, T. & Schinner, F. (1993). Native fungal pellets as a biosorbent for heavy metals. *FEMS Microbiology Reviews,* **11**, 159–164.

Puranik P. R., Modak J. M. & Paknikar K. M. (1999). A comparative study of the mass transfer kinetics of metal biosorption by microbial biomass. *Hydrometallurgy*, **52**, 189–197.

Rapoport, A. I. & Muter, O. A. (1995). Biosorption of hexavalent chromium by yeasts. *Process Biochemistry*, **30**, 145–149.

Sag, Y. & Kutsal, T. (1995). Copper(II) and nickel (II) adsorption by *Rhizopus arrhizus* in batch stirred reactors in series. *Chemical Engineering Journal*, **58**, 265–273.

Sarrat, G., Manceau, A., Spadini, L., Roux, J. R., Hazemann, J. L., Soldo, Y., Eybert-Berard, L. & Monthennex, J. J. (1999). Structural determination of Pb binding sites in *Penicillium chrysogenum* cell walls by EXAFS spectroscopy and solution chemistry. *Journal of Synchroton Radiation*, **6**, 414–416.

Schiewer, S. (1999). Modelling the complexation and electrostatic attraction in heavy metal biosorption by *Sargassum* biomass. *Journal of Applied Phycology*, **1**, 79–87.

Schiewer, S. & Volesky, B. (1995). Modelling of the proton-metal ion exchange in biosorption. *Environmental Science and Technology,* **29**, 3049–3058.

Sharma, D. C. & Forster, C. F. (1993). Removal of hexavalent chromium using sphagnum moss peat. *Water Research*, **27**, 1201–1208.

Siegel, S. M., Galun, M. & Siegel, B. Z. (1990). Filamentous fungi as metal biosorbents: a review. *Water, Air and Soil Pollution*, **53**, 335–344.

Singleton, I. & Tobin, J. M. (1996). Fungal interactions with metals and radionuclides for environmental bioremediation. In *Fungi and Environmental Change,* eds. J. C. Frankland, N. Magan & G. M. Gadd, pp. 282–298. Cambridge, UK: Cambridge University Press.

Stringfellow, W. T. & Alvarez-Cohen, L. (1999). Evaluating the relationship between the sorption of PAH's to bacterial biomass and biodegradation. *Water Research,* **33**, 2535–2544.

Tan, H. K. S. & Spinner, I. H. (1994). Multicomponent ion exchange column dynamics. *Canadian Journal of Chemical Engineering,* **72**, 330–341.

Tobin, J. M. & Cooney, J. J. (1999). Action of tin and organotins on the hydrocarbon-using yeast, *Candida maltosa. Archives of Environmental Contamination and Toxicology,* **36**, 7–12.

Tobin, J. M. & Roux, J. C. (1998). *Mucor* biosorbent for chromium removal from tanning effluent. *Water Research,* **32**, 1407–1416.

Tobin, J. M., Cooper, D. G. & Neufeld, R. J. (1984). Uptake of metal ions by *Rhizopus arrhizus* biomass. *Applied and Environmental Microbiology,* **47**, 821–824.

Tobin, J. M., Cooper, D. G. & Neufeld, R. J. (1987). Influence of ligands on metal adsorption by *Rhizopus arrhizus* biomass. *Biotechnology and Bioengineering,* **30**, 882–886.

Tobin, J. M., Cooper, D. G. & Neufeld, R. J. (1988). The effects of cation competition on metal adsorption by *Rhizopus arrhizus* biomass. *Biotechnology and Bioengineering,* **31**, 282–286.

Tobin, J. M., Cooper, D. G. & Neufeld, R. J. (1990). Investigation of the mechanism of metal adsorption by *Rhizopus arrhizus* biomass. *Enzyme and Microbial Technology,* **12**, 591–595.

Tobin, J. M, l'Homme, B. & Roux, J. C. (1993). Immobilisation protocols and effects on cadmium uptake by *Rhizopus arrhizus* biosorbents. *Biotechnology Techniques,* **7**, 739–744.

Tobin, J. M., White, C. & Gadd. G. M. (1994). Metal accumulation by fungi: applications in environmental biotechnology. *Journal of Industrial Microbiology,* **13**, 126–130.

Treen-Sears, M. E., Martin, S. M. & Volesky, B. (1984). Propagation of *Rhizopus javanicus* biosorbent. *Applied and Environmental Microbiology,* **48**, 137–141.

Tsezos, M. & Bell, J. P. (1989). Comparison of the biosorption and desorption of hazardous organic pollutants by live and dead biomass. *Water Research,* **23**, 561–568.

Tsezos, M. & Volesky, B. (1981). Biosorption of uranium and thorium. *Biotechnology and Bioengineering,* **23**, 583–604.

Tsezos, M. & Volesky, B. (1982). The mechanism of uranium biosorption. *Biotechnology and Bioengineering,* **24**, 385–401.

Tsezos, M. & Wang, X. (1991). Biosorption and biodegradation interactions. *Biotechnology Forum Europe,* **8**, 120–125.

Tsezos, M., Noh, S. H. & Baird, M. H. I. (1988). A batch reactor mass transfer kinetic model for immobilised biomass biosorption. *Biotechnology and Bioengineering,* **32**, 545–553.

Veglio, F. & Beolchini, F. (1997). Removal of metals by biosorption: a review. *Hydrometallurgy,* **44**, 301–316.

Volesky, B. & Holan, Z. R. (1995). Biosorption of heavy metals. *Biotechnology Progress,* **11**, 235–250.

Volesky, B. & Prasetyo, I. (1994). Cadmium removal in a biosorption column. *Biotechnology and Bioengineering,* **43**, 1010–1015.

Weber. T. W. & Chakravorti, R. K. (1974). Pore and solid diffusion models for fixed bed absorbers. *American Institute of Chemical Engineers Journal*, **20**, 228–238.

Yin, P., Yu, Q., Jin, B. & Ling, Z. (1999). Biosorption removal of cadmium from aqueous solution by using pretreated fungal biomass cultured from starch wastewater. *Water Research*, **8**, 1960–1963.

Young, E. & Banks, C. J. (1998). The removal of lindane from aqueous solution using a fungal biosorbent: the influence of pH, temperature, biomass concentration and culture age. *Environmental Technology*, **19**, 619–625.

Zhao, M. & Duncan, J. R. (1998). Bed-depth-service-time analysis on column removal of Zn^{2+} using *Azolla filiculoides*. *Biotechnology Letters*, **20**, 37–39.

Zhou, J. L (1999). Zn biosorption by *Rhizopus arrhizus* and other fungi. *Applied Microbiology and Biotechnology*, **51**, 686–693.

Zhou, J. L. & Kiff, R. J. (1991). The uptake of copper from aqueous solution by immobilized fungal biomass. *Journal of Chemical Technology and Biotechnology*, **52**, 317–330.

16

The potential for utilizing mycorrhizal associations in soil bioremediation

ANDREW A. MEHARG

Introduction

There is intense interest in utilizing plants to facilitate remediation of contaminated soils because 'rhizoremediation' offers a low-cost and ecologically acceptable approach to dissipating pollutants in soils (Anderson, Guthrie & Walton, 1993). The ability of a limited number of plant species, which are normally endemic to naturally metalliferous soils, to hyperaccumulate metals is being explored with a view to remediating metal-contaminated soils; the process is termed phytoremediation (Cunningham et al., 1996). Phytoremediation as a technology has advantages and disadvantages, but as most hyperaccumulating species that are being explored with a view to commercial exploitation are in the Cruciferae and are generally non-mycorrhizal, these will not be considered in this review. The degradation of organic pollutants in the rhizosphere has also received considerable interest with a view to developing in situ remediation technologies (Anderson et al., 1993). It is here that mycorrhizal associations have to be considered (Donnelly & Fletcher, 1994; Meharg & Cairney, 2000a).

Rhizosphere degradation of organic pollutants

A wide range of organic pollutants are degraded more rapidly in the rhizospheres of most plant species tested than in bulk soils (Anderson et al., 1993). This 'rhizosphere effect' varies according to the chemical being degraded, the plant species used and the soil under study. The following explanations are normally put forward to explain enhanced rhizosphere degradation. First, rhizosphere carbon flow greatly stimulates microbial activity in soil surrounding plant roots, and this enhanced microbial activity results in an enhanced pollutant degradation rates. Second, roots may release a range of plant secondary metabolites, which select for microorganisms that can degrade these compounds. Organic pollutants may be analogues of these compounds and are degraded by the microbial enzymic pathways utilized to exploit root-derived carbon sources. The microorganisms may be able to utilize the pollutants as an energy source, or they may co-metabolize the pollutants when growing on plant-derived compounds of which the pollutants are analogues. Third, plasmid transfer is thought to be more rapid and efficient in rhizosphere environments than in bulk soil; when plasmid-borne genes are at a selective advantage in polluted environments, the rhizosphere facilitates their dispersal. Finally, the rhizosphere alters the physical/chemical structure of soil, thereby altering pollutant partitioning and transfer. In particular, the transpiration stream towards plant roots may enhance diffusion of soluble/semisoluble pollutants into the rhizosphere. While laboratory studies have highlighted the potential for rhizoremediation, translating the observed enhanced degradation in laboratory studies into field-suitable technologies offers considerable challenges. It is the purpose of this chapter to outline how exploiting/manipulating the mycorrhizosphere may be the answer to solving many of the limitations of using plants to remediate soils contaminated with organic pollutants.

Limitations of rhizosphere remediation

When considering bioremediation (rhizoremediation and other forms of biological manipulation) of contaminated soils, the remediating organism(s) must, at a minimum, satisfy three important requirements. The pollutant must be bioavailable to the remediating organism(s), the organism(s) must be resistant to the pollutant(s) present at the concentrations encountered on the site(s) to be remediated, and the organisms(s) must possess the enzymic capacity to degrade the pollutant(s) of concern. In practice, these requirements are very difficult to meet when considering rhizoremediation. The spatial constraints affecting bioavailabilty are considerable, especially for root systems. Specific root architectures will define what volume of soil the roots/rhizospheres come into contact with, and the depth to which soil profiles are occupied. Even if roots do come into contact with soil containing the target contaminants, if the contaminants have a high affinity for the soil phase (as exhibited by lipophilic aromatic pollutants) or are sequestered or encapsulated within the soil matrix (in small pores or within clay mineral lattices or organic matter), the rhizosphere organisms will not effectively interact with the pollutants to transform them. The volume of root–soil contact is critical as this will be a crucial factor in the effectiveness and speed of rhizosphere remediation.

A number of higher plants have evolved resistances to pesticides (Purrington & Bergelson, 1999), but the evolution of resistance to industrial chemicals has received little attention. As many sites that require remediation are contaminated with multiple organic and inorganic contaminants, the ideal rhizosphere-remediating species should be resistant to a wide range of pollutants. Metal (including multiple element) resistances have been found in a range of plants (Macnair, 1993), which is significant for remediation of organic pollutants in multiply contaminated sites. Besides organic and metal contaminants, industrial sites may also be very unbalanced (from deficient to toxic levels) in macro- and micronutrients, have very poor soil structures and anomalously high or low pH values. Plants selected for the remediation of such sites must be adapted to the extreme ecological niches typifying many contaminated environments. For sites contaminated with multiple organic pollutants, the enzymic activities of the organisms deployed to facilitate remediation must be able to degrade a wide range of persistent organic pollutants (POPs). Mycorrhizal associations have many attributes that can be used to extend the capabilities of rhizoremediation (Meharg & Cairney, 2000a). Consideration of how ectomycorrhizal (EcM), ericoid mycorrhizal and arbuscular mycorrhizal (AM) associations may be used for remediation purposes will be discussed below. EcM remediation has received the most attention and will be considered in the most detail.

Ectomycorrhizas

The ectomycorrhizosphere occupies virtually the entire surface organic F (fermentation) layer of forest soils (Smith & Read, 1997), and as EcM fungi have been shown to degrade a wide range of POPs (polychlorinated biphenyls (PCBs), chlorinated phenols, nitrobenzenes, pesticides and polycyclic aromatic hydrocarbons (PAHs)), there is potential to exploit the ectomycorrhizosphere for bioremediation purposes (Donnelly & Fletcher, 1994; Meharg & Cairney, 2000a). Importantly, Meharg, Cairney & Maguire (1997a) showed that 2,4-dichlophenol was degraded more effectively by the mycorrhizal fungi when these occurred in symbiosis than by the free-living mycelium. Heinonsalo *et al.* (2000) showed that crude oil was degraded more rapidly in the mycorrhizosphere than in bulk soil.

POP-degrading bacteria may thrive in the ectomycorrhizosphere (Sarand *et al.* 1998, 1999), and bacteria and EcM fungi can work in concert to facilitate POP degradation (Sarand *et al.* 1998). Bacterial numbers were higher in mycorrhizal fungal mats present in oil-contaminated soil planted with Scots pine (*Pinus sylvestris*) than in non-mycorrhizal mat rhizosphere soil or bulk soil (Heinonsalo *et al.*, 2000). EcM fungal mats may alter pollutant bioavailability in soil, and extracellular POP-degrading enzymes released by EcM fungi can diffuse from the root surfaces, effectively expanding the influence on POPs of the rhizosphere (Meharg & Cairney, 2000a). The benefits of utilizing EcM associations for bioremediation are outlined below.

Root structure

EcM fungi increase the surface area of root systems by up to 47-fold, greatly enhancing the volume of soil in contact with roots (Smith & Read, 1997). EcM fungal hyphae have much smaller diameters than roots, enabling them to penetrate microsites inaccessible to roots. As remediation agents, trees also offer other potential benefits such as soil aeration through root-induced soil cracks, which act as conduits for volatile escape to the atmosphere. Trees also have high transpiration rates and the flux of water to roots caused by transpirational demand may carry soluble compounds into the rhizosphere.

Enzymic capabilities

Many EcM fungi produce a suite of POP-degrading enzymes, such as laccases, tyrosinases, catechol oxidases, ascorbate oxygenases, hydroxylases, non-specific phenol oxidases, manganese peroxidases and lignin peroxidases (Meharg & Cairney, 2000a). POP-degrading activities are also expressed in symbiosis (Meharg *et al.*, 1997a). All EcM fungal isolates tested so far that degrade aromatic compounds (see Meharg & Cairney, 2000a) have been obtained from unpolluted soils, suggesting that ECM fungi express POP-degrading activities constitutively in their natural habitats. Therefore, no selection for, or evolution of, POP-degrading activity is required for EcM fungi to degrade a wide range of aromatic pollutants.

Another very important consideration with respect to the enzymic capabilities of EcM fungi is that the enzymes utilized in POP degradation are non-specific with respect to their ability to transform aromatic rings (Meharg & Cairney, 2000a). This means that many EcM fungi can degrade a wide range of aromatic contaminants, including PCBs and other polyhalogenated biphenyls, PAHs, chlorinated phenols, nitrotoluenes and the pesticide chlorpropham (Meharg & Cairney, 2000a).

A range of the POP-degrading enzymes produced by EcM fungi are excreted extracellularly (Gramss, Günther & Fritsche, 1998; Meharg & Cairney, 2000a), greatly increasing the volume of contaminated soil accessed by degrading enzymes. This extracellular activity confers considerable benefits to POP remediation as POPs generally have low mobility in soil. Extracellular degradation of 2,4,6-trinitrotoluene (TNT) has been demonstrated for EcM fungi (Meharg, Dennis & Cairney, 1997b). If degrading enzymes diffuse to sites where POPs are present, this overcomes bioavailability limitations, which many other forms of bioremediation technologies suffer from. Extracellular POP-degrading enzyme production is also the reason why white rot fungi (to which EcM fungi are related) have been identified as efficient bioremediators of POPs (Barr & Aust, 1994).

The only study to identify metabolic pathways of aromatic pollutant degradation by EcM fungi is the investigation by Green *et al.* (1999), which showed that EcM fungi sequentially hydroxylate the non-fluorinated ring of monofluorobiphenyl. The EcM fungi used in this study could not cleave the aromatic rings of monofluorobiphenyl, and the compound was not converted to its inorganic constituents. Meharg *et al.* (1997a) showed that EcM fungi could convert 1,2-dichlorophenol to carbon dioxide. Given that EcM fungi from uncontaminated habitats degrade a wide range of aromatic pollutants, it is pertinent to ask why they do so. Although the answer has not been resolved, the simplest hypothesis is that they naturally inhabit soil environments that are rich in polymerized aromatics and non-polymerized phenolic compounds. As EcM fungi in association obtain carbon from their host (with perhaps the exception of the most proximal hyphae), it is likely that they are degrading aromatic soil organic matter to liberate nitrogen, phosphorus and other macro- and micronutrients as many EcM association habitats are deficient in these nutrients.

Resistance of ectomycorrhizal fungi and their hosts to pollutants

A number of tree species (such as birch, willow, loblolly pine and Scots pine) and their associated EcM fungi are highly resistant to metal and organic pollutants (Hartley, Cairney & Meharg, 1997; Cairney & Meharg, 1999). Consequently, they can colonize contaminated sites and will facilitate POP degradation in the presence of metal co-contaminants.

EcM fungi appear to be able to exhibit both constitutive and adaptive resistances to metals; conflicting studies have suggested that either they possess constitutive resistance to a wide range of toxic metals (regardless of the metal status of their soils of origin) or that EcM fungi isolated from metal-contaminated sites have evolved resistance to the metals concerned (Hartley *et al.*, 1997). Only a limited number of tree species have been screened for metal resistance (see Wilkinson & Dickinson, 1995), but from these studies it appears that tree species colonizing contaminated soils generally display constitutive resistances. Mycorrhizal fungi proliferate on host roots in extremely contaminated sites such as old pesticide and munitions manufacture and storage sites, mine tailings, oil spill sites, chemical accident sites, coal gas sites and oil shales (Hartley *et al.*, 1997; Cairney & Meharg, 1999; Meharg & Cairney, 2000b). For example, Nicolotti & Egli (1998) found that additions of crude oil could stimulate or inhibit EcM infection on *Picea abies* and *Populus nigra*, depending on a range of factors. Oil addition altered EcM fungal community structure, inhibiting some morphotypes, having a neutral effect on others and a stimulatory effect on the remainder, with some morphotypes observed only at the highest concentrations. These morphotypes were dominant at high oil application rates. The *P. sylvestris–Suillus bovinus* association was not impacted by the presence of 2% (w/v) toluene (Sarand *et al.*, 1999). *P. sylvestris–S. bovinus/Paxillus involutus* associations grown in the presence of petroleum hydrocarbons also showed no adverse impacts (Sarand *et al.*, 1998). When EcM fungi were tested under culture conditions, some pollutants were toxic to the tested strains (TNT; Meharg *et al.*, 1997b), while other contaminants (such as organophosphate pesticides) were nontoxic even at their maximum water solubilities (A. A. Meharg, unpublished data).

The taxonomic diversity of EcM fungi is considerable (Molina, Massicotte & Trappe, 1992). This is highly advantageous for bioremediation. If a number of fungal taxa/strains are sensitive to particular suites of pollutants, it appears from a wide number of field observations that other taxa/strains will step in to fill ecological niches vacated because of pollutant stressors (see Hartley *et al.*, 1997; Cairney & Meharg; 1999; Meharg & Cairney 2000b).

Ectomycorrhizosphere microbial consortia

The ectomycorrhizosphere comprises a diverse community, with considerable structural and functional diversity exhibited between the EcM fungal species in addition to the free-living fungi and bacteria harboured within the mycorrhizosphere. As these free-living mycorrhizosphere organisms can also express POP-degrading activities (Sarand *et al.*, 1998, 1999), the mycorrhizosphere comprises an environment ideally suited to POP degradation. *P. sylvestris–S. bovinus/P. involutus* associations grown in soil contaminated with petroleum hydrocarbons formed bacterial biofilms (harbouring plasmid-encoded catabolic genes for hydrocarbon degradation) on the surface of external hyphae (Sarand *et al.*, 1998). Degradation of *m*-toluate has also been observed in *P. sylvestris–S. bovinus* rhizospheres inoculated with a strain of *Pseudomonas fluorescens* that possessed a toluene-degrading plasmid (Sarand *et al.*, 1999).

While EcM fungi may not have the metabolic capabilities to carry out complete degradation of the mineral constituents for a range of POPs, Green *et al.* (1999) demonstrated that EcM fungi could sequentially hydroxylate a fluorinated biphenyl ring. Ring hydroxylation is a thermodynamically limiting step in biphenyl ring degradation. If EcM fungi can initiate metabolism of aromatic rings, free-living rhizosphere microorganisms may facilitate subsequent metabolism. Initial EcM fungal metabolism of biphenyl rings may be particularly important in developing ectomycorrhizosphere remediation technologies in that bacteria generally require biphenyl as a co-substrate to degrade halogenated biphenyls (Donnelly & Fletcher, 1994; Gilbert & Crowley, 1997). The presence of EcM fungi may negate the need for co-substrates. As it is undesirable to add pollutants (i.e. biphenyl) to remediate pollutants, EcM associations may offer a more elegant solution to remediation.

Enhanced degradation of tetrachloroethylene (TCE) and crude oil have also been observed in intact ectomycorrhizospheres of *Pinus taeda* (Anderson & Walton, 1995) and *P. sylvestris* (Heinonsalo *et al.*, 2000), respectively.

Field application of ectomycorrhizosphere technologies

When considering the utility of the ectomycorrhizosphere as a POP remediation technology, the benefits and the limitations of the technique must be explored. Considering first the limitations; like most potential rhizoremediation technologies, ectomycorrhizosphere remediation is untested in the field, and laboratory microcosm studies only demonstrate that there is a potential for developing useful technologies. It is still necesary to examine if field-scale application is feasible. Further, although fast-growing tree species (such as birch and willow) could be deployed, using trees to facilitate remediation would still require years to decades. Consequently, ectomycorrhizosphere remediation can only be used on sites where long-term remediation is an option. In addition, ectomycorrhizosphere remediation is not suitable for sites where it is a requirement to remove the metal contaminant burdens. However, growing trees may have a number of benefits. Economically, it may be viable to turn land over to forestry for timber or fuel (i.e. willow coppicing). Aesthetically, trees may be used to revegetate derelict zones to create amenity or wildlife areas.

The use of forestry to facilitate remediation may involve a number of management strategies. Traditional forestry is one option. However, it may be possible to bring certain types of waste (industrial effluent, sewage sludge) to established forests and incorporate these wastes into surface organic soil horizons. Management options would have to be thought out carefully, with full long-term risk assessments

Ericoid mycorrhizal associations

Many of the attributes of EcM associations with respect to organic pollutant remediation are also applicable to ericoid mycorrhizal associations. The dominant ericoid mycorrhizal fungus *Hymenoscyphus ericae* has many of the enzymic characteristics of EcM fungi (Smith & Read, 1997) and, like EcM fungi, *H. ericae* is capable of degrading a wide range of POPs (Donnelly & Fletcher, 1994; Meharg *et al.*, 1997b; Green *et al.*, 1999). It is likely that EcM fungi and ericoid mycorrhizal fungi possess similar POP-degrading activities as both classes occupy similar ecological niches, such as acid, organic-rich, nutrient-deficient soils (Smith & Read, 1997).

Calluna vulgaris/H. ericae associations are often the dominant vegetation cover on highly metal(loid)-contaminated sites (Meharg & Cairney, 2000b) and, therefore, are ideally suited to colonizing such habitats. Other ericoid associations are also known to colonize such sites (Meharg & Cairney, 2000b). The ability of ericoid mycorrhizal associations to grow on substrates contaminated with organic pollutants has received little or no attention. *H. ericae* growth was unaffected in the presence of organophosphorus pesticides at the maximum water solubilities of the two insecticides tested (A. A. Meharg, unpublished data). However, ericoid associations tend to be slow growing. Hyphal extension from infected roots is limited compared with the much more extensive hyphae of EcM fungi (Smith & Read, 1997). Both of these factors may preclude ericoid associations from bioremediation technologies in view of the advantages of EcM associations.

Arbuscular mycorrhizal associations

The POP-degrading activities of AM fungi have not been explored. This is primarily because they cannot be cultured in the absence of their host, and the extracelluar hyphal network is not as extensive as EcM associations. However, AM fungi may have a role in POP remediation by enabling plants to establish on highly contaminated soils.

AM fungi are found colonizing roots on soils contaminated with crude oil (Cabello, 1997; Nicolottti & Egli, 1998), on oil shale wastes (Stahl & Williams, 1986), coal wastes and on lignite and calcite mine spoils (Ganesan *et al.*, 1991). They are also found on plants colonizing highly metal(loid)-contaminated soils (Meharg & Cairney, 2000b). There is mounting evidence that AM fungi evolve resistances to metal(loid) contaminants, enabling them to colonize mine wastes (Meharg & Cairney, 2000b). Meharg & Cairney (2000b) discuss the ecological significance of metal(loid) resistances in AM fungi. The AM fungi may simply have evolved resistances so that they can carry out their normal ecological functions (such as nutrient acquisition for their host), or the AM fungi may actually confer enhanced resistance to host plants. Evidence for the latter hypothesis had been lacking until recently. Gonzalez-Chavez (2000) showed that AM fungi from arsenic mine spoils further enhanced the arsenate resistance of the already-arsenate resistant grass *Holcus lanatus*. In a study of *Violetum calaminarie*, which is endemic on metaliferous soil, Hildebrandt, Kaldorf & Bothe (1999) showed that its AM fungi considerably enhanced the growth of *Zea mays, Hordeum vulgare, Lupinus luteus* and *Medicago vulgare* on contaminated soil over that seen with a standard AMF inocula. If AM fungal strains can stimulate the growth of plant hosts that are of remediation interest for contaminated soil, the benefits for rhizosphere remediation are obvious.

Conclusions

Rhizosphere remediation technologies are in their infancy. The potential is great, both in terms of the mycorrhizal fungal and host germplasm available for exploitation and in the range of site-management strategies that can be deployed using mycorrhizal associations. The technologies need to be validated at a field scale to ascertain ultimately if rhizoremediation is practical. Notwithstanding the exploitation of enhanced rhizosphere degradation of POPs for remediation purposes, processes underlying POP dynamics in the rhizosphere are also pertinent to pollutant fate and behaviour in ecosystems and must be taken into account when assessing the ecological implications of chemicals in surface soils.

References

Anderson, T. A. & Walton, B. T. (1995). Comparative fate of [^{14}C]trichloroethylene in the root zone of plants from a former solvent disposal site. *Environmental Toxicology and Chemistry,* **14**, 2041–2047.

Anderson, T. A., Guthrie, E. A. & Walton, B. T. (1993). Bioremediation in the rhizosphere. *Environmental Science and Technology*, **27**, 2630–2636.

Barr, D. P. & Aust, S. D. (1994). Mechanisms white rot fungi use to degrade pollutants. *Environmental Science and Technology*, **28**, 79–87.

Cabello, M. N. (1997). Hydrocarbon pollution: its effect on native arbuscular mycorrhizal fungi (AMF). *FEMS Microbiology Ecology*, **22**, 233–236.

Cairney, J. W. G. & Meharg, A. A. (1999). Influences of anthropogenic pollution on mycorrhizal fungal communities. *Environmental Pollution*, **106**, 169–182.

Cunningham, S. D., Anderson, T. A., Schwab, A. P. & Hsu, F. C. (1996). Phytoremediation of soils contaminated with organic pollutants. *Advances in Agronomy,* **56**, 55–114.

Donnelly, P. K. & Fletcher, J. S. (1994). Potential use of mycorrhizal fungi as bioremediation agents. In *Bioremediation Through Rhizosphere Technology*, eds. T. A. Anderson & J. R. Coats, pp. 93–99. Washington DC: American Chemical Society.

Ganesan, V., Ragupathy, S., Parthipan, B., Rajini Rani, D. B. & Mahadevab, A. (1991). Distribution of vesicular-arbuscular mycorrhizal fungi in coal, lignite and calcite mine spoils of India. *Biology and Fertility of Soils*, **12**, 131–136.

Gilbert, E. S. & Crowley, D. E. (1997). Plant compounds that induce polychlorinated biphenyl biodegradation by *Arthrobacter* sp. strain B1B. *Applied and Environmental Microbiology*, **63**, 1933–1938.

Gonzalez-Chavez, C. (2000). *Arbuscular mycorrhizal fungi from As/Cu polluted soils, contribution to plant tolerance and importance of external mycelium.* PhD thesis, University of Reading, UK.

Gramms, G., Günther, T. & Fritsche, W. (1998). Spot tests for oxidative enzymes in ectomycorrhizal, wood and litter decaying fungi. *Mycological Research*, **102**, 67–72.

Green, N. A., Meharg, A. A., Till, C., Troke, J. & Nicholson, J. K. (1999). Degradation of 4-fluorobiphenyl by mycorrhizal fungi as determined by ^{19}F nuclear magnetic resonance spectroscopy and ^{14}C radiolabelling analysis. *Applied and Environmental Microbiology*, **65**, 4021–4027.

Hartley, J., Cairney, J. W. G. & Meharg, A. A. (1997). Do ectomycorrhizal fungi exhibit adaptive tolerance to potentially toxic metals in the rhizosphere? *Plant and Soil*, **189**, 303–319.

Heinonsalo, J., Jorgensen, K. S., Haahtela, K. & Sen, R. (2000). Effect of *Pinus sylvestris* root growth and mycorrhizosphere development on bacterial carbon source utilization and hydrocarbon oxidation in forest and petroleum-contaminated soils. *Canadian Journal of Microbiology*, **46**, 451–464.

Hildebrandt, U., Kaldorf, M. & Bothe, H. (1999). The zinc violet and its colonization by arbuscular mycorrhizal fungi. *Journal of Plant Physiology*, **154**, 709–717.

Macnair, M. R. (1993). Genetics of metal resistance in vascular plants. *New Phytologist*, **124**, 541–559.

Meharg, A. A. & Cairney, J. W. G. (2000a). Ectomycorrhizas – extending the capabilities of rhizosphere remediation? *Soil Biology and Biochemistry*, **32**,

1475–1484.

Meharg, A. A. & Cairney, J. W. G. (2000b). Co-evolution of mycorrhizal symbionts and their hosts to metal contaminated environments. *Advances in Ecological Research*, **30**, 69–112.

Meharg, A. A., Cairney, J. W. G. & Maguire, N. (1997a). Mineralization of 2,4-dichlorophenol by ectomycorrhizal fungi in axenic culture and in symbiosis with pine. *Chemosphere*, **34**, 2495–2504.

Meharg, A. A., Dennis, G. R. & Cairney, J. W. G. (1997b). Biotransformation of 2,4,6-trinitrotoluene (TNT) by ectomycorrhizal basidiomycetes. *Chemosphere*, **35**, 513–521.

Molina, R., Massicotte H. & Trappe, J. M. (1992). Specificity phenomena in mycorrhizal symbiosis: community-ecological consequences and practical implications. In: *Mycorrhizal Functioning*, ed. M. F. Allen, pp. 357–423. London: Chapman & Hall.

Nicolotti, G. & Egli, S. (1998). Soil contamination by crude oil: impact on the mycorrhizosphere and on revegetation potential of forest trees. *Environmental Pollution*, **99**, 37–43.

Purrington, C. B. & Bergelson, J. (1999). Exploring the physiological basis of costs of herbicide resistance in *Arabidopsis thaliana*. *American Naturalist*, **154**, S82–S91.

Sarand, I., Timonen, S., Nurmiaho-Lassila, E.-L., Koivila, T., Haahtela, K., Romantschuk, M. & Sen, R. (1998). Microbial biofilms and catabolic plasmid harbouring degradative fluorescent pseudomonads in Scots pine mycorrhizospheres developed on petroleum contaminated soil. *FEMS Microbiology Ecology*, **27**, 115–126.

Sarand, I., Timonen, S., Koivula, T., Peltola, R., Haahtela, K., Sen, R. & Romantschuk, M. (1999). Tolerance and biodegradation of *m*-toluate by Scots pine, a mycorrhizal fungus and fluorescent pseudomonads individually and under associative conditions. *Journal of Applied Microbiology*, **86**, 817–826.

Smith, S. E. & Read, D. J. (1997). *Mycorrhizal Symbiosis*. London: Academic Press.

Stahl, P. D. & Williams, S. E. (1986). Oil shale process water affects activity of vesicular-arbuscular fungi and *Rhizobium* four years after application to soil. *Soil Biology and Biochemistry*, **18**, 451–455.

Wilkinson, D. M. & Dickinson, N. M. (1995). Metal resistances in trees – the role of mycorrhizae. *Oikos*, **72**, 298–300.

17

Mycorrhizas and hydrocarbons

MARTA NOEMÍ CABELLO

Introduction

Knowledge of plant–microorganism interactions is of great importance for bioremediation and phytoremediation. A wide variety of microbial populations live in natural and agricultural soils, and in marginal soils contaminated with xenobiotics. Plant roots strongly influence the surrounding environment, producing the so-called 'rhizosphere effect' in which microbial populations are qualitatively and quantitatively altered with, reciprocally, their metabolism directly affecting plant biology and the accompanying biota.

Arbuscular mycorrhizal fungi (AMF) belong to the wide spectrum of soil microbiota and are able to improve the growth of the host plant, particularly in soils of low nutritional status or in those modified by human activity. This positive effect can be ascribed to the improvement of nutrient uptake by mycorrhizal colonized plant roots and the increase of soil volume explored for nutrient uptake by the plant, extending from areas in which nutrients have been exhausted to new regions where they are still available. An understanding of the interactions of arbuscular and vesicular–arbuscular mycorrhizas, together with the remaining soil microorganisms naturally associated with plant roots, will provide the basis for development of an important biotechnological tool for bioremediation.

Bioremediation is a managed or spontaneous process in which biological, especially microbiological, catalysis acts on pollutant compounds, thereby reducing or eliminating environmental contamination (Madsen, 1991). Phytoremediation is a natural process involving plants, especially those that are able to survive in contaminated soil and water (Watanabe, 1997), their associated microbiota, appropriate soil amendments and agronomic techniques that remove, contain, or detoxify environmental contaminants (Cunningham *et al.*, 1996). These processes have become important since the expanding use of hydrocarbon derivatives from petroleum, as a source of energy and raw materials for industrial purposes, has resulted in a dramatic increase in contaminants in soil and water. This phenomenon is a global environmental problem. Spillages during the extraction and storage of crude oil and its derivatives regularly contaminate surrounding soils; such pollution is also increased by overflows during crude oil processing owing to failures in cleaning, refrigeration and transport. The deleterious effects of oil derivatives are very serious because they affect all levels of biological hierarchy, from subcellular components to the highest trophic levels of ecosystem organization (Bossert & Bartha, 1984).

Soil pollution with hydrocarbons has strong and negative effects on plant communities. Contact toxicity occurs because the low-boiling-point hydrocarbon components have a solvent action on the lipid membrane structures of cells; this toxicity increases with increasing polarity and decreases with increasing molecular weight (McGill, Rowell & Westlake, 1981). Petroleum products also have indirect effects through interactions with biotic and abiotic soil components (Bossert & Bartha, 1984). Plant roots can be deprived of oxygen because of exhaustion of soil oxygen by hydrocarbon-degrading microorganisms and of nutrients by competition between plants and microorganisms. Intermediate metabolites generated by incomplete microbial degradation of oil can be even more toxic than their precursors and increase the phytotoxicity of contaminated soils if they are not removed by leaching or humification processes. In addition, oil alters the physical structure of soil, for example decreasing its capacity to store moisture and air (de Jong, 1980).

Arbuscular mycorrhizal symbioses are crucial for present-day natural, terrestrial ecosystems (Harley & Smith, 1983); they may have developed with the first appearance of vascular plants on earth and may have strongly influenced plant succession (Pirozynski, 1981). Consequently, AMF may have an important influence on terrestrial plant evolution. The mycorrhizal plants have particular advantages over non-mycorrhizal plant species in marginal ecological conditions. Mycorrhizas are essential for plant survival in coastal sand dunes, which are characterized by their low phosphorus content (Nicolson, 1959, 1960), in desert ecosystems undergoing rapid erosion of surface soil, and in revegetation programmes of mine soils (Daft & Nicolson, 1974; Khan, 1978, 1981). The mycelium of mycorrhizal fungi is more resistant to abiotic agents than the root itself and this may compensate for reduced root growth. AMF increase tolerance to extreme drought conditions (Kothari, Marschner & George, 1990), high soil salinity (Juniper & Abbott, 1993) and heavy metal toxicity (Weissenhorn, Leyval & Berthelin, 1993, 1995). In this chapter, the effect of soil hydrocarbon contamination on the mycorrhizal status of plant species and

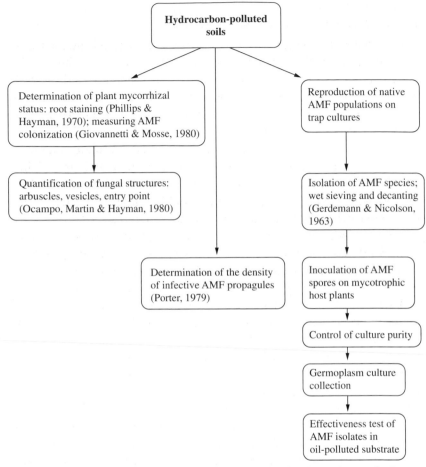

Fig. 17.1. Protocols for studying the effects of hydrocarbon pollution on arbuscular mycorrhizal fungi (AMF).

the native AMF populations growing in those soils is discussed. The benefits of inoculating plants with native AMF isolated from polluted sites can be seen in a variety of plant growth parameters, such as biomass production and phosphorus and zinc accumulation. These are discussed, together with their potential for phytoremediation.

Effect of hydrocarbon pollution on arbuscular mycorrhizal fungi (AMF)

Figure 17.1 shows the steps followed to study the effect of hydrocarbon soil contamination on AMF.

Mycorrhizal status of soil vegetation

Knowledge of the relationships between mycorrhizal fungi and their hosts in hydrocarbon-polluted habitats is limited. Cabello (1997) showed that native AMF propagules survived habitat contamination, although the proportion of mycorrhizal roots was 150% higher than in plants from non-polluted soils. In field work, Cabello (1995) studied the effect of hydrocarbon-containing sludges, incorporated at a rate of $151 m^{-2}$ in soil plots, on alfalfa plants (*Medicago sativa* L.) and on alfalfa plants inoculated with *Glomus mosseae* (Nicol. & Gerd.) Gerd. & Trappe. There were low levels of root colonization in non-inoculated plants, with 5% of the root length colonized after 3 months and 15% after 12 months. Plants inoculated with *G. mosseae* showed 41% and 70% colonization at 3 and 12 months, respectively. The incorporation of hydrocarbon-containing sludges into the soil reduced the capacity of native AMF propagules to initiate colonization. Recovery to levels similar to those found in non-polluted soils was slow and commenced in neighbouring areas.

AMF are in close contact with the roots they colonize by means of abundant arbuscular structures within the cells of the cortex. The external fungal mycelium spreads out from the root and connects the plant to its below-ground environment. The growth of the mycelium could be ecologically significant in helping the plant to develop under stressful situations. Although the proportion of root colonized must be taken into account when evaluating mycorrhizal status, this can lead to an underestimate of the effect of different environmental factors on the symbiosis (Harley & Smith, 1983). The effect of contamination on fungal physiology within the root is not easy to determine. Although the length of the root that is colonized does reflect a general harmful action of hydrocarbons on native AMF, the negative effects are more noticeable on the fungal structures (Cabello, 1997). Pollution does reduce the arbuscular percentage but the symbiosis still remains functional since these fungal structures are the major sites for nutrient exchange between the fungal endophytes and the host plant (Cox *et al.*, 1975; Scannerini & Bonfante Fasolo, 1983). The percentage of arbuscular structures in non-inoculated plants transplanted into polluted sludges was very low, being only 1% 8 months after transplantation, whereas inoculated plants reached higher levels (Cabello, 1995). A high vesicle number is correlated with a decrease in the arbuscular percentage, and under stress situations the fungus metabolizes the lipid stored in vesicles before their degradation. The vesicle number increases in old and/or dead roots, as well as in plants growing under abiotic or biotic stress, suggesting that these fungal structures may have a role in resistance to stress (Cooper, 1984).

Density of native arbuscular mycorrhizal fungal propagules

The density of native AMF propagules in contaminated soils from Ensenada (Argentina) and Rositz (Germany) has been determined by the most probable number method. Overall, propagule densities in non-polluted areas were 7.5-fold and 5-fold higher than those in polluted soils in Argentina and Germany, respectively (Cabello, 1997). This suggests that the mycorrhizal status of plants growing in the polluted soils may be limited by the level of infective propagules. Bioassay of the mycorrhizal inoculum potential in topsoil placed over deposited oil shale revealed colonization levels of approximately half those obtained in undisturbed soil (Call & McKell, 1982). High propagule densities may be associated with rapid fungal spread within roots, which is important in field situations (Smith & Read, 1997). The low levels of AMF inoculum found in contaminated soils could result in failure of revegetation processes in these areas (Miller, 1979). However, if the surviving propagules were highly efficient in the symbiotic association, they could still significantly increase the production of the selected plants.

Isolation of arbuscular mycorrhizal fungi from polluted sites

Soils polluted with different types of hydrocarbon have been processed to isolate native mycorrhizal fungi. These soils were sandy soils from Mar de Ajó (Argentina) contaminated with gas-oil from storage tanks and electricity generators; clay soils from Berisso and Ensenada (Argentina) oil polluted with fuel and crude oil, respectively; and sandy-muddy soils from Rositz (Germany) contaminated with crude oil, polycyclic aromatic hydrocarbons (PAHs) and phenols from asphaltic production residues and bombs from World War II.

AMF and vesicular–arbuscular mycorrhizal fungi from native communities present in contaminated soils were isolated in pot cultures using *M. sativa* L. and *Sorghum vulgare* Pers. as trap plants. All fungal isolates were deposited in the culture collection of the Spegazzini Institute (LPS) and are listed in Table 17.1. The narrow fungal spectrum found in contaminated soils suggests that only some AMF species can survive pollutants, a fact closely related to the low level of colonized roots and the low number of propagules in these soils.

Table 17.1. *Arbuscular mycorrhizal fungi isolated from hydrocarbon-polluted soils*

Fungal species	Substrate	Soil sampled
Glomus aggregatum Schenck & Smith emend. Koske	Sand : soil; 9 : 1	Clay soil, Berisso, Argentina
Glomus claroideum? Schenck & Smith	Sand : soil; 9 : 1	Clay soil, Berisso, Argentina
Glomus deserticola Trappe, Bloss & Menge	Vermiculite : soil; 2 : 3	Sandy soil, Mar de Ajó, Argentina
Glomus geosporum (Nicol. & Gerd.) Walker	Sand : soil; 9 : 1	Clay soil, Berisso, Argentina
Glomus intraradices Schenck & Smith	Vermiculite : soil; 2 : 3	Clay soil, Ensenada, Argentina
G. intraradices Schenck & Smith	Vermiculite : soil; 2 : 3	Sandy-muddy soil, Rositz, Germany
Glomus mosseae (Nicol. & Gerd.) Gerdemann & Nicolson	Sand : soil; 9 : 1	Clay soil, Berisso, Argentina
Glomus tortuosum? Schenck & Smith	Sand : soil; 9 : 1	Clay soil, Ensenada, Argentina
Scutellospora sp.	Sand : soil; 9 : 1	Clay soil, Ensenada, Argentina

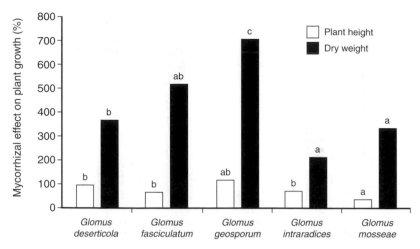

Fig. 17.2. Influence of mycorrhizal infection on plant height and dry mass of alfalfa after 60 days of growth in hydrocarbon-polluted substrate. The mycorrhizal effect was calculated as 100 (infected − noninfected)/non-infected. Bars with the same letter are not significantly different (least significant difference: $P = 0.01$). (Modified from Cabello, 1999.)

Effectiveness of indigenous arbuscular mycorrhizal fungi from polluted soils

The AMF *Glomus deserticola*, *Glomus geosporum* and *Glomus intraradices* have been screened for their symbiotic response with *M. sativa* in a hydrocarbon-polluted substrate under greenhouse conditions and compared with two laboratory AMF species, *Glomus fasciculatum* and *G. mosseae*, and uninoculated plants as controls (Cabello, 1999). With the exception of root length, all measured parameters were significantly higher for inoculated plants. Total biomass production, estimated as dry weight, and plant height were significantly enhanced by AMF inoculation, with the maximal response being achieved by *G. geosporum* (Fig. 17.2). Despite the low colonization percentages, the observed increase in the dry weight and height of the inoculated plants shows that high levels of colonization are not always necessary for a good plant response: efficient biomass production seems to result from physiological interactions between the symbionts.

Both phosphorus and zinc concentrations were higher in plants inoculated with the three fungal strains isolated from contaminated soils (Fig. 17.3). *G. deserticola* and *G. geosporum* inoculation significantly increased the phosphorus concentration. Moderate though significant in-

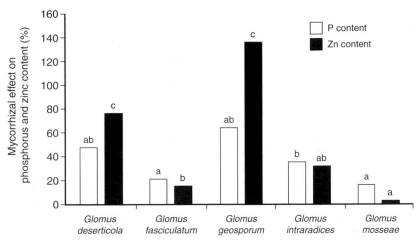

Fig. 17.3. Influence of mycorrhizal infection on phosphorus and zinc content of alfalfa after 60 days of growth in hydrocarbon-polluted substrate. The mycorrhizal effect was calculated as 100 (infected − noninfected)/non-infected. Bars with the same letter are not significantly different (least significant difference: $P = 0.01$). (Modified from Cabello, 1999.)

creases were also observed for the other AMF infections compared with uninoculated plants. The correlation between phosphorus content and biomass production by inoculated and non-inoculated plants confirms the beneficial effects of inoculation with mycorrhizal species that can efficiently adapt to contamination (Cabello, 1999). Call & McKell (1984) reached similar conclusions using container-grown fourwing saltbush plants (*Atriplex canescens* (Pursch) Nutt.) inoculated with native AMF species and transplanted into processed-oil shale and disturbed native soils. Inoculated plants had a higher biomass content, enhanced their percentage of covered area, absorbed more water and increased their phosphorus concentrations compared with non-inoculated controls. These results strongly suggest an ecological adaptation of AMF and this should be taken into account when selecting fungi able to associate with plants suitable for revegetation of contaminated areas. The AMF appear to be able to adapt to changes in soil pH, the presence of heavy metals, the source of inoculum and their host endophyte specificity (Pfleger, Stewart & Noyd, 1994). However, this response cannot be generalized, since Shetty *et al.* (1994) found that mycorrhizal fungi from uncontaminated soils increased plant biomass more than mycorrhizal fungi isolated from soils polluted with heavy metals.

Alfalfa plants growing on hydrocarbon-contaminated substrates show a remarkable mycorrhizal dependence. According to Habte & Manjunath (1991), this plant is highly dependent on its mycorrhizal association with respect to biomass production and phosphorus and zinc uptake and storage. The term mycorrhizal dependency has been defined as 'the degree to which a plant species is dependent on the mycorrhizal condition to produce its maximum growth or yield at a given level of soil fertility' (Gerdemann, 1975). Based on our experiments, the definition could be extended to 'or if the soil is polluted with xenobiotics'. This is because some plant species do not need a symbiotic association with AMF under normal conditions but do require this interaction when subjected to pollution.

Phytoremediation using arbuscular and vesicular–arbuscular fungi: a challenge

The time required for bioremediation processes depends on a number of variables such as the nature of the contaminant and its source, the soil physicochemical characteristics, the climatic conditions and the level of infective inoculum. At the same time, degradation of recalcitrant compounds is generally limited by factors such as a microbial population of low catabolic efficiency, decreased nutrient availability, inadequate pH and other soil properties. To revegetate degraded areas successfully, it is necessary to understand the dynamics of plant–microbe–toxicant interactions in the rhizosphere since vegetation plays an important role in the biodegradation of surface soils contaminated with toxic organic chemicals through stimulation of microorganisms with a capacity for biodegradation (Walton & Anderson, 1990; Walton, Guthrie & Hoylman, 1994; Günther, Dornberger & Fritsche, 1996). Radwan, Sorkhoh & El-Nemr (1995) suggested the use of suitable plants for the bioremediation of oil-polluted desert soil. Günther *et al.* (1996) assayed the effect of ryegrass on the degradation of soil hydrocarbons under controlled laboratory conditions. In artificially loaded soils, they observed rapid hydrocarbon disappearance in the ryegrass rhizosphere, together with a greater microbial population with elevated respiration rates compared with control assays without plants. These authors concluded that improvement in hydrocarbon degradation was caused by an increased rhizosphere microbial community compared with that in root-free soil. The climate and soil nutritional state also should be taken into account when re-introducing vegetation in hydrocarbon-polluted areas. During the revegetation process, nitrogen and phosphorus limitation might favour the selection of plant species with a low requirement for these elements.

Call & McKell (1982) observed that annual non-mycorrhizal weeds form the initial plant successional stages in oil-shale spoils, as well as in other disturbed ecosytems and several studies have shown that mycorrhizas are rare or absent at these early stages. The presence or absence of mycorrhizal fungi in secondary successional sites can determine the composition of the plant population (Reeves *et al.*, 1979). Gudin & Syratt (1975) investigated 15 oil-contaminated sites and found that leguminous plants constituted the prevailing flora, indicating an adaptive advantage from their association with symbiotic nitrogen-fixing microorganisms. However, further studies are necessary to determine if these microorganisms can survive pollution. Westlake, Jobson & Cook (1978) reported that fertilization of oil-contaminated soils with nitrogen and phosphorus improved the revegetation rate.

Taking into account the increase of microorganism biomass in root systems, soil penetration by roots and nutrient excretion, the rhizosphere represents a system ideally suited for the acceleration of biodegradation of pollutants. In the rhizosphere, the soil physicochemical properties change; there is an increase in and selection of microbial communities caused by the supply of root exudates, sloughed off cells and tissue fragments; and there is increased availability of co-metabolites for the degradation processes. Fungi able to form arbuscular and vesicular–arbuscular mycorrhizas might, therefore, be of vital importance for phytoremediation processes. However, soil disturbance can significantly reduce the amount of colonizing propagules which affects their potential to colonize new host plants (Miller, 1979; Reeves *et al.*, 1979; Cabello, 1997). The higher the amount of disturbance, the longer is the time needed to re-establish the mycorrhizal vegetation. The success of revegetation programmes with minimal treatments will depend on the method developed to re-introduce AMF in those sites where they have disappeared. Call & McKell (1984) demonstrated that inoculation with AMF in either nursery beds or containers and the addition of low levels of fertilizers promoted the growth of plants better adapted to the environmental conditions of the outplanting site. In addition, these infected plants provided an inoculum source for other desirable plant species requiring the AMF association for successful establishment and survival. When planted out, the AMF-inoculated plants were better adapted to revegetation programmes with minimal treatments in processed-oil shale and disturbed native soils. After two growing seasons in the field, the biomass increased 2.5 times, the plant height 1.8 times and the vegetation coverage was 2.2 times greater than with uninoculated controls. However, there were significant differences in the growth in disturbed soils or processed oil shale: the evaluated plant parameters were smaller in the latter, which could arise from the low water-holding capability, high pH and nutrient deficiency of the spent shale (Call & McKell, 1984).

AMF have not yet been assayed for degradation of complex organic compounds as they cannot be cultured without a host plant. Nevertheless, laboratory studies have demonstrated that ectomycorrhizal fungi possess enzymatic properties similar to saprotrophic species (Donnelly & Fletcher, 1994). These fungi could play a role in bioremediation when loaded together with the host plant into the contaminated site. Although the capability of AMF to degrade xenobiotics has not been assessed, they should be considered before starting phytoremediation in oil-polluted areas since AMF not only improve the growth, production and survival of plants but also positively influence the environment in several ways, for example, affecting the accumulation and transport of phosphorus, nitrogen and other essential nutrients required by plants (Smith & Read, 1997). AMF can also bind soil particles through the intensively growing mycelium, which improves soil physical properties, and positively interact with other rhizosphere microbiota such as phosphate-solubilizing microorganisms, nitrogen-fixing bacteria, cyanobacteria and actinomycetes, and plant growth-promoting rhizobacteria, constituting the so-called 'mycorrhizosphere effect' (Linderman, 1988). AMF can stimulate plant development through production of growth factors; for example, Allen, Moore & Christensen (1980) detected increases in cytokinin activity in mycorrhizal plants of *Bouteloua*. AMF can also decrease the accumulation of heavy metals, by storing them in mycorrhizal roots (Schüepp & Bodmer, 1991); improve tolerance to stresses caused by cold, flood, drought, nutrient deficiency and salinity; enhance resistance to pathogens; and provide a reservoir of nutrients that improve soil fertility when the biomass is recycled.

Figure 17.4 shows a practical approach for restoration of oil-polluted soil. The questions can be answered using bioassays of the type described above. The last step requires a technological development to allow the production of large amounts of AMF inoculum. AMF inoculation technology is mainly limited by the lack of production and commercial distribution of AMF inoculum; however, private companies, government agencies and several research groups have attempted to develop commercial AMF inoculants with effective AMF strains and different culture systems for these fungi have been developed and patented (Wood, 1992).

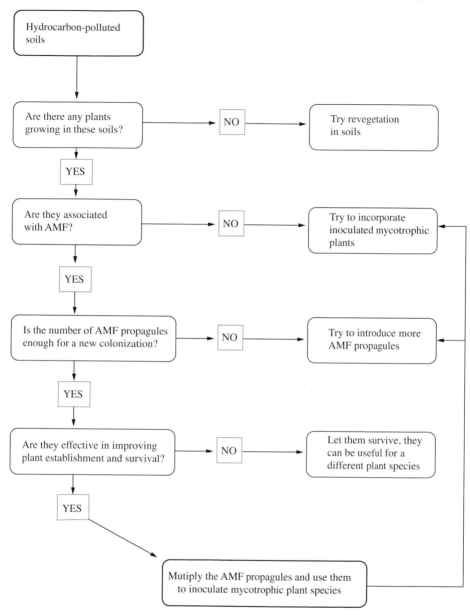

Fig. 17.4. Practical procedures for restoration of vegetation in oil-polluted soils.

Conclusions

Revegetation of oil-polluted areas with arbuscular mycorrhizal plants is challenging since it involves reconstruction of the above-ground flora and the below-ground microbiota. With this technology, the number of microbial communities, including bacteria, actinomycetes, saprotrophic and biotrophic fungi, is increased. These microorganisms contribute to the biological activity of the rhizosphere soil, where they improve toxic chemical degradation, thus reducing the time factor in the remediation process. Re-establishment of mycorrhizal fungi is necessary for successful plant succession, with improved nutrient uptake and the elimination of soil stress being key points for the plant survival. By introducing AMF associated with plant roots, there is an inoculum of mycorrhizal propagules, which will increase in the restored soil with time. In this way, new mycotrophic plants might be encouraged to colonize the area and the invasion of the soil by weed plants, which do not usually form mycorrhizas, will be limited. Increasing the biodiversity of both microorganisms and plants is of crucial importance for high production in a restored ecosystem.

Acknowledgements

The author is grateful to N. Tedesco (CONICET) for technical assistance. This research work was supported in part by grants from the Comisión de Investigaciones Científicas (CIC) de la Provincia de Buenos Aires and Consejo Nacional de Investigaciones Científicas y Tecnológicas (CONICET). M. N. Cabello is a researcher of the CIC.

References

Allen, M. F., Moore, T. S. & Christensen, M. (1980). Phytohormone changes in *Bouteloua gracilis* infected by vesicular-arbuscular mycorrhizae. I. Cytokinin increases in the host plant. *Canadian Journal of Botany*, **58**, 371–374.

Bossert, I. & Bartha, R. (1984). The fate of petroleum in soil ecosystems. In *Petroleum Microbiology*, ed. R. M. Atlas, pp. 435–473. New York: Macmillan.

Cabello, M. N. (1995). Effect of hydrocarbon pollution on vesicular–arbuscular mycorrhizal fungi (VAM). [In Spanish] *Boletín Micológico* (Chile), **10**, 77–83.

Cabello, M. N. (1997). Hydrocarbon pollution: its effect on native arbuscular mycorrhizal fungi (AMF). *FEMS Microbiology Ecology*, **22**, 233–236.

Cabello, M. N. (1999). Effectiveness of indigenous arbuscular mycorrhizal fungi (AMF) isolated from hydrocarbon polluted soils. *Journal of Basic Microbiology*, **39**, 89–95.

Call, C. A. & McKell, C. M. (1982). Vesicular-arbuscular mycorrhizae – a natural revegetation strategy for disposed processed oil shale. *Reclamation and Revegetation Research*, **1**, 337–347.

Call, C. A. & McKell, C. M. (1984). Field establishment of fourwing saltbush in processed oil shale and disturbed native soil as influenced by vesicular–arbuscular mycorrhizae. *Great Basin Naturalist*, **44**, 363–371.

Cooper, K. M. (1984). Physiology of VA mycorrhizal associations. In *VA Mycorrhiza*, eds. C. L. Powell, & D. J. Bagyaraj, pp. 155–186. Boca Raton, FL: CRC Press.

Cox, G., Sanders, F. E., Tinker, P. B. & Wild, J. A. (1975). Ultrastructural evidence relating to host–endophyte transfer in a vesicular-arbuscular mycorrhiza. In *Endomycorrhizas*, eds. F. E. Sanders, B. Mosse & P. B. Tinker, pp. 297–312. London: Academic Press.

Cunningham, S. D., Anderson, T. A., Schwab, A. P. & Hsu, F. C. (1996). Phytoremediation of soils contaminated with organic pollutants. *Advances in Agronomy*, **56**, 55–114.

Daft, M. J. & Nicolson, T. H. (1974). Arbuscular mycorrhizas in plants colonizing coal wastes in Scotland. *New Phytologist*, **73**, 1129–1138.

de Jong, E. (1980). The effect of a crude oil spill on cereals. *Environmental Pollution*, **22**, 187–196.

Donnelly, P. K. & Fletcher, J. S. (1994). Potential use of mycorrhizal fungi as bioremediation agents. In *American Chemical Society Symposium Series*, No. 563. *Bioremediation through Rhizosphere Technology*, eds. T. A. Anderson & J. R. Coats, pp. 93–99. Washington DC: American Chemical Society.

Gerdemann, J. W. (1975). Vesicular-arbuscular mycorrhizae. In *The Development and Function of Roots*, eds. J. C. Torrey & D. T. Clarckson, pp. 575–591. London: Academic Press.

Gerdemann, J. W. & Nicolson, T. H. (1963). Spores of mycorrhiza *Endogone* species extracted from soil by wet sieving and decanting. *Transactions of the British Mycological Society*, **46**, 235–244.

Giovannetti, M. & Mosse, B. (1980). An evaluation of techniques for measuring vesicular–arbuscular mycorrhizal infection in roots. *New Phytologist*, **84**, 489–499.

Gudin, C. & Syratt, W. J. (1975). Biological aspects of land rehabilitation following hydrocarbon contamination. *Environmental Pollution*, **8**, 107–112.

Günther, T., Dornberger, U & Fritsche, W. (1996). Effects of ryegrass on biodegradation of hydrocarbons in soil. *Chemosphere*, **33**, 203–215.

Habte, M. & Manjunath, A. (1991). Categories of vesicular–arbuscular mycorrhizal dependency of host species. *Mycorrhiza*, **1**, 3–12.

Harley, J. L. & Smith, S. E. (1983). *Mycorrhizal symbiosis*. London: Academic Press.

Juniper, S. & Abbott, L. (1993). Vesicular–arbuscular mycorrhizas and soil salinity. *Mycorrhiza*, **4**, 45–57.

Khan, A. G. (1978). Vesicular–arbuscular mycorrhizas in plants colonizing black wastes from bituminous coal mining in the Illawarra region of New South Wales. *New Phytologist*, **81**, 53–63.

Khan, A. G. (1981). Growth responses of endomycorrhizal onions in unsterilized coal waste. *New Phytologist*, **87**, 363–370.

Kothari, S. K., Marschner, H. & George, E. (1990). Effect of VA mycorrhizal fungi and rhizosphere microorganisms on root and shoot morphology, growth and water relations in maize. *New Phytologist*, **116**, 303–311.

Linderman, R. G. (1988). Mycorrhizal interactions with the rhizosphere microflora. The mycorrhizosphere effect. *Phytopathology*, **78**, 366–371.

Madsen, E. L. (1991). Determining *in situ* biodegradation. Facts and challenges. *Environmental Science and Technology*, **25**, 1663–1673.

McGill, W. B., Rowell, M. J. & Westlake, D. W. S. (1981). Biochemistry, ecology and microbiology of petroleum components in soil. In *Soil Biochemistry*, Vol. 5, eds. A. Paul & J. N. Ladd, pp. 229–296. New York: Marcel Dekker.

Miller, R. M. (1979). Some occurrences of vesicular–arbuscular mycorrhiza in natural and disturbed ecosystems of the Red Desert. *Canadian Journal of Botany*, **57**, 619–623.

Nicolson, T. H. (1959). Mycorrhiza in the Gramineae. I. Vesicular–arbuscular endophytes, with special reference to the external phase. *Transactions of the British Mycological Society*, **42**, 421–438.

Nicolson, T. H. (1960). Mycorrhiza in the Gramineae. II. Development in different habitats, particularly sand dunes. *Transactions of the British Mycological Society*, **43**, 132–145.

Ocampo, J. A., Martin, J. & Hayman, D. S. (1980). Influence of plant interactions on vesicular–arbuscular mycorrhizal infection. I. Host and non-host plants grown together. *New Phytologist*, **84**, 27–35.

Pfleger, F. L., Stewart, E. L. & Noyd, R. K. (1994). Role of VAM fungi in mine land revegetation. In *Mycorrhizae and Plant Health*, eds. F. L. Pfleger & R. G. Linderman, pp. 47–81. St Paul, MO: APS Press.

Phillips, J. M. & Hayman, D. S. (1970). Improved procedures for clearing roots and staining parasitic and vesicular–arbuscular mycorrhizal fungi for rapid assessment of infection. *Transactions of the British Mycological Society*, **55**, 158–161.

Pirozynski, K. A. (1981). Interactions between fungi and plants through the ages. *Canadian Journal of Botany*, **59**, 1824–1827.

Porter, W. M. (1979). The 'most probable number' method for enumerating infective propagules of vesicular–arbuscular mycorrhizal fungi in soil. *Australian Journal of Soil Research*, **17**, 515–519.

Radwan, S., Sorkhoh, N. & El-Nemr, I. (1995). Oil biodegradation around roots. *Nature*, **376**, 302.

Reeves, F. B., Wagner, D., Moorman, T. & Kiel, J. (1979). The role of endomycorrizae in revegetation practices in the semi-arid west. I. A comparison of incidence of mycorrhizae in severely disturbed vs. natural environments. *American Journal of Botany*, **66**, 6–13.

Scannerini, S. & Bonfante-Fasolo, P. (1983). Comparative ultrastructural analysis of mycorrhizal associations. *Canadian Journal of Botany*, **61**, 917–943.

Schüepp, H. & Bodmer, M. (1991). Complex response of VA–mycorrhizae to xenobiotic substances. *Toxicological and Environmental Chemistry*, **30**, 193–199.

Shetty, K. G., Hetrick, B. A. D., Figge, D. A. H. & Schwab, A. P. (1994). Effects of mycorrhizae and other soil microbes on revegetation of heavy metal contaminated mine spoil. *Enviromental Pollution*, **86**, 181–188.

Smith, S. E. & Read, D. J. (1997). *Mycorrhizal Symbiosis*, 2nd edn. London: Academic Press.

Walton, B. T. & Anderson, T. A. (1990). Microbial degradation of trichloroethylene in the rhizosphere: potential application to biological remediation of waste sites. *Applied and Environmental Microbiology*, **56**,

1012–1016.

Walton, B. T., Guthrie, E. A. & Hoylman, A. M. (1994). Toxicant degradation in the rhizosphere. In *American Chemical Society, Symposium Series*, No. 563: *Bioremediation through Rhizosphere Technology*, eds. T. A. Anderson & J. R. Coats, pp. 11–26. Washington DC: American Chemical Society.

Watanabe, M. E. (1997). Phytoremediation on the brink of commercialization. *Environmental Science and Technology News*, **31**, 182–186.

Weissenhorn, I., Leyval, C. & Berthelin, J. (1993). Cd-tolerant arbuscular mycorrhizal (AM) fungi from heavy-metal polluted soils. *Plant and Soil*, **157**, 247–256.

Weissenhorn, I., Leyval, C. & Berthelin, J. (1995). Bioavailability of heavy metals and abundance of arbuscular mycorrhiza (AM) in a soil polluted by atmospheric deposition from a smelter. *Biology and Fertility of Soils*, **19**, 22–28.

Westlake, D. W. S., Jobson, A. M. & Cook, F. D. (1978). *In situ* degradation of oil in a soil of the boreal region of the Northwest Territories. *Canadian Journal of Microbiology*, **24**, 254–260.

Wood, T. (1992). VA mycorrhizal fungi: challenges for commercialization. In *Handbook of Applied Mycology*, Vol. 4, *Fungal Biotechnology*, eds. D. K. Arora, R. P. Elander & K. G. Mukerji, pp. 823–847. New York: Marcel Dekker.

Index